**AT&T**

# Elementary Number Theory and Its Applications

## Second Edition

Kenneth H. Rosen

*AT&T Systems Laboratories*
*(formerly part of*
*Bell Laboratories)*

ADDISON-WESLEY PUBLISHING COMPANY
Reading, Massachusetts • Menlo Park, California • New York
Don Mills, Ontario • Wokingham, England • Amsterdam • Bonn
Sydney • Singapore • Tokyo • Madrid • San Juan

*Cover:* Six codes for the same message.
See Chapter 7 for hints on decrypting
these messages.

**Library of Congress Cataloging-in-Publication Data**

Rosen, Kenneth H.5
  Elementary number theory and its applications.

  Bibliography: p.
  Includes index.
  1. Numbers, Theory of.  I. Title.
  QA241.R67  1988      512'.72      87-33278
  ISBN 1-201-11958-7

This book was prepared using the UNIX® TROFF text processing system
and the Autologic APS-5 phototypsetter at AT&T Bell Laboratories in
Murray Hill, New Jersey.

ABCDEFGHIJ-MA-898

# Preface

Number theory has long been a favorite subject for students and teachers of mathematics. It is a classical subject with a reputation for being the "purest" part of mathematics, yet recent developments in cryptology and computer science are based on elementary number theory. This book is the first text to integrate these important applications of elementary number theory with the traditional topics covered in an introductory number theory course.

This book is suitable as a text in an undergraduate number theory course at any level. There are no formal prerequisites for most of the material covered, so that even a bright high-school student could use this book. Also, the text is designed to be useful as a supplement for computer science courses, and as a number theory primer for computer scientists interested in learning about new developments in cryptography.

## FEATURES

There have been many satisfied users of the first edition and favorable reviews in journals. This second edition preserves all the strengths of the first edition while adding many new features requested by users. My overall goal in the preparation of the second edition has been to enhance the teachability and flexibility of the text.

- **A Development of Classical Number Theory**

  A major goal of the text is to present classical number theory in a lucid, comprehensive, and captivating manner. At the core of this book is a solid, carefully conceived development of elementary number theory. I have started with the basic material on each important topic, developed it carefully, and then moved on to more difficult results on this topic.

- **Applications**

  This book is loaded with applications of number theory. These are integrated with the development of the subject to illustrate the usefulness of the different parts of the theory. I have concentrated on two major application areas, cryptography and computer science. Extensive coverage is devoted to such topics as character and block ciphers, public key cryptography, pseudorandom number generation, fast multiplication of integers, and data storage. However, I have also included applications to many other areas, including scheduling, telephony, entomology, and zoology.

- **Unifying Themes**

Primality testing and factorization are important areas in both classical number theory and in applications to cryptography. I have used these topics as unifying themes, returning to them repeatedly. Almost every chapter has material devoted to these topics.

- **Accessibility**

This book has been designed with a minimum of prerequisites. The goal has been to present material so that it is as self-contained as possible. Students only need a solid background in college algebra to understand most of the book. Also, all the computer science applications can be explained to students with a minimal background in this discipline.

- **Accuracy**

Thanks to the reviewers of the first and second editions and the many users of the text, the examples, statements of theorems, tables, and answers to exercises have a high level of accuracy. Being realistic, I realize some errors remain. I would appreciate that these errors, whether they be mathematical or typographical, be brought to my attention.

- **Building Mathematical Maturity**

The book begins with axioms for the integers and examples showing how they are used in proofs. This material gives students an appreciation for the importance of rigor and careful thinking in constructing proofs. Furthermore, the first chapter contains a section devoted entirely to mathematical induction, with many examples and a tremendous variety of exercises whose solutions require inductive proofs.

The text contains many different types of proofs, providing students the opportunity to learn about indirect proofs, proofs by contradiction, proofs by cases, existence proofs, and counterexamples.

- **Extensive Use of Examples**

  Numerous examples are given to illustrate the basic skills and concepts developed in the text. The proofs of many theorems are preceded by numerical examples that illustrate the central ideas in these proofs. Finally, examples are used to show how theorems can be applied.

- **Carefully Motivated Proofs**

  Proofs in the text are presented in a careful, rigorous, and fully explained manner. Students can refer to the numerical examples illustrating the working of the proofs to help understand the more general statements being proved.

- **Algorithmic Thinking**

  I have stressed the algorithmic aspects of number theory in this book. Not only are many algorithms described, but also their complexities are analyzed. Examples of algorithms discussed in this book include algorithms for computing greatest common divisors, for primality testing, and for computer arithmetic.

- **Rich Exercise Sets**

  Student learn mathematics by doing exercises. Consequently this text contains extremely extensive and diverse exercise sets. There are many routine computational exercises that develop basic skills. There are also many exercises and block of exercises that develop new concepts. Challenging exercises are included in large numbers and are so marked, with one star (*) indicating a difficult exercise and two stars (**) indicating an extremely difficult exercise. Exercises containing results used in the text are marked with a hand pointing toward them (☞). These exercises should be assigned whenever possible.

- **Answers to Exercises**

  The back of the book contains a section with answers to selected exercises in the text. These answers are meant to be checks for students' work, and should not be used as a crutch.

- **Computer Projects**

Each section of the text is followed by a selection of computer projects involving concepts or algorithms from that section. These projects are open-ended exercises that give students the opportunity to tie together the mathematics in the text with their computer skills.

- **Biographies**

Contributors to the development of number theory include mathematicians of ancient times, such as Euclid and Eratosthenes, mathematicians of the middle ages, such as Fibonacci, mathematicians of the 17th, 18th, and 19th centuries such as Fermat, Euler, and Gauss, and mathematicians of this century. Biographical footnotes describing the lives of 16 important contributors to number theory have been interspersed in the text to convey the human aspect of the subject.

- **Historical Notes**

There are numerous historical remarks included in the text to show the chronology of the development of the subject. These notes emphasize that number theory has an old and rich history as well as a modern vitality.

- **Unsolved Questions**

I have made it a point to mention unsolved questions in number theory in both the text and exercise sets. This shows students that the subject offers many opportunities for future discoveries. Many of these problems have eluded solution for centuries. Students should be forewarned that attempts to settle such problems are often time-consuming and futile. Often people think they have solved such problems, only to discover some subtle flaw in their reasoning.

- **Bibliography**

An extensive bibliography, split into separate sections for books and articles, is included at the end of the text. Each section is subdivided by subject area. The first section contains lists of number theory texts and references, texts tying together computer science and number theory, books on the history of number theory, books on aspects of computer science dealt with in the text, books on cryptography, and general references. In the articles section there are lists of pertinent expository and research papers in number theory and in cryptography.

- **Appendix**

  A set of five tables is included in the appendix to help students with their computations and experimentation. Students may want to compile tables different than those found in the text and in the appendix; compiling such tables would provide additional computer projects.

- **List of Symbols**

  A list of the symbols used in the text and where they are defined is included.

## CHANGES IN THE SECOND EDITION

The following list summarizes the major improvements in this second edition.

- New sections on basic axioms for the integers and mathematical induction have been added.
- A large number of routine computational exercises have been added.
- Many challenging exercises have been added.
- Difficult exercises have been marked with one or two stars.
- Exercises containing results used in the text have been marked with a special symbol.
- Simplified proofs of several important theorems have been included.
- Misprints and errors in the first edition have been corrected.
- The Fibonacci numbers have been introduced earlier and their treatment has been expanded.
- The coverage of big-$O$ notation has been expanded with more motivation included.

- New sections on sum of squares and on factoring using continued fractions have been added.

- An improved system for numbering theorems, lemmas, corollaries, and examples has been used. For important theorems both a name and a number have been included.

- The latest discoveries of large primes have been included.

- Biographies of mathematicians have been included.

- The page layout and format have been made more attractive.

- Several figures have been added.

- The bibliography has been expanded.


## HOW TO USE THIS BOOK

This text is designed to be extremely flexible. The core material for a course in number theory is presented in Chapters 1, 2, and 5, and in Sections 3.1-3.3 and 6.1.

Section 3.4 requires a background in linear algebra; the material in this section is used in Section 7.2; these two sections can be omitted if desired. Each section in Chapter 4 is optional; Sections 4.1, 4.2, and 4.3 present traditional applications of number theory and Section 4.4 presents an application to computer science. Sections 6.2 and 6.3 discuss arithmetic functions, Mersenne primes, and perfect numbers; some of this material is used in Chapter 8. Chapter 7 covers the applications of number theory to cryptology. Sections 7.1, 7.3, and 7.4, which contain discussions of classical and public-key cryptography, are highly recommended. Chapter 8 deals with primitive roots; Sections 8.1-8.4 should be covered if possible, with Sections 8.5 and 8.6 optional. Section 8.7 will interest computer science students; it deals with pseudo-random numbers. Sections 9.1 and 9.2 are about quadratic residues and reciprocity, a fundamental topic which should be covered if possible; Sections 9.3 and 9.4, which deal with Jacobi symbols and Euler pseudoprimes, are optional. Section 10.1, which covers decimal fractions, and Sections 11.1 and 11.2 which discuss Pythagorean triples and Fermat's last theorem are covered in most number theory courses. Sections 10.2-10.5 and 11.4 involve continued fractions; these sections are optional. Also optional is Section 11.3 which deals with sums of squares.

The dependence of chapters on earlier chapters is as follows. Each of Chapters 2,3, and 4 depends on all previous chapters; Chapter 5 depends on

Chapters 1-3; Chapter 6 depends on Chapters 1-3 and 5; Chapter 7 depends on Chapters 1-3 and 5-6; Chapters 8 and 9 both depend on Chapters 1-3 and 5-6; Chapter 10 depends only on Chapters 1-3; and Chapter 11 depends only on Chapters 1-3 with the exception of Section 11.4 which depends on Chapter 10.

## INSTRUCTOR'S MANUAL

An instructor's manual is available to adopters of this book. This manual contains the solutions of selected exercises. Also included are suggestions for teaching, sample exams, and comments on computer projects.

## ACKNOWLEDGMENTS FOR THE SECOND EDITION

I thank my management at AT&T Information Systems Laboratories, including Len Stier, Mac Mugglin, Mike Taylor, and John Sheehan for their support in the preparation of this edition.

Thanks also go to Tom Taylor, my editor for this edition, for his support and helpful suggestions, and to Jeff Pepper and Wayne Yuhasz, my editors for the first edition. I would also like to thank the following people at Addison-Wesley: Addie Ruggles and Stephanie Botvin who have provided editorial assistance and Karen Garrison who has handled production issues, taking the the output from my phototypesetting system and turning it into a book.

I would like to thank Donna Rycyk of AT&T Bell Laboratories for her assistance in producing a polished camera ready manuscript using the troff phototypesetting system. She has been extremely resourceful and helpful in implementing the page layout and design and making the many changes from the first edition.

I have appreciated all the suggestions, corrections, and accolades from users of the first edition book. Many of their ideas have been incorporated in this edition. My thanks go to the following users and reviewers of the first edition and drafts of the second edition:

| David Bressoud | *Pennsylvania State University* |
| Sydney Bulman-Fleming | *Wilfrid Laurier University* |
| Euda Dean | *Tarleton State University* |
| Daniel Drucker | *Wayne State University* |
| Roy Jordan | *Monmouth College* |
| Neil Koblitz | *University of Washington* |
| Charles Lewis | *Monmouth College* |
| James McKay | *Oakland University* |
| John Mairhuber | *University of Maine-Orono* |
| Evelyn Bender Vaskas | *Clark University* |
| Samuel Wagstaff | *Purdue University* |
| Edward Wong | *Wilfrid Laurier University* |
| Paul Zwier | *Calvin College* |

Finally, I thank in advance all those who send me suggestions and corrections in the future. Keep those card and letters coming!

*Middletown, New Jersey*                                    *Kenneth H. Rosen*
*December, 1987*

# Contents

**Introduction** ............................................................................ 1

**Chapter 1.**  **The Integers**

1.1     Basic properties ........................................................... 4
1.2     Summations and products ........................................... 9
1.3     Mathematical induction .............................................. 15
1.4     Binomial coefficients ................................................. 27
1.5     Divisibility ................................................................. 35
1.6     Representations of integers ........................................ 40
1.7     Computer operations with integers ............................. 48
1.8     Complexity of integer operations ............................... 55
1.9     Prime numbers ........................................................... 61

**Chapter 2.**  **Greatest Common Divisors and Prime Factorization**

2.1     Greatest common divisors ........................................... 70
2.2     The Euclidean algorithm ............................................. 76
2.3     The fundamental theorem of arithmetic ...................... 85
2.4     Factorization of integers and the Fermat numbers ...... 97
2.5     Linear diophantine equations ...................................... 104

**Chapter 3.**  **Congruences**

3.1     Introduction to congruences ....................................... 110
3.2     Linear congruences ..................................................... 122
3.3     The Chinese remainder theorem .................................. 126
3.4     Systems of linear congruences ..................................... 135

**Chapter 4.**  **Applications of Congruences**

4.1     Divisibility tests ......................................................... 147
4.2     The perpetual calendar ............................................... 152
4.3     Round-robin tournaments ........................................... 158
4.4     Computer file storage and hashing functions .............. 160

**Chapter 5.  Some Special Congruences**

5.1     Wilson's theorem and Fermat's little theorem .............................. 166
5.2     Pseudoprimes ...................................................................... 172
5.3     Euler's theorem ................................................................... 180

**Chapter 6.  Multiplicative Functions**

6.1     Euler's phi-function ............................................................. 185
6.2     The sum and number of divisors ........................................... 194
6.3     Perfect numbers and Mersenne primes ................................... 199

**Chapter 7.  Cryptology**

7.1     Character ciphers ................................................................ 208
7.2     Block ciphers ..................................................................... 218
7.3     Exponentiation ciphers ........................................................ 224
7.4     Public-key cryptography ...................................................... 230
7.5     Knapsack ciphers ............................................................... 237
7.6     Some applications to computer science .................................. 244

**Chapter 8.  Primitive Roots**

8.1     The order of an integer and primitive roots .............................. 249
8.2     Primitive roots for primes .................................................... 255
8.3     Existence of primitive roots ................................................. 260
8.4     Index arithmetic ................................................................. 267
8.5     Primality testing using primitive roots .................................... 277
8.6     Universal exponents ........................................................... 281
8.7     Pseudo-random numbers ..................................................... 287
8.8     The splicing of telephone cables ........................................... 292

**Chapter 9.  Quadratic Residues and Reciprocity**

9.1     Quadratic residues .............................................................. 299
9.2     Quadratic reciprocity .......................................................... 315
9.3     The Jacobi symbol .............................................................. 324
9.4     Euler pseudoprimes ............................................................ 334

**Chapter 10.   Decimal Fractions and Continued Fractions**

10.1        Decimal fractions............................................................... 343
10.2        Finite continued fractions ................................................ 353
10.3        Infinite continued fractions.............................................. 363
10.4        Periodic continued fractions ............................................ 375
10.5        Factoring using continued fractions................................. 389

**Chapter 11.   Some Nonlinear Diophantine Equations**

11.1        Pythagorean triples........................................................... 393
11.2        Fermat's last theorem ...................................................... 399
11.3        Sums of squares................................................................ 402
11.4        Pell's equations ............................................................... 413


            **Appendix** .......................................................................... 421
            **Answers to Selected problems** ....................................... 437
            **Bibliography** .................................................................... 451
            **Index** ................................................................................ 460

# Introduction

Number theory is a subject that has interested mankind for thousands of years. In this introduction we give a brief overview of number theory, describing why the subject is both fascinating and important.

Number theory in the most general sense is the study of numbers and their properties. In this book, we primarily deal with the integers, $0, \pm 1, \pm 2, \ldots$ . We will discuss many interesting properties of integers and relationships between integers. In addition, we will study the applications of number theory, particularly those directed towards computer science. The development of modern science and technology depended on the development of adequate methods for representing integers and doing integer arithmetic. Starting as long as 5000 years ago different methods, using different integers as bases, have been devised to denote integers. The ancient Babylonians used 60 as the base for their number system and the ancient Mayans used 20. Our method of expressing integers, the decimal system using 10 as the base, was first developed in India approximately six centuries ago. Today, the binary system, which takes 2 as the base, is used extensively by computing machines.

Once a method for representing integers has been devised, procedures for doing arithmetic using these representations are needed. In fact, the term *algorithm*, which now refers to any procedure for solving a general problem, originally referred specifically to such procedures. The development of efficient procedures for arithmetic has become extremely important as computers work with progressively larger integers. Number theory has been used in many ways to devise efficient algorithms for efficient computer arithmetic. Substantial new results in this area have been discovered in the past decade.

The ancient Greeks in the school of Pythagoras, 2500 years ago, made the distinction between *primes* and *composites*. A *prime* is a positive integer with no positive factors other than one and the integer itself. Questions concerning primes have interested mathematicians from antiquity to modern times. Perhaps the first question about primes that comes to mind is whether there are infinitely many. A demonstration that there are infinitely many primes can be found in the the *Elements* of Euclid, an ancient Greek mathematician. We will prove this result in the text. Another question that arises is: Are there formulae that generate primes? Pierre de Fermat, the great French number theorist of the seventeenth century, thought that all integers of the form $2^{2^n} + 1$ are prime; that this is false was shown, a century after Fermat made this claim, by the renowned Swiss mathematician Leonard Euler, who demonstrated that 641 is a factor of $2^{2^5} + 1$. We will discuss these, and other related results, in the sequel.

The problem of distinguishing primes from composites has been extensively studied. Although originally this problem was studied only out of intellectual curiousity, it is now an extremely important problem in cryptology. The ancient Greek scholar Eratosthenes devised a method, now called the *sieve of Eratosthenes*, that finds all primes less than a specified limit. It first removes all even integers, then removes remaining integers divisible by three, then removes remaining integers divisible by five, and so on, stopping when integers divisible by the largest prime not exceeding the square root of the upper limit are removed. All integers left, other than 1, are prime. We will provide the details of this procedure, and justify its validity, in the text.

Sometimes we need to determine whether a particular integer is prime. Unfortunately, it is inefficient to use the sieve of Eratosthenes to determine whether a particular integer is prime. So mathematicians have sought other methods for establishing the primality of an integer. Ancient Chinese mathematicians thought that the primes were precisely those positive integers $n$ such that $n$ divides $2^n - 2$. Fermat showed that if $n$ is prime, then $n$ does divide $2^n - 2$. However, by the early nineteenth century, it was known that there are composite integers $n$ such that $n$ divides $2^n - 2$, such as $n = 341$. These composite integers are called *pseudoprimes*. Because most composite integers are not pseudoprimes, it is possible to develop primality tests based on the original Chinese idea, together with extra observations. It is now possible to efficiently find primes; in fact, primes with as many as 200 decimal digits can be found in minutes of computer time. There is a tremendous amount of current research devoted to showing integers are prime. We will discuss methods for showing integers are prime throughout the text.

The *fundamental theorem of arithmetic*, known to the ancient Greeks, says that every positive integer can be written uniquely as the product of primes. This factorization can be found by trial division of the integer by primes less than its square-root; unfortunately, this method is very time-consuming. Fermat, Euler, and many other mathematicians have produced imaginative factorization techniques. However, using the most efficient technique yet devised, billions of years of computer time may be required to factor an integer with 200 decimal digits. There are large scale efforts underway today to find new ways to quickly factor large integers.

The German mathematician Carl Friedrich Gauss, considered to be one of the greatest mathematicians of all time, developed the language of *congruences* in the early nineteenth century. When doing certain computations, integers may be replaced by their remainders when divided by a specific integer, using the language of congruences. Many questions can be phrased using the notion of a congruence that can only be awkwardly stated without this terminology. Congruences have diverse applications to computer science, including applications to computer file storage, arithmetic with large integers, and the generation of pseudo-random numbers.

One of the most important applications of number theory to computer science is in the area of cryptography. Congruences can be used to develop various types of ciphers. Recently, a new type of cipher system, called a *public-key cipher system*, has been devised. When a public-key cipher is used, each individual has a public enciphering key and a private deciphering key. Messages are enciphered using the public key of the receiver. Moreover, only the receiver can decipher the message, since an overwhelming amount of computer time is required to decipher when just the enciphering key is known. The most widely used public-key cipher system relies on the disparity in computer time required to find large primes and to factor large integers. In particular, to produce an enciphering key requires that two large primes be found and then multiplied; this can be done in minutes on a computer. When these large primes are known, the deciphering key can be quickly found. To find the deciphering key from the enciphering key requires that a large integer, namely the product of the large primes, be factored. This may take billions of years.

In the following chapters, we discuss these and other topics of elementary number theory and its applications.

# 1

---

# The Integers

---

## 1.1 Basic Properties

In this section we state a collection of fundamental properties for the set of integers $\{...,-2,-1,0,1,2,...\}$ which we will take as axioms. These properties provide the foundations for proving results in number theory. We begin with properties dealing with addition and multiplication As usual, we denote the sum and product of $a$ and $b$ by $a+b$ and $a \cdot b$, respectively. Following convention, we will also write $ab$ for $a \cdot b$.

- **Closure:** $a+b$ and $a \cdot b$ are integers whenever $a$ and $b$ are integers.

- **Commutativite laws:** $a+b = b+a$ and $a \cdot b = b \cdot a$ for all integers $a$ and $b$.

- **Associative laws:** $(a+b)+c = a+(b+c)$ and $(a \cdot b) \cdot c = a \cdot (b \cdot c)$ for all integers $a$, $b$, and $c$.

- **Distributive law:** $(a+b) \cdot c = a \cdot c + b \cdot c$ for all integers $a$, $b$, and $c$.

- **Identity elements:** $a+0 = a$ and $a \cdot 1 = a$ for all integers $a$.

- **Additive inverse:** For every integer $a$ there is a integer solution $x$ to the equation $a + x = 0$; this integer $x$ is called the *additive inverse* of $a$ and is denoted by $-a$. By $b-a$ we mean $b+(-a)$.

- **Cancellation law:** If $a$, $b$, and $c$ are integers with $a \cdot c = b \cdot c$ with $c \neq 0$ then $a = b$.

We can use these axioms and the usual properties of equality to establish properties of integers. An example illustrating how this is done follows. In subsequent sections of the book results that are easily proved from these axioms will be used without comment.

**Example 1.1**. To show that $0 \cdot a = 0$ begin with the equation $0+0=0$; this holds since 0 is an identity element for addition. Next multiply both sides by $a$ to obtain $(0+0) \cdot a = 0 \cdot a$. By the distributive law the left-hand side of this equation equals $(0+0) \cdot a = 0 \cdot a + 0 \cdot a$. Hence $0 \cdot a + 0 \cdot a = 0 \cdot a$. Next subtract $0 \cdot a$ from both sides (which is the same as adding the inverse of $0 \cdot a$). Using the associate law for addition and the fact that 0 is an additive identity element, the left-hand side becomes $0 \cdot a + (0 \cdot a - 0 \cdot a) = 0 \cdot a + 0 = 0 \cdot a$. The right-hand side becomes $0 \cdot a - 0 \cdot a = 0$. We conclude that $0 \cdot a = 0$. □

Ordering of integers is defined using the set of positive integers $\{1,2,3,...\}$. We have the following definition.

**Definition**. If $a$ and $b$ are integers, then $a < b$ if $b-a$ is a positive integer. If $a < b$, we also write $b > a$.

Note that $a$ if a positive integer if and only if $a > 0$.

The fundamental properties of ordering of integers follow.

- **Closure for the Positive Integers**: $a+b$ and $a \cdot b$ are positive integers whenever $a$ and $b$ are positive integers.

- **Trichotomy law**: For every integer $a$, either $a > 0$, $a=0$, or $a < 0$.

The set of integers is said to be an *ordered set* because it has a subset that is closed under addition and multiplication and such that the trichotomy law holds.

Basic properties of ordering of integers can now be proved using our axioms, as the following example shows. We will use without proof properties of ordering that easily follow from our axioms.

**Example 1.2**. Suppose that $a,b$, and $c$ integers with $a < b$ and $c > 0$. We can show that $ac < bc$. First, note that from the definition of $a < b$ we have

$b - a > 0$. Since the set of positive integers is closed under multiplication, $c(b - a) > 0$. Since $c(b - a) = cb - ca$, it follows that $ca < cb$.  □

We need one more property to complete our set of axioms.

- **The Well-Ordering Property.** Every nonempty set of positive integers has a least element.

We say that the set of positive integers is *well ordered*. On the other hand, the set of all integers is not well ordered since there are sets of integers that do not have a smallest element (as the reader should verify).

By the well-ordering property the set of integers less than or equal to a given integer has a largest element (the reader should supply the details of the proof of this statement). This leads to the following definition.

**Definition.** The *greatest integer* in $x$, denoted by $[x]$, is the largest integer less than or equal to $x$. That is, $[x]$ is the integer satisfying

$$[x] \leqslant x < [x] + 1.$$

**Example 1.3.** We have $[5/2] = 2$, $[-5/2] = -3$, $[\pi] = 3$, $[-2] = -2$, and $[0] = 0$.  □

We will also deal with the set of rational numbers in this book. Recall that a real number $x$ is *rational* if and only if there are integers $a$ and $b$, $b \neq 0$, with $x = a/b$. A number that is not rational is called *irrational*. Examples of irrational numbers are $\pi$, $\sqrt{2}$, and $e$. We will now use the well-ordering property to show that $\sqrt{2}$ is irrational; we will prove this result in a different way in Chapter 3. (We refer the reader to Hardy and Wright [20] for proofs that $\pi$ and $e$ are irrational.)

**Theorem 1.1.** $\sqrt{2}$ is irrational.

*Proof.* Suppose that $\sqrt{2}$ were rational. Then there would exist positive integers $a$ and $b$ with $\sqrt{2} = a/b$. Consequently the set $S = \{b\sqrt{2} \mid b \text{ and } b\sqrt{2} \text{ are positive integers}\}$ is a nonempty set of positive integers (it is nonempty since $a = b\sqrt{2}$ is a member of $S$). Therefore by the well-ordering property $S$ has a smallest element, say $s = t\sqrt{2}$. We have $s\sqrt{2} - s = s\sqrt{2} - t\sqrt{2} = (s - t)\sqrt{2}$. Since $s\sqrt{2} = 2t$ and $s$ are both integers, $s\sqrt{2} - s = (s - t)\sqrt{2}$ must also be an integer. Furthermore, it is positive since $s\sqrt{2} - s = s(\sqrt{2} - 1)$ and $\sqrt{2} > 1$. It is less than $s$ since $s = t\sqrt{2}$, $s\sqrt{2} = 2t$, and $\sqrt{2} < 2$. This contradicts the choice of $s$ as the smallest positive integer in $S$. It follows that $\sqrt{2}$ is irrational.  ■

The set of integers, the set of positive integers, the set of rational numbers, and the set of real numbers, are traditionally denoted by $\mathbf{Z}$, $\mathbf{Z}^+$, $\mathbf{Q}$, and $\mathbf{R}$, respectively. Also, we write $x \in S$ to indicate that $x$ belongs to the set $S$. Notation such as this will be used occasionally in this book.

## 1.1 Exercises

1. Use the axioms for the integers to prove the following statements for all integers $a$, $b$, and $c$.
   a) $a \cdot (b + c) = a \cdot b + a \cdot c$      c) $a + (b + c) = (c + a) + b$
   b) $(a + b)^2 = a^2 + 2ab + b^2$      d) $(b - a) + (c - b) + (c - a) = 0$

2. Use the axioms for the integers to prove the following statements for all integers $a$ and $b$.
   a) $-1 \cdot a = -a$      c) $(-a) \cdot (-b) = ab$
   b) $-(a \cdot b) = a \cdot (-b)$      d) $-(a + b) = (-a) + (-b)$

3. What is the value of $-0$? Give a reason for your answer.

4. Use the axioms for the integers to show that if $a$ and $b$ are integers with $ab = 0$, then $a = 0$ or $b = 0$.

5. Show that an integer $a$ is positive if and only if $a > 0$.

6. Use the definition of ordering of integers and properties of the set of positive integers to prove the following statements for integers $a, b$, and $c$ with $a < b$ and $c < 0$.
   a) $a + c < b + c$      b) $a^2 \geqslant 0$
   c) $ac > bc$      d) $c^3 < 0$

7. Show that if $a$, $b$, and $c$ are integers with $a > b$ and $b > c$ then $a > c$.

8. Show that there is no positive integer that is less than 1.

9. Determine whether each of the following sets is well-ordered. Either give a proof using the well-ordering property of the set of positive integers or give an example of a subset of the set with no smallest element.

   a) the set of negative integers
   b) the set of integers greater than 3
   c) the set of even positive integers
   d) the set of positive rational numbers
   e) the set of positive rational numbers that can be written in the form $a/2$ where $a$ is a positive integer

10. Find the following values of the greatest integer function.

   a) $[-3/4]$      c) $[22/7]$      e) $[3]$
   b) $[1/4]$       d) $[-2]$        f) $[-\pi]$

11. Show that a nonempty set of negative integers has a largest element.

12. Let $k$ be an integer. Show that $[x+k] = [x] + k$ for every real number $x$.

13. What is the value of $[x] + [-x]$ where $x$ is a real number?

14. Show that $[x] + [x+\frac{1}{2}] = [2x]$ whenever $x$ is a real number.

15. Show that $[x+y] \geqslant [x] + [y]$ for all real numbers $x$ and $y$.

16. Show that $[2x] + [2y] \geqslant [x] + [y] + [x+y]$ whenever $x$ and $y$ are real numbers.

17. Show that if $x$ and $y$ are positive real numbers then $[xy] > [x][y]$. What is the corresponding inequality when both $x$ and $y$ are negative? When one of $x$ and $y$ is negative and the other positive?

18. Show that $-[-x]$ is the least integer greater than or equal to $x$ when $x$ is a real number.

19. Show that $[x + \frac{1}{2}]$ is the integer nearest to $x$ (when there are two integers equidistant from $x$ it is the larger of the two).

20. Show that $[[x]/n] = [x/n]$ whenever $x$ is a real number and $n$ is a positive integer.

21. Show that if $a$ and $b$ are positive integers, then there is a smallest positive nonnegative integer of the form $a - bk$, $k \in Z$.

22. Prove that both the sum and product of two rational numbers are rational.

23. Prove or disprove each of the following statements.

   a) The sum of a rational and an irrational number is irrational.

   b) The sum of two irrational numbers is irrational.

   c) The product of a rational number and an irrational number is irrational.

   d) The product of two irrational numbers is irrational.

* 24. Use the well-ordering property to show that $\sqrt{3}$ is irrational.

** 25. Suppose that $a$ and $b$ are irrational numbers such that $1/a + 1/b = 1$. Show that every integer can be uniquely expressed as either $[ka]$ or $[kb]$ for some integer $k$.

The *Ulam numbers* $u_n$, $n=1,2,3,...$ are defined as follows. We specify that $u_1 = 1$ and $u_2 = 2$. For each successive integer $m$, $m > 2$, this integer is an Ulam number if and only if it can be written uniquely as the sum of two distinct Ulam numbers.

**26.** Find the first ten Ulam numbers.

**\* 27.** Show that there are infinitely many Ulam numbers.

## 1.1 Computer Projects

Write programs to do the following:

**1.** Find the first $n$ Ulam numbers where $n$ is a positive integer.

---

## 1.2 Summations and Products

In this section we introduce notation for summations and products. The following notation represents the sum of the numbers $a_1, a_2, ..., a_n$.

$$\sum_{k=1}^{n} a_k = a_1 + a_2 + \cdots + a_n.$$

The letter $k$, the *index of summation*, is a "dummy variable" and can be replaced by any letter, so that

$$\sum_{k=1}^{n} a_k = \sum_{j=1}^{n} a_j = \sum_{i=1}^{n} a_i \text{ , and so forth.}$$

**Example 1.4.** We see that

$$\sum_{j=1}^{5} j = 1 + 2 + 3 + 4 + 5 = 15,$$

$$\sum_{j=1}^{5} 2 = 2 + 2 + 2 + 2 + 2 = 10,$$

and

$$\sum_{j=1}^{5} 2^j = 2 + 2^2 + 2^3 + 2^4 + 2^5 = 62.$$

We also note that in summation notation, the index of summation may range between any two integers, as long as the lower limit does not exceed the upper limit. If $m$ and $n$ are integers such that $m \leqslant n$, then

$$\sum_{k=m}^{n} a_k = a_m + a_{m+1} + \cdots + a_n.$$

For instance, we have

$$\sum_{k=3}^{5} k^2 = 3^3 + 4^2 + 5^2 = 50,$$

$$\sum_{k=0}^{2} 3^k = 3^0 + 3^1 + 3^2 = 13,$$

and

$$\sum_{k=-2}^{1} k^3 = (-2)^3 + (-1)^3 + 0^3 + 1^3 = -8.$$

We will use several properties of summation notation in the sequel. We first consider sums where every term is multiplied by a constant. Using the distributive law, we have

$$\sum_{j=m}^{n} ka_j = ka_m + ka_{m+1} + \cdots + ka_n$$

$$= k\,(a_m + a_{m+1} + \cdots + a_n)$$

$$= k \sum_{j=m}^{n} a_j$$

Next we consider sums of sums of terms from two sequences.

$$\sum_{j=m}^{n} (a_j + b_j) = (a_m + b_m) + (a_{m+1} + b_{m+1}) + \cdots + (a_n + b_n)$$

$$= (a_m + a_{m+1} + \cdots + a_n) + (b_m + b_{m+1} + \cdots + b_n)$$

$$= \sum_{j=m}^{n} a_j + \sum_{j=m}^{n} b_j.$$

Finally, we show how that we can change the order of summation in a double sum.

$$\sum_{i=m}^{n} \sum_{j=p}^{q} a_i b_j = \sum_{i=m}^{n} a_i (b_p + \cdots + b_q)$$

$$= a_m (b_p + \cdots + b_q) + \cdots + a_n (b_p + \cdots + b_q)$$

$$= a_m b_p + \cdots + a_m b_q + \cdots + a_n b_p + \cdots + a_n b_q$$

$$= (a_m + a_{m+1} + \cdots + a_n)(b_p + b_{p+1} + \cdots + b_q)$$

$$= \left(\sum_{i=m}^{n} a_i\right)\left(\sum_{j=p}^{q} b_j\right).$$

Summarizing, we have the following three properties for summations.

(1.1)
$$\sum_{j=m}^{n} k a_j = k \sum_{j=m}^{n} a_j$$

(1.2)
$$\sum_{j=m}^{n} (a_j + b_j) = \sum_{j=m}^{n} a_j + \sum_{j=m}^{n} b_j$$

(1.3)
$$\sum_{i=m}^{n} \sum_{j=p}^{q} a_i b_j = \left(\sum_{j=p}^{q} a_i\right)\left(\sum_{i=m}^{n} b_j\right).$$

We also define a notation for products, analogous to that for summations. The product of the numbers $a_1, a_2, \ldots, a_n$ is denoted by

$$\prod_{j=1}^{n} a_j = a_1 a_2 \cdots a_n.$$

The letter $j$ above is a "dummy variable", and can be replaced arbitrarily.

**Example 1.5.** To illustrate the notation for products we have

$$\prod_{j=1}^{5} j = 1 \cdot 2 \cdot 3 \cdot 4 \cdot 5 = 120,$$

$$\prod_{j=1}^{5} 2 = 2 \cdot 2 \cdot 2 \cdot 2 \cdot 2 = 2^5 = 32,$$

$$\prod_{j=1}^{5} 2^j = 2 \cdot 2^2 \cdot 2^3 \cdot 2^4 \cdot 2^5 = 2^{15}. \qquad \square$$

The *factorial function* arises throughout number theory.

**Definition.** Let $n$ be a positive integer. Then $n!$, read as $n$ factorial, is the product of the integers $1, 2, \ldots, n$. We also specify that $0! = 1$. In terms of product notation, we have $n! = \prod\limits_{j=1}^{n} j.$

**Example 1.6.** We    have    $1! = 1,$    $4! = 1 \cdot 2 \cdot 3 \cdot 4 = 24,$    and    $12! = 1 \cdot 2 \cdot 3 \cdot 4 \cdot 5 \cdot 6 \cdot 7 \cdot 8 \cdot 9 \cdot 10 \cdot 11 \cdot 12 = 479,001,600.$ □

---

## 1.2  Exercises

1.  Find the value of each of the following sums.

   a) $\sum\limits_{j=1}^{10} 2$                                   c) $\sum\limits_{j=1}^{10} j^2$

   b) $\sum\limits_{j=1}^{10} j$                                   d) $\sum\limits_{j=1}^{10} 2^j$

2.  Find the value of each of the following sums.

   a) $\sum\limits_{j=3}^{6} (j-1)$                                c) $\sum\limits_{j=-2}^{2} j^2$

   b) $\sum\limits_{j=0}^{5} j!$                                   d) $\sum\limits_{j=-8}^{-2} j(j+1)$

3.  Find the value of each of the following products.

   a) $\prod\limits_{j=1}^{5} 3$                                   c) $\prod\limits_{j=1}^{5} j^2$

   b) $\prod\limits_{j=1}^{5} (j+1)$                               d) $\prod\limits_{j=1}^{5} 2^j$

4.  Find the value of each of the following products.

   a) $\prod\limits_{j=3}^{7} 5$                                   c) $\prod\limits_{j=-1}^{4} (j+1)$

   b) $\prod\limits_{j=0}^{5} j!$                                  d) $\prod\limits_{j=-5}^{-1} j^2$

**5.** Express each of the following sums in terms of $\sum\limits_{i=m}^{n} a_i$ and $\sum\limits_{i=m}^{n} b_i$.

a) $\sum\limits_{i=m}^{n} 5a_i$         c) $\sum\limits_{i=m}^{n} (a_i - b_i)$

b) $\sum\limits_{i=m}^{n} (-b_i)$       d) $\sum\limits_{i=m}^{n} (3 a_i + 4 b_i)$

**6.** Find $n!$ for $n$ equal to each of the first ten positive integers.

**7.** List the integers $100!$, $100^{100}$, $2^{100}$, and $(50!)^2$ in order of increasing size. Justify your answer.

**8.** Express each of the following products in terms of $\prod\limits_{i=1}^{n} a_i$ where $k$ is a constant.

a) $\prod\limits_{i=1}^{n} ka_i$     b) $\prod\limits_{i=1}^{n} ia_i$     c) $\prod\limits_{i=1}^{n} a_i^k$

**9.** Find the value of each of the following double sums.

a) $\sum\limits_{i=1}^{5} \sum\limits_{j=1}^{6} ij$          c) $\sum\limits_{i=1}^{5} \sum\limits_{j=1}^{6} i$

b) $\sum\limits_{i=1}^{5} \sum\limits_{j=1}^{6} (i+1)(j+1)$          d) $\sum\limits_{i=1}^{5} \sum\limits_{j=1}^{6} (i+j)$

**10.** Find the value of each of the following double products.

a) $\prod\limits_{i=1}^{4} \prod\limits_{j=1}^{3} ij$          c) $\prod\limits_{i=1}^{4} \prod\limits_{j=1}^{3} i$

b) $\prod\limits_{i=1}^{4} \prod\limits_{j=1}^{3} (i+j)$          d) $\prod\limits_{i=1}^{4} \prod\limits_{j=1}^{3} (i+1)(j-1)$

**11.** Find the value of each of the following quantities.

a) $\sum\limits_{i=1}^{3} \prod\limits_{j=1}^{4} ij$          c) $\sum\limits_{i=1}^{3} \prod\limits_{j=1}^{4} (i+j)$

b) $\prod\limits_{i=1}^{3} \sum\limits_{j=1}^{4} ij$          d) $\prod\limits_{i=1}^{3} \sum\limits_{j=1}^{4} (i+j)$

**12.** Show that $\sum\limits_{j=1}^{n} (a_j - a_{j-1}) = a_n - a_0$ where $a_0, a_1, \ldots, a_n$ is a sequence of numbers. Such a sum is said to be *telescoping*.

**13.** Use the identity $\dfrac{1}{k(k+1)} = \dfrac{1}{k} - \dfrac{1}{k+1}$ and Exercise 12 to evaluate $\displaystyle\sum_{k=1}^{n} \dfrac{1}{k(k+1)}$.

**14.** Use the identity $\dfrac{1}{k^2-1} = \dfrac{1}{2}\left[\dfrac{1}{k-1} - \dfrac{1}{k+1}\right]$ and Exercise 12 to evaluate $\displaystyle\sum_{k=2}^{n} \dfrac{1}{k^2-1}$.

**15.** Sum both sides of the identity $(k+1)^2 - k^2 = 2k + 1$ from $k=1$ to $k=n$, together with Exercise 12 to find a formula for $\displaystyle\sum_{k=1}^{n} k$.

**16.** Find a formula for $\displaystyle\sum_{k=1}^{n} k^2$ using a technique analogous to that outlined in Exercise 15 and the result of that exercise.

**17.** Find a formula for $\displaystyle\sum_{k=1}^{n} k^3$ using a technique analogous to that outlined in Exercise 15 and the results of Exercises 15 and 16.

**18.** Find the values of the following products.

a) $\displaystyle\prod_{j=1}^{n}(1 - 1/j)$  

b) $\displaystyle\prod_{j=1}^{n}(1 - 1/j^2)$.

**19.** Without multiplying all the terms, show that

a) $10! = 6!\ 7!$    c) $16! = 14!\ 5!\ 2!$
b) $10! = 7!\ 5!\ 3!$    d) $9! = 7!\ 3!\ 3!\ 2!$.

**20.** Let $a_1, a_2, \ldots, a_n$ be positive integers. Let $b = (a_1!\ a_2! \cdots a_n!) - 1$, and $c = a_1!\ a_2! \cdots a_n!$. Show that $c! = a_1!\ a_2! \cdots a_n!b!$.

**21.** Find all positive integers $x$, $y$, and $z$ such that $x! + y! = z!$.

**\* 22.** Show that $\displaystyle\sum_{j=0}^{n-1}[x + j/n] = [nx]$ whenever $x$ is a real number and $n$ is a positive integer.

## 1.2 Computer Projects

Write programs to do the following:

**1.** Find $n!$ for a given nonnegative integer $n$.

## 1.3 Mathematical Induction

The *principle of mathematical induction* is a valuable tool for proving results about the integers. We will first state this principle and show how it is used. Afterwards, we will use the well-ordering principle to show that mathematical induction is a valid technique. In our study of number theory, we will use both the well-ordering property and the principle of mathematical induction many times.

**Theorem 1.2. The Principle of Mathematical Induction.** A set of positive integers that contains 1, and for each positive integer $n$ contains $n+1$ if it contains $n$ must be the set of all positive integers.

We need to do two things to prove by mathematical induction that a result holds for every positive integer. First, we must show the result is true for the integer 1. This is called the *basis step*. Second, we must show for each positive integer $n$ that the result is true for the positive integer $n + 1$ if it is true for the positive integer $n$. This is called the *inductive step*. Once these two steps are completed, by the principle of mathematical induction we can conclude that the set $S$ of all positive integers for which the statement is true is the set of all positive integers.

We illustrate how mathematical induction works with several examples. First we prove an inequality.

**Example 1.7.** We can show by mathematical induction that $n! \leqslant n^n$ for every positive integer $n$. The basis step, namely the case where $n=1$, holds since $1!=1 \leqslant 1^1 = 1$. Now assume that $n! \leqslant n^n$; this is the inductive hypothesis. To complete the proof we must show, under the assumption that the inductive hypothesis is valid, that $(n+1)! \leqslant (n+1)^{n+1}$. Using the inductive hypothesis we have

$$\begin{aligned}
(n+1)! &= (n+1)\cdot n! \\
&\leqslant (n+1)n^n \\
&< (n+1)(n+1)^n \\
&< (n+1)^{n+1}.
\end{aligned}$$

This completes both the inductive step and the proof.                    □

One of the most common applications of the mathematical induction is in proofs of explicit formulae for sums. Often we can conjecture such a formula from the first few cases, and then use mathematical induction to confirm (or contradict) our conjecture. (Note that mathematical induction can only be used to prove formulae for sums and not find such formulae.) Consider the following example.

**Example 1.8.** By examining the sums of the first $n$ odd positive integers for small values of $n$ we can conjecture a formula for this sum. We have

$$1 = 1,$$
$$1+3 = 4,$$
$$1+3+5 = 9,$$
$$1+3+5+7 = 16,$$
$$1+3+5+7+9 = 25,$$
$$1+3+5+7+9+11 = 36.$$

From these values, we conjecture that $\sum_{j=1}^{n} (2j-1) = 1+3+5+7+...+2n-1 = n^2$ for every positive integer $n$.

We now attempt to prove this conjecture using mathematical induction. (Clearly this attempt will fail if our guess is wrong.) The basis step follows since $\sum_{j=1}^{2 \cdot 1-1} (2j - 1) = 1 = 1^2$. For the inductive step, we assume that the formula holds for $n$. We must show that under this assumption the formula holds for $n+1$. We assume that $\sum_{j=1}^{n} (2j - 1) = n^2$. Using the inductive hypothesis we have

$$\sum_{j=1}^{n+1} (2j - 1) = \sum_{j=1}^{n} (2j - 1) + 2(n+1) - 1$$
$$= n^2 + 2(n+1) - 1$$
$$= n^2 + 2n + 1$$
$$= (n+1)^2.$$

Since both the basis step and inductive step have been completed, we know that the result holds.                                                                 □

Geometric progressions are sequences of integers where each term is a constant multiple of the previous term. We will often need to sum terms of such progressions. After giving the formal definition of a geometric progression, we will use mathematical induction to establish a formula for the sum of the terms of such a sequence.

**Definition.** Given real numbers $a$ and $r$, the real numbers

$$a, \ ar, \ ar^2, \ ar^3,...$$

are said to form a *geometric progression*. Also, $a$ is called the *initial term* and $r$ is called the *common ratio*.

**Example 1.9.** The numbers $5, -15, 45, -135,...$ form a geometric progression with initial term 5 and common ratio $-3$.                         □

We give an explicit formula for the sum of the first $n$ consecutive terms of a geometric progression in the following example.

**Example 1.10.** We will prove that if $a$ and $r$ are real numbers and $r \neq 1$, then for each positive integer $n$

$$(1.4) \qquad \sum_{j=0}^{n} ar^j = a + ar + ar^2 + \cdots + ar^n = \frac{ar^{n+1}-a}{r-1}.$$

To prove that this formula is valid we complete the basis step, that is, we must first show that the formula holds for $n = 1$. Then, we must complete the inductive step, that is, we must show that if the formula is valid for the positive integer $n$, it must also be true for the positive integer $n + 1$.

To start things off, let $n = 1$. Then, the left side of (1.4) is $a + ar$, while on the right side of (1.4) we have

$$\frac{ar^2-a}{r-1} = \frac{a(r^2-1)}{r-1} = \frac{a(r+1)(r-1)}{r-1} = a(r+1) = a + ar.$$

So the formula is valid when $n = 1$.

Now we assume that (1.4) holds for the positive integer $n$. That is, we assume that

$$(1.5) \qquad a + ar + ar^2 + \cdots + ar^n = \frac{ar^{n+1}-a}{r-1}.$$

We must show that the formula also holds for the positive integer $n + 1$. What we must show is that

$$(1.6) \quad a + ar + ar^2 + \cdots + ar^n + ar^{n+1} = \frac{ar^{(n+1)+1}-a}{r-1} = \frac{ar^{n+2}-a}{r-1}.$$

To show that (1.6) is valid, we add $ar^{n+1}$ to both sides of (1.5), to obtain

$$(1.7) \qquad (a+ar+ar^2+\cdots+ar^n) + ar^{n+1} = \frac{ar^{n+1}-a}{r-1} + ar^{n+1},$$

The left-hand side of (1.7) is identical to that of (1.6). To show that the right sides are equal, we note that

$$\frac{ar^{n+1}-a}{r-1} + ar^{n+1} = \frac{ar^{n+1}-a}{r-1} + \frac{ar^{n+1}(r-1)}{r-1}$$

$$= \frac{ar^{n+1} - a + ar^{n+2} - ar^{n+1}}{r-1}$$

$$= \frac{ar^{n+2}-a}{r-1}.$$

Since we have shown that (1.5) implies (1.6), we can conclude that (1.4) holds for all positive integers $n$.                                          □

The following example illustrates the usefulness of the formula proved in Example 1.10.

**Example 1.11**. Let $n$ be a positive integer. To find the sum

$$\sum_{k=0}^{n} 2^k = 1 + 2 + 2^2 + \cdots + 2^n,$$

we use Theorem 1.1 with $a = 1$ and $r = 2$, to obtain

$$1 + 2 + 2^2 + \cdots + 2^n = \frac{2^{n+1}-1}{2-1} = 2^{n+1}-1.$$

Hence, the sum of consecutive nonnegative powers of 2 is one less than the next largest power of 2.                                          □

We now present the proof that the principle of mathematical induction follows from the well-ordering principle.

*Proof.* Let $S$ be a set of positive integers containing the integer 1 and the integer $n + 1$ whenever it contains $n$. Assume that $S$ is not the set of all positive integers. Therefore, there are some positive integers not contained in $S$. By the well-ordering property, since the set of positive integers not contained in $S$ is nonempty, there is a least positive integer $n$ which is not in $S$. Note that $n \neq 1$, since 1 is in $S$. Now since $n > 1$ (since there is no positive integer $n$ with $n < 1$; see Exercise 8 in Section 1.1), the integer $n - 1$ is a positive integer smaller than $n$, and hence must be in $S$. But since $S$ contains $n - 1$, it must also contain $(n-1) + 1 = n$, which is a contradiction, since $n$ is supposedly the smallest positive integer not in $S$. This shows that $S$ must be the set of all positive integers.                                          ■

A slight variant of the principle of mathematical induction is also sometimes useful in proofs.

**Theorem 1.3. The Second Principle of Mathematical Induction**. A set of positive integers which contains the integer 1, and which has the property that for every positive integer $n$ if it contains all the positive integers 1, 2, ..., $n$, then it also contains the integer $n + 1$, must be the set of all positive integers.

Before proving that the second principle of mathematical induction is valid, we will give an example to illustrate its use.

**Example 1.12**. We will show that any postage more than one cent can be formed using just 2-cent and 3-cent stamps. For the basis step, note that postage of two cents can be formed using one two cent stamp and postage of three cents can be formed using one three cent stamp. For the inductive step, assume that every postage not exceeding $n$ cents, $n \geqslant 3$, can be formed using 2-cent and 3-cent stamps. Then a postage of $n+1$ cents can be formed by forming a postage of $n-1$ cents together with a 2-cent stamp. This completes the proof. (This result can also be proved using the principle of mathematical induction, as the reader can verify.) □

We will now prove that the second principle of mathematical induction is a valid technique.

*Proof.* Let $T$ be a set of integers containing 1 and such that for every positive integer $n$ if it contains 1, 2, ..., $n$, also contains $n+1$. Let $S$ be the set of all positive integers $n$ such that all the positive integers less than or equal to $n$ are in $T$. Then 1 is in $S$, and by the hypotheses, we see that if $n$ is in $S$, then $n + 1$ is in $S$. Hence, by the principle of mathematical induction, $S$ must be the set of all positive integers, so clearly $T$ is also the set of all positive integers, since $S$ is a subset of $T$.                                     ■

## Recursive Definitions

The principle of mathematical induction provides a method for defining the values of functions at positive integers. Instead of explicitly specifying the value of the function at $n$ we give the value of the function at 1 and give a rule for finding for each positive integer $n$ the value of the function at $n+1$ from the value of the function at $n$.

**Definition.** We say the function $f$ is *defined recursively* if the value of $f$ at 1 is specified and if for each positive integer $n$ a rule is provided for determining $f(n+1)$ from $f(n)$.

The principle of mathematical induction can be used to show that a function that is defined recursively is defined uniquely at each positive integer. (See Exercise 29 at the end of this section.) We illustrate how to define a function recursively with the following definition.

**Example 1.13.** We will recursively define the *factorial function* $f(n) = n!$. First, we specify that

$$f(1) = 1.$$

Then we give a rule for finding $f(n+1)$ from $f(n)$ for each positive integer, namely

$$f(n+1) = (n+1){\cdot}f(n).$$

These two statements uniquely define $n!$.

To find the value of $f(6) = 6!$ from the recursive definition use the second property successively, as follows:

$$f(6) = 6 \cdot f(5) = 6 \cdot 5 \cdot f(4) = 6 \cdot 5 \cdot 4 \cdot f(3) = 6 \cdot 5 \cdot 4 \cdot 3 \cdot f(2) = 6 \cdot 5 \cdot 4 \cdot 3 \cdot 2f(1).$$

Then use the first statement of the definition to replace $f(1)$ by its stated value 1, to conclude that

$$6! = 6 \cdot 5 \cdot 4 \cdot 3 \cdot 2 \cdot 1 = 720.$$

□

The second principle of mathematical induction also serves as a basis for recursive definitions. We can define a function whose domain is the set of positive integer by specifying its value at 1 and giving a rule for each positive integer $n$ for finding $f(n)$ from the values $f(j)$ for $1 \leqslant j \leqslant n-1$. (Exercise 32 at the end of this section asks for a proof that such a function is well defined.)

We illustrate how such an inductive definition works with the definition of the *Fibonacci sequence,* named after the Fibonacci*, an Italian mathematician of the thirteenth century who used this sequence to solve a problem about rabbit populations (see Exercise 46 at the end of this section.)

**Example 1.14.** The Fibonacci sequence $f_1, f_2, \ldots, f_n$ is defined recursively by $f_1 = 1$, $f_2 = 1$, and $f_n = f_{n-1} + f_{n-2}$ for $n \geqslant 3$. From this definition we see that

$$\begin{aligned}
f_3 &= f_2 + f_1 = 1+1 = 2, \\
f_4 &= f_3 + f_2 = 2+1 = 3, \\
f_5 &= f_4 + f_3 = 3+2 = 5, \\
f_6 &= f_5 + f_4 = 5+3 = 8,
\end{aligned}$$

and so on. (*Note*: A sequence is just a function from the set of positive integers to the set of real numbers. Consequently it makes sense to recursively define sequences.)               □

The Fibonacci numbers satisfy many identities. We prove one of these in the following example and give several others as exercises at the end of the section.

**Example 1.15.** We can conjecture a formula for $\sum_{j=1}^{n} f_j$ by examining this sum for small values of $n$. We have

---

\* **FIBONACCI (c. 1180-1228),** (short for filus Bonacci, son of Bonacci), also known as Leonardo of Pisa, was born in the Italian commercial center of Pisa. Fibonacci was an merchant who traveled extensively throughout the Mideast, where he came into contact with mathematical works from the Arabic world. In his book *Liber Abaci* Fibonacci introduced Arabic notation for numerals and their algorithms for arithmetic into the European world. It was in this book that his famous rabbit problem appeared. Fibonacci also wrote *Practica geometriae*, a treatise on geometry and trigonometry, and *Liber quadratorum*, a book on Diophantine equations.

$$f_1 = 1,$$
$$f_1 + f_2 = 1+1 = 2,$$
$$f_1 + f_2 + f_3 = 1+1+2 = 4,$$
$$f_1 + f_2 + f_3 + f_4 = 1+1+2+3 = 7,$$
$$f_1 + f_2 + f_3 + f_4 + f_5 = 1+1+2+3+5 = 12.$$

These sums are all one less than the Fibonacci number that is two terms out in the sequence. Hence it is reasonable to guess that

$$\sum_{j=1}^{n} f_i = f_{n+2} - 1.$$

We now attempt to prove this conjecture by mathematical induction. The basis step follows since $\sum_{j=1}^{1} f_i = 1$ and this is the same as $f_{1+2} - 1 = f_3 - 1 = 2 - 1 = 1$. The inductive hypothesis is $\sum_{j=1}^{n} f_i = f_{n+2} - 1$. We must show that under this assumption that $\sum_{j=1}^{n+1} f_i = f_{n+3} - 1$. To prove this, by the inductive hypothesis it follows that

$$\sum_{j=1}^{n+1} f_i = (\sum_{j=1}^{n} f_i) + f_{n+1}$$
$$= (f_{n+2} - 1) + f_{n+1}$$
$$= (f_{n+1} + f_{n+2}) - 1$$
$$= f_{n+3} - 1.$$

$\square$

The following inequality for the Fibonacci numbers will be used in Chapter 2.

**Example 1.16.** We can use the second principle of mathematical induction to prove that $f_n > \alpha^{n-2}$ for $n \geqslant 3$ where $\alpha = (1+\sqrt{5})/2$. The basis step consists of verifying this inequality for $n=3$ and $n=4$. We have $\alpha < 2 = f_3$, so that the theorem is true for $n=3$. Since $\alpha^2 = (3 + \sqrt{5})/2 < 3 = f_4$, the theorem is true for $n=4$.

The inductive hypothesis consists of assuming that $\alpha^{k-2} < f_k$ for all integers $k$ with $k \leqslant n$. Since $\alpha = (1+\sqrt{5})/2$ is a solution of $x^2 - x - 1 = 0$, we have $\alpha^2 = \alpha + 1$. Hence

$$\alpha^{n-1} = \alpha^2 \cdot \alpha^{n-3} = (\alpha+1) \cdot \alpha^{n-3} = \alpha^{n-2} + \alpha^{n-3}.$$

By the induction hypothesis, we have the inequalities

$$\alpha^{n-2} < f_{n-2}, \; \alpha^{n-3} < f_{n-1}.$$

Therefore, we conclude that

$$\alpha^{n-1} < f_n + f_{n-1} = f_{n+1}.$$

This finishes the proof.                                                                    □

We have presented a few of the most important results involving the Fibonacci numbers. There is a vast literature concerning these numbers and their many applications to botany, computer science, geography, physics, and other areas. There is even a scholarly journal, *The Fibonacci Quarterly*, devoted to the study of these numbers.

## 1.3 Exercises

1.  Use mathematical induction to prove that $n < 2^n$ whenever $n$ is a positive integer.

2.  Conjecture a formula for the sum of the first $n$ even positive integers. Prove your result using mathematical induction.

3.  Use mathematical induction to prove that $\displaystyle\sum_{k=1}^{n} \frac{1}{k^2} =$ $\dfrac{1}{1^2} + \dfrac{1}{2^2} + ... + \dfrac{1}{n^2} \geq 2 - \dfrac{1}{n}$ whenever $n$ is a positive integer.

4.  Conjecture a formula for $\displaystyle\sum_{j=1}^{n} \frac{1}{k(k+1)} = \dfrac{1}{1\cdot2} + \dfrac{1}{2\cdot3} + ... + \dfrac{1}{n(n+1)}$ from the value of this sum for small integers $n$. Prove your conjecture is correct using mathematical induction.

5.  Conjecture a formula for $\mathbf{A}^n$ where $\mathbf{A} = \begin{pmatrix} 1 & 1 \\ 0 & 1 \end{pmatrix}$. Prove your conjecture using mathematical induction.

6.  Use mathematical induction to prove that $\displaystyle\sum_{j=1}^{n} j = 1 + 2 + 3 + ... + n = n(n+1)/2$ for every positive integer $n$.

7.  Use mathematical induction to prove that $\displaystyle\sum_{j=1}^{n} j^2 = 1^2 + 2^2 + 3^2 + ... + n^2 = n(n+1)(2n+1)/6$ for every positive integer $n$.

8.  Use mathematical induction to prove that $\displaystyle\sum_{j=1}^{n} j^3 = 1^3 + 2^3 + 3^3 + ... + n^3 = [n(n+1)/2]^2$ for every positive integer $n$.

9.  Use mathematical induction to prove that $\sum\limits_{j=1}^{n} j\,(j+1) =$
    $1\cdot2 + 2\cdot3 + ... + n\cdot(n+1) = n\,(n+1)\,(n+2)/3$ for every positive integer $n$.

10. Use mathematical induction to prove that $\sum\limits_{j=1}^{n} (-1)^{j-1} j^2 =$
    $1^2 - 2^2 + 3^2 - ... + (-1)^{n-1} n^2 = (-1)^{n-1} n\,(n+1)/2$ for every positive integer $n$.

11. Find a formula for $\prod\limits_{j=1}^{n} 2^j$.

12. Find $1 - 2 + 2^2 - 2^3 + \cdots + 2^{100}$.

13. Find $3 + 3\cdot5^2 + 3\cdot5^4 + ... + 3\cdot5^{1000}$.

14. Find $7^3\cdot11^2 + 7^5\cdot11^6 + ... + 7^{99}\cdot11^{194}$.

15. Find $1 + \dfrac{1}{2} + \dfrac{1}{2^2} + ... + \dfrac{1}{2^{100}}$.

16. Show that $\sum\limits_{j=1}^{n} j\cdot j! = 1\cdot1!+2\cdot2!+...+n\cdot n! = (n+1)!-1$ for every positive integer $n$.

17. Show that any postage that is an integer number of cents greater than 11 cents can be formed using just 4-cent and 5-cent stamps.

18. Show that any postage that is an integer number of cents greater than 53 cents can be formed using just 7-cent and 10-cent stamps.

Let $H_n$ be the $n$th partial sum of the harmonic series, that is, $H_n = \sum\limits_{j=1}^{n} 1/j$.

* 19. Use mathematical induction to show that $H_{2^n} \geqslant 1 + n/2$.

* 20. Use mathematical induction to show that $H_{2^n} \leqslant 1 + n$.

21. Show by mathematical induction that if $n$ is a positive integer, then $(2n)! < 2^{2n}(n!)^2$.

22. Use mathematical induction to prove that $x - y$ is a factor of $x^n - y^n$ where $x$ and $y$ are variables.

☛ 23. Use the principle of mathematical induction to show that a set of integers that contains the integer $k$ and such that this set contains $n+1$ whenever it contains $n$ is the set of integers that are greater than or equal to $k$.

24. Use mathematical induction to prove that $2^n < n!$ for $n \geqslant 4$.

25. Use mathematical induction to prove that $n^2 < n!$ for $n \geqslant 4$.

26. Show by mathematical induction that if $h > -1$ then $1+nh \leqslant (1+h)^n$ for all nonnegative integers $n$.

**27.** A jigsaw puzzle is solved by putting its pieces together in the correct way. Show that exactly $n-1$ moves are required to solve a jigsaw puzzle with $n$ pieces, where a move consists of putting together two blocks of pieces, with a block consisting of one or more assembled pieces. (*Hint:* Use the second principle of mathematical induction.)

**28.** What is wrong with the following proof by mathematical induction that all horses are the same color? Clearly all horses in any set of 1 horse are all the same color. This completes the basis step. Now assume that all horses in any set of $n$ horses are the same color. Consider a set of $n+1$ horses, labeled with the integers $1,2,...,n+1$. By the induction hypothesis horses $1,2,...,n$ are all the same color, as are horses $2,3,...,n,n+1$. Since these two sets of horses have common members, namely horses $2,3,4,...,n$ all $n+1$ horses must be the same color. This completes the induction argument.

**29.** Use the principle of mathematical induction to show that the value at each positive integer of a function defined recursively is uniquely determined.

**30.** What function $f(n)$ is defined recursively by $f(1) = 2$ and $f(n+1) = 2f(n)$ for $n \geq 1$? Prove your answer using mathematical induction.

**31.** If $g$ is defined recursively by $g(1) = 2$ and $g(n) = 2^{g(n-1)}$ for $n \geq 2$, what is $g(4)$?

**32.** Use the second principle of mathematical induction to show that if $f(1)$ is specified and a rule for finding $f(n+1)$ from the values of $f$ at the first $n$ positive integers is given, then $f(n)$ is uniquely determined for every positive integer $n$.

**33.** We define a function recursively for all positive integers $n$ by $f(1) = 1$, $f(2) = 5$, and for $n > 2$, $f(n+1) = f(n) + 2f(n-1)$. Show that $f(n) = 2^n + (-1)^n$, using the second principle of mathematical induction.

The *triangular number* $t_n$ is the number of dots in the triangular array with $n$ rows that has $j$ dots in the $j$th row. For instance $t_1 = 1$, $t_2 = 3$, $t_3 = 6$ , and $t_4 = 10$.

$t_1 = 1$     $t_2 = 3$     $t_3 = 6$     $t_4 = 10$

Exercises 34 - 36 deal with triangular numbers.

34. Show that $t_n$ satisfies the recursive defintion $t_1 = 1$ and $t_{n+1} = t_n + (n+1)$ for $n > 1$. From this, show that $t_n = \sum_{j=1}^{n} j$.

35. By putting together two triangular arrays of this kind, one with $n$ rows and one with $n-1$ rows (as illustrated for $n=4$) to form a square, show that $t_{n-1} + t_n = n^2$.

36. By putting together two triangular arrays of this kind, each with $n$ rows, to form a rectangular array of dots of size $n$ by $n+1$ (as illustrated for $n=4$) show that $2t_n = n(n+1)$. From this conclude that $t_n = n(n+1)/2$.

In Exercises 37-50 $f_n$ denotes the $n$th Fibonacci number.

37. Find the following Fibonacci numbers.

    a) $f_{10}$      c) $f_{15}$      e) $f_{20}$
    b) $f_{13}$      d) $f_{18}$      f) $f_{25}$

38. Use the fact that $f_k = f_{k+2} - f_{k+1}$ for every positive integer $k$ to find $\sum_{k=1}^{n} f_k$.

39. Determine a formula for $\sum_{j=1}^{n} f_{2j-1} = f_1 + f_3 + \cdots + f_{2n-1}$ where $n$ is a positive integer by examining the value of this sum for small positive integers. Prove this formula using mathematical induction.

40. Prove that $\sum_{j=1}^{n} f_j^2 = f_1^2 + f_2^2 + \cdots + f_n^2 = f_n f_{n+1}$ for every positive integer $n$.

41. Prove that $f_{n+1} f_{n-1} - f_n^2 = (-1)^n$ for every positive integer $n$.

42. Prove that $f_{n+1} f_n - f_{n-1} f_{n-2} = f_{2n-1}$ for every positive integer $n$, $n > 2$.

43. Prove that $f_1 f_2 + f_2 f_3 + \cdots + f_{2n-1} f_{2n} = f_{2n}^2$ whenever $n$ is a positive integer.

44. Let $\alpha = (1+\sqrt{5})/2$ and $\beta = (1-\sqrt{5})/2$. Prove that $f_n = (\alpha^n - \beta^n)/\sqrt{5}$ for every positive integer $n$.

45. Show that $f_n \leqslant \alpha^{n-1}$ for every integer $n$ with $n \geqslant 2$ where $\alpha = (1+\sqrt{5})/2$.

46. Suppose that on January 1 a pair of baby rabbits is left on an island. These rabbits take two months to mature, and on March 1 they produce another pair of rabbits. They continually produce a new pair the first of every succeeding month. Each newborn pair takes two months to mature and produces a new pair on the first day of the third month of its life, and on the first day of every succeeding month. Show that the number of pairs of rabbits on the island after $n$ months is $f_n$, assuming that no rabbits ever die.

* **47.** Show that every positive integer can be written as the sum of distinct Fibonacci numbers.

48. Let $\mathbf{F} = \begin{pmatrix} 1 & 1 \\ 1 & 0 \end{pmatrix}$. Show that $\mathbf{F}^n = \begin{pmatrix} f_{n+1} & f_n \\ f_n & f_{n-1} \end{pmatrix}$ when $n \in \mathbf{Z}^+$.

49. By taking determinants of both sides of the result of Exercise 48 prove the identity in Exercise 41.

50. Define the *generalized Fibonacci numbers* recursively by $g_1 = a$, $g_2 = b$, and $g_n = g_{n-1} + g_{n-2}$ for $n \geqslant 3$. Show that $g_n = a f_{n-2} + b f_{n-1}$ for $n \geqslant 3$.

51. Suppose that $a_0 = 1$, $a_1 = 3$, $a_2 = 9$, and $a_n = a_{n-1} + a_{n-2} + a_{n-3}$ for $n \geqslant 3$. Show that $a_n \leqslant 3^n$ for every nonnegative integer $n$.

52. The tower of Hanoi was a popular puzzle of the late nineteenth century. The puzzle includes three pegs and eight rings of different sizes placed in order of size, with the largest on the bottom, on one of the pegs. The goal of the puzzle is to move all the rings, one at a time without ever placing a larger ring on top of a smaller ring, from the first peg to the second, using the third peg as an auxiliary peg.

    a) Use mathematical induction to show that the minimum number of moves to transfer $n$ rings, with the rules we have described, from one peg to another is $2^n - 1$.

    b) An ancient legend tells of the monks in a tower with 64 gold rings and 3 diamond pegs. They started moving the rings, one move per second, when the world was created. When they finish transferring the rings to the second peg, the world ends. How long will the world last?

** **53.** The *arithmetic mean* of the positive real numbers $a_1, a_2, \ldots, a_n$ is $A = (a_1 + a_2 + \ldots + a_n)/n$ and the *geometric mean* of these number is $G = (a_1 a_2 \cdots a_n)^{1/n}$. Use mathematical induction to prove that $A \geqslant G$ for every finite sequence of positive real numbers. When does equality hold?

&ast; **54.**  Use mathematical induction to show that a $2^n \times 2^n$ chessboard with one square missing can be covered with L-shaped pieces, where each L-shaped piece covers three squares.

## 1.3 Computer Projects

Write programs to do the following:

**1.**  Find the sum of the terms of a geometric series.

**2.**  Find the first $n$ terms of the Fibonacci sequence.

&ast; **3.**  List the moves in the Tower of Hanoi puzzle (see Exercise 52).

&ast;&ast; **4.**  Cover a $2^n \times 2^n$ chessboard missing one square with L-shaped pieces (see Exercise 54).

---

## 1.4 Binomial Cofficients

Sums of two terms are called *binomial expressions*. Powers of binomial expressions are used throughout number theory and thoroughout mathematics. In this section we will define the *binomial coefficients* and show that these are precisely the coefficients that arise in expansions of powers of binomial expressions.

**Definition.**  Let $m$ and $k$ be nonnegative integers with $k \leqslant m$. The *binomial coefficient* $\begin{pmatrix} m \\ k \end{pmatrix}$ is defined by

$$\begin{pmatrix} m \\ k \end{pmatrix} = \frac{m!}{k!(m-k)!}.$$

When $k$ and $m$ are positive integers with $k > m$ we define $\begin{pmatrix} m \\ k \end{pmatrix} = 0$.

In computing $\begin{pmatrix} m \\ k \end{pmatrix}$, we see that there is a good deal of cancellation, because

$$\begin{pmatrix} m \\ k \end{pmatrix} = \frac{m!}{k!(m-k)!} = \frac{1 \cdot 2 \cdot 3 \ \cdots \ (m-k)(m-k+1) \ \cdots \ (m-1)m}{k! \ \ 1 \cdot 2 \cdot 3 \ \cdots \ (m-k)}$$

$$= \frac{(m-k+1) \ \cdots \ (m-1)m}{k!}.$$

**Example  1.17.**  To evaluate the binomial coefficient $\begin{pmatrix} 7 \\ 3 \end{pmatrix}$, we note that

$$\binom{7}{3} = \frac{7!}{3!4!} = \frac{1\cdot2\cdot3\cdot4\cdot5\cdot6\cdot7}{1\cdot2\cdot3\cdot1\cdot2\cdot3\cdot4} = \frac{5\cdot6\cdot7}{1\cdot2\cdot3} = 35.$$ □

We now prove some simple properties of binomial coefficients.

**Theorem 1.4.** Let $n$ and $k$ be nonnegative integers with $k \leqslant n$. Then

(i) $$\binom{n}{0} = \binom{n}{n} = 1$$

(ii) $$\binom{n}{k} = \binom{n}{n-k}.$$

*Proof.* To see that (i) is true, note that

$$\binom{n}{0} = \frac{n!}{0!n!} = \frac{n!}{n!} = 1$$

and

$$\binom{n}{n} = \frac{n!}{n!0!} = \frac{n!}{n!} = 1.$$

To verify (ii), we see that

$$\binom{n}{k} = \frac{n!}{k!(n-k)!} = \frac{n!}{(n-k)!(n-(n-k))!} = \binom{n}{n-k}.$$ ∎

An important property of binomial coefficients is the following identity.

**Theorem 1.5.** Let $n$ and $k$ be positive integers with $n \geqslant k$. Then

$$\binom{n}{k} + \binom{n}{k-1} = \binom{n+1}{k}.$$

*Proof.* We perform the addition

$$\binom{n}{k} + \binom{n}{k-1} = \frac{n!}{k!(n-k)!} + \frac{n!}{(k-1)!(n-k+1)!}$$

by using the common denominator $k!(n-k+1)!$. This gives

$$\binom{n}{k} + \binom{n}{k-1} = \frac{n!(n-k+1)}{k!(n-k+1)!} + \frac{n!k}{k!(n-k+1)!}$$

$$= \frac{n!((n-k+1)+k)}{k!(n-k+1)!}$$

$$= \frac{n!(n+1)}{k!(n-k+1)!}$$

$$= \frac{(n+1)!}{k!(n-k+1)!}$$

$$= \binom{n+1}{k}. \qquad \blacksquare$$

Using Theorem 1.5 we can construct *Pascal's triangle*, named after the French mathematician Blaise Pascal* who used the binomial coefficients in his analysis of gambling games. In Pascal's triangle, the binomial coefficient $\binom{n}{k}$ is the $(k+1)$th number in the $(n+1)$th row. The first nine rows of Pascal's triangle are displayed in Figure 1.1.

We see that the exterior numbers in the triangle are all 1. To find an interior number, we simply add the two numbers in the positions above, and to either side, of the position being filled. From Theorem 1.5, this yields the correct integer.

---

\* **BLAISE PASCAL** (1623-1662) exhibited his mathematical talents early even though his father, who had made discoveries in analytic geometry, kept mathematical books from him to encourage his other interests. At 16 Pascal discovered an important result concerning conic sections. At 18 he designed a calculating machine, which he had built and successfully sold. Later Pascal made substantial contributions to hydrostatics. Pascal, together with Fermat, laid the foundations for the modern theory of probability. It was in his work on probability where Pascal made new discoveries concerning what is now called Pascal's triangle. In this work he gave what is considered the first lucid description of the principle of mathematical induction. In 1654, catalyzed by an intense religious experience, Pascal abandoned his mathematical and scientific pursuits to devote himself to theology. After this he returned to mathematics only once. One night he had insomnia caused by the discomfort of a toothache and, as a distraction, he studied the mathematical properties of the cycloid. Miraculously, his pain subsided, which he took as a signal of divine approval of the study of mathematics.

$$1$$
$$1 \ 1$$
$$1 \ 2 \ 1$$
$$1 \ 3 \ 3 \ 1$$
$$1 \ 4 \ 6 \ 4 \ 1$$
$$1 \ 5 \ 10 \ 10 \ 5 \ 1$$
$$1 \ 6 \ 15 \ 20 \ 15 \ 6 \ 1$$
$$1 \ 7 \ 21 \ 35 \ 35 \ 21 \ 7 \ 1$$
$$1 \ 8 \ 28 \ 56 \ 70 \ 56 \ 28 \ 8 \ 1$$

$$\cdots$$

**Figure 1.1.**   Pascal's triangle.

Binomial coefficients occur in the expansions of powers of sums. Exactly how they occur is described by the *binomial theorem*.

**Theorem 1.6  The Binomial Theorem.**  Let $x$ and $y$ be variables and $n$ a positive integer. Then

$$(x+y)^n = \binom{n}{0}x^n + \binom{n}{1}x^{n-1}y + \binom{n}{2}x^{n-2}y^2 + \cdots$$
$$+ \binom{n}{n-2}x^2y^{n-2} + \binom{n}{n-1}xy^{n-1} + \binom{n}{n}y^n,$$

or using summation notation,

$$(x+y)^n = \sum_{j=0}^{n} \binom{n}{j}x^{n-j}y^j.$$

*Proof.*  We use mathematical induction. When $n = 1$, according to the binomial theorem, the formula becomes

$$(x+y)^1 = \binom{1}{0}x^1y^0 + \binom{1}{1}x^0y^1.$$

But because $\binom{1}{0} = \binom{1}{1} = 1$, this states that $(x+y)^1 = x + y$, which is obviously true.

We now assume the theorem is valid for the positive integer $n$, that is, we assume that

$$(x+y)^n = \sum_{j=0}^{n} \binom{n}{j}x^{n-j}y^j.$$

We must now verify that the corresponding formula holds with $n$ replaced by $n + 1$, assuming the result holds for $n$. Hence, we have

$$(x+y)^{n+1} = (x+y)^n(x+y)$$

$$= \left[ \sum_{j=0}^{n} \binom{n}{j} x^{n-j} y^j \right] (x+y)$$

$$= \sum_{j=0}^{n} \binom{n}{j} x^{n-j+1} y^j + \sum_{j=0}^{n} \binom{n}{j} x^{n-j} y^{j+1}.$$

We see that by removing terms from the sums and consequently shifting indices, that

$$\sum_{j=0}^{n} \binom{n}{j} x^{n-j+1} y^j = x^{n+1} + \sum_{j=1}^{n} \binom{n}{j} x^{n-j+1} y^j$$

and

$$\sum_{j=0}^{n} \binom{n}{j} x^{n-j} y^{j+1} = \sum_{j=0}^{n-1} \binom{n}{j} x^{n-j} y^{j+1} + y^{n+1}$$

$$= \sum_{j=1}^{n} \binom{n}{j-1} x^{n-j+1} y^j + y^{n+1}.$$

Hence, we find that

$$(x+y)^{n+1} = x^{n+1} + \sum_{j=1}^{n} \left[ \binom{n}{j} + \binom{n}{j-1} \right] x^{n-j+1} y^j + y^{n+1}.$$

By Pascal's Identity we have

$$\binom{n}{j} + \binom{n}{j-1} = \binom{n+1}{j},$$

so we conclude that

$$(x+y)^{n+1} = x^{n+1} + \sum_{j=1}^{n} \binom{n+1}{j} x^{n-j+1} y^j + y^{n+1}$$

$$= \sum_{j=0}^{n+1} \binom{n+1}{j} x^{n+1-j} y^j.$$

This establishes the theorem. ∎

The binomial theorem shows that the coefficients of $(x+y)^n$ are the numbers in the $(n+1)$st row of Pascal's triangle.

We now illustrate one use of the binomial theorem.

**Corollary 1.6.1.** Let $n$ be a nonnegative integer. Then

$$2^n = (1+1)^n = \sum_{j=0}^{n} \binom{n}{j} 1^{n-j} 1^j = \sum_{j=0}^{n} \binom{n}{j}.$$

*Proof.* Let $x=1$ and $y=1$ in the binomial theorem.    ∎

Corollary 1.6.1 shows that if we add all elements of the $(n+1)$th row of Pascal's triangle, we get $2^n$. For instance, for the fifth row, we find that

$$\binom{4}{0} + \binom{4}{1} + \binom{4}{2} + \binom{4}{3} + \binom{4}{4} = 1 + 4 + 6 + 4 + 1 = 16 = 2^4.$$

## 1.4 Exercises

1.  Find the value of each of the following binomial coefficients.

    a) $\binom{100}{0}$          d) $\binom{11}{5}$

    b) $\binom{50}{1}$           e) $\binom{10}{7}$

    c) $\binom{20}{3}$           f) $\binom{70}{70}$

2.  Find the binomial coefficients $\binom{9}{3}$, $\binom{9}{4}$, and $\binom{10}{4}$, and verify that

    $$\binom{9}{3} + \binom{9}{4} = \binom{10}{4}.$$

3.  Use the binomial theoreom to find the write out all terms in the expansions of the following expressions.

    a) $(a+b)^5$                 d) $(2a + 3b)^4$
    b) $(x+y)^{10}$              e) $(3x - 4y)^5$
    c) $(m-n)^7$                 f) $(5x + 7)^8$

4.  What is the coefficient of $x^{99}y^{101}$ in $(2x + 3y)^{200}$ ?

5.  Let $n$ be a positive integer. Using the binomial theorem to expand $(1+(-1))^n$ show that

    $$\sum_{k=0}^{n} (-1)^k \binom{n}{k} = 0.$$

6.  Use Corollary 1.6.1 and Exercise 5 to find

$$\binom{n}{0} + \binom{n}{2} + \binom{n}{4} + \cdots$$

and

$$\binom{n}{1} + \binom{n}{3} + \binom{n}{5} + \cdots .$$

7. Show that if $n, r$, and $k$ are integers with $0 \leqslant k \leqslant r \leqslant n$ then

$$\binom{n}{r}\binom{r}{k} = \binom{n}{k}\binom{n-k}{r-k}.$$

* 8. What is the largest value of $\binom{m}{n}$ where $m$ is a positive integer and $n$ is an integer such that $0 \leqslant n \leqslant m$? Justify your answer.

9. Show that

$$\binom{r}{r} + \binom{r+1}{r} + \cdots + \binom{n}{r} = \binom{n+1}{r+1}$$

where $n$ and $r$ are integers with $1 \leqslant r \leqslant n$.

10. Show that

$$\binom{n}{0} + \binom{n-1}{1} + \binom{n-2}{2} + \cdots = f_{n+1}$$

where $n$ is a nonnegative integer and $f_{n+1}$ is the $(n+1)$st Fibonacci number.

11. Prove that whenever $n$ is a nonegative integer that $\sum_{j=1}^{n} \binom{n}{j} f_j = f_{2n}$ where $f_j$ is the $j$th Fibonacci number.

The binomial coefficients $\binom{x}{n}$, where $x$ is a real number and $n$ is a positive integer, can be defined recursively by the equations $\binom{x}{1} = x$ and

$$\binom{x}{n+1} = \frac{x-n}{n+1}\binom{x}{n}.$$

12. Show from the recursive definition that if $x$ is a positive integer, then $\binom{x}{k} = \frac{x!}{k!(x-k)!}$, where $k$ is an integer with $1 \leqslant k \leqslant x$.

13. Show from the recursive definition that if $x$ is a positive integer, then $\binom{x}{n} + \binom{x}{n+1} = \binom{x+1}{n+1}$, whenever $n$ is a positive integer.

**14.** Show that the binomial coefficient $\binom{n}{k}$, where $n$ and $k$ are integers with $0 \leqslant k \leqslant n$, gives the number of subsets with $k$ elements of a set with $n$ elements.

**15.** Use Exercise 14 to give an alternate proof of the binomial theorem.

**16.** Let $S$ be a set with $n$ elements and let $P_1$ and $P_2$ be two properties that an element of $S$ may have. Show that the number of elements of $S$ possessing neither properties $P_1$ nor property $P_2$ is

$$n - [n(P_1) + n(P_2) - n(P_1,P_2)]$$

where $n(P_1), n(P_2)$, and $n(P_1,P_2)$ are the number of elements of $S$ with property $P_1$, with properties $P_2$, and both properties $P_1$ and $P_2$.

**17.** Let $S$ be a set with $n$ elements and let $P_1$, $P_2$, and $P_3$ be three properties that an element $S$ may have. Show that the number of elements of $S$ possessing none of the properties $P_1$, $P_2$, and $P_3$ is

$$n - [n(P_1) + n(P_2) + n(P_3)$$
$$- n(P_1P_2) + n(P_1P_3) + n(P_2P_3) - n(P_1P_2P_3)]$$

where $n(P_{i_1},...P_{i_k})$ is the number of elements of $S$ with properties $P_{i_1}, \ldots, P_{i_k}$.

* **18.** In this exercise we develop the *principle of inclusion-exclusion*. Suppose that $S$ is a set with $n$ elements and let $P_1$, $P_2$, ..., $P_t$ be $t$ different properties that an element of $S$ may have. Show that the number of elements of $S$ possessing *none* of the $t$ properties is

$$n - [n(P_1) + n(P_2) + \cdots + n(P_t)]$$
$$+ [n(P_1,P_2) + n(P_1,P_3) + \cdots + n(P_{t-1},P_t)]$$
$$- [n(P_1,P_2,P_3) + n(P_1,P_2,P_4) + \cdots + n(P_{t-2},P_{t-1},P_t)]$$
$$+ \cdots + (-1)^t n(P_1,P_2,...,P_t),$$

where $n(P_{i_1},P_{i_2},..., P_{i_j})$ is the number of elements of $S$ possessing all of the properties $P_{i_1},P_{i_2},...,P_{i_j}$. The first expression in brackets contains a term for each property, the second expression in brackets contains terms for all combinations of two properties, the third expression contains terms for all combinations of three properties, and so forth. (*Hint:* For each element of $S$ determine the number of times it is counted in the above expression. If an element has $k$ of the properties, show it is counted

$$1 - \binom{k}{1} + \binom{k}{2} - \cdots + (-1)^k \binom{k}{k}$$

times. This is 0 by Exercise 5.)

## 1.4 Computer Projects

Write programs to do the following:

**1.** Evaluate binomial coefficients.

2.  Print out Pascal's triangle.

3.  Expand $(x+y)^n$, where $n$ is a positive integer, using the binomial theorem.

## 1.5 Divisibility

When an integer is divided by a second nonzero integer, the quotient may or may not be an integer. For instance, $24/8 = 3$ is an integer, while $17/5 = 3.4$ is not. This observation leads to the following definition.

**Definition.** If $a$ and $b$ are integers, we say that $a$ *divides* $b$ if there is an integer $c$ such that $b = ac$. If $a$ divides $b$, we also say that $a$ is a *divisor* or *factor* of $b$.

If $a$ divides $b$ we write $a \mid b$, while if $a$ does not divide $b$, we write $a \nmid b$.

**Example 1.18.** The following examples illustrate the concept of divisibility of integers: $13 \mid 182$, $-5 \mid 30$, $17 \mid 289$, $6 \nmid 44$, $7 \nmid 50$, $-3 \mid 33$, and $17 \mid 0$.    □

**Example 1.19.** The divisors of 6 are $\pm 1$, $\pm 2$, $\pm 3$, and $\pm 6$. The divisors of 17 are $\pm 1$ and $\pm 17$. The divisors of 100 are $\pm 1$, $\pm 2$, $\pm 4$, $\pm 5$, $\pm 10$, $\pm 20$, $\pm 25$, $\pm 50$, and $\pm 100$.    □

In subsequent sections, we will need some simple properties of divisibility. We now state and prove these properties.

**Theorem 1.7.** If $a$, $b$, and $c$ are integers with $a \mid b$ and $b \mid c$, then $a \mid c$.

*Proof.* Since $a \mid b$ and $b \mid c$, there are integers $e$ and $f$ with $ae = b$ and $bf = c$. Hence, $bf = (ae)f = a(ef) = c$, and we conclude that $a \mid c$.    ∎

**Example 1.20.** Since $11 \mid 66$ and $66 \mid 198$, Theorem 1.7 tells us that $11 \mid 198$.    □

**Theorem 1.8.** If $a$, $b$, $m$, and $n$ are integers, and if $c \mid a$ and $c \mid b$, then $c \mid (ma+nb)$.

*Proof.* Since $c \mid a$ and $c \mid b$, there are integers $e$ and $f$ such that $a = ce$ and $b = cf$. Hence, $ma + nb = mce + ncf = c(me+nf)$. Consequently, we see that $c \mid (ma+nb)$.    ∎

**Example 1.21.** Since $3 \mid 21$ and $3 \mid 33$, Theorem 1.8 tells us that 3 divides

$$5 \cdot 21 - 3 \cdot 33 = 105 - 99 = 6.$$    □

The following theorem states an important fact about division.

**Theorem 1.9. The Division Algorithm.** If $a$ and $b$ are integers such that $b > 0$, then there are unique integers $q$ and $r$ such that $a = bq + r$ with $0 \leqslant r < b$.

In the equation given in the division algorithm, we call $q$ the *quotient* and $r$ the *remainder*. (*Note*: We use the traditional name for this theorem even though the division algorithm is not an algorithm. We discuss algorithms, which are procedures for solving general problems, in Section 7 of this chapter.)

We note that $a$ is divisible by $b$ if and only if the remainder in the division algorithm is zero. Before we prove the division algorithm, consider the following examples.

**Example 1.22.** If $a = 133$ and $b = 21$, then $q = 6$ and $r = 7$, since $133 = 21 \cdot 6 + 7$. Likewise, if $a = -50$ and $b = 8$, then $q = -7$ and $r = 6$, since $-50 = 8(-7) + 6$. $\qquad\qquad\square$

We now prove the division algorithm. We prove this theorem using the well-ordering property.

*Proof.* Consider the set $S$ of all integers of the form $a - bk$ where $k$ is an integer, that is, $S = \{a - bk \mid k \in \mathbf{Z}\}$. Let $T$ be the set of all nonnegative integers in $S$. $T$ is nonempty, since $a - bk > 0$ is positive whenever $k$ is an integer with $k < b/a$. By the well-ordering property $T$ has a least element, say with $k = q$. We set $r = a - bq$. (These are the values for $q$ and $r$ specified in the theorem). We know that $r \geq 0$ by construction, and it is easy to see that $r < b$. If $r \geq b$ then $r > r - b = a - bq - b = a - b(q+1) \geq 0$, which contradicts the choice of $r = a - bq$ as the least nonnegative integer of the form $a - bk$. Hence $0 \leq r < b$.

To show that these values for $q$ and $r$ are unique, assume that we have two equations $a = bq_1 + r_1$ and $a = bq_2 + r_2$, with $0 \leq r_1 < b$ and $0 \leq r_2 < b$. By subtracting the second of these from the first, we find that

$$0 = b(q_1 - q_2) + (r_1 - r_2).$$

Hence, we see that

$$r_2 - r_1 = b(q_1 - q_2).$$

This tells us that $b$ divides $r_2 - r_1$. Since $0 \leq r_1 < b$ and $0 \leq r_2 < b$, we have $-b < r_2 - r_1 < b$. Hence $b$ can divide $r_2 - r_1$ only if $r_2 - r_1 = 0$, or, in other words, if $r_1 = r_2$. Since $bq_1 + r_1 = bq_2 + r_2$ and $r_1 = r_2$ we also see that $q_1 = q_2$. This shows that the quotient $q$ and the remainder $r$ are unique. $\qquad\blacksquare$

We can use the greatest integer function to give explicit formulae for the quotient and remainder in the division algorithm. Since the quotient $q$ is the largest integer such that $bq \leq a$, and $r = a - bq$, it follows that

$$q = [a/b], \quad r = a - b[a/b].$$

The following examples illustrate how to find the quotient and remainder of a division.

**Example 1.23.** Let $a = 1028$ and $b = 34$. Then $a = bq + r$ with $0 \leqslant r < b$, where $q = [1028/34] = 30$ and $r = 1028 - [1028/34]\cdot34 = 1028 - 30\cdot34 = 8$. □

**Example 1.24.** Let $a = -380$ and $b = 75$. Then $a = bq + r$ with $0 \leqslant r < b$, where $q = [-380/75] = -6$ and $r = -380 - [-380/75] = -380 - (-6)75 = 70$. □

Given a positive integer $d$, we can classify integers according to their remainders when divided by $d$. For example, with $d = 2$, we see from the division algorithm that every integer when divided by 2 leaves a remainder of either 0 or 1. This leads to the following defintion of some common terminology.

**Definition.** If the remainder when $n$ is divided by 2 is 0, then $n = 2k$ for some positive integer $k$, and we say $n$ is *even*, while if the remainder when $n$ is divided by 2 is 1, then $n = 2k + 1$ for some integer $k$, and we say $n$ is *odd*.

Similarly, when $d = 4$, we see from the division algorithm that when an integer $n$ is divided by 4, the remainder is either 0,1,2, or 3. Hence, every integer is of the form $4k$, $4k + 1$, $4k + 2$, or $4k + 3$, where $k$ is a positive integer.

We will pursue these matters further in Chapter 3.

---

## 1.5 Exercises

1. Show that $3 \mid 99$, $5 \mid 145$, $7 \mid 343$, and $888 \mid 0$.

2. Decide which of the following integers are divisible by 22

   a) 0
   b) 444
   c) 1716
   d) 192544
   e) −32516
   f) −195518.

3. Find the quotient and remainder in the division algorithm with divisor 17 and dividend

   a) 100
   b) 289
   c) −44
   d) −100.

4. What can you conclude if $a$ and $b$ are nonzero integers such that $a \mid b$ and $b \mid a$?

5. Show that if $a$, $b$, $c$, and $d$ are integers with $a$ and $c$ nonzero such that $a \mid b$ and $c \mid d$, then $ac \mid bd$.

6. Are there integers $a$, $b$, and $c$ such that $a \mid bc$, but $a \nmid b$ and $a \nmid c$?

7.  Show that if $a$, $b$, and $c \neq 0$ are integers, then $a \mid b$ if and only if $ac \mid bc$.

8.  Show that if $a$ and $b$ are positive integers and $a \mid b$, then $a \leqslant b$.

☞ 9.  Show that if $a$ and $b$ are integers such that $a \mid b$ then $a^k \mid b^k$ for every positive integer $k$.

10.  Show that the sum of two even or of two odd integers is even, while the sum of an odd and an even integer is odd.

11.  Show that the product of two odd integers is odd, while the product of two integers is even if either of the integers is even.

12.  Show that if $a$ and $b$ are odd positive integers and $b$ does not divide $a$, then there are integers $s$ and $t$ such that $a = bs + t$, where $t$ is odd and $|t| < b$.

13.  When the integer $a$ is divided by the integer $b$ where $b > 0$, the division algorithm gives a quotient of $q$ and a remainder of $r$. Show that if $b \nmid a$, when $-a$ is divided by $b$, the division algorithm gives a quotient of $-(q+1)$ and a remainder of $b - r$, while if $b \mid a$, the quotient is $-q$ and the remainder is zero.

14.  Show that if $a$, $b$, and $c$ are integers with $b > 0$ and $c > 0$, such that when $a$ is divided by $b$ the quotient is $q$ and the remainder is $r$, and when $q$ is divided by $c$ the quotient is $t$ and the remainder is $s$, then when $a$ is divided by $bc$, the quotient is $t$ and the remainder is $bs + r$.

15.  a)  Extend the division algorithm by allowing negative divisors. In particular, show that whenever $a$ and $b \neq 0$ are integers, there are integers $q$ and $r$ such that $a = bq + r$, where $0 \leqslant r < |b|$.

   b)  Find the remainder when 17 is divided by $-7$.

16.  Show that if $a$ and $b$ are positive integers, then there are integers $q, r$ and $e = \pm 1$ such that $a = bq + er$ where $-b/2 < r \leqslant b/2$.

☞ 17.  Show that if $m$ and $n > 0$ are integers, then

$$\left[\frac{m+1}{n}\right] = \begin{cases} \left[\dfrac{m}{n}\right] & \text{if } m \neq kn - 1 \text{ for some integer } k \\ \left[\dfrac{m}{n}\right] + 1 & \text{if } m = kn - 1 \text{ for some integer } k. \end{cases}$$

18.  Show that the integer $n$ is even if and only if $n - 2[n/2] = 0$.

19.  Show that the number of positive integers less than or equal to $x$, where $x$ is a number real number, that are divisible by the positive integer $d$ equals $[x/d]$.

20.  Find the number of positive integers not exceeding 1000 that are divisible by 5, by 25, by 125, and by 625.

21. How many integers between 100 and 1000 are divisible by 7? by 49?

22. Find the number of integers not exceeding 1000 that are not divisible by 3 or 5.

23. Find the number of integers not exceeding 1000 that are not divisible by 3, 5, or 7.

24. Find the number of integers not exceeding 1000 that are divisible by 3 but not by 4.

25. To mail a letter in the U.S.A. it costs 22 cents for the first ounce and 17 cents for each additional ounce or fraction thereof. Find a formula involving the greatest integer function for the cost of mailing a letter. Could it possibly cost $ 1.08 or $ 1.58 to mail a letter?

26. Show that if $a$ is an integer, then 3 divides $a^3 - a$ .

27. Show that the product of two integers of the form $4k + 1$ is again of this form, while the product of two integers of the form $4k + 3$ is of the form $4k + 1$.

28. Show that the square of every odd integer is of the form $8k + 1$.

29. Show that the fourth power of every odd integer is of the form $16k + 1$.

30. Show that the product of two integers of the form $6k + 5$ is of the form $6k + 1$.

31. Show that the product of any three consecutive integers is divisible by 6.

32. Use mathematical induction to show that $n^5 - n$ is divisible by 5 for every positive integer $n$.

33. Use mathematical induction to show that the sum of the cubes of three consecutive integers is divisible by 9.

In Exercises 34 - 38 let $f_n$ denote the $n$th Fibonacci number.

34. Show that $f_n$ is even if and only if $n$ is divisible by 3.

35. Show that $f_n$ is divisible by 3 if and only if $n$ is divisible by 4.

36. Show that $f_n$ is divisible by 4 if and only if $n$ is divisible by 6.

37. Show that $f_n = 5f_{n-4} + 3f_{n-5}$ whenever $n$ is a positive integer with $n > 5$. Use this result to show that $f_n$ is divisible by 5 whenever $n$ is divisible by 5.

* 38. Show that $f_{n+m} = f_m f_{n+1} + f_{m-1} f_n$ whenever $m$ and $n$ are positive integers with $m > 1$. Use this result to show that $f_n \mid f_m$ when $m$ and $n$ are positive integers with $m \mid n$.

Let $n$ be a positive integer. We define

$$T(n) = \begin{cases} n/2 & \text{if } n \text{ is even} \\ (3n+1)/2 & \text{if } n \text{ is odd.} \end{cases}$$

We then form the sequence obtained by iterating $T$; $n$, $T(n)$, $T(T(n))$, $T(T(T(n)))$,.... For instance, starting with $n = 7$ we have $7,11,17,26,13,20,10,5,8,4,2,1,2,1,2,1,...$ . A well-known conjecture, sometimes called the *Collatz conjecture*, asserts that the sequence obtained by iterating $T$ always reaches the integer 1 no matter which positive integer $n$ begins the sequence.

**39.** Find the sequence obtained by iterating $T$ starting with $n = 39$.

**40.** Show that the sequence obtained by iterating $T$ starting with $n = (2^{2k}-1)/3$, where $k$ is a positive integer, $k > 1$, always reaches the integer 1.

**41.** Show that the Collatz conjecture is true if it can be shown that for every positive integer $n$ with $n \geqslant 2$ there is a term in the sequence obtained by iterating $T$ that is less than $n$.

**42.** Verify that there is a term in the sequence obtained by iterating $T$, starting with the positive integer $n$, that is less than $n$ for all positive integers $n$ with $2 \leqslant n \leqslant 100$. (*Hint*: Begin by considering sets of positive integers for which it is easy to show this is true.)

**\* 43.** Show that $[(2+\sqrt{3})^n]$ is odd whenever $n$ is a nonnegative integer.

## 1.5 Computer Projects

Write programs to do the following:

**1.** Decide whether an integer is divisible by a given integer.

**2.** Find the quotient and remainder in the division algorithm.

**3.** Find the quotient, remainder, and sign in the modified division algorithm given in Exercise 15.

**4.** Compute the terms of the sequence $n$, $T(n)$, $T(T(n))$, $T(T(T(n)))$,... for a given positive integer $n$ defined in the preamble to Exercise 39.

---

## 1.6 Representations of Integers

The conventional manner of expressing numbers is by decimal notation. We write out numbers using digits to represent multiples of powers of ten. For instance, when we write the integer 34765 we mean

$$3 \cdot 10^4 + 4 \cdot 10^3 + 7 \cdot 10^2 + 6 \cdot 10^1 + 5 \cdot 10^0.$$

There is no particular reason for the use of ten as the base of notation, other than the fact that we have ten fingers. Other civilizations have used different bases, including the Babylonians, who used base sixty, and the Mayans, who used base twenty. Electronic computers use two as a base for internal representation of integers, and either eight or sixteen for display purposes.

We now show that every positive integer greater than one may be used as a base.

**Theorem 1.10.**   Let $b$ be a positive integer with $b > 1$. Then every positive integer $n$ can be written uniquely in the form

$$n = a_k b^k + a_{k-1} b^{k-1} + \cdots + a_1 b + a_0,$$

where $a_j$ is an integer with $0 \leqslant a_j \leqslant b-1$ for $j = 0, 1, ..., k$ and the initial coefficient $a_k \neq 0$.

*Proof.*   We obtain an expression of the desired type by successively applying the division algorithm in the following way. We first divide $n$ by $b$ to obtain

$$n = bq_0 + a_0, \quad 0 \leqslant a_0 \leqslant b-1.$$

If $q_0$ , we continue by dividing $q_0$ by $b$ to find that

$$q_0 = bq_1 + a_1, \quad 0 \leqslant a_1 \leqslant b-1.$$

We continue this process to obtain

$$q_1 = bq_2 + a_2, \quad 0 \leqslant a_2 \leqslant b-1,$$
$$q_2 = bq_3 + a_3, \quad 0 \leqslant a_3 \leqslant b-1,$$
$$\vdots$$
$$q_{k-2} = bq_{k-1} + a_{k-1}, \quad 0 \leqslant a_{k-1} \leqslant b-1,$$
$$q_{k-1} = b \cdot 0 + a_k, \ 0 \leqslant a_k \leqslant b-1.$$

The last step of the process occurs when a quotient of 0 is obtained. This is guaranteed to occur, because the sequence of quotients satisfies

$$n > q_0 > q_1 > q_2 > \cdots \geqslant 0,$$

and any decreasing sequence of nonnegative integers must eventually terminate with a term equaling 0.

From the first equation above we find that

$$n = bq_0 + a_0.$$

We next replace $q_0$ using the second equation, to obtain

$$n = b(bq_1 + a_1) + a_0 = b^2 q_1 + a_1 b + a_0,$$

Successively substituting for $q_1, q_2, ..., q_{k-1}$, we have

$$n = b^3 q_2 + a_2 b^2 + a_1 b + a_0,$$

.

.

.

$$n = b^{k-1} q_{k-2} + a_{k-2} b^{k-2} + \cdots + a_1 b + a_0,$$
$$n = b^k q_{k-1} + a_{k-1} b^{k-1} + \cdots + a_1 b + a_0$$
$$= a_k b^k + a_{k-1} b^{k-1} + \cdots + a_1 b + a_0,$$

where $0 \leqslant a_j \leqslant b-1$ for $j = 0,1,...,k$ and $a_k \neq 0$, since $a_k = q_{k-1}$ is the last nonzero quotient. Consequently, we have found an expansion of the desired type.

To see that the expansion is unique, assume that we have two such expansions equal to $n$, i.e.

$$n = a_k b^k + a_{k-1} b^{k-1} + \cdots + a_1 b + a_0$$
$$= c_k b^k + c_{k-1} b^{k-1} + \cdots + c_1 b + c_0,$$

where $0 \leqslant a_k < b$ and $0 \leqslant c_k < b$ (and if necessary we add initial terms with zero coefficients to have the number of terms agree). Subtracting one expansion from the other, we have

$$(a_k-c_k)b^k + (a_{k-1} - c_{k-1})b^{k-1} + \cdots + (a_1-c_1)b + (a_0-c_0) = 0.$$

If the two expansions are different, there is a smallest integer $j$, $0 \leqslant j \leqslant k$, such that $a_j \neq c_j$. Hence,

$$b^j \left[ (a_k-c_k)b^{k-j} + \cdots + (a_{j+1}-c_{j+1})b + (a_j-c_j) \right] = 0,$$

so that

$$(a_k-c_k)b^{k-j} + \cdots + (a_{j+1}-c_{j+1})b + (a_j-c_j) = 0.$$

Solving for $a_j-c_j$ we obtain

$$a_j-c_j = (c_k-a_k)b^{k-j} + \cdots + (c_{j+1}-a_{j+1})b$$
$$= b \left[ (c_k-a_k)b^{k-j-1} + \cdots + (c_{j+1}-a_{j+1}) \right].$$

Hence, we see that

$$b \mid (a_j-c_j).$$

But since $0 \leqslant a_j < b$ and $0 \leqslant c_j < b$, we know that $-b < a_j-c_j < b$. Consequently, $b \mid (a_j-c_j)$ implies that $a_j = c_j$. This contradicts the assumption that the two expansions are different. We conclude that our base $b$ expansion of $n$ is unique. ∎

For $b = 2$, we see from Theorem 1.10 that the following corollary holds.

**Corollary 1.10.1.** Every positive integer may be represented as the sum of distinct powers of two.

*Proof.*   Let n be a positive integer. From Theorem 1.10 with $b = 2$, we know that $n = a_k 2^k + a_{k-1} 2^{k-1} + \cdots + a_1 2 + a_0$ where each $a_j$ is either 0 or 1. Hence, every positive integer is the sum of distinct powers of 2.   ∎

In the expansions described in Theorem 1.10 $b$ is called the *base* or *radix* of the expansion. We call base 10 notation, our conventional way of writing integers, *decimal* notation. Base 2 expansions are called *binary* expansions, base 8 expansions are called *octal* expansions, and base 16 expansions are called *hexadecimal*, or *hex* for short, expansions. The coefficients $a_j$ are called the *digits* of the expansion. Binary digits are called *bits* (*binary digits*) in computer terminology.

To distinguish representations of integers with different bases, we use a special notation. We write $(a_k a_{k-1} ... a_1 a_0)_b$ to represent the expansion $a_k b^k + a_{k-1} b^{k-1} + \cdots + a_1 b + a_0$.

**Example 1.25.** To illustrate base $b$ notation, note that $(236)_7 = 2 \cdot 7^2 + 3 \cdot 7 + 6$ and $(10010011)_2 = 1 \cdot 2^7 + 1 \cdot 2^4 + 1 \cdot 2^1 + 1$.   □

Note that the proof of Theorem 1.10 gives us a method of finding the base $b$ expansion of a given positive integer. We simply perform the division algorithm successively, replacing the dividend each time with the quotient, and stop when we come to a quotient which is zero. We then read up the list of remainders to find the base $b$ expansion.

**Example 1.26.** To find the base 2 expansion of 1864, we use the division algorithm successively:

$$
\begin{aligned}
1864 &= 2 \cdot 932 + 0, \\
932 &= 2 \cdot 466 + 0, \\
466 &= 2 \cdot 233 + 0 \\
233 &= 2 \cdot 116 + 1, \\
116 &= 2 \cdot 58 + 0, \\
58 &= 2 \cdot 29 + 0, \\
29 &= 2 \cdot 14 + 1, \\
14 &= 2 \cdot 7 + 0, \\
7 &= 2 \cdot 3 + 1, \\
3 &= 2 \cdot 1 + 1, \\
1 &= 2 \cdot 0 + 1.
\end{aligned}
$$

To obtain the base 2 expansion of 1984, we simply take the remainders of these divisions. This shows that $(1864)_{10} = (11101001000)_2$.   □

Computers represent numbers internally by using a series of "switches" which may be either "on" or "off". (This may be done mechanically using magnetic tape, electrical switches, or by other means.) Hence, we have two possible states for each switch. We can use "on" to represent the digit 1 and "off" to represent the digit 0. This is why computers use binary expansions to represent integers internally.

Computers use base 8 or base 16 for display purposes. In base 16, or hexadecimal, notation there are 16 digits, usually denoted by 0,1,2,3,4,5,6,7,8,9, A, B, C, D, E and F. The letters A, B, C, D, E, and F are used to represent the digits that correspond to 10,11,12,13,14 and 15 (written in decimal notation). We give the following example to show how to convert from hexadecimal notation to decimal notation.

**Example 1.27.** To convert $(A35B0F)_{16}$ from hexadecimal notation to decimal notation we write

$$(A35B0F)_{16} = 10 \cdot 16^5 + 3 \cdot 16^4 + 5 \cdot 16^3 + 11 \cdot 16^2 + 0 \cdot 16 + 15$$
$$= (10705679)_{10}. \qquad \square$$

A simple conversion is possible between binary and hexadecimal notation. We can write each hex digit as a block of four binary digits according to the correspondence given in Table 1.1.

| Hex Digit | Binary Digits | Hex Digit | Binary Digits |
|:---:|:---:|:---:|:---:|
| 0 | 0000 | 8 | 1000 |
| 1 | 0001 | 9 | 1001 |
| 2 | 0010 | A | 1010 |
| 3 | 0011 | B | 1011 |
| 4 | 0100 | C | 1100 |
| 5 | 0101 | D | 1101 |
| 6 | 0110 | E | 1110 |
| 7 | 0111 | F | 1111 |

**Table 1.1.** Conversion from hex digits to blocks of binary digits.

**Example 1.28.** An example of conversion from hex to binary is $(2FB3)_{16} = (10111110110011)_2$. Each hex digit is converted to a block of four binary digits (the initial zeros in the initial block $(0010)_2$ corresponding to the digit $(2)_{16}$ are omitted).

To convert from binary to hex, consider $(11110111101001)_2$. We break this into blocks of four starting from the right. The blocks are, from right to left, 1001, 1110, 1101, and 0011 (we add the initial zeros). Translating each block to hex, we obtain $(3DE9)_{16}$. $\qquad \square$

We note that a conversion between two different bases is as easy as binary hex conversion, whenever one of the bases is a power of the other.

## 1.6 Exercises

1. Convert $(1999)_{10}$ from decimal to base 7 notation. Convert $(6105)_7$ from base 7 to decimal notation.

2. Convert $(101001000)_2$ from binary to decimal notation and $(1984)_{10}$ from decimal to binary notation.

3. Convert $(100011110101)_2$ and $(11101001110)_2$ from binary to hexadecimal.

4. Convert $(ABCDEF)_{16}$, $(DEFACED)_{16}$, and $(9A0B)_{16}$ from hexadecimal to binary.

5. Explain why we really are using base 1000 notation when we break large decimal integers into blocks of three digits, separated by commas.

6. Show that if $b$ is a negative integer less than $-1$, then every integer $n$ can be uniquely written in the form

$$n = a_k b^k + a_{k-1} b^{k-1} + \cdots + a_1 b + a_0,$$

where $a_k \neq 0$ and $0 \leqslant a_j < |b|$ for $j = 0,1,2,..., k$. We write $n = (a_k a_{k-1}...a_1 a_0)_b$, just as we do for positive bases.

7. Find the decimal representation of $(101001)_{-2}$ and $(12012)_{-3}$.

8. Find the base $-2$ representations of the decimal numbers $-7, -17$, and 61.

9. Show that any weight not exceeding $2^k - 1$ may be measured using weights of $1, 2, 2^2,..., 2^{k-1}$, when all the weights are placed in one pan.

10. Show that every integer can be uniquely represented in the form

$$e_k 3^k + e_{k-1} 3^{k-1} + \cdots + e_1 3 + e_0$$

where $e_j = -1, 0$, or 1 for $j=0,1,2, ..., k$. This expansion is called a *balanced ternary expansion.*

11. Use Exercise 10 to show that any weight not exceeding $(3^k - 1)/2$ may be measured using weights of $1, 3, 3^2, ..., 3^{k-1}$, when the weights may be placed in either pan.

12. Explain how to convert from base 3 to base 9 notation, and from base 9 to base 3 notation.

13. Explain how to convert from base $r$ to base $r^n$ notation, and from base $r^n$ notation to base $r$ notation, when $r > 1$ and $n$ are positive integers.

14. Show that if $r = (a_k a_{k-1}...a_1 a_0)_b$, then the quotient and remainder when $n$ is divided by $b^j$ are $q = (a_k a_{k-1}...a_j)_b$ and $r = (a_{j-1}...a_1 a_0)_b$, respectively.

15. If the base $b$ expansion of $n$ is $n = (a_k a_{k-1}...a_1 a_0)_b$, what is the base $b$ expansion of $b^m n$?

*One's complement* representations of integers are used to simplify computer arithmetic. To represent positive and negative integers with absolute value less than $2^n$, a total of $n+1$ bits are used.

The leftmost bit is used to represent the sign. A zero bit in this position is used for positive integers and a one bit in this position is used for negative integers.

For positive integers the remaining bits are identical to the binary expansion of the integer. For negative integers, the remaining bits are obtained by first finding the binary expansion of the absolute value of the integer, and then taking the complement of each of these bits, where the complement of a 1 is a 0 and the complement of a 0 is a 1.

**16.** Find the one's complement representations, using bit strings of length six, of the following integers

a) 22    c) -7
b) 31    d) -19

**17.** What integer do each of the following one's complement representations of length five represent?

a) 11001    c) 10001
b) 01101    d) 11111

**18.** How is the one's complement representation of $-m$ obtained from the one's complement of $m$, when bit strings of length $n$ are used?

**19.** Sometimes integers are encoded by using four digit binary expansions to represent each decimal digit. This produces the *binary coded decimal* form of the integer. For instance, 791 is encoded in this way by 011110010001. How many bits are required to represent a number with $n$ decimal digits using this type of encoding?

A *Cantor expansion* of a positive integer $n$ is a sum

$$n = a_m m! + a_{m-1}(m-1)! + \cdots + a_2 2! + a_1 1!$$

where each $a_j$ is an integer with $0 \leqslant a_j \leqslant j$.

**20.** Find Cantor expansions of 14, 56, and 384.

* **21.** Show that every positive integer has a unique Cantor expansion. (*Hint*: For each positive integer $n$ there is a positive integer $m$ such that $m! \leqslant n < (m+1)!$. For $a_m$ take the quotient from the division algorithm when $n$ is divided by $m!$. Then iterate.)

The Chinese game of *nim* is played as follows. There are a number of piles of matches, each containing an arbitrary number of matches at the start of the game. A move consists of a player removing one or more matches from one of the piles. The players take turns, with the player removing the last match winning the game.

A *winning position* is an arrangement of matches in piles so that if a player can move to this position, then, no matter what the second player does, the first player can continue to play in a way that will win the game. An example is the position where there are two piles each containing one match; this is a winning position, because the second player must remove a match leaving the first player the opportunity to win by removing the last match.

22. Show that the position in nim where there are two piles, each with two matches, is a winning position.

23. For each arrangement of matches into piles, write the number of matches in each pile in binary notation, and then line up the digits of these numbers into columns (adding initial zeroes if necessary to some of the numbers). Show that a position is a winning one if and only if the number of ones in each column is even (*Example*: Three piles of 3, 4, and 7 give

$$
\begin{array}{ccc}
0 & 1 & 1 \\
1 & 1 & 1 \\
1 & 0 & 0
\end{array}
$$

where each column has exactly two ones). (*Hint*: Show that any move from a winning position produces a non-winning one. Show that there is a move from a non-winning position to a winning one.)

Let $a$ be an integer with a four-digit decimal expansion, with not all digits the same. Let $a'$ be the integer with a decimal expansion obtained by writing the digits of $a$ in descending order, and let $a''$ be the integer with a decimal expansion obtained by writing the digits of $a$ in ascending order. Define $T(a) = a' - a''$. For instance, $T(7318) = 8731 - 1378 = 7358$.

* 24. Show that the only integer with a four-digit decimal expansion with not all digits the same such that $T(a) = a$ is $a = 6174$.

** 25. Show that if $a$ is a positive integer with a four-digit decimal expansion with not all digits the same, then the sequence $a$, $T(a)$, $T(T(a))$, $T(T(T(a)))$,..., obtained by iterating $T$, eventually reaches the integer 6174. Because of this property, 6174 is called *Kaprekar's constant*.

Let $b$ be a positive integer and let $a$ be an integer with a four-digit base $b$ expansion, with not all digits the same. Define $T_b(a) = a' - a''$, where $a'$ is the integer with base $b$ expansion obtained by writing the base $b$ digits of $a$ in descending order, and let $a''$ is the integer with base $b$ expansion obtained by writing the base $b$ digits of $a$ in ascending order.

** 26. Let $b = 5$. Find the unique integer $a_0$ with a four-digit base 5 expansion such that $T_5(a_0) = a_0$. Show that this integer $a_0$ is a Kaprekar constant for the base 5, *i.e.*, $a$, $T(a)$, $T(T(a))$, $T(T(T(a)))$,... eventually reaches $a_0$, whenever $a$ is an integer which a four-digit base 5 expansion with not all digits the same.

* **27.**  Show that no Kaprekar constant exists for the base 6.

## 1.6  Computer Projects

Write programs to do the following:

**1.**  Find the binary expansion of an integer from the decimal expansion of this integer and *vice versa*.

**2.**  Convert from base $b_1$ notation to base $b_2$ notation, where $b_1$ and $b_2$ are arbitrary positive integers greater than one.

**3.**  Convert from binary notation to hexadecimal notation and *vice versa*.

**4.**  Find the base $(-2)$ notation of an integer from its decimal notation (see Exercise 6).

**5.**  Find the balanced ternary expansion of an integer from its decimal expansion (see Exercise 10).

**6.**  Find the Cantor expansion of an integer from its decimal expansion (see the preamble to Exercise 20).

**7.**  Play a winning strategy in the game of nim (see the preamble to Exercise 22).

* **8.**  Find the sequence $a$, $T(a)$, $T(T(a))$, $T(T(T(a)))$,... defined in the preamble to Exercise 24), where $a$ is a positive integer, to discover how many iterations are needed to reach 6174.

* **9.**  Let $b$ be a positive integer. Find the Kaprekar constant to the base $b$, when it exists (see Exercise 25-27).

---

## 1.7  Computer Operations with Integers

We have mentioned that computers internally represent numbers using bits, or binary digits. Computers have a built-in limit on the size of integers that can be used in machine arithmetic. This upper limit is called the *word size*, which we denote by $w$. The word size is usually a power of 2, such as $2^{35}$, although sometimes the word size is a power of 10.

To do arithmetic with integers larger than the word size, it is necessary to devote more than one word to each integer. To store an integer $n > w$, we express $n$ in base $w$ notation, and for each digit of this expansion we use one computer word. For instance, if the word size is $2^{35}$, using ten computer words we can store integers as large as $2^{350}-1$, since integers less than $2^{350}$ have no more than ten digits in their base $2^{35}$ expansions. Also note that to find the base $2^{35}$ expansion of an integer, we need only group together blocks of 35 bits.

The first step in discussing computer arithmetic with large integers is to describe how the basic arithmetic operations are methodically performed.

We will describe the classical methods for performing the basic arithmetic operations with integers in base $r$ notation where $r > 1$ is an integer. These methods are examples of *algorithms*\*.

**Definition.** An *algorithm* is a specified set of rules for obtaining a desired result from a set of input.

We will describe algorithms for performing addition, subtraction, and multiplication of two $n$-digit integers $a = (a_{n-1}a_{n-2}...a_1a_0)_r$ and $b = (b_{n-1}b_{n-2}...b_1b_0)_r$, where initial digits of zero are added if necessary to make both expansions the same length. The algorithms described are used both for binary arithmetic with integers less than the word size of a computer, and for *multiple precision* arithmetic with integers larger than the word size $w$, using $w$ as the base.

We first discuss the algorithm for addition. When we add $a$ and $b$, we obtain the sum

$$a + b = \sum_{j=0}^{n-1} a_j r^j + \sum_{j=0}^{n-1} b_j r^j = \sum_{j=0}^{n-1} (a_j + b_j)r^j.$$

To find the base $r$ expansion of the $a + b$, first note that by the division algorithm, there are integers $C_0$ and $s_0$ such that

$$a_0 + b_0 = C_0 r + s_0, \; 0 \leqslant s_0 < r.$$

Because $a_0$ and $b_0$ are positive integers not exceeding $r$, we know that $0 \leqslant a_0 + b_0 \leqslant 2r - 2$, so that $C_0 = 0$ or 1; here $C_0$ is the *carry* to the next place. Next, we find that there are integers $C_1$ and $s_1$ such that

$$a_1 + b_1 + C_0 = C_1 r + s_1, \; 0 \leqslant s_1 < r.$$

Since $0 \leqslant a_1 + b_1 + C_0 \leqslant 2r - 1$, we know that $C_1 = 0$ or 1. Proceeding inductively, we find integers $C_i$ and $s_i$ for $1 \leqslant i \leqslant n - 1$ by

---

\* The word "algorithm" has an interesting history, as does the evolution of the concept of an algorithm. "Algorithm" is a corruption of the original term "algorism" which originally comes from the book *Kitab al jabr w'al-muqabala (Rules of Restoration and Reduction)* written by Abu Ja´far Mohammed ibn Musa al-Khowarizmi in the ninth century. The word "algorism" originally referred only to the rules of performing arithmetic using Arabic numerals. The word "algorism" evolved into "algorithm" by the eighteenth century. With growing interest in computing machines, the concept of an algorithm became more general, to include all definite procedures for solving problems, not just the procedures for performing arithmetic with integers expressed in Arabic notation.

$$a_i + b_i + C_{i-1} = C_i r + s_i, \ 0 \leqslant s_i < r,$$

with $C_i = 0$ or 1. Finally, we let $s_n = C_{n-1}$ , since the sum of two integers with $n$ digits has $n + 1$ digits when there is a carry in the $n$th place. We conclude that the base $r$ expansion for the sum is $a + b = (s_n s_{n-1}...s_1 s_0)_r$ .

When performing base $r$ addition by hand, we can use the same familiar technique as is used in decimal addition.

**Example 1.29.** To add $(1101)_2$ and $(1011)_2$ we write

$$
\begin{array}{ccccc}
\textit{1} & & & \textit{1} & \\
& 1 & 1 & 0 & 1 \\
+ & 1 & 0 & 0 & 1 \\
\hline
1 & 0 & 1 & 1 & 0 \\
\end{array}
$$

where we have indicated carries by 1's in italics written above the appropriate column. We found the binary digits of the sum by noting that $1 - 01 + 1 = 1 \cdot 2 + 0$, $0 + 0 + 1 = 0 \cdot 2 + 1$, $1 + 0 + 0 = 0 \cdot 2 + 1$, and $1 + 1 = 1 \cdot 2 + 0$. $\square$

We now turn our attention to subtraction. We consider

$$a - b = \sum_{j=0}^{n-1} a_j r^j - \sum_{j=0}^{n-1} b_j r^j = \sum_{j=0}^{n-1} (a_j - b_j) r^j,$$

where we assume that $a > b$. Note that by the division algorithm, there are integers $B_0$ and $d_0$ such that

$$a_0 - b_0 = B_0 r + d_0, \quad 0 \leqslant d_0 < r,$$

and since $a_0$ and $b_0$ are positive integers less than $r$, we have

$$-(r - 1) \leqslant a_0 - b_0 \leqslant r - 1.$$

When $a_0 - b_0 \geqslant 0$, we have $B_0 = 0$. Otherwise, when $a_0 - b_0 < 0$ , we have $B_0 = -1$; $B_0$ is the *borrow* from the next place of the base $r$ expansion of $a$. We use the division algorithm again to find integers $B_1$ and $d_1$ such that

$$a_1 - b_1 + B_0 = B_1 r + d_1, \quad 0 \leqslant d_1 < r.$$

From this equation, we see that the borrow $B_1 = 0$ as long as $a_1 - b_1 + B_0 \geqslant 0$, and $B_1 = -1$ otherwise, since $-r \leqslant a_1 - b_1 + B_0 \leqslant r - 1$. We proceed inductively to find integers $B_i$ and $d_i$, such that

$$a_i - b_i + B_{i-1} = B_i r + d_i, \quad 0 \leqslant d_i < r$$

with $B_i = 0$ or $-1$, for $1 \leqslant i \leqslant n - 2$. We see that $B_{n-1} = 0$, since $a > b$. We can conclude that

$$a - b = (d_{n-1} d_{n-2}...d_1 d_0)_r.$$

When performing base $r$ subtraction by hand, we use the same familiar technique as is used in decimal subtraction.

**Example 1.30.** To subtract $(10110)_2$ from $(11011)_2$, we have

$$
\begin{array}{r}
\textit{-1}\phantom{00000} \\
1\ 1\ 0\ 1\ 1 \\
-\ 1\ 0\ 1\ 1\ 0 \\
\hline
1\ 0\ 1
\end{array}
$$

where the $-1$ in italics above a column indicates a borrow. We found the binary digits of the difference by noting that $1 - 0 = 0{\cdot}2 + 1$, $1 - 1 = 0{\cdot}2 + 0$, $0 - 1 = -1{\cdot}2 + 1$, $1 - 0 - 1 = 0{\cdot}2 + 0$, and $1 - 1 = 0{\cdot}2 + 0$. $\qquad\square$

Before discussing multiplication, we describe *shifting*. To multiply $(a_{n-1}...a_1 a_0)_r$ by $r^m$, we need only shift the expansion left $m$ places, appending the expansion with $m$ zero digits.

**Example 1.31.** To multiply $(101101)_2$ by $2^5$, we shift the digits to the left five places and append the expansion with five zeros, obtaining $(10110100000)_2$. $\qquad\square$

To deal with multiplication, we first discuss the multiplication of an $n$-place integer by a one-digit integer. To multiply $(a_{n-1}...a_1 a_0)_r$ by $(b)_r$, we first note that

$$ a_0 b = q_0 r + p_0, \quad 0 \leqslant p_0 < r, $$

and $0 \leqslant q_0 \leqslant r-2$, since $0 \leqslant a_0 b \leqslant (r-1)^2$. Next, we have

$$ a_1 b + q_0 = q_1 r + p_1, \quad 0 \leqslant p_1 < r, $$

and $0 \leqslant q_1 \leqslant r-1$. In general, we have

$$ a_i b + q_{i-1} = q_i r + p_i, \quad 0 \leqslant p_i < r $$

and $0 \leqslant q_i \leqslant r - 1$. Furthermore, we have $p_n = q_{n-1}$. This yields $(a_{n-1}...a_1 a_0)_r (b)_r = (p_n p_{n-1}...p_1 p_0)_r$.

To perform a multiplication of two n-place integers we write

$$ ab = a\left(\sum_{j=1}^{n-1} b_j r^j\right) = \sum_{j=0}^{n-1} (ab_j)r^j. $$

For each $j$, we first multiply $a$ by the digit $b_j$, then shift to the left $j$ places, and finally add all of the $n$ integers we have obtained to find the product.

When multiplying two integers with base $r$ expansions, we use the familiar method of multiplying decimal integers by hand.

**Example 1.32.** To multiply $(1101)_2$ and $(1110)_2$ we write

$$
\begin{array}{r}
1\ 1\ 0\ 1 \\
\times\ 1\ 1\ 1\ 0 \\
\hline
0\ 0\ 0\ 0 \\
1\ 1\ 0\ 1 \\
1\ 1\ 0\ 1 \\
1\ 1\ 0\ 1 \\
\hline
1\ 0\ 1\ 1\ 0\ 1\ 1\ 0
\end{array}
$$

Note that we first multiplied $(1101)_2$ by each digit of $(1110)_2$, shifting each time by the appropriate number of places, and then we added the appropriate integers to find our product.  □

We now discuss integer division. We wish to find the quotient $q$ in the division algorithm

$$a = bq + R, \quad 0 \leqslant R < b.$$

If the base $r$ expansion of $q$ is $q = (q_{n-1}q_{n-2}...q_1q_0)_r$, , then we have

$$a = b \left[ \sum_{j=0}^{n-1} q_j r^j \right] + R, \quad 0 \leqslant R < b.$$

To determine the first digit $q_{n-1}$ of $q$, notice that

$$a - bq_{n-1}r^{n-1} = b \left[ \sum_{j=0}^{n-2} q_j r^j \right] + R.$$

The right-hand side of this equation is not only positive, but also it is less than $br^{n-1}$, since $\displaystyle\sum_{j=0}^{n-2} q_j r^j \leqslant \sum_{j=0}^{n-2} (r-1)r^j = \sum_{j=1}^{n-1} r^j - \sum_{j=0}^{n-2} r^j = r^{n-1} - 1$  Therefore, we know that

$$0 \leqslant a - bq_{n-1}r^{n-1} < br^{n-1}.$$

This tells us that

$$q_{n-1} = [a/br^{n-1}].$$

We can obtain $q_{n-1}$ by successively subtracting $br^{n-1}$ from $a$ until a negative result is obtained, and then $q_{n-1}$ is one less than the number of subtractions.

To find the other digits of $q$, we define the sequence of *partial remainders* $R_i$ by

$$R_0 = a$$

and

$$R_i = R_{i-1} - bq_{n-i}r^{n-i}$$

for $i = 1, 2, ..., n$. By mathematical induction, we show that

$$(1.8) \qquad R_i = \left[ \sum_{j=0}^{n-i-1} q_j r^j \right] b + R.$$

For $i = 0$, this is clearly correct, since $R_0 = a = qb + R$. Now assume that

$$R_k = \left[ \sum_{j=0}^{n-k-1} q_j r^j \right] b + R.$$

Then

$$R_{k+1} = R_k - bq_{n-k-1} r^{n-k-1}$$

$$= \left[ \sum_{j=0}^{n-k-1} q_j r^j \right] b + R - bq_{n-k-1} r^{n-k-1}$$

$$= \left[ \sum_{j=0}^{n-(k+1)-1} q_j r^j \right] b + R,$$

establishing (1.8).

By (1.8) we see that $0 \leqslant R_i < r^{n-i} b$, for $i = 1, 2, ..., n$, since $\sum_{j=0}^{n-i-1} q_j r^j \leqslant r_{n-i} - 1$. Consequently, since $R_i = R_{i-1} - bq_{n-i} r^{n-i}$ and $0 \leqslant R_i < r^{n-1} b$, we see that the digit $q_{n-i}$ is given by $[R_{i-1}/br^{n-i}]$ and can be obtained by successively subtracting $br^{n-i}$ from $R_{i-1}$ until a negative result is obtained, and then $q_{n-i}$ is one less than the number of subtractions. This is how we find the digits of $q$.

**Example 1.33.** To divide $(11101)_2$ by $(111)_2$, we let $q = (q_2 q_1 q_0)_2$. We subtract $2^2(111)_2 = (11100)_2$ once from $(11101)_2$ to obtain $(1)_2$, and once more to obtain a negative result, so that $q_2 = 1$. Now $R_1 = (11101)_2 - (11100)_2 = (1)_2$. We find that $q_1 = 0$, since $R_1 - 2(111)_2$ is less than zero, and likewise $q_0 = 0$. Hence the quotient of the division is $(100)_2$ and the remainder is $(1)_2$. □

---

## 1.7 Exercises

1. Add $(101111011)_2$ and $(1100111011)_2$.

2. Subtract $(101110101)_2$ from $(1101101100)_2$.

3. Multiply $(11101)_2$ and $(110001)_2$.

4. Find the quotient and remainder when $(110100111)_2$ is divided by $(11101)_2$.

5. Add $(1234321)_5$ and $(2030104)_5$.

6.  Subtract $(434421)_5$ from $(4434201)_5$.

7.  Multiply $(1234)_5$ and $(3002)_5$.

8.  Find the quotient and remainder when $(14321)_5$ is divided by $(334)_5$.

9.  Add $(ABAB)_{16}$ and $(BABA)_{16}$.

10. Subtract $(CAFE)_{16}$ from $(FEED)_{16}$.

11. Multiply   $(FACE)_{16}$ and $(BAD)_{16}$.

12. Find the quotient and remainder when $(BEADED)_{16}$ is divided by $(ABBA)_{16}$.

13. Explain how to add, subtract, and multiply the integers 18235187 and 22135674 on a computer with word size 1000.

14. Write algorithms for the basic operations with integers in base $(-2)$ notation (see Exercise 6 of Section 1.6).

15. How is the one's complement representation of the sum of two integers obtained from the one's complement representations of these integers?

16. How is the one's complement representation of the difference of two integers obtained from the one's complement representations of these integers?

17. Give an algorithm for adding and an algorithm for subtracting *Cantor expansions* (see preamble to Exercise 20 of Section 1.6).

18. A *dozen* equals 12 and a *gross* equals $12^2$. Using base 12, or *duodecimal*, arithmetic answer the following questions.

    a)  If 3 gross, 7 dozen, and 4 eggs are removed from a total of 11 gross and 3 dozen eggs, how many eggs are left?

    b)  If 5 truckloads of 2 gross, 3 dozen, and 7 eggs each are delivered to the supermarket, how many eggs were delivered?

    c)  If 11 gross, 10 dozen and 6 eggs are divided in 3 groups of equal size, how many eggs are in each group?

19. A well-known rule used to find the square of an integer with decimal expansion $(a_{n-1}...a_1 a_0)_{10}$ with final digit $a_0 = 5$ is to find the decimal expansion of the product $(a_n a_{n-1}...a_1)_{10} [(a_n a_{n-1}...a_1)_{10} + 1]$ and append this with the digits $(25)_{10}$. For instance, we see that the decimal expansion of $(165)^2$ begins with $16 \cdot 17 = 272$, so that $(165)^2 = 27225$. Show that the rule just described is valid.

20. In this exercise we generalize the rule given in Exercise 19 to find the squares of integers with final base *2B* digit *B*, where *B* is a positive integer. Show that the base *2B* expansion of the integer $(a_n a_{n-1}...a_1 a_0)_{2B}$

starts with the digits of the base $2B$ expansion of the integer $(a_n a_{n-1}...a_1 a_0)_{2B}$ $[(a_n a_{n-1}...a_1 a_0)_{2B} + 1]$ and ends with the digits $B/2$ and 0 when $B$ is even, and the digits $(B-1)/2$ and $B$ when $B$ is odd.

## 1.7 Computer Projects

Write programs to do the following:

1.  Perform addition with arbitrarily large integers.

2.  Perform subtraction with arbitrarily large integers.

3.  Multiply two arbitrarily large integers using the conventional algorithm.

4.  Divide arbitrarily large integers, finding the quotient and remainder.

---

## 1.8 Complexity of Integer Operations

Once an algorithm is specified for an operation we can consider the amount of time required to perform this algorithm on a computer. We will measure the amount of time needed in terms of *bit operations*. By a bit operation we mean the addition, subtraction, or multiplication of two binary digits, the division of a two-bit integer by one-bit, or the shifting of a binary integer one place. (The actual amount of time required to carry out a bit operation on a computer varies depending on the computer architecture and capacity.) When we describe the number of bit operations needed to perform an algorithm, we are describing the *computational complexity* of this algorithm.

In describing the number of bit operations needed to perform calculations we will use *big-O* notation. To motivate the definition of this notation, consider the following situation. Suppose that performing a specified operation on an integer $n$ requires at most $n^3 + 8n^2 \log n$ bit operations. Since $8n^2 \log n < 8n^3$ for every positive integer, less that $9n^3$ bit operations are required for this operation for every integer $n$. Since the number of bit operations required is always less than a constant times $n^3$, namely $9n^3$, we say that $O(n^3)$ bit operations are needed. In general we have the following definition.

**Definition.** If $f$ and $g$ are functions taking positive values, defined for all $x \in S$, where $S$ is a specified set of real numbers, then $f$ is $O(g)$ on $S$ if there is a positive constant $K$ such that $f(x) < Kg(x)$ for all $x \in S$. (Normally we take $S$ to be the set of positive integers and we drop all reference to $S$.)

We illustrate this concept with several examples.

**Example 1.34.** We can show on the set of positive integers $n^4 + 2n^3 + 5$ is $O(n^4)$. To do this, note that $n^4 + 2n^3 + 5 \leqslant n^4 + 2n^4 + 5n^4 = 8n^4$ for all positive integers. (We take $K = 8$ in the definition.) The reader should also note that $n^4$ is $O(n^4 + 2n^3 + 5)$. □

**Example 1.35.** We can easily give a big-$O$ estimate for $\sum_{j=1}^{n} j$. Noting that each summand is less than $n$ tells us that $\sum_{j=1}^{n} j \leqslant \sum_{j=1}^{n} n = n \cdot n = n^2$. Note that we could also easily derive this estimate from the formula $\sum_{j=1}^{n} j = n(n+1)/2$. □

We now will give some useful results for working with big-$O$ estimates for combinations of functions.

**Theorem 1.11.** If $f$ is $O(g)$ and $c$ is a positive constant, then $cf$ is $O(g)$.

*Proof.* If $f$ is $O(g)$, then there is a constant $K$ with $f(x) < Kg(x)$ for all $x$ under consideration. Hence $cf(x) < (cK)g(x)$, so that $cf$ is $O(g)$. ∎

**Theorem 1.12.** If $f_1$ is $O(g_1)$ and $f_2$ is $O(g_2)$, then $f_1 + f_2$ is $O(g_1+g_2)$ and $f_1 f_2$ is $O(g_1 g_2)$.

*Proof.* If $f$ is $O(g_1)$ and $f_2$ is $O(g_2)$, then there are constants $K_1$ and $K_2$ such that $f_1(x) < K_1 g_1(x)$ and $f_2(x) < K_2 g_2(x)$ for all $x$ under consideration. Hence

$$f_1(x) + f_2(x) < K_1 g_1(x) + K_2 g_2(x)$$
$$\leqslant K(g_1(x) + g_2(x))$$

where $K$ is the maximum of $K_1$ and $K_2$. Hence $f_1 + f_2$ is $O(g_1 + g_2)$.

Also

$$f_1(x)f_2(x) < K_1 g_1(x) K_2 g_2(x)$$
$$= (K_1 K_2)(g_1(x)g_2(x)),$$

so that $f_1 f_2$ is $O(g_1 g_2)$. ∎

**Corollary 1.12.1.** If $f_1$ and $f_2$ are $O(g)$, then $f_1 + f_2$ is $O(g)$.

*Proof.* Theorem 1.12 tells us that $f_1 + f_2$ is $O(2g)$. But if $f_1 + f_2 < K(2g)$, then $f_1 + f_2 < (2K)g$, so that $f_1 + f_2$ is $O(g)$. ∎

We illustrate how to use these theorems with the following examples.

**Example 1.36.** To give a big-$O$ estimate for $(n + 8 \log n)$ $(10n \log n + 17n^2)$, first note that $n + 8 \log n$ is $O(n)$ and $10n \log n + 17n^2$ is $O(n^2)$ (since $\log n$ is $O(n)$ and $n \log n$ is $O(n^2)$ ) by Theorems 1.11 and 1.12 and Corollary 1.12.1. By Theorem 1.12 we see that $(n + 8 \log n)(10n \log n + 17n^2)$ is $O(n^3)$. □

Using big-$O$ notation we can see that to add or subtract two $n$-bit integers takes $O(n)$ bit operations, while to multiply two $n$-bit integers in the conventional way takes $O(n^2)$ bit operations (see Exercises 12 and 13 at the end of this section). Surprisingly, there are faster algorithms for multiplying large integers. To develop one such algorithm, we first consider the multiplication of two $2n$-bit integers, say $a = (a_{2n-1}a_{2n-2}...a_1a_0)_2$ and $b = (b_{2n-1}b_{2n-2}...b_1b_0)_2$. We write $a = 2^nA_1 + A_0$ and $b = 2^nB_1 + B_0$, where $A_1 = (a_{2n-1}a_{2n-2}...a_{n+1}a_n)_2$, $A_0 = (a_{n-1}a_{n-2}...a_1a_0)_2$, $B_1 = (b_{2n-1}b_{2n-2}...b_{n+1}\ b_n)_2$, and $B_0 = (b_{n-1}\ b_{n-2}...b_1b_0)_2$. We will use the identity

$$(1.9) \qquad ab = (2^{2n}+2^n)A_1B_1 + 2^n(A_1-A_0)(B_0-B_1) + (2^n+1)A_0B_0.$$

To find the product of $a$ and $b$ using (1.9), requires that we perform three multiplications of $n$-bit integers (namely $A_1B_1$, $(A_1 - A_0)(B_0 - B_1)$, and $A_0B_0$), as well as a number of additions and shifts. If we let $M(n)$ denote the number of bit operations needed to multiply two $n$-bit integers, we find from (1.9) that

$$(1.10) \qquad M(2n) \leqslant 3M(n) + Cn,$$

where $C$ is a constant, since each of the three multiplications of $n$-bit integers takes $M(n)$ bit operations, while the number of additions and shifts needed to compute $a \cdot b$ via (1.9) does not depend on $n$, and each of these operations takes $O(n)$ bit operations.

From (1.10), using mathematical induction, we can show that

$$(1.11) \qquad M(2^k) \leqslant c(3^k - 2^k),$$

where $c$ is the maximum of the quantities $M(2)$ and $C$ (the constant in (1.10)). To carry out the induction argument, we first note that with $k = 1$, we have $M(2) \leqslant c(3^1 - 2^1) = c$, since $c$ is the maximum of $M(2)$ and $C$.

As the induction hypothesis, we assume that

$$M(2^k) \leqslant c(3^k - 2^k).$$

Then, using (1.10), we have

$$\begin{aligned} M(2^{k+1}) &\leqslant 3M(2^k) + C2^k \\ &\leqslant 3c(3^k - 2^k) + C2^k \\ &\leqslant c\,3^{k+1} - c\cdot3\cdot2^k + c\,2^k \\ &\leqslant c(3^{k+1} - 2^{k+1}). \end{aligned}$$

This establishes that (1.11) is valid for all positive integers $k$.

Using inequality (1.11), we can prove the following theorem.

**Theorem 1.13.** Multiplication of two $n$-bit integers can be performed using $O(n^{\log_2 3})$ bit operations. (*Note:* $\log_2 3$ is approximately 1.585, which is considerably less than the exponent 2 that occurs in the estimate of the number of bit operations needed for the conventional multiplication algorithm.)

*Proof.* From (1.11) we have

$$M(n) = M(2^{\log_2 n}) \leqslant M(2^{[\log_2 n]+1})$$
$$\leqslant c(3^{[\log_2 n]+1} - 2^{[\log_2 n]+1})$$
$$\leqslant 3c \cdot 3^{[\log_2 n]} \leqslant 3c \cdot 3^{\log_2 n} = 3cn^{\log_2 3}$$
$$(\text{since } 3^{\log_2 n} = n^{\log_2 3}).$$

Hence, $M(n) = O(n^{\log_2 3})$.                                                     ■

We now state, without proof, two pertinent theorems. Proofs may be found in Knuth [80] or Kronsjö [82].

**Theorem 1.14.** Given a positive number $\epsilon > 0$, there is an algorithm for multiplication of two $n$-bit integers using $O(n^{1+\epsilon})$ bit operations.

Note that Theorem 1.4 is a special case of Theorem 1.5 with $\epsilon = \log_2 3 - 1$, which is approximately 0.585.

**Theorem 1.15.** There is an algorithm to multiply two $n$-bit integers using $O(n \log_2 n \log_2 \log_2 n)$ bit operations.

Since $\log_2 n$ and $\log_2 \log_2 n$ are much smaller than $n^\epsilon$ for large numbers $n$, Theorem 1.15 is an improvement over Theorem 1.14. Although we know that $M(n) = O(n \log_2 n \log_2 \log_2 n)$, for simplicity we will use the obvious fact that $M(n) = O(n^2)$ in our subsequent discussions.

The conventional algorithm described above performs a division of a $2n$-bit integer by an $n$-bit integer with $O(n^2)$ bit operations. However, the number of bit operations needed for integer division can be related to the number of bit operations needed for integer multiplication. We state the following theorem, which is based on an algorithm which is discussed in Knuth [80].

**Theorem 1.16.** There is an algorithm to find the quotient $q = [a/b]$, when the $2n$-bit integer $a$ is divided by the integer $b$ having no more than $n$ bits, using $O(M(n))$ bit operations, where $M(n)$ is the number of bit operations needed to multiply two $n$-bit integers.

## 1.8 Exercises

1. Determine whether each of the following functions is $O(n)$ on the set of positive integers.

   a) $2n + 7$      d) $\log(n^2 + 1)$
   b) $n^2/3$       e) $\sqrt{n^2 + 1}$
   c) $10$         f) $(n^2 + 1)/(n + 1)$

2. Show that $2n^4 + 3n^3 + 17$ is $O(n^4)$ on the set of positive integers.

3. Show that $(n^3 + 4n^2 \log n + 101n^2)(14n \log n + 8n)$ is $O(n^4 \log n)$.

4. Show that $n! = O(n^n)$ on the set of positive integers.

5. Show that $(n! + 1)(n + \log n) + (n^3 + n^n)((\log n)^3 + n + 7)$ is $O(n^{n+1})$.

6. Suppose that $m$ is a positive real number. Show that $\sum_{j=1}^{n} j^m = O(n^{m+1})$.

\* 7. Show that $n \log n = O(\log n!)$ on the set of positive integers.

8. Show that if $f_1$ and $f_2$ are $O(g_1)$ and $O(g_2)$, respectively, and $c_1$ and $c_2$ are constants, then $c_1 f_1 + c_2 f_2$ is $O(g_1 + g_2)$.

9. Show that if $f$ is $O(g)$, then $f^k$ is $O(g^k)$ for all positive integers $k$.

10. Show that a function $f$ is $O(\log_2 n)$ if and only if $f$ is $O(\log_r n)$ whenever $r > 1$. (*Hint*: Recall that $\log_a n/\log_b n = \log_a b$.)

11. Show that the base $b$ expansion of a positive integer $n$ has $[\log_b n]+1$ digits.

12. Analyzing the conventional algorithms for subtraction and addition, show that with $n$-bit integers these operations require $O(n)$ bit operations.

13. Show that to multiply an $n$-bit and an $m$-bit integer in the conventional manner requires $O(nm)$ bit operations.

14. Estimate the number of bit operations needed to find $1+2+ \cdots + n$

   a) by performing all the additions.

   b) by using the identity $1+2+ \cdots + n = n(n+1)/2$, and multiplying and shifting.

15. Give an estimate for the number of bit operations needed to find each of the following quantities.

   a) $n!$        b) $\dbinom{n}{k}$

16. Give an estimate of the number of bit operations needed to find the binary expansion of an integer from its decimal expansion.

17. a) Show there is an identity analogous to (1.9) for decimal expansions.

    b) Using part (a), multiply 73 and 87 performing only three multiplications of one-digit integers, plus shifts and additions.

    c) Using part (a), reduce the multiplication of 4216 and 2733 to three multiplications of two-digit integers, plus shifts and additions, and then using part (a) again, reduce each of the multiplications of two-digit integers into three multiplications of one-digit integers, plus shifts and additions. Complete the multiplication using only nine multiplications of one-digit integers, and shifts and additions.

18. If $A$ and $B$ are $n \times n$ matrices, with entries $a_{ij}$ and $b_{ij}$ for $1 \leqslant i \leqslant n$, $1 \leqslant j \leqslant n$, then $AB$ is the $n \times n$ matrix with entries $c_{ij} = \sum_{k=1}^{n} a_{ik} b_{kj}$. Show that $n^3$ multiplications of integers are used to find $AB$ directly from its definition.

19. Show it is possible to multiply two $2 \times 2$ matrices using only seven multiplications of integers by using the identity

$$
\begin{pmatrix} a_{11} & a_{12} \\ a_{21} & a_{22} \end{pmatrix} \begin{pmatrix} b_{11} & b_{12} \\ b_{21} & b_{22} \end{pmatrix} =
$$

$$
\begin{pmatrix} a_{11}b_{11} + a_{12}b_{21} & x + (a_{21} + a_{22})(b_{12}-b_{11}) + \\ & (a_{11}+a_{12}-a_{21}-a_{22})b_{22} \\ \\ x + (a_{11}-a_{21})(b_{22}-b_{12}) - & x + (a_{11}-a_{21})(b_{22}-b_{12}) + \\ a_{22}(b_{11}-b_{21}-b_{12}+b_{22}) & (a_{21} + a_{22})(b_{12}-b_{11}) \end{pmatrix}
$$

where $x = a_{11}b_{11} - (a_{11} - a_{21} - a_{22})(b_{11} - b_{12} + b_{22})$.

* 20. Using an inductive argument, and splitting $2n \times 2n$ matrices into four $n \times n$ matrices, use Exercise 19 to show that it is possible to multiply two $2^k \times 2^k$ matrices using only $7^k$ multiplications, and less than $7^{k+1}$ additions.

21. Conclude from Exercise 20 that two $n \times n$ matrices can be multiplied using $O(n^{\log_2 7})$ bit operations when all entries of the matrices have less than $c$ bits, where $c$ is a constant.

## 1.8 Computer Projects

Write programs to do the following:

* 1. Multiply two arbitrarily large integers using the identity (1.9).

** **2.** Multiply two $n \times n$ matrices using the algorithm discussed in Exercise 18-20.

## 1.9 Prime Numbers

The positive integer 1 has just one positive divisor. Every other positive integer has at least two positive divisors, because it is divisible by 1 and by itself. Integers with exactly two positive divisors are of great importance in number theory; they are called *primes*.

**Definition.** A *prime* is a positive integer greater than 1 that is divisible by no positive integers other than 1 and itself.

**Example 1.37.** The integers 2,3,5,13,101 and 163 are primes. □

**Definition.** A positive integer which is not prime, and which is not equal to 1, is called *composite*.

**Example 1.38.** The integers $4 = 2 \cdot 2$, $8 = 4 \cdot 2$, $33 = 3 \cdot 11$, $111 = 3 \cdot 37$, and $1001 = 7 \cdot 11 \cdot 13$ are composite. □

The primes are the building blocks of the integers. Later, we will show that every positive integer can be written uniquely as the product of primes.

Here, we briefly discuss the distribution of primes and mention some conjectures about primes. We start by showing that there are infinitely many primes. The following lemma is needed.

**Lemma 1.1.** Every positive integer greater than one has a prime divisor.

*Proof.* We prove the lemma by contradiction; we assume that there is a positive integer having no prime divisors. Then, since the set of positive integers with no prime divisors is non-empty, the well-ordering property tells us that there is a least positive integer $n$ with no prime divisors. Since $n$ has no prime divisors and $n$ divides $n$, we see that $n$ is not prime. Hence, we can write $n=ab$ with $1 < a < n$ and $1 < b < n$. Because $a < n$, $a$ must have a prime divisor. By Theorem 1.7 any divisor of $a$ is also a divisor of $n$, so that $n$ must have a prime divisor, contradicting the fact that $n$ has no prime divisors. We can conclude that every positive integer has at least one prime divisor. ∎

We now show that the number of primes is infinite. This result was known to the ancient Greek mathematician Euclid. We give a proof similar to that given by Euclid (the proof from his *Elements* is outlined in Exercise 6).

**Theorem 1.17.** There are infinitely many primes.

*Proof.* Consider the integer

$$Q_n = n! + 1, \qquad n \geqslant 1.$$

Lemma 1.1. tells us that $Q_n$ has at least one prime divisor, which we denote by $q_n$. Thus, $q_n$ must be larger than $n$; for if $q_n \leqslant n$, it would follow that $q_n \mid n!$, and then, by Thereom 1.8 $q_n \mid (Q_n - n!) = 1$, which is impossible.

Since we have found a prime larger than $n$, for every positive integer $n$, there must be infinitely many primes. ∎

Later we will be interested in finding, and using, extremely large primes. We will be concerned throughout this book with the problem of determining whether a given integer is prime. We first deal with this question by showing that by trial divisions of $n$ by primes not exceeding the square root of n, we can determine whether $n$ is prime.

**Theorem 1.18**. If $n$ is a composite integer, then $n$ has a prime factor not exceeding $\sqrt{n}$.

*Proof.* Since $n$ is composite, we can write $n = ab$, where $a$ and $b$ are integers with $1 < a \leqslant b < n$. We must have $a \leqslant \sqrt{n}$, since otherwise $b \geqslant a > \sqrt{n}$ and $ab > \sqrt{n} \cdot \sqrt{n} = n$. Now, by Lemma 1.1 $a$ must have a prime divisor, which by Theorem 1.7 is also a divisor of $n$ and which is clearly less than or equal to $\sqrt{n}$. ∎

We can use Theorem 1.18 to find all the primes less than or equal to a given positive integer $n$. This procedure is called the *sieve of Eratosthenes,* since it was invented by the ancient Greek mathematician Eratosthenes*. We illustrate its use in Figure 1.2 by finding all primes less than 100. We first note that every composite integer less than 100 must have a prime factor less than $\sqrt{100} = 10$. Since the only primes less than 10 are 2,3,4, and 7, we only need to check each integer less than 100 for divisibility by these primes. We first cross out, below by a horizontal slash —, all multiples of 2. Next we cross out with a slash / those integers remaining that are multiples of 3. Then all multiples of 5 that remain are crossed out, below by a backslash \. Finally, all multiples of 7 that are left are crossed out, below with a vertical

---

\* **ERATOSTHENES** (276 - 194 B.C.E.) was born in Cyrene, which was a Greek colony west of Egypt. It is known that he spent some time studying at Plato's school in Athens. King Ptolemy II invited Eratosthenes to Alexandria to tutor his son. Later Eratosthenes became the chief librarian of the famous library at Alexandria which was a central repository of ancient works of literature, art, and science. He was an extremely versatile scholar, having written on mathematics, geography, astronomy, history, philosophy, and literary criticism. Besides his work in mathematics, Eratosthenes was most noted for his chronology of ancient history and for his geographical measurements, including his famous measurement of the size of the earth.

slash |. All remaining integers (other than 1) must be prime.

| 1 | 2 | 3 | ~~4~~ | 5 | ~~6~~ | 7 | ~~8~~ | ~~9~~ | ~~10~~ |
|---|---|---|---|---|---|---|---|---|---|
| 11 | ~~12~~ | 13 | ~~14~~ | ~~15~~ | ~~16~~ | 17 | ~~18~~ | 19 | ~~20~~ |
| ~~21~~ | ~~22~~ | 23 | ~~24~~ | ~~25~~ | ~~26~~ | ~~27~~ | ~~28~~ | 29 | ~~30~~ |
| 31 | ~~32~~ | ~~33~~ | ~~34~~ | ~~35~~ | ~~36~~ | 37 | ~~38~~ | ~~39~~ | ~~40~~ |
| 41 | ~~42~~ | 43 | ~~44~~ | ~~45~~ | ~~46~~ | 47 | ~~48~~ | ~~49~~ | ~~50~~ |
| ~~51~~ | ~~52~~ | 53 | ~~54~~ | ~~55~~ | ~~56~~ | ~~57~~ | ~~58~~ | 59 | ~~60~~ |
| 61 | ~~62~~ | ~~63~~ | ~~64~~ | ~~65~~ | ~~66~~ | 67 | ~~68~~ | ~~69~~ | ~~70~~ |
| 71 | ~~72~~ | 73 | ~~74~~ | ~~75~~ | ~~76~~ | ~~77~~ | ~~78~~ | 79 | ~~80~~ |
| ~~81~~ | ~~82~~ | 83 | ~~84~~ | ~~85~~ | ~~86~~ | ~~87~~ | ~~88~~ | 89 | ~~90~~ |
| ~~91~~ | ~~92~~ | ~~93~~ | ~~94~~ | ~~95~~ | ~~96~~ | 97 | ~~98~~ | ~~99~~ | ~~100~~ |

**Figure 1.2.** Finding the Primes Less Than 100 Using the Sieve of Eratosthenes.

Although the sieve of Eratosthenes produces all primes less than or equal to a fixed integer, to determine whether a particular integer $n$ is prime in this manner, it is necessary to check $n$ for divisibility by all primes not exceeding $\sqrt{n}$. This is quite inefficient; later on we will have better methods for deciding whether or not an integer is prime.

We know that there are infinitely many primes, but can we estimate how many primes there are less than a positive real number $x$? One of the most famous theorems of number theory, and of all mathematics, is the *prime number theorem* which answers this question. To state this theorem, we introduce some notation.

**Definition.** The function $\pi(x)$, where $x$ is a positive real number, denotes the number of primes not exceeding $x$.

**Example 1.39.** From our example illustrating the sieve of Eratosthenes, we see that $\pi(10) = 4$ and $\pi(100) = 25$. □

We now state the prime number theorem.

**Theorem 1.19. The Prime Number Theorem.** The ratio of $\pi(x)$ to $x/\log x$ approaches one as $x$ grows without bound. (Here $\log x$ denotes the natural logarithm of $x$. In the language of limits, we have $\lim_{x \to \infty} \pi(x) / \dfrac{x}{\log x} = 1$).

The prime number theorem was conjectured by Gauss in 1793, but it was not proved until 1896, when a French mathematician J. Hadamard and a Belgian mathematician C. J. de la Vallée-Poussin produced independent proofs. We will not prove the prime number theorem here; the various proofs

known are either quite complicated or rely on advanced mathematics. In table
1.1 we give some numerical evidence to indicate the validity of the theorem.

| $x$ | $\pi(x)$ | $x/\log x$ | $\pi(x)/\dfrac{x}{\log x}$ | $li(x)$ | $\pi(x)/li(x)$ |
|---|---|---|---|---|---|
| $10^3$ | 168 | 144.8 | 1.160 | 178 | 0.9438202 |
| $10^4$ | 1229 | 1085.7 | 1.132 | 1246 | 0.9863563 |
| $10^5$ | 9592 | 8685.9 | 1.104 | 9630 | 0.9960540 |
| $10^6$ | 78498 | 72382.4 | 1.085 | 78628 | 0.9983466 |
| $10^7$ | 664579 | 620420.7 | 1.071 | 664918 | 0.9998944 |
| $10^8$ | 5761455 | 5428681.0 | 1.061 | 5762209 | 0.9998691 |
| $10^9$ | 50847534 | 48254942.4 | 1.054 | 50849235 | 0.9999665 |
| $10^{10}$ | 455052512 | 434294481.9 | 1.048 | 455055614 | 0.9999932 |
| $10^{11}$ | 4118054813 | 3948131663.7 | 1.043 | 4118165401 | 0.9999731 |
| $10^{12}$ | 37607912018 | 36191206825.3 | 1.039 | 37607950281 | 0.9999990 |
| $10^{13}$ | 346065535898 | 334072678387.1 | 1.036 | 346065645810 | 0.9999997 |

**Table 1.1.** Approximations to $\pi(x)$.

The prime number theorem tells us that $x/\log x$ is a good approximation to
$\pi(x)$ when $x$ is large. It has been shown that an even better approximation is
given by

$$li(x) = \int_2^x \frac{dt}{\log t}$$

(where $\int_2^x \dfrac{dt}{\log t}$ represents the area under the curve $y = 1/\log$ t, and above
the $t$-axis from $t = 2$ to $t = x$). In Table 1.1 we see evidence that $li(x)$ is an
excellent approximation of $\pi(x)$.

We can now estimate the number of bit operations needed to show that an
integer $n$ is prime by trial divisions of $n$ by all primes not exceeding $\sqrt{n}$. The
prime number theorem tells us that there are approximately
$\sqrt{n}/\log\sqrt{n} = 2\sqrt{n}/\log n$ primes not exceeding $\sqrt{n}$. To divide $n$ by an integer
$m$ takes $O(\log_2 n \cdot \log_2 m)$ bit operations. Therefore, the number of bit
operations needed to show that $n$ is prime by this method is at least
$(2\sqrt{n}/\log n)(c \log_2 n) = c\sqrt{n}$ (where we have ignored the $\log_2 m$ term since it
is at least 1, even though it sometimes is as large as $(\log_2 n)/2$). This method
of showing that an integer $n$ is prime is very inefficient, for not only is it

necessary to know all the primes not larger than $\sqrt{n}$, but it is also necessary to do at least a constant multiple of $\sqrt{n}$ bit operations. Later on we will have more efficient methods of showing that an integer is prime.

We remark here that it is not necessary to find all primes not exceeding $x$ in order to compute $\pi(x)$. One way that $\pi(x)$ can be evaluated without finding all the primes less then $x$ is to use a counting argument based on the sieve of Eratosthenes (see Exercise 16). (Recently, very efficient ways of finding $\pi(x)$ using $O(x^{3/5+\epsilon})$ bit operations have been devised by Lagarias and Odlyzko [69].)

We have shown that there are infinitely many primes and we have discussed the abundance of primes below a given bound $x$, but we have yet to discuss how regularly primes are distributed throughout the positive integers. We first give a result that shows that there are arbitrarily long runs of integers containing no primes.

**Theorem 1.20.** For any positive integer n, there are at least $n$ consecutive composite positive integers.

*Proof.* Consider the n consecutive positive integers

$$(n + 1)! + 2, \ (n + 1)! + 3, ..., (n + 1)! + n + 1.$$

When $2 \leqslant j \leqslant n + 1$, we know that $j \mid (n + 1)!$. By Theorem 1.8 it follows that $j \mid (n + 1)! + j$. Hence, these $n$ consecutive integers are all composite. ∎

**Example 1.40.** The seven consecutive integers beginning with $8! + 2 = 40322$ are all composite. (However, these are much larger than the smallest seven consecutive composites, 90, 91, 92, 93, 94, 95, and 96.) □

Theorem 1.20 shows that the gap between consecutive primes is arbitrarily long. On the other hand, primes may often be close together. The only consecutive primes are 2 and 3, because 2 is the only even prime. However, many pairs of primes differ by two; these pairs of primes are called *twin primes*. Examples are the primes 5 and 7, 11 and 13, 101 and 103, and 4967 and 4969. A famous unsettled conjecture asserts that there are infinitely many twin primes.

There are a multitude of conjectures concerning the number of primes of various forms. For instance, it is unknown whether there are infinitely many primes of the form $n^2 + 1$ where n is a positive integer. Questions such as this may be easy to state, but are sometimes extremely difficult to resolve.

We conclude this section by discussing perhaps the most notorious conjecture about primes.

**Goldbach's Conjecture.** Every even positive integer greater than two can be written as the sum of two primes.

This conjecture was stated by Christian Goldbach* in a letter to Leonhard Euler in 1742. It has been verified for all even integers less than a million. One sees by experimentation, as the following example illustrates, that usually there are many sums of two primes equal to a particular integer, but a proof that there always is at least one such sum has not yet been found.

**Example 1.41.** The integers 10,24, and 100 can be written as the sum of two primes in the following ways:

$$10 = 3 + 7 = 5 + 5,$$
$$24 = 5 + 19 = 7 + 17 = 11 + 13,$$
$$100 = 3 + 97 = 11 + 89 = 17 + 83$$
$$= 29 + 71 = 41 + 59 = 47 + 53. \qquad \square$$

---

## 1.9 Exercises

**1.** Determine which of the following integers are primes

    a)   101        c)   107        e)   113
    b)   103        d)   111        f)   121.

**2.** Use the sieve of Eratosthenes to find all primes less than 200.

**3.** Find all primes that are the difference of the fourth powers of two integers.

**4.** Show that no integer of the form $n^3 + 1$ is a prime, other than $2 = 1^3 + 1$.

**5.** Show that if $a$ and $n$ are positive integers such that $a^n - 1$ is prime, then $a = 2$ and $n$ is prime. (*Hint*: Use the identity $a^{kl} - 1 = (a^k - 1)(a^{k(l-1)} + a^{k(l-2)} + \cdots + a^k + 1)$).

---

* **CHRISTIAN GOLDBACH** (1690-1764) was born in Königsberg, Prussia (the city noted in mathematical circles for its famous bridge problem). He became professor of mathematics at the Imperial Academy of St. Petersburg in 1725. In 1728 Goldbach went to Moscow to tutor Tsarevich Peter II. In 1742 he entered the Russian Ministry of Foreign Affairs as a staff member. Goldbach is most noted for his correspondence with eminent mathematicians, in particular Leonhard Euler and Daniel Bernoulli. Besides his well known conjectures that every even positive integer in the sum of two primes and that every odd positive integer greater than five is the sum of three primes, Goldbach made several notable contributions to analysis.

6. In this exercise, another proof of the infinitude of primes is given. Assume there are only finitely many primes $p_1, p_2, ..., p_n$. Form the integer $Q = p_1 p_2 \cdots p_n + 1$. Show that $Q$ has a prime factor not in the above list. Conclude that there are infinitely many primes.

7. Let $Q_n = p_1 p_2 \cdots p_n + 1$ where $p_1, p_2, ..., p_n$ are the $n$ smallest primes. Determine the smallest prime factor of $Q_n$ for $n = 1, 2, 3, 4, 5$, and 6. Do you think $Q_n$ is prime infinitely often? (*Note*: This is an unresolved question.)

8. Let $p_1, p_2, ..., p_n$ be the first $n$ primes and let $m$ be an integer with $1 < m < n$. Let $Q$ be the product of a set of $m$ primes in the list and let $R$ be the product of the remaining primes. Show that $Q + R$ is not divisible by any primes in the list, and hence must have a prime factor not in the list. Conclude that there are infinitely many primes.

9. Show that if the smallest prime factor $p$ of the positive integer $n$ exceeds $\sqrt[3]{n}$ then $n/p$ must be prime or 1.

10. a) Find the smallest five consecutive composite integers.
    b) Find one million consecutive composite integers.

11. Show that there are no "prime triplets", *i.e.* primes $p$, $p + 2$, and $p + 4$, other than 3,5, and 7.

12. Verify Goldbach's conjecture for each of the following values of $n$.

    a) 50    c) 102    e) 200
    b) 98    d) 144    f) 222

13. Goldbach also conjectured that every odd positive integer greater than five is the sum of three primes. Verify this conjecture for each of the following odd integers.

    a) 7    c) 27    e) 101
    b) 17   d) 97    f) 199

14. Show that every integer greater than 11 is the sum of two composite integers.

15. Use the second principle of mathematical induction to prove that every integer greater than one is either prime or the product of two or more primes.

* 16. Use the principle of inclusion-exclusion (Exercise 18 of Section 1.4) to show that

$$\pi(n) = (\pi(\sqrt{n})-1) + n - \left(\left[\frac{n}{p_1}\right] + \left[\frac{n}{p_2}\right] + \cdots + \left[\frac{n}{p_r}\right]\right)$$

$$+ \left(\left[\frac{n}{p_1 p_2}\right] + \left[\frac{n}{p_1 p_3}\right] + \cdots + \left[\frac{n}{p_{r-1} p_r}\right]\right)$$

$$- \left(\left[\frac{n}{p_1 p_2 p_3}\right] + \left[\frac{n}{p_1 p_2 p_4}\right] + \cdots + \left[\frac{n}{p_{r-2} p_{r-1} p_r}\right]\right) + \cdots,$$

where $p_1, p_2, \ldots, p_r$ are the primes less than or equal to $\sqrt{n}$ (with $r = \pi(\sqrt{n})$). (*Hint*: Let property $P_i$ be the property that an integer is divisible by $p_i$.

17. Use Exercise 16 to find $\pi(250)$.

18. Show that the polynomial $x^2 - x + 41$ is prime for all integers $x$ with $0 \leq x \leq 40$. Show, however, that it is composite for $x = 41$.

* 19. Show that if $f(x) = a_n x^n + a_{n-1} x^{n-1} + \cdots + a_1 x + a_0$ where the coefficients are integers, then there is an integer $y$ such that $f(y)$ is composite. (*Hint*: Assume that $f(x) = p$ is prime, and show $p$ divides $f(x+kp)$ for all integers $k$. Conclude from the fact that a polynomial of degree $n$ takes on each value at most $n$ times, that there is an integer $y$ such that $f(y)$ is composite.)

The *lucky numbers* are generated by the following sieving process. Start with the positive integers. Begin the process by crossing out every second integer in the list, starting your count with the integer 1. Other than 1 the smallest integer left is 3, so we continue by crossing out every third integer left, starting the count with the integer 1. The next integer left is 7, so we cross out every seventh integer left. Continue this process, where at each stage we cross out every $k$th integer left where $k$ is the smallest integer left other than one. The integers that remain are the lucky numbers.

20. Find all lucky numbers less than 100.

21. Show that there are infinitely many lucky numbers.

22. Show that if $p$ is prime and $1 \leq k < p$, then the binomial coefficient $\binom{p}{k}$ is divisible by $p$.

* 23. Find all pairs of prime powers that differ by 1. Prove your answer is correct.

## 1.9 Computer Projects

Write programs to do the following:

1. Decide whether an integer is prime using trial division of the integer by all primes not exceeding its square root.

* **2.** Use the sieve of Eratosthenes to find all primes less than 10000.

** **3.** Find $\pi(n)$, the number of primes less than or equal to $n$, using Exercise 16.

    **4.** Verify Goldbach's conjecture for all even integers less than 10000.

    **5.** Find all twin primes less than 10000.

    **6.** Find the first 100 primes of the form $n^2 + 1$.

* **7.** Find the lucky numbers less than 10000 (see the preamble to Exercise 19).

# 2

---

# Greatest Common Divisors and Prime Factorization

---

## 2.1 Greatest Common Divisors

If $a$ and $b$ are integers, not both zero, then the set of common divisors of $a$ and $b$ is a finite set of integers, always containing the integers $+1$ and $-1$. We are interested in the largest integer among the common divisors of the two integers.

**Definition.** The *greatest common divisor* of two integers $a$ and $b$, that are not both zero, is the largest integer which divides both $a$ and $b$.

The greatest common divisor of $a$ and $b$ is written as $(a,b)$. We also define $(0,0) = 0$.

**Example 2.1.** The common divisors of 24 and 84 are $\pm 1$, $\pm 2$, $\pm 3$, $\pm 4$, $\pm 6$, and $\pm 12$. Hence $(24,84) = 12$. Similarly, looking at sets of common divisors, we find that $(15,81) = 3$, $(100,5) = 5$, $(17,25) = 1$, $(0,44) = 44$, $(-6,-15) = 3$, and $(-17,289) = 17$. $\qquad\square$

We are particularly interested in pairs of integers sharing no common divisors greater than 1. Such pairs of integers are called *relatively prime*.

**Definition.** The integers $a$ and $b$ are called *relatively prime* if $a$ and $b$ have greatest common divisor $(a,b) = 1$.

**70**

**Example 2.2.** Since $(25,42) = 1$, 25 and 42 are relatively prime.

Note that since the divisors of $-a$ are the same as the divisors of $a$, it follows that $(a,b) = (|a|,|b|)$ (where $|a|$ denotes the absolute value of $a$ which equals $a$ if $a \geqslant 0$ and equals $-a$ if $a < 0$). Hence, we can restrict our attention to greatest common divisors of pairs of positive integers. □

We now prove some properties of greatest common divisors.

**Theorem 2.1.** Let $a$, $b$, and $c$ be integers with $(a,b) = d$. Then

$$(i) \quad (a/d,b/d) = 1$$
$$(ii) \quad (a+cb,b) = (a,b).$$

*Proof.* (i) Let $a$ and $b$ be integers with $(a,b) = d$. We will show that $a/d$ and $b/d$ have no common positive divisors other than 1. Assume that $e$ is a positive integer so that $e \mid (a/d)$ and $e \mid (b/d)$. Then, there are integers $k$ and $l$ with $a/d = ke$ and $b/d = le$, such that $a = dek$ and $b = del$. Hence, $de$ is a common divisor of $a$ and $b$. Since $d$ is the greatest common divisor of $a$ and $b, de \leqslant d$ so that $e$ must be 1. Consequently, $(a/d,b/d) = 1$.

(ii) Let $a$, $b$, and $c$ be integers. We will show that the common divisors of $a$ and $b$ are exactly the same as the common divisors of $a + cb$ and $b$. This will show that $(a+cb,b) = (a,b)$. Let $e$ be a common divisor of $a$ and $b$. By Theorem 1.8 we see that $e \mid (a+cb)$, so that $e$ is a common divisor of $a + cb$ and $b$. If $f$ is a common divisor of $a + cb$ and $b$, then by Theorem 1.8, we see that $f$ divides $(a+cb) - cb = a$, so that $f$ is a common divisor of $a$ and $b$. Hence $(a+cb,b) = (a,b)$. ∎

We will show that the greatest common divisor of the integers $a$ and $b$, that are not both zero, can be written as a sum of multiples of $a$ and $b$. To phrase this more succinctly, we use the following definition.

**Definition.** If $a$ and $b$ are integers, then a *linear combination* of $a$ and $b$ is a sum of the form $ma + nb$, where both $m$ and $n$ are integers.

We can now state and prove the following theorem about greatest common divisors.

**Theorem 2.2.** The greatest common divisor of the integers $a$ and $b$, that are not both zero, is the least positive integer that is a linear combination of $a$ and $b$.

*Proof.* Let $d$ be the least positive integer which is a linear combination of $a$ and $b$. (There is a *least* such positive integer, using the well-ordering property, since at least one of two linear combinations $1 \cdot a + 0 \cdot b$ and $(-1)a + 0 \cdot b$, where $a \neq 0$, is positive.) We write

(2.1)                            $d = ma + nb,$

where $m$ and $n$ are integers. We will show that $d \mid a$ and $d \mid b$.

By the division algorithm, we have

$$a = dq + r, \qquad 0 \leqslant r < d.$$

From this equation and (2.1), we see that

$$r = a - dq = a - q(ma + nb) = (1 - qm)a - qnb.$$

This shows that the integer $r$ is a linear combination of $a$ and $b$. Since $0 \leqslant r < d$, and $d$ is the least positive linear combination of $a$ and $b$, we conclude that $r = 0$, and hence $d \mid a$. In a similar manner, we can show that $d \mid b$.

We now demonstrate that $d$ is the *greatest* common divisor of $a$ and $b$. To show this, all we need to show is that any common divisor $c$ of $a$ and $b$ must divide $d$. Since $d = ma + nb$, if $c \mid a$ and $c \mid b$, Theorem 1.8 tells us that $c \mid d$.  ∎

We have shown that the greatest common divisor of the integers $a$ and $b$, that are not both zero, is a linear combination of $a$ and $b$. How to find a particular linear combination of $a$ and $b$ equal to $(a,b)$ will be discussed in the next section.

We can also define the greatest common divisor of more than two integers.

**Definition.** Let $a_1, a_2,..., a_n$ be integers, that are not all zero. The *greatest common divisor* of these integers is the largest integer which is a divisor of all of the integers in the set. The greatest common divisor of $a_1, a_2,..., a_n$ is denoted by $(a_1, a_2,..., a_n)$.

**Example 2.3.** We easily see that $(12, 18, 30) = 6$ and $(10, 15, 25) = 5$.  □

We can use the following lemma to find the greatest common divisor of a set of more than two integers.

**Lemma 2.1.** If $a_1, a_2,..., a_n$ are integers, that are not all zero, then $(a_1, a_2,..., a_{n-1}, a_n) = (a_1, a_2,..., (a_{n-1}, a_n))$.

*Proof.* Any common divisor of the $n$ integers $a_1, a_2,..., a_{n-1}, a_n$ is, in particular, a divisor of $a_{n-1}$ and $a_n$, and therefore a divisor of $(a_{n-1}, a_n)$. Also, any common divisor of the $n-2$ integers $a_1, a_2,..., a_{n-2}$, and $(a_{n-1}, a_n)$ must be a common divisor of all $n$ integers, for if it divides $(a_{n-1}, a_n)$, it must divide both $a_{n-1}$ and $a_n$. Since the set of $n$ integers and the set of the first $n-2$ integers together with the greatest common divisor of the last two integers have exactly the same divisors, their greatest common divisors are equal.  ∎

**Example 2.4.** To find the greatest common divisor of the three integers 105, 140, and 350, we use Lemma 2.1 to see that $(105,140,350) = (105,(140,350)) = (105,70) = 35$.  □

**Definition.** We say that the integers $a_1,a_2,...,a_n$ are *mutually relatively prime* if $(a_1,a_2,...,a_n) = 1$. These integers are called *pairwise relatively prime* if for each pair of integers $a_i$ and $a_j$ from the set, $(a_i,a_j) = 1$, that is, if each pair of integers from the set is relatively prime.

It is easy to see that if integers are pairwise relatively prime, they must be mutually relatively prime. However, the converse is false as the following example shows.

**Example 2.5.** Consider the integers 15, 21, and 35. Since

$$(15,21,35) = (15,(21,35)) = (15,7) = 1,$$

we see that the three integers are mutually relatively prime. However, any two of these integers are not relatively prime because $(15,21) = 3$, $(15,35) = 5$, and $(21,35) = 7$.  □

---

## 2.1 Exercises

1. Find the greatest common divisor of each of the following pairs of integers.

   a) 15, 35
   b) 0, 111
   c) −12, 18
   d) 99, 100
   e) 11, 121
   f) 100, 102

2. Let $a$ be a positive integer. What is the greatest common divisor of $a$ and $a^2$?

3. Let $a$ be a positive integer. What is the greatest common divisor of $a$ and $a+1$?

4. Let $a$ be a positive integer. What is the greatest common divisor of $a$ and $a+2$?

5. Show that if $a$ and $b$ are integers that are not both zero, and $c$ is a nonzero integer, then $(ca,cb) = |c|(a,b)$.

6. Show that if $a$ and $b$ are integers with $(a,b) = 1$, then $(a+b,a-b) = 1$ or 2.

7. What is $(a^2+b^2,a+b)$, where $a$ and $b$ are relatively prime integers that are not both zero?

8. Show that if $a$ and $b$ are both even integers that are not both zero, then $(a,b) = 2(a/2,b/2)$.

9. Show that if $a$ is an even integer and $b$ is an odd integer, then $(a,b) = (a/2,b)$.

10. Show that if $a,b$, and $c$ are integers such that $(a,b) = 1$ and $c \mid (a+b)$, then $(c,a) = (c,b) = 1$.

11. Show that if $a$, $b$, and $c$ are integers with $c \mid ab$, then $c \mid (a,c)(b,c)$.

12. a) Show that if $a$ and $b$ are positive integers with $(a,b) = 1$, then $(a^n,b^n) = 1$ for all positive integers $n$.

    b) Use part (a) to prove that if $a$ and $b$ are integers such that $a^n \mid b^n$ where $n$ is a positive integer, then $a \mid b$.

13. Show that if $a$, $b$ and $c$ are mutually relatively prime nonzero integers, then $(a,bc) = (a,b)(a,c)$.

☛ 14. a) Show that if $a$, $b$, and $c$ are integers with $(a,b) = (a,c) = 1$, then $(a,bc) = 1$.

    b) Use mathematical induction to show that if $a_1,a_2,...,a_n$ are integers, and $b$ is another integer such that $(a_1,b) = (a_2,b) = ... = (a_n,b) = 1$, then $(a_1a_2 \cdots a_n,b) = 1$.

15. Find a set of three integers that are mutually relatively prime, but any two of which are not relatively prime. Do not use examples from the text.

16. Find four integers that are mutually relatively prime such that any two of these integers are not relatively prime.

17. Find the greatest common divisor of each of the following sets of integers

    a) 8, 10, 12          d) 6, 15, 21
    b) 5, 25, 75          e) $-7$, 28, $-35$
    c) 99, 9999, 0        f) 0, 0, 1001

18. Find three mutually relatively prime integers from among the integers 66, 105, 42, 70, and 165.

19. Show that $a_1$, $a_2$,..., $a_n$ are integers that are not all zero and $c$ is a positive integer, then $(ca_1,ca_2,...,ca_n) = c(a_1, a_2..., a_n)$.

20. Show that the greatest common divisor of the integers $a_1$, $a_2$,..., $a_n$, that are not all zero, is the least positive integer that is a linear combination of $a_1, a_2,..., a_n$.

21. Show that if $k$ is an integer, then the integers $6k-1$, $6k+1$, $6k+2$, $6k+3$, and $6k+5$ are pairwise relatively prime.

22. Show that if $k$ is a positive integer then $3k+2$ and $5k+3$ are relatively prime.

23. Show that every positive integer greater than six is the sum of two relatively prime integers greater than one.

The *Farey series* $\mathscr{F}_n$ *of order n* is the set of fractions $h/k$ where $h$ and $k$ are integers, $0 \leqslant h \leqslant k \leqslant n$, and $(h,k) = 1$, in ascending order. We include 0 and 1 in the forms $\dfrac{0}{1}$ and $\dfrac{1}{1}$, respectively. For instance, the Farey series of order 4 is

$$\frac{0}{1}, \frac{1}{4}, \frac{1}{3}, \frac{1}{2}, \frac{2}{3}, \frac{3}{4}, \frac{1}{1}.$$

Exercises 24 - 27 deal with the Farey series.

24. Find the Farey series of order 7.

* 25. Show that if $a/b$ and $c/d$ are successive terms of a Farey series, then $ad - bc = 1$.

* 26. Show that if $a/b$, $c/d$, and $e/f$ are successive terms of a Farey series, then

$$\frac{c}{d} = \frac{a+e}{b+f}.$$

* 27. Show that if $a/b$ and $c/d$ are successive terms of the Farey series of order $n$, then $b+d > n$.

* 28. Show that if $a$ and $b$ are relatively prime positive integers, then $((a^n-b^n)/(a-b),a-b) = 1$ or $n$.

* 29. Show that if $a$ and $b$ are positive integers, then $((a^n-b^n)/(a-b), a-b) = (n(a, b)^{n-1},a-b)$.

## 2.1 Computer Projects

Write programs to do the following things.

1. Find the greatest common divisor of two integers from the lists of their divisors.

2. Print out the Farey series of order $n$ for a given positive integer $n$.

## 2.2 The Euclidean Algorithm

We are going to develop a systematic method, or *algorithm*, to find the greatest common divisor of two positive integers. This method is called the *Euclidean algorithm*. It is named after the ancient Greek mathematician Euclid* who describes this algorithm in his book *The Elements*.

Before we discuss the algorithm in general, we demonstrate its use with an example. We find the greatest common divisor of 30 and 72. First, we use the division algorithm to write $72 = 30 \cdot 2 + 12$, and we use Theorem 2.1 to note that $(30,72) = (30,72 - 2 \cdot 30) = (30,12)$. Another way to see that $(30,72) = (30,12)$ is to notice that any common divisor of 30 and 72 must also divide 12 because $12 = 72 - 30 \cdot 2$, and conversely, any common divisor of 12 and 30 must also divide 72, since $72 = 30 \cdot 2 + 12$. Note we have replaced 72 by the smaller number 12 in our computations since $(72,30) = (30,12)$. Next, we use the division algorithm again to write $30 = 2 \cdot 12 + 6$. Using the same reasoning as before, we see that $(30,12) = (12,6)$. Because $12 = 6 \cdot 2 + 0$, we now see that $(12,6) = (6,0) = 6$. Consequently, we can conclude that $(72,30) = 6$, without finding all the common divisors of 30 and 72.

We now set up the general format of the Euclidean algorithm for computing the greatest common divisor of two positive integer.

**Theorem 2.3. The Euclidean Algorithm.** Let $r_0 = a$ and $r_1 = b$ be integers such that $a \geqslant b > 0$. If the division algorithm is successively applied to obtain $r_j = r_{j+1}q_{j+1} + r_{j+2}$ with $0 < r_{j+2} < r_{j+1}$ for $j = 0,1,2,...,n-2$ and $r_{n+1} = 0$, then $(a,b) = r_n$, the last nonzero remainder.

From this theorem, we see that the greatest common divisor of $a$ and $b$ is the last nonzero remainder in the sequence of equations generated by successively using the division algorithm continuing until a remainder is 0, where at each step, the dividend and divisor are replaced by smaller numbers, namely the divisor and remainder.

---

\*  EUCLID (c. 350 B.C.E) was the author of the most successful mathematics textbook ever written, namely his *Elements,* which appeared in over a thousand different additions from ancient to modern times. Very little is known about Euclid's life other than that he taught at the famed academy at Alexandria. Evidently he did not stress the applications of mathematics for it is reputed that when asked by a student for the use of geometry, Euclid had his slave give the student some coins, "since he must needs make gain of what he learns." Euclid's *Elements* provides an introduction to plane and solid geometry and to number theory. The Euclidean algorithm is found in Book VII of the 13 books in the *Elements* and his proof of the infinitude of primes in found in Book IX. Euclid also wrote books on a variety of other topics, including astronomy, optics, music, and mechanics.

To prove that the Euclidean algorithm produces greatest common divisors, the following lemma will be helpful.

**Lemma 2.2.** If $c$ and $d$ are integers and $c = dq + r$ where $q$ and $r$ are integers, then $(c,d) = (d,r)$.

*Proof.* If an integer $e$ divides both $c$ and $d$, then since $r = c - dq$, Theorem 1.8 shows that $e \mid r$. If $e \mid d$ and $e \mid r$, then since $c = dq + r$, from Theorem 1.8 we see that $e \mid c$. Since the common divisors of $c$ and $d$ are the same as the common divisors of $d$ and $r$, we see that $(c,d) = (d,r)$. ∎

We now prove that the Euclidean algorithm works.

*Proof.* Let $r_0 = a$ and $r_1 = b$ be positive integers with $a \geqslant b$. By successively applying the division algorithm, we find that

$$
\begin{aligned}
r_0 &= r_1 q_1 + r_2 & 0 &\leqslant r_2 < r_1, \\
r_1 &= r_2 q_2 + r_3 & 0 &\leqslant r_3 < r_2,
\end{aligned}
$$

$$\vdots$$

$$
r_{j-2} = r_{j-1} q_{j-1} + r_j \qquad 0 \leqslant r_j < r_{j-1},
$$

$$\vdots$$

$$
\begin{aligned}
r_{n-4} &= r_{n-3} q_{n-3} + r_{n-2} & 0 &\leqslant r_{n-2} < r_{n-3}, \\
r_{n-3} &= r_{n-2} q_{n-2} + r_{n-1} & 0 &\leqslant r_{n-1} < r_{n-2}, \\
r_{n-2} &= r_{n-1} q_{n-1} + r_n & 0 &\leqslant r_n < r_{n-1} \\
r_{n-1} &= r_n q_n.
\end{aligned}
$$

We can assume that we eventually obtain a remainder of zero since the sequence of remainders $a = r_0 > r_1 > r_2 > \cdots \geqslant 0$ cannot contain more than $a$ terms. By Lemma 2.2 we see that $(a,b) = (r_0,r_1) = (r_1,r_2) = (r_2,r_3)$ $= \ldots = (r_{n-3},r_{n-2}) = (r_{n-2}, r_{n-1}) = (r_{n-1},r_n) = (r_n,0) = r_n$. Hence $(a,b) = r_n$, the last nonzero remainder. ∎

We illustrate the use of the Euclidean algorithm with the following example.

**Example 2.6.** To find $(252,198)$, we use the division algorithm successively to obtain

$$
\begin{aligned}
252 &= 1 \cdot 198 + 54 \\
198 &= 3 \cdot 54 + 36 \\
54 &= 1 \cdot 36 + 18 \\
36 &= 2 \cdot 18.
\end{aligned}
$$

Hence $(252,198) = 18$. □

Later in this section, we give estimates for the maximum number of divisions used by the Euclidean algorithm to find the greatest common divisor of two positive integers. However, we first show that given any positive integer $n$, there are integers $a$ and $b$ such that exactly $n$ divisions are required to find $(a,b)$ using the Euclidean algorithm. We can find such numbers by taking successive terms of the Fibonacci sequence.

The reason that the Euclidean algorithm operates most slowly when it finds the greatest common divisor of successive Fibonacci numbers is that the quotient in all but the last step is 1, as illustrated in the following example.

**Example 2.7.** We apply the Euclidean algorithm to find $(34,55)$. Note that $f_9 = 34$ and $f_{10} = 55$. We have

$$
\begin{aligned}
55 &= 34 \cdot 1 + 21 \\
34 &= 21 \cdot 1 + 13 \\
21 &= 13 \cdot 1 + 8 \\
13 &= 8 \cdot 1 + 5 \\
8 &= 5 \cdot 1 + 3 \\
5 &= 3 \cdot 1 + 2 \\
3 &= 2 \cdot 1 + 1 \\
2 &= 1 \cdot 2.
\end{aligned}
$$

Observe that when the Euclidean algorithm is used to find the greatest common divisor of $f_9 = 34$ and $f_{10} = 55$, a total of eight divisions are required. Furthermore, $(34,55) = 1$.  □

The following theorem tells us how many divisions are needed to find the greatest common divisor of successive Fibonacci numbers.

**Theorem 2.4.** Let $f_{n+1}$ and $f_{n+2}$ be successive terms of the Fibonacci sequence, with $n > 1$. Then the Euclidean algorithm takes exactly $n$ divisions to show that $(f_{n+1}, f_{n+2}) = 1$.

*Proof.* Applying the Euclidean algorithm, and using the defining relation for the Fibonacci numbers $f_j = f_{j-1} + f_{j-2}$ in each step, we see that

$$
\begin{aligned}
f_{n+2} &= f_{n+1} \cdot 1 + f_n, \\
f_{n+1} &= f_n \cdot 1 + f_{n-1},
\end{aligned}
$$

.

.

.

$$
\begin{aligned}
f_4 &= f_3 \cdot 1 + f_2, \\
f_3 &= f_2 \cdot 2.
\end{aligned}
$$

Hence, the Euclidean algorithm takes exactly $n$ divisions, to show that $(f_{n+2}, f_{n+1}) = f_2 = 1$.  ■

We can now prove a theorem first proved by Gabriel Lamé, a French mathematician of the nineteenth century, which gives an estimate for the number of divisions needed to find the greatest common divisor using the Euclidean algorithm.

**Theorem 2.5. Lamé's Theorem.** The number of divisions needed to find the greatest common divisor of two positive integers using the Euclidean algorithm does not exceed five times the number of decimal digits in the smaller of the two integers.

*Proof.* When we apply the Euclidean algorithm to find the greatest common divisor of $a = r_0$ and $b = r_1$ with $a > b$, we obtain the following sequence of equations:

$$r_0 = r_1 q_1 + r_2, \quad 0 \leqslant r_2 < r_1,$$
$$r_1 = r_2 q_2 + r_3, \quad 0 \leqslant r_3 < r_2,$$
$$\cdot$$
$$\cdot$$
$$\cdot$$
$$r_{n-2} = r_{n-1} q_{n-1} + r_n, \quad 0 \leqslant r_n < r_{n-1},$$
$$r_{n-1} = r_n q_n.$$

We have used $n$ divisions. We note that each of the quotients $q_1, q_2,..., q_{n-1}$ is greater than or equal to 1, and $q_n \geqslant 2$, since $r_n < r_{n-1}$. Therefore,

$$r_n \geqslant 1 = f_2,$$
$$r_{n-1} \geqslant 2r_n \geqslant 2f_2 = f_3,$$
$$r_{n-2} \geqslant r_{n-1} + r_n \geqslant f_3 + f_2 = f_4,$$
$$r_{n-3} \geqslant r_{n-2} + r_{n-1} \geqslant f_4 + f_3 = f_5,$$
$$\cdot$$
$$\cdot$$
$$\cdot$$
$$r_2 \geqslant r_3 + r_4 \geqslant f_{n-1} + f_{n-2} = f_n,$$
$$b = r_1 \geqslant r_2 + r_3 \geqslant f_n + f_{n-1} = f_{n+1}.$$

Thus, for there to be $n$ divisions used in the Euclidean algorithm we must have $b \geqslant f_{n+1}$. By Example 1.16, we know that $f_{n+1} > \alpha^{n-1}$ for $n > 2$ where

---

*    **GABRIEL LAMÉ** (1795 -1870) was a graduate of the Ecole Polytechnique. Lamé was a civil and railway engineer. He advanced the mathematical theory of elasticity and invented curvilinear coordinates. Although his main contributions were to mathematical physics, he made several discoveries in number theory, including the estimate of the number of steps required by the Euclidean algorithm, and the proof that Fermat's last theorem holds for $n=7$ (see Section 11.2). It is interesting to note that Gauss considered Lamé to be the foremost French mathematician of his time.

$\alpha = (1+\sqrt{5})/2$. Hence, $b > \alpha^{n-1}$. Now, since $\log_{10}\alpha > 1/5$, we see that

$$\log_{10}b > (n-1)\log_{10}\alpha > (n-1)/5.$$

Consequently,

$$n - 1 < 5\cdot\log_{10}b.$$

Let $b$ have $k$ decimal digits, so that $b < 10^k$ and $\log_{10}b < k$. Hence, we see that $n - 1 < 5k$ and since $k$ is an integer, we can conclude that $n \leqslant 5k$. This establishes Lamé's theorem.   ■

The following result is a consequence of Lamé's theorem.

**Corollary 2.5.1.** The number of bit operations needed to find the greatest common divisor of two positive integers $a$ and $b$ with $a < b$ is $O((\log_2 a)^3)$.

*Proof.* We know from Lamé's theorem that $O(\log_2 a)$ divisions, each taking $O((\log_2 a)^2)$ bit operations, are needed to find $(a, b)$. Hence, by Theorem 1.12 $(a,b)$ may be found using a total of $O((\log_2 a)^3)$ bit operations.   ■

The Euclidean algorithm can be used to express the greatest common divisor of two integers as a linear combination of these integers. We illustrate this by expressing $(252,198) = 18$ as a linear combination of 252 and 198. Referring to the steps of the Euclidean algorithm used to find $(252,198)$, by the next to the last step we see that

$$18 = 54 - 1\cdot 36.$$

By the second to the last step it follows that

$$36 = 198 - 3\cdot 54,$$

which implies that

$$18 = 54 - 1\cdot(198-3\cdot 54) = 4\cdot 54 - 1\cdot 198.$$

Likewise, by the first step we have

$$54 = 252 - 1\cdot 198,$$

so that

$$18 = 4(252-1\cdot 198) - 1\cdot 198 = 4\cdot 252 - 5\cdot 198.$$

This last equation exhibits $18 = (252,198)$ as a linear combination of 252 and 198.

In general, to see how $d = (a,b)$ may be expressed as a linear combination of $a$ and $b$, refer to the series of equations that is generated by use of the Euclidean algorithm. By the penultimate equation we have

$$r_n = (a,b) = r_{n-2} - r_{n-1}q_{n-1}.$$

This expresses $(a,b)$ as a linear combination of $r_{n-2}$ and $r_{n-1}$. The second to the last equation can be used to express $r_{n-1}$ as $r_{n-3} - r_{n-2}q_{n-2}$. Using this last equation to eliminate $r_{n-1}$ in the previous expression for $(a,b)$, we find that

$$r_n = r_{n-3} - r_{n-2}q_{n-2},$$

so that

$$(a, b) = r_{n-2} - (r_{n-3} - r_{n-2}q_{n-2})q_{n-1}$$
$$= (1 + q_{n-1}q_{n-2})r_{n-2} - q_{n-1}r_{n-3},$$

which expresses $(a,b)$ as a linear combination of $r_{n-2}$ and $r_{n-3}$. We continue working backwards through the steps of the Euclidean algorithm to express $(a, b)$ as a linear combination of each preceding pair of remainders until we have found $(a,b)$ as a linear combination of $r_0 = a$ and $r_1 = b$. Specifically, if we have found at a particular stage that

$$(a,b) = sr_j + tr_{j-1},$$

then, since

$$r_j = r_{j-2} - r_{j-1}q_{j-1},$$

we have

$$(a,b) = s(r_{j-2} - r_{j-1}q_{j-1}) + tr_{j-1}$$
$$= (t - sq_{j-1})r_{j-1} + sr_{j-2}.$$

This shows how to move up through the equations that are generated by the Euclidean algorithm so that, at each step, the greatest common divisor of $a$ and $b$ may be expressed as a linear combination of $a$ and $b$.

This method for expressing $(a,b)$ as a linear combination of $a$ and $b$ is somewhat inconvenient for calculation, because it is necessary to work out the steps of the Euclidean algorithm, save all these steps, and then proceed backwards through the steps to write $(a,b)$ as a linear combination of each successive pair of remainders. There is another method for finding $(a,b)$ which requires working through the steps of the Euclidean algorithm only once. The following theorem gives this method.

**Theorem 2.6.** Let $a$ and $b$ be positive integers. Then

$$(a,b) = s_n a + t_n b,$$

for $n = 0,1,2,...$, where $s_n$ and $t_n$ are the $n$th terms of the sequences defined recursively by

$$s_0 = 1, \ t_0 = 0,$$
$$s_1 = 0, \ t_1 = 1,$$

and

$$s_j = s_{j-2} - q_{j-1}s_{j-1}, \ t_j = t_{j-2} - q_{j-1}t_{j-1}$$

for $j = 2, 3, ..., n$, where the $q_j$'s are the quotients in the divisions of the Euclidean algorithm when it is used to find $(a,b)$.

*Proof.* We will prove that

(2.2)                          $r_j = s_j a + t_j b$

for $j = 0, 1,..., n$. Since $(a,b) = r_n$, once we have established (2.2), we will know that

$$(a,b) = s_n a + t_n b.$$

We prove (2.2) using the second principle of mathematical induction. For $j = 0$, we have $a = r_0 = 1 \cdot a + 0 \cdot b = s_0 a + t_0 b$. Hence, (2.2) is valid for $j = 0$. Likewise, $b = r_1 = 0 \cdot a + 1 \cdot b = s_1 a + t_1 b$, so that (2.2) is valid for $j = 1$.

Now, assume that

$$r_j = s_j a + t_j b$$

for $j = 1, 2,..., k-1$. Then, from the $k$th step of the Euclidean algorithm, we have

$$r_k = r_{k-2} - r_{k-1}q_{k-1}.$$

Using the induction hypothesis, we find that

$$\begin{aligned} r_k &= (s_{k-2}a + t_{k-2}b) - (s_{k-1}a + t_{k-1}b)q_{k-1} \\ &= (s_{k-2} - s_{k-1}q_{k-1})a + (t_{k-2} - t_{k-1}q_{k-1})b \\ &= s_k a + t_k b. \end{aligned}$$

This finishes the proof.                                                        ∎

The following example illustrates the use of this algorithm for expressing $(a,b)$ as a linear combination of $a$ and $b$.

**Example 2.8.** Let $a = 252$ and $b = 198$. Then

| | |
|---|---|
| $s_0 = 1,$ | $t_0 = 0,$ |
| $s_1 = 0,$ | $t_1 = 1,$ |
| $s_2 = s_0 - s_1 q_1 = 1 - 0 \cdot 1 = 1,$ | $t_2 = t_0 - t_1 q_1 = 0 - 1 \cdot 1 = -1,$ |
| $s_3 = s_1 - s_2 q_2 = 0 - 1 \cdot 3 = -3,$ | $t_3 = t_1 - t_2 q_2 = 1 - (-1)3 = 4,$ |
| $s_4 = s_2 - s_3 q_3 = 1 - (-3) \cdot 1 = 4,$ | $t_4 = t_2 - t_3 q_3 = -1 - 4 \cdot 1 = -5.$ |

Since $r_4 = 18 = (252, 198)$ and $r_4 = s_4 a + t_4 b$, we have

$$18 = (252, 198) = 4 \cdot 252 - 5 \cdot 198.$$                               □

It should be noted that the greatest common divisor of two integers may be expressed in an infinite number of different ways as a linear combination of these integers. To see this, let $d = (a,b)$ and let $d = sa + tb$ be one way to write $d$ as a linear combination of $a$ and $b$, guaranteed to exist by the previous discussion. Then

$$d = (s + k(b/d))a + (t - k(a/d))b$$

for all integers $k$.

**Example 2.9.** With $a = 252$ and $b = 198$, $18 = (252, 198) = (4 + 11k)252 + (-5 - 14k)198$ whenever $k$ is an integer. $\square$

---

## 2.2 Exercises

1. Use the Euclidean algorithm to find each of the following greatest common divisors.

   a) (45, 75)     c) (666, 1414)
   b) (102, 222)    d) (20785, 44350)

2. For each pair of integers in Exercise 1, express the greatest common divisor of the integers as a linear combination of these integers.

3. Find the greatest common divisor of each of the following sets of integers.

   a) 6, 10, 15
   b) 70, 98, 105
   c) 280, 330, 405, 490

4. Express the greatest common divisor of each set of numbers in Exercise 3 as a linear combination of the numbers in that set.

The greatest common divisor of two integers can be found using only subtractions, parity checks, and shifts of binary expansions, without using any divisions. The algorithm proceeds recursively using the following reduction

$$(a,b) = \begin{cases} a & \text{if } a = b \\ 2(a/2,b/2) & \text{if } a \text{ and } b \text{ are even} \\ (a/2,b) & \text{if } a \text{ is even and } b \text{ is odd} \\ (a-b,b) & \text{if } a \text{ and } b \text{ are odd}. \end{cases}$$

(*Note:* Reverse the roles of $a$ and $b$ when necessary).

5. Find (2106,8318) using this algorithm.

6. Show that this algorithm always produces the greatest common divisor of a pair of positive integers.

\* 7. How many steps does this algorithm use to find $(a,b)$ if $a = (2^n - (-1)^n)/3$ and $b = 2(2^{n-1} - (-1)^{n-1})/3$ when $n$ is a positive integer?

* **8.**  Show that to find $(a,b)$ this algorithm uses the subtraction step in the reduction no more than $1 + [\log_2 \max(a,b)]$ times.

* **9.**  Devise an algorithm for finding the greatest common divisor of two positive integers using their balanced ternary expansions.

In Exercise 16 of Section 1.5, a modified division algorithm is given which states that if $a$ and $b > 0$ are integers, then there exist unique integers $q$, $r$, and $e$ such that $a = bq + er$, where $e = \pm 1$, $r \geq 0$, and $-b/2 < er \leq b/2$. We can set up an algorithm, analogous to the Euclidean algorithm, based on this modified division algorithm, called the *least-remainder algorithm*. It works as follows. Let $r_0 = a$ and $r_1 = b$, where $a > b > 0$. Using the modified division algorithm repeatedly, obtain the greatest common divisor of $a$ and $b$ as the last nonzero remainder $r_n$ in the sequence of divisions

$$r_0 = r_1 q_1 + e_2 r_2, \qquad -r_1/2 < e_2 r_2 \leq r_1/2$$

$$\cdot$$
$$\cdot$$
$$\cdot$$

$$r_{n-2} = r_{n-1} q_{n-1} + e_n r_n, \qquad -r_{n-1}/2 < e_n r_n \leq r_{n-1}/2$$
$$r_{n-1} = r_n q_n.$$

**10.**  Use the least-remainder algorithm to find $(384,226)$.

**11.**  Show that the least-remainder algorithm always produces the greatest common divisor of two integers.

** **12.**  Show that the least-remainder algorithm is always faster, or as fast, as the Euclidean algorithm. (*Hint*: First show that if $a$ and $b$ are positive integers with $2b < a$, then the least-remainder algorithm can find $(a,b)$ with no more steps than it uses to find $(a,a-b)$.)

* **13.**  Find a sequence of integers $v_0$, $v_1$, $v_2$,... such that the least-remainder algorithm takes exactly $n$ divisions to find $(v_{n+1}, v_{n+2})$.

* **14.**  Show that the number of divisions needed to find the greatest common divisor of two positive integers using the least-remainder algorithm is less than $8/3$ times the number of digits in the smaller of the two numbers, plus $4/3$.

* **15.**  Let $m$ and $n$ be positive integers and let $a$ be an integer greater than one. Show that $(a^m-1, a^n-1) = a^{(m,n)} - 1$.

* **16.**  Show that if $m$ and $n$ are positive integers then $(f_m, f_n) = f_{(m,n)}$.

The next two exercises deal with the *game of Euclid*. Two players begin with a pair of positive integers and take turns making moves of the following type. A player can move from the pair of positive integers $\{x,y\}$ with $x \geq y$, to any of the pairs $\{x-ty,y\}$, where $t$ is a positive integer and $x-ty \geq 0$. A *winning move* consists of moving to a pair with one element equal to 0.

17. Show that every sequence of moves starting with the pair $\{a,b\}$ must eventually end with the pair $\{0, (a,b)\}$.

* 18. Show that in a game beginning with the pair $\{a,b\}$, the first player may play a winning strategy if $a = b$ or if $a > b(1+\sqrt{5})/2$; otherwise the second player may play a winning strategy. (*Hint*: First show that if $y < x \leqslant y(1+\sqrt{5})/2$ then there is a unique move from $\{x,y\}$ that goes to a pair $\{z,y\}$ with $y > z(1+\sqrt{5})/2$.)

## 2.2 Computer Projects

Write programs to do the following:

1. Find the greatest common divisor of two integers using the Euclidean algorithm.

2. Find the greatest common divisor of two integers using the modified Euclidean algorithm given in the preamble to Exercise 10.

3. Find the greatest common divisor of two integers using no divisions (see the preamble to Exercise 5).

4. Find the greatest common divisor of a set of more than two integers.

5. Express the greatest common divisor of two integers as a linear combination of these integers.

6. Express the greatest common divisor of a set of more than two integers as a linear combination of these integers.

* 7. Play the game of Euclid described in the preamble to Exercise 17.

---

## 2.3 The Fundamental Theorem of Arithmetic

The fundamental theorem of arithmetic is an important result that shows that the primes are the building blocks of the integers. Here is what the theorem says.

**Theorem 2.7. The Fundamental Theorem of Arithmetic.** Every positive integer greater than one can be written uniquely as a product of primes, with the prime factors in the product written in order of nondecreasing size.

**Example 2.10.** The factorizations of some positive integers are given by

$$240 = 2 \cdot 2 \cdot 2 \cdot 2 \cdot 3 \cdot 5 = 2^4 \cdot 3 \cdot 5, \ 289 = 17 \cdot 17 = 17^2, \ 1001 = 7 \cdot 11 \cdot 13. \qquad \square$$

Note that it is convenient to combine all the factors of a particular prime into a power of this prime, such as in the previous example. There, for the factorization of 240, all the factors of 2 were combined to form $2^4$.

Factorizations of integers in which the factors of primes are combined to form powers are called *prime-power factorizations*.

To prove the fundamental theorem of arithmetic, we need the following lemma concerning divisibility.

**Lemma 2.3**. If $a$, $b$, and $c$ are positive integers such that $(a,b) = 1$ and $a \mid bc$, then $a \mid c$.

*Proof.* Since $(a,b) = 1$, there are integers $x$ and $y$ such that $ax + by = 1$. Multiplying both sides of this equation by $c$, we have $acx + bcy = c$. By Theorem 1.8, $a$ divides $acx + bcy$, since this is a linear combination of $a$ and $bc$, both of which are divisible by $a$. Hence $a \mid c$.  ∎

The following consequence of this lemma will be needed in the proof the Fundamental Theorem of Arithmetic.

**Lemma 2.4**. If $p$ divides $a_1 a_2 \cdots a_n$ where $p$ is a prime and $a_1, a_2,..., a_n$ are positive integers, then there is an integer $i$ with $1 \leqslant i \leqslant n$ such that $p$ divides $a_i$.

*Proof.* We prove this result by induction. The case where $n = 1$ is trivial. Assume that the result is true for $n$. Consider a product of $n + 1$, integers, $a_1 a_2 \cdots a_{n+1}$ that is divisible by the prime $p$. Since $p \mid a_1 a_2 \cdots a_{n+1} = (a_1 a_2 \cdots a_n)a_{n+1}$, we know from Lemma 2.3 that $p \mid a_1 a_2 \cdots a_n$ or $p \mid a_{n+1}$. Now, if $p \mid a_1 a_2 \cdots a_n$, from the induction hypothesis there is an integer $i$ with $1 \leqslant i \leqslant n$ such that $p \mid a_i$. Consequently $p \mid a_i$ for some $i$ with $1 \leqslant i \leqslant n + 1$. This establishes the result.  ∎

We now begin the proof of the fundamental theorem of arithmetic. First, we will show that every positive integer greater than one can be written as the product of primes in at least one way. Then we will show that this product is unique up to the order of primes that appear.

*Proof.* We use proof by contradiction. Assume that some positive integer cannot be written as the product of primes. Let $n$ be the smallest such integer (such an integer must exist from the well-ordering property). If $n$ is prime, it is obviously the product of a set of primes, namely the one prime $n$. So $n$ must be composite. Let $n = ab$, with $1 < a < n$ and $1 < b < n$. But since $a$ and $b$ are smaller than $n$ they must be the product of primes. Then, since $n = ab$, we conclude that $n$ is also a product of primes. This contradiction shows that every positive integer can be written as the product of primes.

We now finish the proof of the fundmental theorem of arithmetic by showing that the factorization is unique. Suppose that there is an integer $n$ that has two different factorizations into primes:

$$n = p_1 p_2 \cdots p_s = q_1 q_2 \cdots q_t,$$

where $p_1, p_2, ..., p_s, q_1, ..., q_t$ are all primes, with $p_1 \leqslant p_2 \leqslant \cdots \leqslant p_s$ and $q_1 \leqslant q_2 \leqslant \cdots \leqslant q_t$.

Remove all common primes from the two factorizations to obtain

$$p_{i_1} p_{i_2} \cdots p_{i_u} = q_{j_1} q_{j_2} \cdots q_{j_v}$$

where the primes on the left-hand side of this equation differ from those on the right-hand side, $u \geqslant 1$, and $v \geqslant 1$ (since the two original factorizations were presumed to differ). However, this leads to a contradiction of Lemma 2.4; by this lemma $p_{i_1}$ must divide $q_{i_j}$ for some $j$, which is impossible, since each $q_{i_j}$ is prime and is different from $p_{i_1}$. This shows that the prime factorization of an integer $n$ is unique. ∎

The prime factorization of an integer is often useful. As an example, let us find all the divisors of an integer from its prime factorization.

**Example 2.11.** The positive divisors of $120 = 2^3 \cdot 3 \cdot 5$ are those positive integers with prime power factorizations containing only the primes 2, 3, and 5, to powers less than or equal to 3, 1, and 1, respectively. These divisors are

| | | | |
|---|---|---|---|
| 1 | 3 | 5 | $3 \cdot 5 = 15$ |
| 2 | $2 \cdot 3 = 6$ | $2 \cdot 5 = 10$ | $2 \cdot 3 \cdot 5 = 30$ |
| $2^2 = 4$ | $2^2 \cdot 3 = 12$ | $2^2 \cdot 5 = 20$ | $2^2 \cdot 3 \cdot 5 = 60$ |
| $2^3 = 8$ | $2^3 \cdot 3 = 24$ | $2^3 \cdot 5 = 40$ | $2^3 \cdot 3 \cdot 5 = 120.$ |

□

Another way in which we can use prime factorizations is to find greatest common divisors, as illustrated in the following example.

**Example 2.12.** To be a common divisor of $720 = 2^4 \cdot 3^2 \cdot 5$ and $2100 = 2^2 \cdot 3 \cdot 5^2 \cdot 7$, a positive integer can contain only the primes 2, 3, and 5 in its prime-power factorization, and the power to which one of these primes appears cannot be larger than either of the powers of that prime in the factorizations of 720 and 2100. Consequently, to be a common divisor of 720 and 2100, a positive integer can contain only the primes 2, 3, and 5 to powers no larger than 2, 1, and 1, respectively. Therefore, the greatest common divisor of 720 and 2100 is $2^2 \cdot 3 \cdot 5 = 60$. □

To describe, in general, how prime factorizations can be used to find greatest common divisors, let $\min(a, b)$ denote the smaller or minimum, of the two numbers $a$ and $b$. Now let the prime factorizations of $a$ and $b$ be

$$a = p_1^{a_1} p_2^{a_2} \cdots p_n^{a_n}, \quad b = p_1^{b_1} p_2^{b_2} \cdots p_n^{b_n},$$

where each exponent is a nonnegative integer and where all primes occurring in the prime factorizations of $a$ and of $b$ are included in both products, perhaps with zero exponents. We note that

$$(a,b) = p_1^{\min(a_1,b_1)} p_2^{\min(a_2,b_2)} \cdots p_n^{\min(a_n,b_n)},$$

since for each prime $p_i$, $a$ and $b$ share exactly $\min(a_i,b_i)$ factors of $p_i$.

Prime factorizations can also be used to find the smallest integer that is a multiple of two positive integers. The problem of finding this integer arises when fractions are added.

**Definition.** The *least common multiple* of two positive integers $a$ and $b$ is the smallest positive integer that is divisible by $a$ and $b$.

The least common multiple of $a$ and $b$ is denoted by $[a,b]$.

**Example 2.13.** We have the following least common multiples: $[15,21] = 105$, $[24,36] = 72$, $[2,20] = 20$, and $[7,11] = 77$.  □

Once the prime factorizations of $a$ and $b$ are known, it is easy to find $[a,b]$. If $a = p_1^{a_1} p_2^{a_2} \cdots p_n^{a_n}$ and $b = p_1^{a_1} p_2^{b_2} \cdots p_n^{b_n}$, where $p_1,p_2,...,p_n$ are the primes occurring in the prime-power factorizations of $a$ and $b$, then for an integer to be divisible by both $a$ and $b$, it is necessary that in the factorization of the integer, each $p_j$ occurs with a power at least as large as $a_j$ and $b_j$. Hence, $[a,b]$, the smallest positive integer divisible by both $a$ and $b$ is

$$[a,b] = p_1^{\max(a_1,b_1)} p_2^{\max(a_2,b_2)} \cdots p_n^{\max(a_n,b_n)}$$

where $\max(x,y)$ denotes the larger, or maximum, of $x$ and $y$.

Finding the prime factorization of large integers is time-consuming. Therefore, we would prefer a method for finding the least common multiple of two integers without using the prime factorizations of these integers. We will show that we can find the least common multiple of two positive integers once we know the greatest common divisor of these integers. The latter can be found via the Euclidean algorithm. First, we prove the following lemma.

**Lemma 2.5.** If $x$ and $y$ are real numbers, then $\max(x,y) + \min(x,y) = x + y$.

*Proof.* If $x \geqslant y$, then $\min(x,y) = y$ and $\max(x,y) = x$, so that $\max(x,y) + \min(x,y) = x + y$. If $x < y$, then $\min(x,y) = x$ and $\max(x,y) = y$, and again we find that $\max(x,y) + \min(x,y) = x + y$. ∎

We use the following theorem to find $[a,b]$ once $(a,b)$ is known.

**Theorem 2.8.** If $a$ and $b$ are positive integers, then $[a,b] = ab/(a,b)$, where $[a,b]$ and $(a,b)$ are the least common multiple and greatest common divisor of $a$ and $b$, respectively.

*Proof.* Let $a$ and $b$ have prime-power factorizations $a = p_1^{a_1} p_2^{a_2} \cdots p_n^{a_n}$ and $b = p_1^{b_1} p_s^{b_2} \cdots p_n^{b_n}$, where the exponents are nonnegative integers and all primes occurring in either factorization occur in both, perhaps with zero exponents. Now let $M_j = \max(a_j, b_j)$ and $m_j = \min(a_j, b_j)$. Then, we have

$$
\begin{aligned}
[a,b](a,b) &= p_1^{M_1} p_2^{M_2} \cdots p_n^{M_n} p_1^{m_1} p_2^{m_2} \cdots p_n^{m_n} \\
&= p_1^{M_1 + m_1} p_2^{M_2 + m_2} \cdots p_n^{M_n + m_n} \\
&= p_1^{a_1 + b_1} p_2^{a_2 + b_2} \cdots p_n^{a_n + b_n} \\
&= p_1^{a_1} p_2^{a_2} \cdots p_n^{a_n} p_1^{b_1} \cdots p_n^{b_n} \\
&= ab,
\end{aligned}
$$

since $M_j + m_j = \max(a_j, b_j) + \min(a_j, b_j) = a_j + b_j$ by Lemma 2.5.   ■

The following consequence of the fundamental theorem of arithmetic will be needed later.

**Lemma 2.6.** Let $m$ and $n$ be relatively prime positive integers. Then if $d$ is a positive divisor of $mn$, there is a unique pair of positive divisors $d_1$ of $m$ and $d_2$ of $n$ such that $d = d_1 d_2$. Conversely, if $d_1$ and $d_2$ are positive divisors of $m$ and $n$, respectively, then $d = d_1 d_2$ is a positive divisor of $mn$.

*Proof.* Let the prime-power factorizations of $m$ and $n$ be $m = p_1^{m_1} p_2^{m_2} \cdots p_s^{m_s}$ and $n = q_1^{n_1} q_2^{n_2} \cdots q_t^{n_t}$. Since $(m,n) = 1$, the set of primes $p_1, p_2, ..., p_s$ and the set of primes $q_1, q_2, ..., q_t$ have no common elements. Therefore, the prime-power factorization of $mn$ is

$$
mn = p_1^{m_1} p_2^{m_2} \cdots p_s^{m_s} q_1^{n_1} q_2^{n_2} \cdots q_t^{n_t}.
$$

Hence, if $d$ is a positive divisor of $mn$, then

$$
d = p_1^{e_1} p_2^{e_2} \cdots p_s^{e_s} q_1^{f_1} q_2^{f_2} \cdots q_t^{f_t}
$$

where $0 \leqslant e_i \leqslant m_i$ for $i = 1, 2, ..., s$ and $0 \leqslant f_j \leqslant n_j$ for $j = 1, 2, ..., t$. Now let $d_1 = (d,m)$ and $d_2 = (d,n)$, so that

$$
d_1 = p_1^{e_1} p_2^{e_2} \cdots p_s^{e_s}
$$

and

$$
d_2 = q_1^{f_1} q_2^{f_2} \cdots q_t^{f_t}.
$$

Clearly $d = d_1 d_2$ and $(d_1, d_2) = 1$. This is the decomposition of $d$ we desire. Furthermore this decomposition is unique. To see this, note that every prime power in the factorization of $d$ must occur in either $d_1$ or $d_2$, and prime powers in the factorization of $d$ that are powers of primes dividing $m$ must appear in $d_1$, while prime powers in the factorization of $d$ that are powers of primes dividing $n$ must appear in $d_2$. It follows that $d_1$ must be $(d,m)$ and $d_2$ must be $(d,n)$.

Conversely, let $d_1$ and $d_2$ be positive divisors of m and n, respectively. Then

$$d_1 = p_1^{e_1} p_2^{e_2} \cdots p_s^{e_s}$$

where $0 \leqslant e_i \leqslant m_i$ for $i = 1, 2, \dots, s$, and

$$d_2 = q_1^{f_1} q_2^{f_2} \cdots q_t^{f_t}$$

where $0 \leqslant f_j \leqslant n_j$ for $j = 1, 2, \dots, t$. The integer

$$d = d_1 d_2 = p_1^{e_1} p_2^{e_2} \cdots p_s^{e_s} q_1^{f_1} q_2^{f_2} \cdots q_t^{f_t}$$

is clearly a divisor of

$$mn = p_1^{m_1} p_2^{m_2} \cdots p_s^{m_s} q_1^{n_1} q_2^{n_2} \cdots q_t^{n_t},$$

since the power of such prime occurring in the prime-power factorization of $d$ is less than or equal to the power of that prime in the prime-power factorization of $mn$.     ∎

A famous result of number theory deals with primes in arithmetic progressions.

**Theorem 2.9. Dirichlet's Theorem on Primes in Arithmetic Progressions.** Let $a$ and $b$ be relatively prime positive integers. Then the arithmetic progression $an + b$, $n = 1, 2, 3, \dots,$ contains infinitely many primes.

G. Lejeune Dirichlet*, proved this theorem in 1837. Since proofs of Dirichlet's Theorem are complicated and rely on advanced techniques, we do not present a proof here. However, it is not difficult to prove special cases of Dirichlet's theorem, as the following theorem illustrates.

**Theorem 2.10.** There are infinitely many primes of the form $4n + 3$, where $n$ is a positive integer.

Before we prove this result, we first prove a useful lemma.

**Lemma 2.7.** If $a$ and $b$ are integers both of the form $4n + 1$, then the product $ab$ is also of this form.

---

* **G. LEJEUNE DIRICHLET** (1805-1859) was born into a French family living in the vicinity of Cologne, Germany. He studied at the University of Paris when this was an important world center of mathematics. He held positions at the University of Breslau and the University of Berlin, and in 1855, was chosen to succeed Gauss at the University of Göttingen. Dirichelt is said to be the first person to master Gauss' *Disquisitiones Arithmeticae* which had appeared 20 years earlier. He is said to have kept a copy of this book at his side even when he traveled. His book on number theory, *Vorlesungen über Zahlentheorie,* helped make Gauss' discoveries accessible to other mathematicians. Besides his fundamental work in number theory, Dirichlet made many important contributions to analysis. His famous "drawer principle", also called the pigeonhole principle, is used extensively in combinatorics and in number theory.

*Proof.* Since $a$ and $b$ are both of the form $4n + 1$, there exist integers $r$ and $s$ such that $a = 4r + 1$ and $b = 4s + 1$. Hence,

$$ab = (4r+1)(4s+1) = 16rs + 4r + 4s + 1 = 4(4rs + r + s) + 1,$$

which is again of the form $4n + 1$.    ∎

We now prove the desired result.

*Proof.* Let us assume that there are only a finite number of primes of the form $4n + 3$, say $p_0 = 3, p_1, p_2, ..., p_r$. Let

$$Q = 4p_1 p_2 \cdots p_r + 3.$$

Then, there is at least one prime in the factorization of $Q$ of the form $4n + 3$. Otherwise, all of these primes would be of the form $4n + 1$, and by Lemma 2.7, this would imply that $Q$ would also be of this form, which is a contradiction. However, none of the primes $p_0, p_1,..., p_n$ divides $Q$. The prime 3 does not divide $Q$, for if $3 \mid Q$, then $3 \mid (Q-3) = 4p_1 p_2 \cdots p_r$, which is a contradiction. Likewise, none of the primes $p_j$ can divide $Q$, because $p_j \mid Q$ implies $p_j \mid (Q-4p_1 p_2 \cdots p_r) = 3$ which is absurd. Hence, there are infinitely many primes of the form $4n + 3$.    ∎

We conclude this section by proving some results about irrational numbers. If $\alpha$ is a rational number then we may write $\alpha$ as the quotient of two integers in infinitely many ways, for if $\alpha = a/b$, where $a$ and $b$ are integers with $b \neq 0$, then $\alpha = ka/kb$ whenever $k$ is a nonzero integer. It is easy to see that a positive rational number may be written uniquely as the quotient of two relatively prime positive integers; when this is done we say that the rational number is in *lowest terms*. We note that the rational number $11/21$ is in lowest terms. We also see that

$$\cdots = -33/-63 = -22/-42 = -11/-21 = 11/21 = 22/42 = 33/63 = \cdots.$$

The next two results show that certain numbers are irrational. We start by giving another proof that $\sqrt{2}$ is irrational (we proved this originally in Section 1.1).

**Example 2.14.** Suppose that $\sqrt{2}$ is rational. Then $\sqrt{2} = a/b$, where $a$ and $b$ are relatively prime integers with $b \neq 0$. It follows that $2 = a^2/b^2$, so that $2b^2 = a^2$. Since $2 \mid a^2$, it follows from the Fundamental Theorem of Arithmetic that (see Exercise 28 at the end of this section) $2 \mid a$. Let $a = 2c$, so that $b^2 = 2c^2$. Hence, $2 \mid b^2$, and by Exercise 28 2 also divides $b$. However, since $(a,b) = 1$, we know that 2 cannot divide both $a$ and $b$. This contradiction shows that $\sqrt{2}$ is irrational.    □

We can also use the following more general result to show that $\sqrt{2}$ is irrational.

**Theorem 2.11.** Let $\alpha$ be a root of the polynomial $x^n + c_{n-1}x^{n-1} + \cdots + c_1 x + c_0$ where the coefficients $c_0, c_1, \ldots, c_{n-1}$, are integers with $c_0 \neq 0$. Then $\alpha$ is either an integer or an irrational number.

*Proof.* Suppose that $\alpha$ is rational. Then we can write $\alpha = a/b$ where $a$ and $b$ are relatively prime integers with $b \neq 0$. Since $\alpha$ is a root of $x^n + c_{n-1}x^{n-1} + \cdots + c_1 x + c_0$, we have

$$(a/b)^n + c_{n-1}(a/b)^{n-1} + \cdots + c_1(a/b) + c_0 = 0.$$

Multiplying by $b^n$, we find that

$$a^n + c_{n-1}a^{n-1}b + \cdots + c_1 ab^{n-1} + c_0 b^n = 0.$$

Since

$$a^n = b(-c_{n-1}a^{n-1} - \cdots - c_1 ab^{n-2} - c_0 b^{n-1}),$$

we see that $b \mid a^n$. Assume that $b \neq \pm 1$. Then, $b$ has a prime divisor $p$. Since $p \mid b$ and $b \mid a^n$, we know that $p \mid a^n$. Hence, by Exercise 29 we see that $p \mid a$. However, since $(a, b) = 1$, this is a contradiction which shows that $b = \pm 1$. Consequently, if $\alpha$ is rational then $\alpha = \pm a$, so that $\alpha$ must be an integer. ∎

We illustrate the use of Theorem 2.11 with the following example.

**Example 2.15.** Let $a$ be a positive integer that is not the $m$th power of an integer, so that $\sqrt[m]{a}$ is not an integer. Then $\sqrt[m]{a}$ is irrational by Theorem 2.11, since $\sqrt[m]{a}$ is a root of $x^m - a$. Consequently, such numbers as $\sqrt{2}$, $\sqrt[3]{5}$, $\sqrt[10]{17}$, *etc* are irrational. □

---

## 2.3 Exercises

1. Find the prime factorizations of each of the following integers.

   a) 36       e) 222      i) 5040
   b) 39       f) 256      j) 8000
   c) 100      g) 515      k) 9555
   d) 289      h) 989      l) 9999

2. Show that all the powers in the prime-power factorization of an integer $n$ are even if and only if $n$ is a perfect square.

3.  Which positive integers have exactly three positive divisors?  Which have exactly four positive divisors?

4.  Show that every positive integer can be written as the product of possibly a square and a square-free integer. A *square-free integer* is an integer that is not divisible by any perfect squares other than 1.

5.  An integer $n$ is called *powerful* if whenever a prime $p$ divides $n$, $p^2$ divides $n$. Show that every powerful number can be written as the product of a perfect square and a perfect cube.

6.  Show that if $a$ and $b$ are positive integers and $a^3 \mid b^2$, then $a \mid b$.

Let $p$ be a prime and $n$ a positive integer. If $p^a \mid n$, but $p^{a+1} \nmid n$, we say that $p^a$ *exactly divides* $n$, and we write $p^a \parallel n$.

7.  Show that if $p^a \parallel m$ and $p^b \parallel n$, then $p^{a+b} \parallel mn$.

8.  Show that if $p^a \parallel m$, then $p^{ka} \parallel m^k$.

9.  Show that if $p^a \parallel m$ and $p^b \parallel n$ with $a \neq b$, then $p^{\min(a,b)} \parallel (m+n)$.

10.  Let $n$ be a positive integer. Show that the power of the prime $p$ occurring in the prime power factorization of $n!$ is

$$[n/p] + [n/p^2] + [n/p^3] + \cdots .$$

11.  Use Exercise 10 to find the prime-power factorization of 20!.

12.  How many zeros are there at the end of 1000! in decimal notation? How many in base eight notation?

13.  Find all positive integers $n$ such that $n!$ ends with exactly 74 zeros in decimal notation.

14.  Show that if $n$ is a positive integer it is impossible for $n!$ to end with exactly 153, 154, or 155 zeros when it is written in decimal notation.

The next four exercises present an example of a system where unique factorization into primes fails. Let $H$ be the set of all positive integers of the form $4k+1$, where $k$ is a positive integer.

15.  Show that the product of two elements of $H$ is also in $H$.

16.  An element $h \neq 1$ in $H$ is called a "*Hilbert prime*" if the only way it can be written as the product of two integers in $H$ is $h = h \cdot 1 = 1 \cdot h$, Find the 20 smallest Hilbert primes.

17.  Show every element of $H$ can be factored into Hilbert primes.

18.  Show that factorization of elements of $H$ into Hilbert primes is not necessarily unique by finding two different factorizations of 693 into Hilbert primes.

19. Which positive integers $n$ are divisible by all integers not exceeding $\sqrt{n}$?

20. Find the least common multiple of each of the following pairs of integers

    a)  8, 12          d)   111, 303
    b)  14, 15         e)   256, 5040
    c)  28, 35         f)   343, 999.

21. Find the greatest common divisor and least common multiple of the following pairs of integers

    a)  $2^2 3^3 5^5 7^7, 2^7 3^5 5^3 7^2$
    b)  $2 \cdot 3 \cdot 5 \cdot 7 \cdot 11 \cdot 13, 17 \cdot 19 \cdot 23 \cdot 29$
    c)  $2^3 5^7 11^{13}, 2 \cdot 3 \cdot 5 \cdot 7 \cdot 11 \cdot 13$
    d)  $47^{11} 79^{111} 101^{1001}, 41^{11} 83^{111} 101^{1000}$.

22. Show that every common multiple of the positive integers $a$ and $b$ is divisible by the least common multiple of $a$ and $b$.

23. Periodical cicadas are insects with very long larval periods and brief adult lives. For each species of periodical cicada with larval period of 17 years, there is a similar species with a larval period of 13 years. If both the 17-year and 13-year species emerged in a particular location in 1900, when will they next both emerge in that location?

24. Which pairs of integers $a$ and $b$ have greatest common divisor 18 and least common multiple 540?

25. Show that if $a$ and $b$ are positive integers, then $(a,b) \mid [a,b]$. When does $(a,b) = [a,b]$?

26. Show that if $a$ and $b$ are positive integers, then there are divisors $c$ of $a$ and $d$ of $b$ with $(c,d) = 1$ and $cd = [a,b]$.

☞ 27. Show that if $a,b$, and $c$ are integers, then $[a,b] \mid c$ if and only if $a \mid c$ and $b \mid c$.

☞ 28. Show that if $p$ is a prime and $a$ is a positive integer with $p \mid a^2$, then $p \mid a$.

☞ 29. Show that if $p$ is a prime, $a$ is an integer, and $n$ is a positive integer such that $p \mid a^n$, then $p \mid a$.

30. Show that $\sqrt[3]{5}$ is irrational

    a)  by an argument similar to that given in Example 2.14
    b)  using Theorem 2.11

31. Show that $\sqrt{2} + \sqrt{3}$ is irrational.

32. Show that $\log_2 3$ is irrational.

33. Show that $\log_p b$ is irrational, where $p$ is a prime and $b$ is a positive integer which is not a power of $p$.

* **34.** Let $n$ be a positive integer greater than 1. Show that $1 + \dfrac{1}{2} + \dfrac{1}{3} + ... + \dfrac{1}{n}$ is not an integer.

**35.** Show that if $a$ and $b$ are positive integers then $(a,b) = (a+b,[a,b])$.

**36.** Find the two positive integers with sum 798 and least common multiple 10780. (*Hint*: Use Exercise 35.)

**37.** Show that if $a,b,$ and $c$ are positive integers, then $([a,b],c) = [(a,c),(b,c)]$ and $[(a,b),c] = ([a,c],[b,c])$.

The *least common multiple* of the integers $a_1,a_2,...,a_n$, that are not all zero, is the smallest positive integer that is divisible by all the integers $a_1,a_2,...,a_n$; it is denoted by $[a_1,a_2,...,a_n]$.

**38.** Find $[6,10,15]$ and $[7,11,13]$.

**39.** Show that $[a_1,a_2,...,a_{n-1},a_n] = [[a_1,a_2,...,a_{n-1}],a_n]$.

**40.** Let $n$ be a positive integer. How many pairs of positive integers satisfy $[a,b] = n$?

**41.** a) Show that if $a,b,$ and $c$ are positive integers, then

$$\max(a,b,c) = a + b + c - \min(a,b) - \min(a,c) - \min(b,c) + \min(a,b,c).$$

b) Use part (a) to show that
$$[a,b,c] = \frac{abc\,(a,b,c)}{(a,b)\,(a,c)\,(b,c)}.$$

**42.** Generalize Exercise 41 to find a formula for $(a_1,a_2,...,a_n) \cdot [a_1,a_2,...,a_n]$ where $a_1,a_2,...,a_n$ are positive integers.

**43.** Show that if $a,b,$ and $c$ are positive integers then $(a,b,c)[ab,ac,bc] = abc$.

**44.** Show that if $a,b,$ and $c$ are positive integers then $[a,b,c](ab,ac,bc) = abc$.

**45.** Show that if $a,b,$ and $c$ are positive integers then $([a,b],[a,c],[b,c]) = [(a,b),(a,c),(b,c)]$.

**46.** Prove that there are infinitely many primes of the form $6k + 5$, where $k$ is a positive integer.

**47.** Show that if $a$ and $b$ are integers, then the arithmetic progression $a$, $a+b$, $a+2b$,... contains an arbitrary number of consecutive composite terms.

**48.** Find the prime factorizations of each of the following integers.

a) $10^6 - 1$

b) $10^8 - 1$

c) $2^{15} - 1$

d) $2^{24} - 1$

e) $2^{30} - 1$

f) $2^{36} - 1$

49. A discount store sells a camera at a price less than its usual retail price of $99. If they sell $8137 worth of this camera and the discounted dollar price is an integer, how many cameras did they sell?

50. Show that if $a$ and $b$ are positive integers, then $a^2 \mid b^2$ implies that $a \mid b$.

51. Show that if $a, b$, and $c$ are positive integers with $(a, b) = 1$ and $ab = c^n$, then there are positive integers $d$ and $e$ such that $a = d^n$ and $b = e^n$.

☞ 52. Show that if $a_1, a_2, ..., a_n$ are pairwise relatively prime integers, then $[a_1, a_2, ..., a_n] = a_1 a_2 \cdots a_n$.

53. Show that among any set of $n+1$ positive integers not exceeding $2n$ there is an integer that divides a different integer in the set.

54. Show that $\dfrac{(m+n)!}{m!n!}$ is an integer whenever $m$ and $n$ are positive integers.

\* 55. Find all solutions of the equation $m^n = n^m$ where $m$ and $n$ are integers.

The next six exercises establish some estimates for the size of $\pi(x)$, the number of primes less than or equal to $x$. These results were originally proved in the 19th century by the Russian mathematician Chebyshev.

56. Let $p$ be a prime and let $n$ be a positive integer. Show that $p$ divides $\binom{2n}{n}$ exactly

$$([2n/p] - 2[n/p]) + ([2n/p^2] - 2[n/p^2]) + ... + ([2n/p^t] - 2[n/p^t])$$

times, where $t = [\log_p 2n]$. Conclude that if $p^r$ divides $\binom{2n}{n}$ then $p^r \leqslant 2n$.

57. Use Exercise 56 to show that

$$\binom{2n}{n} \leqslant (2n)^{\pi(2n)}.$$

58. Show that the product of all primes between $n$ and $2n$ is between $\binom{2n}{n}$ and $n^{\pi(2n) - \pi(n)}$. (*Hint:* Use the fact that every prime between $n$ and $2n$ divides $(2n)!$ but not $(n!)^2$.)

59. Use Exercise 57 and 58 to show that

$$\pi(2n) - \pi(n) < n \log 4 / \log n$$

**60.**  Use Exercise 57 and 58 to show that

$$\pi(2n) = (\pi(2n) - \pi(n)) + (\pi(n) - \pi(n/2)) + (\pi(n/2) - \pi(n/4))$$
$$+ \cdots \leqslant n \log 64/\log n.$$

**61.**  Use Exercises 59 and 60 to show that are constants $c_1$ and $c_2$ such that

$$c_1 x/log\ x < \pi(x) < c_2 x/log\ x.$$

(Compare this to the strong statement given in the Prime Number Theorem, stated as Theorem 1.19 in Section 1.9)

## 2.3  Computer Projects

Write programs to do the following:

**1.**  Find all positive divisors of a positive integer from its prime factorization.

**2.**  Find the greatest common divisor of two positive integers from their prime factorizations.

**3.**  Find the least common multiple of two positive integers from their prime factorizations.

**4.**  Find the number of zeros at the end of the decimal expansion of $n!$ where $n$ is a positive integer.

**5.**  Find the prime factorization of $n!$ where $n$ is a positive integer.

---

## 2.4  Factorization of Integers and the Fermat Numbers

From the fundamental theorem of arithmetic, we know that every positive integer can be written uniquely as the product of primes. In this section, we discuss the problem of determining this factorization. The most direct way to find the factorization of the positive integer $n$ is as follows. Recall from Theorem 1.18 that $n$ either is prime, or else has a prime factor not exceeding $\sqrt{n}$. Consequently, when we divide $n$ by the primes $2,3,5,...$ successively not exceeding $\sqrt{n}$, we either find a prime factor $p_1$ of $n$ or else we conclude that $n$ is prime. If we have located a prime factor $p_1$ of $n$, we next look for a prime factor of $n_1 = n/p_1$, beginning our search with the prime $p_1$, since $n_1$ has no prime factor less than $p_1$, and any factor of $n_1$ is also a factor of $n$. We continue, if necessary, determining whether any of the primes not exceeding $\sqrt{n_1}$ divide $n_1$. We continue in this manner, proceeding recursively, to find the prime factorization of $n$.

**Example 2.16.**  Let $n = 42833$. We note that $n$ is not divisible by 2,3 and 5, but that $7 \mid n$. We have

$$42833 = 7 \cdot 6119.$$

Trial divisions show that 6119 is not divisible by any of the primes 7,11,13,17,19, and 23. However, we see that

$$6119 = 29 \cdot 211.$$

Since $29 > \sqrt{211}$, we know that 211 is prime. We conclude that the prime factorization of 42833 is $42833 = 7 \cdot 29 \cdot 211$.        □

Unfortunately, this method for finding the prime factorization of an integer is quite inefficient. To factor an integer $N$, it may be necessary to perform as many as $\pi(\sqrt{N})$ divisions, altogether requiring on the order of $\sqrt{N} \log N$ bit operations, since from the prime number theorem $\pi(\sqrt{N})$ is approximately $\sqrt{N}/\log\sqrt{N} = 2\sqrt{N}/\log N$, and from Theorem 1.16, these divisions take $O(\log^2 N)$ bit operations each. More efficient algorithms for factorization have been developed, requiring fewer bit operations than the direct method of factorization previously described. In general, these algorithms are complicated and rely on ideas that we have not yet discussed. For information about these algorithms we refer the reader to Dixon [92], Guy [95], Knuth [80], Riesel [107], Rumely [108], and Wagstaff and Smith [111]. We note that the quickest method yet devised can factor an integer $N$ in approximately

$$\exp( \sqrt{\log N \cdot \log \log N} )$$

bit operations, where exp stands for the exponential function.

In Table 2.1, we give the time required to factor integers of various sizes using the most efficient algorithm known, where the time for each bit operation has been estimated as one microsecond (one microsecond is $10^{-6}$ seconds).

| Number of decimal digits | Number of bit operations | Time |
|:---:|:---:|:---:|
| 50 | $1.4 \times 10^{10}$ | 3.9 hours |
| 75 | $9.0 \times 10^{12}$ | 104 days |
| 100 | $2.3 \times 10^{15}$ | 74 years |
| 200 | $1.2 \times 10^{23}$ | $3.8 \times 10^9$ years |
| 300 | $1.5 \times 10^{29}$ | $4.9 \times 10^{15}$ years |
| 500 | $1.3 \times 10^{39}$ | $4.2 \times 10^{25}$ years |

**Table 2.1.** Time Required For Factorization of Large Integers.

Later we will show that it is far easier to decide whether an integer is prime, then it is to factor the integer. This difference is the basis of a cryptographic system discussed in Chapter 7.

We now describe a factorization technique which is interesting, although it is not always efficient. This technique is known as *Fermat factorization*, since it was discovered by the French mathematician Fermat* in the 17th century. This factorization method is based on the following lemma.

**Lemma 2.7.** If $n$ is an odd positive integer, then there is a one-to-one correspondence between factorizations of $n$ into two positive integers and differences of two squares that equal $n$.

*Proof.* Let $n$ be an odd positive integer and let $n = ab$ be a factorization of $n$ into two positive integers. Then $n$ can be written as the difference of two squares, since

$$n = ab = s^2 - t^2 \, ,$$

where $s = (a+b)/2$ and $t = (a-b)/2$ are both integers since $a$ and $b$ are both odd.

Conversely, if $n$ is the difference of two squares, say $n = s^2 - t^2$, then we can factor $n$ by noting that $n = (s-t)(s+t)$.

We leave it to the reader to show this is a one-to-one correspondence.   ∎

To carry out the method of Fermat factorization, we look for solutions of the equation $n = x^2 - y^2$ by searching for perfect squares of the form $x^2 - n$. Hence, to find factorizations of $n$, we search for a square among the sequence of integers

$$t^2-n, \ (t+1)^2-n, \ (t+2)^2-n,...$$

where $t$ is the smallest integer greater than $\sqrt{n}$. This procedure is guaranteed to terminate, since the trivial factorization $n = n\cdot1$ leads to the equation

---

* **PIERRE DE FERMAT** (1601 - 1665), was a lawyer by profession. He was a noted jurist at the provincial parliament in the French city of Toulouse. Fermat was probably the most famous amateur mathematician in history. He published almost none of his mathematical discoveries, but did correspond with contemporary mathematicians about these discoveries. From his correspondents, especially the French monk Mersenne (discussed in Chapter 6), the world learned about his many contributions to mathematics. Fermat was one of inventors of analytic geometry. Furthermore, he laid the foundations of calculus. Fermat, along with Pascal, gave a mathematical basis to the concept of probability. Some of Fermat's discoveries come to us only because he made notes in the margins of his copy of the work of Diophantus. His son found his copy with these notes and published this so that other mathematicians would be aware of Fermat's results and claims.

$$n = \left(\frac{n+1}{2}\right)^2 - \left(\frac{n-1}{2}\right)^2.$$

**Example 2.17.** We factor 6077 using the method of Fermat factorization. Since $77 < \sqrt{6077} < 78$, we look for a perfect square in the sequence

$$78^2 - 6077 = 7$$
$$79^2 - 6077 = 164$$
$$80^2 - 6077 = 323$$
$$81^2 - 6077 = 484 = 22^2.$$

Since $6077 = 81^2 - 22^2$, we see that $6077 = (81-22)(81+22) = 59 \cdot 103$. ◻

Unfortunately, Fermat factorization can be very inefficient. To factor $n$ using this technique, it may be necessary to check as many as $(n + 1)/2 - \sqrt{n}$ integers to determine whether they are perfect squares. Fermat factorization works best when it is used to factor integers having two factors of similar size.

The integers $F_n = 2^{2^n} + 1$ are called the *Fermat numbers*. Fermat conjectured that these integers are all primes. Indeed, the first few are primes, namely $F_0 = 3$, $F_1 = 5$, $F_2 = 17$, $F_3 = 257$, and $F_4 = 65537$. Unfortunately, $F_5 = 2^{2^5} + 1$ is composite as we will now demonstrate.

**Example 2.18.** The Fermat number $F_5 = 2^{2^5} + 1$ is divisible by 641. We can show that $641 \mid F_5$ without actually performing the division. Note that

$$641 = 5 \cdot 2^7 + 1 = 2^4 + 5^4.$$

Hence,

$$2^{2^5} + 1 = 2^{32} + 1 = 2^4 \cdot 2^{28} + 1 = (641 - 5^4)2^{28} + 1$$
$$= 641 \cdot 2^{28} - (5 \cdot 2^7)^4 + 1 = 641 \cdot 2^{28} - (641 - 1)^4 + 1$$
$$= 641(2^{28} - 641^3 + 4 \cdot 641^2 - 6 \cdot 641 + 4).$$

Therefore, we see that $641 \mid F_5$. ◻

The following result is a valuable aid in the factorization of Fermat numbers.

**Theorem 2.11.** Every prime divisor of the Fermat number $F_n = 2^{2^n} + 1$ is of the form $2^{n+2}k + 1$.

The proof of Theorem 2.11 is left until later. It is presented as a problem in Chapter 9. Here, we indicate how Theorem 2.11 is useful in determining the factorization of Fermat numbers.

**Example 2.19.** From Theorem 2.11, we know that every prime divisor of $F_3 = 2^{2^3} + 1 = 257$ must be of the form $2^5 k + 1 = 32 \cdot k + 1$. Since there are no primes of this form less than or equal to $\sqrt{257}$, we can conclude that $F_3 = 257$ is prime. $\square$

**Example 2.20.** When factoring $F_6 = 2^{2^6} + 1$, we use Thereom 2.11 to see that all its prime factors are of the form $2^8 k + 1 = 256 \cdot k + 1$. Hence, we need only perform trial divisions of $F_6$ by primes of the form $256 \cdot k + 1$ that do not exceed $\sqrt{F_6}$. After considerable computation, we find that a prime divisor is obtained with $k = 1071$, *i.e.* $274177 = (256 \cdot 1071 + 1) \mid F_6$. $\square$

A great deal of effort has been devoted to the factorization of Fermat numbers. As yet, no new Fermat primes have been found, and many people believe that no additional Fermat primes exist. We will develop a primality test for Fermat numbers in Chapter 9.

It is possible to prove that there are infinitely many primes using Fermat numbers. We begin by showing that any two distinct Fermat numbers are relatively prime. The following lemma will be used.

**Lemma 2.8.** Let $F_k = 2^{2^k} + 1$ denote the $k$th Fermat number, where $k$ is a nonnegative integer. Then for all positive integers $n$, we have

$$F_0 F_1 F_2 \cdots F_{n-1} = F_n - 2.$$

*Proof.* We will prove the lemma using mathematical induction. For $n = 1$, the identity reads

$$F_0 = F_1 - 2.$$

This is obviously true since $F_0 = 3$ and $F_1 = 5$. Now let us assume that the identity holds for the positive integer $n$, so that

$$F_0 F_1 F_2 \cdots F_{n-1} = F_n - 2.$$

With this assumption we can easily show that the identity holds for the integer $n + 1$, since

$$\begin{aligned}
F_0 F_1 F_2 \cdots F_{n-1} F_n &= (F_0 F_1 F_2 \cdots F_{n-1}) F_n \\
&= (F_n - 2) F_n = (2^{2^n} - 1)(2^{2^n} + 1) \\
&= (2^{2^n})^2 - 1 = 2^{2^{n+1}} - 2 = F_{n+1} - 2.
\end{aligned}$$ ■

This leads to the following theorem.

**Theorem 2.12.** Let $m$ and $n$ be distinct nonnegative integers. Then the Fermat numbers $F_m$ and $F_n$ are relatively prime.

*Proof.* Let us assume that $m < n$. By Lemma 2.8 we know that

$$F_0 F_1 F_2 \cdots F_m \cdots F_{n-1} = F_n - 2.$$

Assume that $d$ is a common divisor of $F_m$ and $F_n$. Then, Thereom 1.8 tells us that

$$d \mid (F_n - F_0 F_1 F_2 \cdots F_m \cdots F_{n-1}) = 2.$$

Hence, either $d=1$ or $d=2$. However, since $F_m$ and $F_n$ are odd, $d$ cannot be 2. Consequently, $d=1$ and $(F_m, F_n) = 1$.  ∎

Using Fermat numbers we can give another proof that there are infinitely many primes. First, we note that from Lemma 1.1 from Section 1.9, every Fermat number $F_n$ has a prime divisor $p_n$. Since $(F_m, F_n) = 1$, we know that $p_m \neq p_n$ whenever $m \neq n$. Hence, we can conclude that there are infinitely many primes.

The Fermat primes are also important in geometry. The proof of the following famous theorem may be found in Ore [30].

**Theorem 2.13.** A regular polygon of $n$ sides can be constructed using a ruler and compass if and only if $n$ is of the form $n = 2^a p_1 \cdots p_t$ where $p_i$, $i=1,2,...,t$ are distinct Fermat primes and $a$ is a nonnegative integer.

---

## 2.4 Exercises

1. Find the prime factorization of each of the following positive integers.

   a) 33776925     b) 210733237     c) 1359170111.

2. Using the Fermat's factorization method factor each of the following positive integers.

   a)  8051       d)  11021
   b)  73         e)  3200399
   c)  10897      f)  24681023

3. Show that the last two decimal digits of a perfect square must be one of the following pairs: 00, $e1$, $e4$, 25, $o6$, $e9$, where $e$ stands for any even digit and $o$ stands for any odd digit. (*Hint*: Show that $n^2$, $(50+n)^2$, and $(50-n)^2$ all have the same final decimal digits, and then consider those integers $n$ with $0 \leqslant n \leqslant 25$.)

4. Explain how the result of Exercise 3 can be used to speed up Fermat's factorization method.

5. Show that if the smallest prime factor of $n$ is $p$, then $x^2-n$ will not be a perfect square for $x > (n+p^2)/2p$ with the single exception $x = (n+1)/2$.

Exercises 6-8 involve the method of *Draim factorization*. To use this technique to search for a factor of the positive integer $n = n_1$, we start by using the division algorithm, to obtain

$$n_1 = 3q_1 + r_1, \quad 0 \leqslant r_1 < 3.$$

Setting $m_1 = n_1$, we let

$$m_2 = m_1 - 2q_1, \quad n_2 = m_2 + r_1.$$

We use the division algorithm again, to obtain

$$n_2 = 5q_2 + r_2, \quad 0 \leqslant r_2 < 5,$$

and we let

$$m_3 = m_2 - 2q_2, \quad n_3 = m_3 + r_2.$$

We proceed recursively, using the division algorithm, to write

$$n_k = (2k+1)q_k + r_k, \quad 0 \leqslant r_k < 2k+1,$$

and we define

$$m_k = m_{k-1} - 2q_{k-1}, \quad n_k = m_k + r_{k-1}.$$

We stop when we obtain a remainder $r_k = 0$.

6. Show that $n_k = kn_1 - (2k+1)(q_1 + q_2 + \cdots + q_{k-1})$ and $m_k = n_1 - 2 \cdot (q_1 + q_2 + \cdots + q_{k-1})$.

7. Show that if $(2k+1) \mid n$, then $(2k+1) \mid n_k$ and $n = (2k+1)m_{k+1}$.

8. Factor 5899 using the method of Draim factorization.

In Exercises 9-11 we develop a factorization technique known as *Euler's method*. It is applicable when the integer being factored is odd and can be written as the sum of two squares in two different ways. Let $n$ be odd and let $n = a^2 + b^2 = c^2 + d^2$, where $a$ and $c$ are odd positive integers, and $b$ and $d$ are even positive integers.

9. Let $u = (a-c, b-d)$. Show that $u$ is even and that if $r = (a-c)/u$ and $s = (d-b)/u$, then $(r,s) = 1$, $r(a+c) = s(d+b)$, and $s \mid (a+c)$.

10. Let $sv = a+c$. Show that $rv = d + b, v = (a+c, d+b)$, and $v$ is even.

11. Conclude that $n$ may be factored as $n = [(u/2)^2 + (v/2)^2](r^2 + s^2)$.

12. Use Euler's method to factor each of the following integers.

   a) $221 = 10^2 + 11^2 = 5^2 + 14^2$
   b) $2501 = 50^2 + 1^2 = 49^2 + 10^2$
   c) $1000009 = 1000^2 + 3^2 = 972^2 + 235^2$

13. Show that any number of the form $2^{4n+2} + 1$ can be easily factored by the use of the identity $4x^4 + 1 = (2x^2+2x+1)(2x^2-2x+1)$. Factor $2^{18}+1$ using this identity.

14. Show that if $a$ is a positive integer and $a^m+1$ is a prime, then $m=2^n$ for some positive integer $n$. (*Hint*: Recall the identity $a^m+1 = (a^k + 1)(a^{k(l-1)}-a^{k(l-2)} + \cdots -a^k+1)$ where $m=kl$ and $l$ is odd).

15. Show that the last digit in the decimal expansion of $F_n = 2^{2^n} + 1$ is 7 if $n \geqslant 2$. (*Hint*: Using mathematical induction, show that the last decimal digit of $2^{2^n}$ is 6.)

16. Use the fact that every prime divisor of $F_4 = 2^{2^4} + 1 = 65537$ is of the form $2^6 k + 1 = 64k + 1$ to verify that $F_4$ is prime. (You should need only one trial division.)

17. Use the fact that every prime divisor of $F_5 = 2^{2^5} + 1$ is of the form $2^7 k + 1 = 128k + 1$ to demonstrate that the prime factorization of $F_5$ is $F_5 = 641 \cdot 6700417$.

18. Find all primes of the form $2^{2^n} + 5$, where $n$ is a nonnegative integer.

19. Estimate the number of decimal digits in the Fermat number $F_n$.

## 2.4 Computer Projects

Write programs to do the following:

   1. Find the prime factorization of a positive integer.

   2. Perform Fermat factorization.

   3. Perform Draim factorization (see preamble to Exercise 6).

   4. Check a Fermat number for prime factors, using Theorem 2.11.

## 2.5 Linear Diophantine Equations

Consider the following problem. A man wishes to purchase $510 of travelers checks. The checks are available only in denominations of $20 and

$50. How many of each denomination should he buy? If we let $x$ denote the number of $20 checks and $y$ the number of $50 checks that he should buy, then the equation $20x + 50y = 510$ must be satisfied. To solve this problem, we need to find all solutions of this equation, where both $x$ and $y$ are nonnegative integers.

A related problem arises when a woman wishes to mail a package. The postal clerk determines the cost of postage to be 83 cents but only 6-cent and 15-cent stamps are available. Can some combination of these stamps be used to mail the package? To answer this, we first let $x$ denote the number of 6-cent stamps and $y$ the number of 15-cent stamps to be used. Then we must have $6x + 15y = 83$, where both $x$ and $y$ are nonnegative integers.

When we require that solutions of a particular equation come from the set of integers, we have a *diophantine equation*. Diophantine equations get their name from the ancient Greek mathematician Diophantus*, who wrote extensively on such equations. The type of diophantine equation $ax + by = c$, where $a$, $b$, and $c$ are integers is called a *linear diophantine equations in two variables*. We now develop the theory for solving such equations. The following theorem tells us when such an equation has solutions, and when there are solutions, explicitly describes them.

**Theorem 2.14.** Let $a$ and $b$ be integers with $d = (a,b)$. The equation $ax + by = c$ has no integral solutions if $d \nmid c$. If $d \mid c$, then there are infinitely many integral solutions. Moreover, if $x = x_0$, $y = y_0$ is a particular solution of the equation, then all solutions are given by

$$x = x_0 + (b/d)n, \quad y = y_0 - (a/d)n,$$

where $n$ is an integer.

*Proof.* Assume that $x$ and $y$ are integers such that $ax + by = c$. Then, since $d \mid a$ and $d \mid b$, by Theorem 1.8 $d \mid c$ as well. Hence, if $d \nmid c$, there are no integral solutions of the equation.

Now assume that $d \mid c$. From Theorem 2.2, there are integers $s$ and $t$ with

---

* **DIOPHANTUS (c. 250 C.E.)** wrote the *Arithmetica* which is the earliest book we know of on algebra. This book contains the first systematic use of mathematical notation to represent unknowns in equations and powers of these unknowns. Almost nothing is known about Diophantus' life, other than that he lived in Alexandria around 250 C.E. The only source of details about this life come from an epigram found in a collection called the *Greek Anthology:* "Diophantus passed one sixth of his life in childhood, one twelfth in youth, and one seventh as a bachelor. Five years after his marriage was born a son who died four years before his father, at half his father's age." From this the reader can infer that Diophantus lived to the age of 84.

(2.3)                                $d = as + bt.$

Since $d \mid c$, there is an integer $e$ with $de = c$. Multiplying both sides of (2.3) by $e$, we have

$$c = de = (as + bt)e = a(se) + b(te).$$

Hence, one solution of the equation is given by $x = x_0$ and $y = y_0$, where $x_0 = se$ and $y_0 = te$.

To show that there are infinitely many solutions, let $x = x_0 + (b/d)n$ and $y = y_0 - (a/d)n$, where $n$ is an integer. We see that this pair $(x,y)$ is a solution, since

$$ax + by = ax_0 + a(b/d)n + by_0 - b(a/d)n = ax_0 + by_0 = c.$$

We now show that every solution of the equation $ax + by = c$ must be of the form described in the theorem. Suppose that $x$ and $y$ are integers with $ax + by = c$. Since

$$ax_0 + by_0 = c,$$

by subtraction we find that

$$(ax + by) - (ax_0 + by_0) = 0,$$

which implies that

$$a(x - x_0) + b(y - y_0) = 0.$$

Hence,

$$a(x - x_0) = b(y_0 - y).$$

Dividing both sides of this last equality by $d$, we see that

$$(a/d)(x - x_0) = (b/d)(y_0 - y).$$

By Theorem 2.1 we know that $(a/d, b/d) = 1$. Using Lemma 2.3, it follows that $(a/d) \mid (y_0 - y)$. Hence, there is an integer $n$ with $(a/d)n = y_0 - y$; this means that $y = y_0 - (a/d)n$. Now putting this value of $y$ into the equation $a(x - x_0) = b(y_0 - y)$, we find that $a(x - x_0) = b(a/d)n$, which implies that $x = x_0 + (b/d)n$.                                   ■

We now demonstrate how Theorem 2.14 is used to find the solutions of particular linear diophantine equations in two variables.

**Example 2.21.** Consider the problems of finding all the integral solutions of the two diophantine equations described at the beginning of this section. We first consider the equation $6x + 15y = 83$. The greatest common divisor of 6 and 15 is $(6,15) = 3$. Since $3 \nmid 83$, we know that there are no integral solutions. Hence, no combination of 6- and 15-cent stamps gives the correct postage.                                   □

**Example 2.22.** Consider the equation $20x + 50y = 510$. The greatest common divisor of 20 and 50 is $(20,50) = 10$, and since $10 \mid 510$, there are infinitely many integral solutions. Using the Euclidean algorithm, we find that $20(-2) + 50 = 10$. Multiplying both sides by 51, we obtain $20(-102) + 50(51) = 510$. Hence, a particular solution is given by $x_0 = -102$ and $y_0 = 51$. Theorem 2.14 tells us that all integral solutions are of the form $x = -102 + 5n$ and $y = 51 - 2n$. Since we want both $x$ and $y$ to be nonnegative, we must have $-102 + 5n \geqslant 0$ and $51 - 2n \geqslant 0$; thus, $n \geqslant 20\ 2/5$ and $n \leqslant 25\ 1/2$. Since $n$ is an integer, it follows that $n = 21, 22, 23, 24$, or 25. Hence, we have the following 5 solutions: $(x,y) = (3,9)$, $(8,7)$, $(13,5)$, $(18,3)$, and $(23,1)$. $\square$

## 2.5 Exercises

1. For each of the following linear diophantine equations, either find all solutions, or show that there are no integral solutions.

    a) $2x + 5y = 11$
    b) $17x + 13y = 100$
    c) $21x + 14y = 147$
    d) $60x + 18y = 97$
    e) $1402x + 1969y = 1$

2. A student returning from Europe changes his French francs and Swiss francs into U.S. money. If he receives $\$11.91$ and has received $17¢$ for each French franc and $48¢$ for each Swiss franc, how much of each type of currency did he exchange?

3. A grocer orders apples and oranges at a total cost of $\$8.39$. If apples cost him $25¢$ each and oranges cost him $18¢$ each, how many of each type of fruit did he order?

4. A shopper spends a total of $\$5.49$ for oranges, which cost $18¢$ each, and grapefruits, which cost $33¢$ each. What is the minimum number of pieces of fruit the shopper could have bought?

5. A postal clerk has only 14-cent and 21-cent stamps to sell. What combinations of these may be used to mail a package requiring postage of exactly each of the following amounts?

    a)   $\$3.50$      b)   $\$4.00$      c)   $\$7.77$

6. At a clambake, the total cost of a lobster dinner is $\$11$ and of a chicken dinner is $\$8$. What can you conclude if the total bill is each of the following amounts?

   a)   $\$777$     b)   $\$96$     c)   $\$69$

7. Show that the linear diophantine equation $a_1x_1 + a_2x_2 + ... + a_nx_n = b$ has no solutions if $d \nmid b$, where $d = (a_1,a_2,...,a_n)$, and has infinitely many solutions if $d \mid b$.

8. Find all integer solutions of each of the following linear diophantine equations

   a)   $2x + 3y + 4z = 5$
   b)   $7x + 21y + 35z = 8$
   c)   $101x + 102y + 103z = 1$

9. Which combinations of pennies, dimes, and quarters have a total value 99¢?

10. How many ways can change be made for one dollar using each of the following coins?

    a)   dimes and quarters
    b)   nickels, dimes, and quarters
    c)   pennies, nickels, dimes, and quarters

In Exercises 11-14 we consider simultaneous linear diophantine equations. To solve these first eliminate all but one variable and then solve the resulting equation in one variable.

11. Find all integer solutions of the following systems of linear diophantine equations.

    a)   $x + y + z = 100$
         $x + 8y + 50z = 156$

    b)   $x + y + z = 100$
         $x + 6y + 21z = 121$

    c)   $x + y + z + w = 100$
         $x + 2y + 3z + 4w = 300$
         $x + 4y + 9z + 16w = 1000$

12. A piggy bank contains 24 coins, all nickels, dimes, and quarters. If the total value of the coins is two dollars, what combinations of coins are possible?

13. Nadir Airways offers three types of tickets on their Boston to New York flights. First-class tickets are $\$70$, second-class tickets are $\$55$, and

stand-by tickets are $39. If 69 passengers pay a total of $3274 for their tickets on a particular flight, how many of each type of tickets were sold?

14. Is it possible to have 50 coins, all pennies, dimes, and quarters worth $3?

Let $a$ and $b$ be relatively prime positive integers and let $n$ be a positive integer. We call a solution $x,y$ of the linear diophantine equation $ax + by = n$ *nonnegative* when both $x$ and $y$ are nonnegative.

* 15. Show that whenever $n \geqslant (a-1)(b-1)$ there is a nonnegative solution of $ax + by = n$.

* 16. Show that if $n = ab - a - b$, then there are no nonnegative solutions of $ax + by = n$.

* 17. Show that there are exactly $(a-1)(b-1)/2$ nonnegative integers $n < ab-a-b$ such that the equation has a nonnegative solution.

18. The post office in a small Maine town is left with stamps of only two values. They discover that there are exactly 33 postage amounts that cannot be made up using these stamps, including 46¢. What are the values of the remaining stamps?

* 19. An ancient Chinese puzzle found in the sixth century work of the mathematician Chang Ch'iu-chien, called the "hundred fowls" problem, asks: If a cock is worth five coins, a hen three coins, and three chickens together are worth 1 coin, how many cocks, hens, and chickens, totaling 100, can be bought for 100 coins? Solve this problem.

* 20. Find all solutions where $x$ and $y$ are integers to the diophantine equation

$$\frac{1}{x} + \frac{1}{y} = \frac{1}{14}.$$

## 2.5 Computer Projects

Write programs to do the following:

1. Find the solutions of a linear diophantine equation in two variables.

2. Find the positive solutions of a linear diophantine equation in two variables.

3. Find the solutions of a linear diophantine equation in an arbitrary number of variables.

* 4. Find all positive integers $n$ for which the linear diophantine equation $ax + by = n$ has no positive solutions (See the preamble to Exercise 15).

# 3

---

# Congruences

## 3.1 Introduction to Congruences

The special language of congruences that we introduce in this chapter is extremely useful in number theory. This language of congruences was developed at the beginning of the nineteenth century by Karl Friedrich Gauss*, one of the most famous mathematicians in history.

---

* **KARL FRIEDRICH GAUSS** (1777-1785) was the son of a bricklayer. It was quickly apparent that he was child prodigy. In fact, at the age of three he corrected an error in his father's payroll. In his first arithmetic class the teacher gave an assignment designed to keep the class busy, namely to find the sum of the first 100 positive integers. Gauss, who was eight at the time, realized that this sum is $50 \cdot 101 = 5050$ because the terms can be grouped as $1+100=101$, $2+99=101,...,49+52=101$, and $50+51=101$. In 1796 Gauss made an important discovery in an area of geometry that had not progressed since ancient times. In particular, he showed that a heptadecagon, a 17-sided regular polygon, could be drawn using just a ruler and a compass. In 1799 he presented the first rigorous proof of the Fundamental Theorem of Algebra, which states that a polynomial of degree $n$ with real coefficients has exactly $n$ roots. Gauss made fundamental contributions to astronomy including calculating the orbit of the asteroid Ceres. On the basis of this calculation, Gauss was appointed director of the Göttingen Observatory. He laid the foundations of modern number theory with his book *Disquitiones Arithmeticae* in 1801. Gauss was called Princeps Mathematicorum (the Prince of Mathematicians), by his contemporary mathematician. Although Gauss is noted for his many discoveries in geometry, algebra, analysis, astronomy, and mathematical physics, he had a special interest in number theory. This can be seen from his statement: "Mathematics is the queen of sciences, and the theory of numbers is the queen of mathematics." Gauss made most of his important discoveries early in his life, and spent his later years refining them. Gauss made many fundamental discoveries that he did not reveal and when other mathematician made the same discoveries, it was found Gauss had discovered the same things years earlier.

**Definition.** Let $m$ be a positive integer. If $a$ and $b$ are integers, we say that $a$ is *congruent to b modulo m* if $m \mid (a-b)$.

If $a$ is congruent to $b$ modulo $m$, we write $a \equiv b \pmod{m}$. If $m \nmid (a-b)$, we write $a \not\equiv b \pmod{m}$, and say that $a$ and $b$ are *incongruent modulo m*.

**Example 3.1.** We have $22 \equiv 4 \pmod{9}$, since $9 \mid (22-4) = 18$. Likewise $3 \equiv -6 \pmod{9}$ and $200 \equiv 2 \pmod{9}$. On the other hand, $13 \not\equiv 5 \pmod{9}$ since $9 \nmid (13-5) = 8$. □

Congruences often arise in everyday life. For instance, clocks work either modulo 12 or 24 for hours, and modulo 60 for minutes and seconds, calendars work modulo 7 for days of the week and modulo 12 for months. Utility meters often operate modulo 1000, and odometers usually work modulo 100000.

In working with congruences, it is often useful to translate them into equalities. To do this, the following theorem is needed.

**Theorem 3.1.** If $a$ and $b$ are integers, then $a \equiv b \pmod{m}$ if and only if there is an integer $k$ such that $a = b + km$.

*Proof.* If $a \equiv b \pmod{m}$, then $m \mid (a-b)$. This means that there is an integer $k$ with $km = a - b$, so that $a = b + km$.

Conversely, if there is an integer $k$ with $a = b + km$, then $km = a - b$. Hence $m \mid (a-b)$, and consequently, $a \equiv b \pmod{m}$. ∎

**Example 3.2.** We have $19 \equiv -2 \pmod{7}$ and $19 = -2 + 3 \cdot 7$. □

The following proposition establishes some important properties of congruences.

**Theorem 3.2.** Let $m$ be a positive integer. Congruences modulo $m$ satisfy the following properties:

(i) *Reflexive property.* If $a$ is an integer, then $a \equiv a \pmod{m}$.

(ii) *Symmetric property.* If $a$ and $b$ are integers such that $a \equiv b \pmod{m}$, then $b \equiv a \pmod{m}$.

(iii) *Transitive property.* If $a$, $b$, and $c$ are integers with $a \equiv b \pmod{m}$ and $b \equiv c \pmod{m}$, then $a \equiv c \pmod{m}$.

*Proof.*

(i)   We see that $a \equiv a \pmod{m}$, since $m \mid (a-a) = 0$.

(ii)  If $a \equiv b \pmod{m}$, then $m \mid (a-b)$. Hence, there is an integer $k$ with $km = a - b$. This shows that $(-k)m = b - a$, so that $m \mid (b-a)$. Consequently, $b \equiv a \pmod{m}$.

(iii) If $a \equiv b \pmod{m}$ and $b \equiv c \pmod{m}$, then $m \mid (a-b)$ and $m \mid (b-c)$. Hence, there are integers $k$ and $l$ with $km = a - b$ and $lm = b - c$. Therefore, $a - c = (a-b) + (b-c) = km + lm = (k+l)m$. It follows that $m \mid (a-c)$ and $a \equiv c \pmod{m}$.   ∎

By Theorem 3.2 we see that the set of integers is divided into $m$ different sets called *congruence classes modulo m*, each containing integers which are mutually congruent modulo $m$.

**Example 3.3.** The four congruence classes modulo 4 are given by

$$\cdots \equiv -8 \equiv -4 \equiv 0 \equiv 4 \equiv 8 \equiv \cdots \pmod 4$$
$$\cdots \equiv -7 \equiv -3 \equiv 1 \equiv 5 \equiv 9 \equiv \cdots \pmod 4$$
$$\cdots \equiv -6 \equiv -2 \equiv 2 \equiv 6 \equiv 10 \equiv \cdots \pmod 4$$
$$\cdots \equiv -5 \equiv -1 \equiv 3 \equiv 7 \equiv 11 \equiv \cdots \pmod 4.$$   □

Let $a$ be an integer. Given the positive integer $m$, $m > 1$, by the division algorithm, we have $a = bm + r$ where $0 \leqslant r \leqslant m - 1$. From the equation $a = bm + r$, we see that $a \equiv r \pmod{m}$. Hence, every integer is congruent modulo $m$ to one of the integers of the set $0, 1,...,m - 1$, namely the remainder when it is divided by $m$. Since no two of the integers $0, 1,...,m - 1$ are congruent modulo $m$, we have $m$ integers such that every integer is congruent to exactly one of these $m$ integers.

**Definition.** A *complete system of residues modulo m* is a set of integers such that every integer is congruent modulo $m$ to exactly one integer of the set.

**Example 3.4.** The division algorithm shows that the set of integers $0, 1, 2,...,m - 1$ is a complete system of residues modulo $m$. This is called the set of *least nonnegative residues modulo m*.   □

**Example 3.5.** Let $m$ be an odd positive integer. Then the set of integers

$$-\frac{m-1}{2}, -\frac{m-3}{2},...,-1, 0, 1,..., \frac{m-3}{2}, \frac{m-1}{2}$$

is a complete system of residues called the set of *absolute least residues modulo m*.   □

We will often do arithmetic with congruences. Congruences have many of the same properties that equalities do. First, we show that an addition, subtraction, or multiplication to both sides of a congruence preserves the congruence.

**Theorem 3.3.** If $a$, $b$, $c$, and $m$ are integers with $m > 0$ such that $a \equiv b \pmod{m}$, then

    (i)   $a + c \equiv b + c \pmod{m}$,

    (ii)  $a - c \equiv b - c \pmod{m}$,

    (iii) $ac \equiv bc \pmod{m}$.

*Proof.* Since $a \equiv b \pmod{m}$, we know that $m \mid (a-b)$. From the identity $(a+c) - (b+c) = a - b$, we see $m \mid [(a+c) - (b+c)]$, so that (i) follows. Likewise, (ii) follows from the fact that $(a-c) - (b-c) = a - b$. To show that (iii) holds, note that $ac - bc = c(a-b)$. Since $m \mid (a-b)$, it follows that $m \mid c(a-b)$, and hence, $ac \equiv bc \pmod{m}$. ∎

**Example 3.6.** Since $19 \equiv 3 \pmod 8$, it follows from Theorem 3.3 that $26 = 19 + 7 \equiv 3 + 7 = 10 \pmod 8$, $15 = 19 - 4 \equiv 3 - 4 \equiv -1 \pmod 8$, and $38 = 19 \cdot 2 \equiv 3 \cdot 2 = 6 \pmod 8$. ∎

What happens when both sides of a congruence are divided by an integer? Consider the following example.

**Example 3.7.** We have $14 = 7 \cdot 2 \equiv 4 \cdot 2 = 8 \pmod 6$. But we cannot carry the common factor of 2 since $7 \not\equiv 4 \pmod 6$. □

This example shows that it is not necessarily true that we preserve a congruence when we divide both sides by an integer. However, the following theorem gives a valid congruence when both sides of a congruence are divided by the same integer.

**Theorem 3.4.** If $a$, $b$, $c$ and $m$ are integers such that $m > 0$, $d = (c,m)$, and $ac \equiv bc \pmod{m}$, then $a \equiv b \pmod{m/d}$.

*Proof.* If $ac \equiv bc \pmod{m}$, we know that $m \mid (ac-bc) = c(a-b)$. Hence, there is an integer $k$ with $c(a-b) = km$. By dividing both sides by $d$, we have $(c/d)(a-b) = k(m/d)$. Since $(m/d, c/d) = 1$, by Lemma 2.3 it follows that $m/d \mid (a-b)$. Hence, $a \equiv b \pmod{m/d}$. ∎

**Example 3.8.** Since $50 \equiv 20 \pmod{15}$ and $(10,15) = 5$, we see that $50/10 \equiv 20/10 \pmod{15/5}$, or $5 \equiv 2 \pmod 3$. □

The following corollary, which is a special case of Theorem 3.4, is used often.

**Corollary 3.4.1.** If $a$, $b$, $c$, and $m$ are integers such that $m > 0$, $(c,m) = 1$, and $ac \equiv bc \pmod{m}$, then $a \equiv b \pmod{m}$.

**Example 3.9.** Since $42 \equiv 7 \pmod{5}$ and $(5,7) = 1$, we can conclude that $42/7 \equiv 7/7 \pmod{5}$, or that $6 \equiv 1 \pmod{5}$.                                 □

The following theorem which is more general than Theorem 3.3 is also useful.

**Theorem 3.5.** If $a$, $b$, $c$, $d$, and $m$ are integers such that $m > 0$, $a \equiv b \pmod{m}$, and $c \equiv d \pmod{m}$, then

   (i)   $a + c \equiv b + d \pmod{m}$,

   (ii)   $a - c \equiv b - d \pmod{m}$,

   (iii)   $ac \equiv bd \pmod{m}$.

*Proof.* Since $a \equiv b \pmod{m}$ and $c \equiv d \pmod{m}$, we know that $m \mid (a-b)$ and $m \mid (c-d)$. Hence, there are integers $k$ and $l$ with $km = a - b$ and $lm = c - d$.

To prove (i), note that $(a+c) - (b+d) = (a-b) + (c-d) = km + lm = (k+l)m$. Hence, $m \mid [(a+c) - (b+d)]$. Therefore, $a + c \equiv b + d \pmod{m}$.

To prove (ii), note that $(a-c) - (b-d) = (a-b) - (c-d) = km - lm = (k-l)m$. Hence, $m \mid [(a-c)-(b-d)]$, so that $a - c \equiv b - d \pmod{m}$.

To prove (iii), note that $ac - bd = ac - bc + bc - bd = c(a-b) + b(c-d) = ckm + blm = m(ck+bl)$. Hence, $m \mid (ac - bd)$. Therefore, $ac \equiv bd \pmod{m}$.                                 ■

**Example 3.10.** Since $13 \equiv 8 \pmod{5}$ and $7 \equiv 2 \pmod{5}$, using Theorem 3.5 we see that $20 = 13 + 7 \equiv 8 + 2 \equiv 0 \pmod{5}$, $6 = 13 - 7 \equiv 8 - 7 = 1 \pmod{5}$, and $91 = 13 \cdot 7 \equiv 8 \cdot 2 = 16 \pmod{5}$.                                 □

**Theorem 3.6.** If $r_1, r_2, \ldots, r_m$ is a complete system of residues modulo $m$, and if $a$ is a positive integer with $(a,m) = 1$, then

$$ar_1 + b, ar_2 + b, \ldots, ar_m + b$$

is a complete system of residues modulo $m$ for any integer $b$.

*Proof.* First, we show that no two of the integers

$$ar_1 + b, ar_2 + b, \ldots, ar_m + b$$

are congruent modulo $m$. To see this, note that if

$$ar_j + b \equiv ar_k + b \pmod{m},$$

then, by (ii) of Theorem 3.3, we know that

$$ar_j \equiv ar_k \pmod{m}.$$

Because $(a,m) = 1$, Corollary 3.4.1 shows that

$$r_j \equiv r_k \pmod{m}.$$

Since $r_j \not\equiv r_k \pmod{m}$ if $j \neq k$, we conclude that $j = k$.

Since the set of integers in question consists of $m$ incongruent integers modulo $m$, these integers form a complete system of residues modulo $m$. ∎

The following theorem shows that a congruence is preserved when both sides are raised to the same positive integral power.

**Theorem 3.7.** If $a$, $b$, $k$, and $m$ are integers such that $k > 0$, $m > 0$, and $a \equiv b \pmod{m}$, then $a^k \equiv b^k \pmod{m}$.

*Proof.* Because $a \equiv b \pmod{m}$, we have $m \mid (a - b)$. Since

$$a^k - b^k = (a-b)(a^{k-1}+a^{k-2}b+ \cdots +ab^{k-2}+b^{k-1}),$$

we see that $(a - b) \mid (a^k - b^k)$. Therefore, by Theorem 1.7 it follows that $m \mid (a^k - b^k)$. Hence, $a^k \equiv b^k \pmod{m}$. ∎

**Example 3.11.** Since $7 \equiv 2 \pmod{5}$, Theorem 3.7 tells us that $343 = 7^3 \equiv 2^3 \equiv 8 \pmod{5}$. □

The following result shows how to combine congruences of two numbers to different moduli.

**Theorem 3.8.** If $a \equiv b \pmod{m_1}$, $a \equiv b \pmod{m_2}$,..., $a \equiv b \pmod{m_k}$ where $a,b,m_1, m_2,...,m_k$ are integers with $m_1,m_2,...,m_k$ positive, then

$$a \equiv b \pmod{[m_1,m_2,...,m_k]},$$

where $[m_1,m_2,...,m_k]$ is the least common multiple of $m_1,m_2,...,m_k$.

*Proof.* Since $a \equiv b \pmod{m_1}$, $a \equiv b \pmod{m_2}$,..., $a \equiv b \pmod{m_k}$, we know that $m_1 \mid (a - b), m_2 \mid (a - b),..., m_k \mid (a-b)$. By Exercise 27 of Section 2.3 we see that

$$[m_1,m_2,...,m_k] \mid (a - b).$$

Consequently,

$$a \equiv b \pmod{[m_1,m_2,...,m_k]}.$$ ∎

An immediate and useful consequence of this theorem is the following result.

**Corollary 3.8.1.** If $a \equiv b \pmod{m_1}$, $a \equiv b \pmod{m_2}$ ,..., $a \equiv b \pmod{m_k}$ where $a$ and $b$ are integers and $m_1, m_2, ..., m_k$ are pairwise relatively prime positive integers, then

$$a \equiv b \pmod{m_1 m_2 \cdots m_k}.$$

*Proof.* Since $m_1, m_2, ..., m_k$ are pairwise relatively prime, Exercise 52 of Section 2.3 tells us that

$$[m_1, m_2, ..., m_k] = m_1 m_2 \cdots m_k.$$

Hence, by Theorem 3.8 we know that

$$a \equiv b \pmod{m_1 m_2 \cdots m_k}. \qquad \blacksquare$$

In our subsequent studies, we will be working with congruences involving large powers of integers. For example, we will want to find the least positive residue of $2^{644}$ modulo 645. If we attempt to find this least positive residue by first computing $2^{644}$, we would have an integer with 194 decimal digits, a most undesirable thought. Instead, to find $2^{644}$ modulo 645 we first express the exponent 644 in binary notation:

$$(644)_{10} = (1010000100)_2.$$

Next, we compute the least positive residues of $2, 2^2, 2^4, 2^8, ..., 2^{512}$ by successively squaring and reducing modulo 645. This gives us the congruences

$$
\begin{array}{rcll}
2      & \equiv & 2   & \pmod{645}, \\
2^2    & \equiv & 4   & \pmod{645}, \\
2^4    & \equiv & 16  & \pmod{645}, \\
2^8    & \equiv & 256 & \pmod{645}, \\
2^{16} & \equiv & 391 & \pmod{645}, \\
2^{32} & \equiv & 16  & \pmod{645}, \\
2^{64} & \equiv & 256 & \pmod{645}, \\
2^{128}& \equiv & 391 & \pmod{645}, \\
2^{256}& \equiv & 16  & \pmod{645}, \\
2^{512}& \equiv & 256 & \pmod{645}.
\end{array}
$$

We can now compute $2^{644}$ modulo 645 by multiplying the least positive residues of the appropriate powers of 2. This gives

$$2^{644} = 2^{512+128+4} = 2^{512} 2^{128} 2^4 \equiv 256 \cdot 391 \cdot 16 = 1601536 \equiv 1 \pmod{645}.$$

We have just illustrated a general procedure for *modular exponentiation*, that is, for computing $b^N$ modulo $m$ where $b$, $m$, and $N$ are positive integers. We first express the exponent $N$ in binary notation, as $N = (a_k a_{k-1} ... a_1 a_0)_2$. We then find the least positive residues of $b, b^2, b^4, ..., b^{2^k}$ modulo $m$, by successively squaring and reducing modulo $m$. Finally, we multiply the least positive residues modulo $m$ of $b^{2^j}$ for those $j$ with $a_j = 1$, reducing modulo $m$ after each multiplication.

In our subsequent discussions, we will need an estimate for the number of bit operations needed for modular exponentiation. This is provided by the following proposition.

**Theorem 3.9.** Let $b$, $m$, and $N$ be positive integers with $b < m$. Then the least positive residue of $b^N$ modulo $m$ can be computed using $O((\log_2 m)^2 \log_2 N)$ bit operations.

*Proof.* To find the least positive residue of $b^N \pmod{m}$, we can use the algorithm just described. First, we find the least positive residues of $b, b^2, b^4, \ldots, b^{2^k}$ modulo $m$, where $2^k \leqslant N < 2^{k+1}$, by successively squaring and reducing modulo $m$. This requires a total of $O((\log_2 m)^2 \log_2 N)$ bit operations, because we perform $[\log_2 N]$ squarings modulo $m$, each requiring $O((\log_2 m)^2)$ bit operations. Next, we multiply together the least positive residues of the integers $b^{2^j}$ corresponding to the binary digits of $N$ which are equal to one, and we reduce modulo $m$ after each multiplication. This also requires $O((\log_2 m)^2 \log_2 N)$ bit operations, because there are at most $\log_2 N$ multiplications, each requiring $O((\log_2 m)^2)$ bit operations. Therefore, a total of $O((\log_2 m)^2 \log_2 N)$ bit operations are needed. ∎

## 3.1 Exercises

1. Show that each of the following congruences holds.

   a) $13 \equiv 1 \pmod{2}$      e) $-2 \equiv 1 \pmod{3}$
   b) $22 \equiv 7 \pmod{5}$      f) $-3 \equiv 30 \pmod{11}$
   c) $91 \equiv 0 \pmod{13}$     g) $111 \equiv -9 \pmod{40}$
   d) $69 \equiv 62 \pmod{7}$     h) $666 \equiv 0 \pmod{37}$

2. Determine whether each of the following pairs of integers are congruent modulo 7.

   a) $1,15$      d) $-1,8$
   b) $0,42$      e) $-9,5$
   c) $2,99$      f) $-1,699$

3. For which positive integers $m$ is each of the following statements true?

   a) $27 \equiv 5 \pmod{m}$
   b) $1000 \equiv 1 \pmod{m}$
   c) $1331 \equiv 0 \pmod{m}$

4. Show that if $a$ is an even integer, then $a^2 \equiv 0 \pmod{4}$, and if $a$ is an odd integer, then $a^2 \equiv 1 \pmod{4}$.

5. Show that if $a$ is an odd integer, then $a^2 \equiv 1 \pmod 8$.

6. Find the least nonnegative residue modulo 13 of each of the following integers.

   a) 22
   b) 100
   c) 1001
   d) $-1$
   e) $-100$
   f) $-1000$

7. Find the least positive residue of $1! + 2! + 3! + \cdots + 100!$ modulo each of the following integers.

   a) 2      c) 12
   b) 7      d) 25

8. Show that if $a$, $b$, $m$, and $n$ are integers such that $m > 0$, $n > 0$, $n \mid m$, and $a \equiv b \pmod m$, then $a \equiv b \pmod n$.

9. Show that if $a$, $b$, $c$, and $m$ are integers such that $c > 0$, $m > 0$, and $a \equiv b \pmod m$, then $ac \equiv bc \pmod{mc}$.

10. Show that if $a$, $b$, and $c$ are integers with $c > 0$ such that $a \equiv b \pmod c$, then $(a,c) = (b,c)$.

11. Show that if $a_j \equiv b_j \pmod m$ for $j = 1,2,...,n$, where $m$ is a positive integer and $a_j, b_j$, $j = 1,2,...,n$, are integers, then

    a) $\displaystyle\sum_{j=1}^{n} a_j \equiv \sum_{j=1}^{n} b_j \pmod m$

    b) $\displaystyle\prod_{j=1}^{n} a_j \equiv \prod_{j=1}^{n} b_j \pmod m$

In Exercises 12-14 construct tables for arithmetic modulo 6 using the least nonnegative residues modulo 6 to represent the congruence classes.

12. Construct a table for addition modulo 6.

13. Construct a table for subtraction modulo 6.

14. Construct a table for multiplication modulo 6.

15. What time does a clock read

    a)    29 hours after it reads 11 o'clock
    b)    100 hours after it reads 2 o'clock
    c)    50 hours before it reads 6 o'clock?

16. Which decimal digits occur as the final digit of a fourth power of an integer?

17. What can you conclude if $a^2 \equiv b^2 \pmod p$, where $a$ and $b$ are integers and $p$ is prime?

18. Show that if $a^k \equiv b^k \pmod{m}$ and $a^{k+1} \equiv b^{k+1} \pmod{m}$, where $a$, $b$, $k$, and $m$ are integers with $k > 0$ and $m > 0$ such that $(a,m) = 1$, then $a \equiv b \pmod{m}$. If the condition $(a,m) = 1$ is dropped, is the conclusion that $a \equiv b \pmod{m}$ still valid?

19. Show that if $n$ is an odd positive integer, then

$$1 + 2 + 3 + \cdots + (n-1) \equiv 0 \pmod{n}.$$

    Is this statement true if $n$ is even?

20. Show that if $n$ is an odd positive integer or if $n$ is divisible by 4, then

$$1^3 + 2^3 + 3^3 + \cdots + (n-1)^3 \equiv 0 \pmod{n}.$$

    Is this statement true if $n$ is even but not divisible by 4?

21. For which positive integers $n$ is it true that

$$1^2 + 2^2 + 3^2 + \cdots + (n-1)^2 \equiv 0 \pmod{n}?$$

22. Show by mathematical induction that if $n$ is a positive integer then $4^n \equiv 1 + 3n \pmod{9}$.

23. Show by mathematical induction that if $n$ is a positive integer then $5^n \equiv 1 + 4n \pmod{16}$.

24. Give a complete system of residues modulo 13 consisting entirely of odd integers.

25. Show that if $n \equiv 3 \pmod{4}$, then $n$ cannot be the sum of the squares of two integers.

26. Show that if $p$ is prime, then the only solutions of the congruence $x^2 \equiv x \pmod{p}$ are those integers $x$ with $x \equiv 0$ or $1 \pmod{p}$.

27. Show that if $p$ is prime and $k$ is a positive integer, then the only solutions of $x^2 \equiv x \pmod{p^k}$ are those integers $x$ such that $x \equiv 0$ or $1 \pmod{p^k}$.

28. Find the least positive residues modulo 47 of each of the following integers.

    a)  $2^{32}$     b)  $2^{47}$     c)  $2^{200}$

29. Let $m_1, m_2, ..., m_k$ be pairwise relatively prime positive integers. Let $M = m_1 m_2 \cdots m_k$ and $M_j = M/m_j$ for $j = 1, 2, ..., k$. Show that

$$M_1 a_1 + M_2 a_2 + \cdots + M_k a_k$$

    runs through a complete system of residues modulo $M$ when $a_1, a_2, ..., a_k$ run through complete systems of residues modulo $m_1, m_2, ..., m_k$, respectively.

30. Explain how to find the sum $u + v$ from the least positive residue of $u + v$ modulo $m$, where $u$ and $v$ are positive integers less than $m$. (*Hint*: Assume that $u \leqslant v$ and consider separately the cases where the least positive residue of $u + v$ is less than $u$, and where it is greater than $v$.)

**31.** On a computer with word size $w$, multiplication modulo $n$, where $n < w/2$, can be performed as outlined. Let $T = [\sqrt{n} + \frac{1}{2}]$, and $t = T^2 - n$. For each computation, show that all the required computer arithmetic can be done without exceeding the word size. (This method was described by Head [96]).

a) Show that $|t| \leqslant T$.

b) Show that if $x$ and $y$ are nonnegative integers less than $n$, then

$$x = aT + b, \quad y = cT + d$$

where $a, b, c,$ and $d$ are integers such that $0 \leqslant a \leqslant T, 0 \leqslant b < T, 0 \leqslant c < T,$ and $0 \leqslant d < T$.

c) Let $z \equiv ad + bc \pmod{n}$, with $0 \leqslant z < n$. Show that

$$xy \equiv act + zT + bd \pmod{n}.$$

d) Let $ac = eT + f$ where $e$ and $f$ are integers with $0 \leqslant e < T$ and $0 \leqslant f \leqslant T$. Show that

$$xy \equiv (z + et)T + ft + bd \pmod{n}.$$

e) Let $v = z + et \pmod{n}$, with $0 \leqslant v < n$. Show that we can write

$$v = gT + h,$$

where $g$ and $h$ are integers with $0 \leqslant g \leqslant T, 0 \leqslant h < T,$ and such that

$$xy \equiv hT + (f + g)t + bd \pmod{n}.$$

f) Show that the right-hand side of the congruence of part (e) can be computed without exceeding the word size by first finding $j$ with

$$j \equiv (f + g)t \pmod{n}$$

and $0 \leqslant j < n$, and then finding $k$ with

$$k \equiv j + bd \pmod{n}$$

and $0 \leqslant k < n$, so that

$$xy \equiv hT + k \pmod{n}.$$

This gives the desired result.

**32.** Develop an algorithm for modular exponentiation from the base three expansion of the exponent.

**33.** Find the least positive residue of

a) $3^{10}$ modulo 11
b) $2^{12}$ modulo 13
c) $5^{16}$ modulo 17
d) $3^{22}$ modulo 23.
e) Can you propose a theorem from the above congruences?

**34.** Find the least positive residues of

    a)  6! modulo 7
    b)  10! modulo 11
    c)  12! modulo 13
    d)  16! modulo 17.
    e)  Can you propose a theorem from the above congruences?

**\* 35.** Show that for every positive integer $m$ there are infinitely many Fibonacci numbers $f_n$ such that $m$ divides $f_n$. (*Hint*: Show that the sequence of least positive residues modulo $m$ of the Fibonacci numbers is a repeating sequence.)

**36.** Prove Theorem 3.7 using mathematical induction.

**37.** Show that the least nonnegative residue modulo $m$ of the product of two positive integers less than $m$ can be computed using $O(\log^2 m)$ bit operations.

**\* 38.** Five men and a monkey are shipwrecked on an island. The men have collected a pile of coconuts which they plan to divide equally among themselves the next morning. Not trusting the other men, one of the group wakes up during the night and divides the coconuts into five equal parts with one left over, which he gives to the monkey. He then hides his portion of the pile. During the night, each of the other four men does exactly the same thing by dividing the pile they find into five equal parts leaving one coconut for the monkey and hiding his portion. In the morning, the men gather and split the remaining pile of coconuts into five parts and one is left over for the monkey. What is the minimum number of coconuts the men could have collected for their original pile?

**\* 39.** Answer the same question as in Exercise 38 if instead of five men and one monkey, there are $n$ men and $k$ monkeys, and at each stage the monkeys receive one coconut each.

## 3.1 Computer Projects

Write computer programs to do the following:

**1.** Find the least nonnegative residue of an integer with respect to a fixed modulus.

**2.** Perform modular addition and subtraction when the modulus is less than half of the word size of the computer.

**3.** Perform modular multiplication when the modulus is less than half of the word size of the computer using Exercise 31.

**4.** Perform modular exponentiation using the algorithm described in the text.

## 3.2  Linear Congruences

A congruence of the form

$$ax \equiv b \pmod{m},$$

where $x$ is an unknown integer, is called a *linear congruence in one variable*.  In this section we will see that the study of such congruences is similar to the study of linear diophantine equations in two variables.

We first note that if $x = x_0$ is a solution of the congruence $ax \equiv b \pmod{m}$, and if $x_1 \equiv x_0 \pmod{m}$, then $ax_1 \equiv ax_0 \equiv b \pmod{m}$, so that $x_1$ is also a solution.  Hence, if one member of a congruence class modulo $m$ is a solution, then all members of this class are solutions.  Therefore, we may ask how many of the $m$ congruence classes modulo $m$ give solutions; this is exactly the same as asking how many incongruent solutions there are modulo $m$.  The following theorem tells us when a linear congruence in one variable has solutions, and if it does, tells exactly how many incongruent solutions there are modulo $m$.

**Theorem 3.10**.  Let $a$, $b$, and $m$ be integers with $m > 0$ and $(a,m) = d$.  If $d \nmid b$, then $ax \equiv b \pmod{m}$ has no solutions.  If $d \mid b$, then $ax \equiv b \pmod{m}$ has exactly $d$ incongruent solutions modulo $m$.

*Proof.*  By Theorem 3.1 the linear congruence $ax \equiv b \pmod{m}$ is equivalent to the linear diophantine equation in two variables $ax - my = b$.  The integer $x$ is a solution of $ax \equiv b \pmod{m}$ if and only if there is an integer $y$ with $ax - my = b$.  By Theorem 2.14 we know that if $d \nmid b$, there are no solutions, while if $d \mid b$, $ax - my = b$ has infinitely many solutions, given by

$$x = x_0 + (m/d)t, \quad y = y_0 + (a/d)t,$$

where $x = x_0$ and $y = y_0$ is a particular solution of the equation.  The values of $x$ given above,

$$x = x_0 + (m/d)t,$$

are the solutions of the linear congruence; there are infinitely many of these.

To determine how many incongruent solutions there are, we find the condition that describes when two of the solutions $x_1 = x_0 + (m/d)t_1$ and $x_2 = x_0 + (m/d)t_2$ are congruent modulo $m$.  If these two solutions are congruent, then

$$x_0 + (m/d)t_1 \equiv x_0 + (m/d)t_2 \pmod{m}.$$

Subtracting $x_0$ from both sides of this congruence, we find that

$$(m/d)t_1 \equiv (m/d)t_2 \pmod{m}.$$

Now $(m, m/d) = m/d$ since $(m/d) \mid m$, so that by Theorem 3.4 we see that

$$t_1 \equiv t_2 \pmod{d}.$$

This shows that a complete set of incongruent solutions is obtained by taking $x = x_0 + (m/d)t$, where $t$ ranges through a complete system of residues modulo $d$. One such set is given by $x = x_0 + (m/d)t$ where $t = 0,1,2,...,d - 1$. ∎

We now illustrate the use of Theorem 3.10.

**Example 3.12.** To find all solutions of $9x \equiv 12 \pmod{15}$, we first note that since $(9,15) = 3$ and $3 \mid 12$, there are exactly three incongruent solutions. We can find these solutions by first finding a particular solution and then adding the appropriate multiples of $15/3 = 5$.

To find a particular solution, we consider the linear diophantine equation $9x - 15y = 12$. The Euclidean algorithm shows that

$$15 = 9 \cdot 1 + 6$$
$$9 = 6 \cdot 1 + 3$$
$$6 = 3 \cdot 2,$$

so that $3 = 9 - 6 \cdot 1 = 9 - (15 - 9 \cdot 1) = 9 \cdot 2 - 15$. Hence $9 \cdot 8 - 15 \cdot 4 = 12$, and a particular solution of $9x - 15y = 12$ is given by $x_0 = 8$ and $y_0 = 4$.

From the proof of Theorem 3.10, we see that a complete set of 3 incongruent solutions is given by $x = x_0 \equiv 8 \pmod{15}$, $x = x_0 + 5 \equiv 13 \pmod{15}$, and $x = x_0 + 5 \cdot 2 \equiv 18 \equiv 3 \pmod{15}$. □

We now consider congruences of the special form $ax \equiv 1 \pmod{m}$. By Theorem 3.10 there is a solution to this congruence if and only if $(a,m) = 1$, and then all solutions are congruent modulo $m$.

**Definition.** Given an integer $a$ with $(a,m) = 1$, a solution of $ax \equiv 1 \pmod{m}$ is called an *inverse of a modulo m*.

**Example 3.13.** Since the solutions of $7x \equiv 1 \pmod{31}$ satisfy $x \equiv 9 \pmod{31}$, 9, and all integers congruent to 9 modulo 31, are inverses of 7 modulo 31. Analogously, since $9 \cdot 7 \equiv 1 \pmod{31}$, 7 is an inverse of 9 modulo 31. □

When we have an inverse of $a$ modulo $m$, we can use it to solve any congruence of the form $ax \equiv b \pmod{m}$. To see this, let $\bar{a}$ be an inverse of $a$ modulo $m$, so that $a\bar{a} \equiv 1 \pmod{m}$. Then, if $ax \equiv b \pmod{m}$, we can multiply both sides of this congruence by $\bar{a}$ to find that $\bar{a}(ax) \equiv \bar{a}b \pmod{m}$, so that $x \equiv \bar{a}b \pmod{m}$.

**Example 3.14.** To find the solutions of $7x \equiv 22 \pmod{31}$, we multiply both sides of this congruence by 9, an inverse of 7 modulo 31, to obtain $9 \cdot 7x \equiv 9 \cdot 22 \pmod{31}$. Hence, $x \equiv 198 \equiv 12 \pmod{31}$. □

We note here that if $(a,m) = 1$, then the linear congruence $ax \equiv b \pmod{m}$ has a unique solution modulo $m$.

**Example 3.15.** To find all solutions of $7x \equiv 4 \pmod{12}$, we note that since $(7,12) = 1$, there is a unique solution modulo 12. To find this, we need only obtain a solution of the linear diophantine equation $7x - 12y = 4$. The Euclidean algorithm gives

$$12 = 7 \cdot 1 + 5$$
$$7 = 5 \cdot 1 + 2$$
$$5 = 2 \cdot 2 + 1$$
$$2 = 1 \cdot 2.$$

Hence $1 = 5 - 2 \cdot 2 = 5 - (7 - 5 \cdot 1) \cdot 2 = 5 \cdot 3 - 2 \cdot 7 = (12 - 7 \cdot 1) \cdot 3 - 2 \cdot 7 = 12 \cdot 3 - 5 \cdot 7$. Therefore, a particular solution to the linear diophantine equation is $x_0 = -20$ and $y_0 = 12$. Hence, all solutions of the linear congruences are given by $x \equiv -20 \equiv 4 \pmod{12}$.                                                                 □

Later we will want to know which integers are their own inverses modulo $p$ where $p$ is prime. The following theorem tells us which integers have this property.

**Theorem 3.11.** Let $p$ be prime. The positive integer $a$ is its own inverse modulo $p$ if and only if $a \equiv 1 \pmod{p}$ or $a \equiv -1 \pmod{p}$.

*Proof.* If $a \equiv 1 \pmod{p}$ or $a \equiv -1 \pmod{p}$, then $a^2 \equiv 1 \pmod{p}$, so that $a$ is its own inverse modulo $p$.

Conversely, if $a$ is its own inverse modulo $p$, then $a^2 = a \cdot a \equiv 1 \pmod{p}$. Hence, $p \mid (a^2 - 1)$. Since $a^2 - 1 = (a - 1)(a + 1)$, either $p \mid (a - 1)$ or $p \mid (a + 1)$. Therefore, either $a \equiv 1 \pmod{p}$ or $a \equiv -1 \pmod{p}$.                                        ■

---

## 3.2 Exercises

1.  Find all solutions of each of the following linear congruences.

    a)   $3x \equiv 2 \pmod{7}$          d)   $15x \equiv 9 \pmod{25}$
    b)   $6x \equiv 3 \pmod{9}$          e)   $128x \equiv 833 \pmod{1001}$
    c)   $17x \equiv 14 \pmod{21}$       f)   $987x \equiv 610 \pmod{1597}$

2.  Let $a$, $b$, and $m$ be positive integers with $a > 0$, $m > 0$, and $(a,m) = 1$. The following method can be used to solve the linear congruence $ax \equiv b \pmod{m}$.

    a)   Show that if the integer $x$ is a solution of $ax \equiv b \pmod{m}$, then $x$ is also a solution of the linear congruence

$$a_1 x \equiv -b[m/a] \pmod{m}.$$

where $a_1$ is the least positive residue of $m$ modulo $a$. Note that this congruence is of the same type as the original congruence, with a positive integer smaller than $a$ as the coefficient of $x$.

b) When the procedure of part (a) is iterated, one obtains a sequence of linear congruences with coefficients of $x$ equal to $a_0 = a > a_1 > a_2 > \cdots$. Show that there is a positive integer $n$ with $a_n = 1$, so that at the $n$th stage, one obtains a linear congruence $x \equiv B \pmod{m}$.

c) Use the method described in part (b) to solve the linear congruence $6x \equiv 7 \pmod{23}$.

3. An astronomer knows that a satellite orbits the earth in a period that is an exact multiple of 1 hour that is less than 1 day. If the astronomer notes that the satellite completes 11 orbits in an interval starting when a 24-hour clock reads 0 hours and ending when the clock reads 17 hours, how long is the orbital period of the satellite?

4. For which integers $c$ with $0 \leqslant c < 30$ does the congruence $12x \equiv c \pmod{30}$ have solutions? When there are solutions, how many incongruent solutions are there?

5. Find an inverse modulo 17 of each of the following integers.

a) 4                   c) 7
b) 5                   d) 16

6. Show that if $\bar{a}$ is an inverse of $a$ modulo $m$ and $\bar{b}$ is an inverse of $b$ modulo $m$, then $\bar{a}\,\bar{b}$ is an inverse of $ab$ modulo $m$.

7. Show that the linear congruence in two variables $ax + by \equiv c \pmod{m}$, where $a$, $b$, $c$, and $m$ are integers, $m > 0$, with $d = (a,b,m)$, has exactly $dm$ incongruent solutions if $d \mid c$, and no solutions otherwise.

8. Find all solutions of each of the following linear congruences in two variables.

a)   $2x + 3y \equiv 1 \pmod{7}$        c)   $6x + 3y \equiv 0 \pmod{9}$
b)   $2x + 4y \equiv 6 \pmod{8}$        d)   $10x + 5y \equiv 9 \pmod{15}$

9. Let $p$ be an odd prime and $k$ a positive integer. Show that the congruence $x^2 \equiv 1 \pmod{p^k}$ has exactly two incongruent solutions, namely $x \equiv \pm 1 \pmod{p^k}$.

10. Show that the congruence $x^2 \equiv 1 \pmod{2^k}$ has exactly four incongruent solutions, namely $x \equiv \pm 1$ or $\pm(1+2^{k-1}) \pmod{2^k}$, when $k > 2$. Show that when $k = 1$ there is one solution and when $k = 2$ there are two incongruent solutions.

11. Show that if $a$ and $m$ are relatively prime positive integers with $a < m$, then an inverse of $a$ modulo $m$ can be found using $O(\log^3 m)$ bit operations.

12. Show that if $p$ is an odd prime and $a$ is a positive integer not divisible by $p$, then the congruence $x^2 \equiv a \pmod{p}$ has either no solution or exactly two incongruent solutions.

## 3.2 Computer Projects

Write programs to do the following:

1. Solve linear congruence using the method given in the text.

2. Solve linear congruences using the method given in Exercise 2.

3. Find inverses modulo $m$ of integers relatively prime to $m$ where $m$ is a positive integer.

4. Solve linear congruences using inverses.

5. Solve linear congruences in two variables.

## 3.3 The Chinese Remainder Theorem

In this section and in the one following, we discuss systems of simultaneous congruences. We will study two types of such systems. In the first type, there are two or more linear congruences in one variable, with different moduli (moduli is the plural of modulus). The second type consists of more than one simultaneous congruence in more than one variable, where all congruences have the same modulus.

First, we consider systems of congruences that involve only one variable, but different moduli. Such systems arose in ancient Chinese puzzles such as the following: Find a number that leaves a remainder of 1 when divided by 3, a remainder of 2 when divided by 5, and a remainder of 3 when divided by 7. This puzzle leads to the following system of congruences:

$$x \equiv 1 \pmod{3}, \ x \equiv 2 \pmod{5}, \ x \equiv 3 \pmod{7}.$$

We now give a method for finding all solutions of systems of simultaneous congruences such as this. The theory behind the solution of systems of this type is provided by the following theorem, which derives its name from the ancient Chinese heritage of the problem*.

**Theorem 3.12. The Chinese Remainder Theorem.** Let $m_1, m_2, ..., m_r$ be pairwise relatively prime positive integers. Then the system of congruence

$$x \equiv a_1 \pmod{m_1},$$
$$x \equiv a_2 \pmod{m_2},$$
$$\cdot$$
$$\cdot$$
$$\cdot$$
$$x \equiv a_r \pmod{m_t},$$

has a unique solution modulo $M = m_1 m_2 \cdots m_r$.

*Proof.* First, we construct a simultaneous solution to the system of congruences. To do this, let $M_k = M/m_k = m_1 m_2 \cdots m_{k-1} m_{k+1} \cdots m_r$. We know that $(M_k, m_k) = 1$ by Exercise 14 of Section 2.1, since $(m_j, m_k) = 1$ whenever $j \neq k$. Hence, by Theorem 3.10 we can find an inverse $y_k$ of $M_k$ modulo $m_k$, so that $M_k y_k \equiv 1 \pmod{m_k}$. We now form the sum

$$x = a_1 M_1 y_1 + a_2 M_2 y_2 + \cdots + a_r M_r y_r.$$

The integer $x$ is a simultaneous solution of the $r$ congruences. To demonstrate this, we must show that $x \equiv a_k \pmod{m_k}$ for $k = 1, 2, ..., r$. Since $m_k \mid M_j$ whenever $j \neq k$, we have $M_j \equiv 0 \pmod{m_k}$. Therefore, in the sum for $x$, all terms except the $k$th term are congruent to $0 \pmod{m_k}$. Hence, $x \equiv a_k M_k y_k \equiv a_k \pmod{m_k}$, since $M_k y_k \equiv 1 \pmod{m_k}$.

We now show that any two solutions are congruent modulo $M$. Let $x_0$ and $x_1$ both be simultaneous solutions to the system of $r$ congruences. Then, for each $k$, $x_0 \equiv x_1 \equiv a_k \pmod{m_k}$, so that $m_k \mid (x_0 - x_1)$. Using Theorem 3.8, we see that $M \mid (x_0 - x_1)$. Therefore, $x_0 \equiv x_1 \pmod{M}$. This shows that the simultaneous solution of the system of $r$ congruences is unique modulo $M$. ∎

We illustrate the use of the Chinese remainder theorem by solving the system that arises from the ancient Chinese puzzle.

**Example 3.16.** To solve the system

---

* Problems involving systems of linear congruences arose as early as the first century. Such problems can be found in the *Arithmetic* of the Chinese mathematician Sun-Tsu who lived at that time. Problems involving such systems and methods for solving them can be found throughout the works of ancient Chinese and Hindu mathematicians. For more information on this topic consult the references on the history of number theory listed in the Bibliography which appears at the end of the text.

$$x \equiv 1 \pmod 3$$
$$x \equiv 2 \pmod 5$$
$$x \equiv 3 \pmod 7,$$

we have $M = 3 \cdot 5 \cdot 7 = 105$, $M_1 = 105/3 = 35$, $M_2 = 105/5 = 21$, and $M_3 = 105/7 = 15$. To determine $y_1$, we solve $35y_1 \equiv 1 \pmod 3$, or equivalently, $2y_1 \equiv 1 \pmod 3$. This yields $y_1 \equiv 2 \pmod 3$. We find $y_2$ by solving $21y_2 \equiv 1 \pmod 5$; this immediately gives $y_2 \equiv 1 \pmod 5$. Finally, we find $y_3$ by solving $15y_3 \equiv 1 \pmod 7$. This gives $y_3 \equiv 1 \pmod 7$. Hence,

$$x \equiv 1 \cdot 35 \cdot 2 + 2 \cdot 21 \cdot 1 + 3 \cdot 15 \cdot 1$$
$$\equiv 157 \equiv 52 \pmod{105}.$$

There is also an iterative method for solving simultaneous systems of congruences. We illustrate this method with an example. Suppose we wish to solve the system

$$x \equiv 1 \pmod 5$$
$$x \equiv 2 \pmod 6$$
$$x \equiv 3 \pmod 7.$$

We use Theorem 3.1 to rewrite the first congruence as an equality, namely $x = 5t + 1$, where $t$ is an integer. Inserting this expression for $x$ into the second congruence, we find that

$$5t + 1 \equiv 2 \pmod 6,$$

which can easily be solved to show that $t \equiv 5 \pmod 6$. Using Theorem 3.1 again, we write $t = 6u + 5$ where $u$ is an integer. Hence, $x = 5(6u+5) + 1 = 30u + 26$. When we insert this expression for $x$ into the third congruence, we obtain

$$30u + 26 \equiv 3 \pmod 7.$$

When this congruence is solved, we find that $u \equiv 6 \pmod 7$. Consequently, Theorem 3.1 tells us that $u = 7v + 6$, where $v$ is an integer. Hence,

$$x = 30(7v+6) + 26 = 210v + 206.$$

Translating this equality into a congruence, we find that

$$x \equiv 206 \pmod{210},$$

and this is the simultaneous solution.                                                    □

Note that the method we have just illustrated shows that a system of simultaneous questions can be solved by successively solving linear congruences. This can be done even when the moduli of the congruences are not relatively prime as long as congruences are consistent. (See Exercises 11-14 at the end of this section.)

The Chinese remainder theorem provides a way to perform computer arithmetic with large integers. To store very large integers and do arithmetic with them requires special techniques. The Chinese remainder theorem tells us that given pairwise relatively prime moduli $m_1, m_2, ..., m_r$, a positive integer $n$ with $n < M = m_1 m_2 \cdots m_r$, is uniquely determined by its least positive residues moduli $m_j$ for $j = 1, 2, ..., r$. Suppose that the word size of a computer is only 100, but that we wish to do arithmetic with integers as large as $10^6$. First, we find pairwise relatively prime integers less than 100 with a product exceeding $10^6$; for instance, we can take $m_1 = 99$, $m_2 = 98$, $m_3 = 97$, and $m_4 = 95$. We convert integers less than $10^6$ into 4-tuples consisting of their least positive residues modulo $m_1$, $m_2$, $m_3$, and $m_4$. (To convert integers as large as $10^6$ into their list of least positive residues, we need to work with large integers using multiprecision techniques. However, this is done only once for each integer in the input and once for the output.) Then, for instance, to add integers, we simply add their respective least positive residues modulo $m_1$, $m_2$, $m_3$, and $m_4$, making use of the fact that if $x \equiv x_i \pmod{m_i}$ and $y \equiv y_i \pmod{m_i}$, then $x + y \equiv x_i + y_i \pmod{m_i}$. We then use the Chinese remainder theorem to convert the set of four least positive residues for the sum back to an integer.

The following example illustrates this technique.

**Example 3.17.** We wish to add $x = 123684$ and $y = 413456$ on a computer of word size 100. We have

$$x \equiv 33 \pmod{99}, \qquad y \equiv 32 \pmod{99},$$
$$x \equiv 8 \pmod{98}, \qquad y \equiv 92 \pmod{98},$$
$$x \equiv 9 \pmod{97}, \qquad y \equiv 42 \pmod{97},$$
$$x \equiv 89 \pmod{95}, \qquad y \equiv 16 \pmod{95},$$

so that

$$x + y \equiv 65 \pmod{99}$$
$$x + y \equiv 2 \pmod{98}$$
$$x + y \equiv 51 \pmod{97}$$
$$x + y \equiv 10 \pmod{95}.$$

We now use the Chinese remainder theorem to find $x + y$ modulo $99 \cdot 98 \cdot 97 \cdot 95$. We have $M = 99 \cdot 98 \cdot 97 \cdot 95 = 89403930$, $M_1 = M/99 = 903070$, $M_2 = M/98 = 912288$, $M_3 = M/97 = 921690$, and $M_4 = M/95 = 941094$. We need to find the inverse of $M_i \pmod{y_i}$ for $i = 1, 2, 3, 4$. To do this, we solve the following congruences (using the Euclidean algorithm):

$$903070y_1 \equiv 91y_1 \equiv 1 \pmod{99},$$
$$912285y_2 \equiv 3y_2 \equiv 1 \pmod{98},$$
$$921690y_3 \equiv 93y_3 \equiv 1 \pmod{97},$$
$$941094y_4 \equiv 24y_4 \equiv 1 \pmod{95}.$$

We find that $y_1 \equiv 37 \pmod{99}$, $y_2 \equiv 38 \pmod{98}$, $y_3 \equiv 24 \pmod{97}$, and $y_4 \equiv 4 \pmod{95}$. Hence,

$$x + y \equiv 65 \cdot 903070 \cdot 37 + 2 \cdot 912285 \cdot 33 + 51 \cdot 921690 \cdot 24 + 10 \cdot 941094 \cdot 4$$
$$= 3397886480$$
$$\equiv 537140 \pmod{89403930}.$$

Since $0 < x + y < 89403930$, we conclude that $x + y = 537140$.                    □

On most computers the word size is a large power of 2, with $2^{35}$ a common value. Hence, to use modular arithmetic and the Chinese remainder theorem to do computer arithmetic, we need integers less than $2^{35}$ that are pairwise relatively prime which multiply together to give a large integer. To find such integers, we use numbers of the form $2^m - 1$, where $m$ is a positive integer. Computer arithmetic with these numbers turns out to be relatively simple (see Knuth [80]). To produce a set of pairwise relatively prime numbers of this form, we first prove some lemmata (*lemmata* is the plural of lemma).

**Lemma 3.1.** If $a$ and $b$ are positive integers, then the least positive residue of $2^a - 1$ modulo $2^b - 1$ is $2^r - 1$, where $r$ is the least positive residue of $a$ modulo $b$.

*Proof.* From the division algorithm, $a = bq + r$ where $r$ is the least positive residue of $a$ modulo $b$. We have $(2^a-1) = (2^{bq+r}-1) = (2^b-1)(2^{b(q-1)+r} + \cdots + 2^{b+r}+2^r) + (2^r-1)$, which shows that the remainder when $2^a - 1$ is divided by $2^b - 1$ is $2^r - 1$; this is the least positive residue of $2^a - 1$ modulo $2^b - 1$.                    ∎

We use Lemma 3.1 to prove the following result.

**Lemma 3.2.** If $a$ and $b$ are positive integers, then the greatest common divisor of $2^a - 1$ and $2^b - 1$ is $2^{(a,b)} - 1$.

*Proof.* When we perform the Euclidean algorithm with $a = r_0$ and $b = r_1$, we obtain

$$r_0 = r_1 q_1 + r_2 \qquad 0 \leqslant r_2 < r_1$$
$$r_1 = r_2 q_2 + r_3 \qquad 0 \leqslant r_3 < r_2$$
$$\cdot$$
$$\cdot$$
$$\cdot$$
$$r_{n-3} = r_{n-2} q_{n-2} + r_{n-1} \qquad 0 \leqslant r_{n-1} < r_{n-2}$$
$$r_{n-2} = r_{n-1} q_{n-1}.$$

where the last remainder, $r_{n-1}$, is the greatest common divisor of $a$ and $b$.

Using Lemma 3.1, and the steps of the Euclidean algorithm with $a = r_0$ and $b = r_1$, when we perform the Euclidean algorithm on the pair $2^a - 1 = R_0$ and $2^b - 1 = R_1$, we obtain

$$R_0 = R_1 Q_1 + R_2 \qquad R_2 = 2^{r_2} - 1$$
$$R_1 = R_2 Q_2 + R_3 \qquad R_3 = 2^{r_3} - 1$$
$$\cdot$$
$$\cdot$$
$$\cdot$$
$$R_{n-3} = R_{n-2} Q_{n-2} + R_{n-1} \qquad R_{n-1} = 2^{r_{n-1}} - 1$$
$$R_{n-2} = R_{n-1} Q_{n-1}.$$

Here the last non-zero remainder, $R_{n-1} = 2^{r_{n-1}} - 1 = 2^{(a,b)} - 1$, is the greatest common divisor of $R_0$ and $R_1$. ∎

Using Lemma 3.2 we have the following theorem.

**Theorem 3.13.** The positive integers $2^a - 1$ and $2^b - 1$ are relatively prime if and only if $a$ and $b$ are relatively prime.

We can now use Theorem 3.13 to produce a set of pairwise relatively prime integers, each of which is less than $2^{35}$, with product greater than a specified integer. Suppose that we wish to do arithmetic with integers as large as $2^{184}$. We pick $m_1 = 2^{35} - 1$, $m_2 = 2^{34} - 1$, $m_3 = 2^{33} - 1$, $m_4 = 2^{31} - 1$, $m_5 = 2^{29} - 1$, and $m_6 = 2^{23} - 1$. Since the exponents of 2 in the expressions for the $m_j$ are relatively prime, by Theorem 3.13 the $M_j$'s are pairwise relatively prime. Also, we have $M = m_1 m_2 m_3 m_4 m_5 m_6 > 2^{184}$. We can now use modular arithmetic and the Chinese remainder theorem to perform arithmetic with integers as large as $2^{184}$.

Although it is somewhat awkward to do computer operations with large integers using modular arithmetic and the Chinese remainder theorem, there are some definite advantages to this approach. First, on many high-speed computers, operations can be performed simultaneously. So, reducing an operation involving two large integers to a set of operations involving smaller integers, namely the least positive residues of the large integers with respect to the various moduli, leads to simultaneous computations which may be performed more rapidly than one operation with large integers. Second, even without taking into account the advantages of simultaneous computations, multiplication of large integers may be done faster using these ideas than with many other multiprecision methods. The interested reader should consult Knuth [80].

---

## 3.3 Exercises

1. Which integers leave a remainder of one when divided by either 2 or 3?

2. Find an integer that leaves a remainder of one when divided by either 2 or 5, but that is divisible by 3.

3. Find an integer that leaves a remainder of 2 when divided by either 3 or 5 but that is divisible by 4.

4. Find all the solutions of each of the following systems of linear congruences.

   a)  $x \equiv 4 \pmod{11}$        c)  $x \equiv 0 \pmod 2$
       $x \equiv 3 \pmod{17}$             $x \equiv 0 \pmod 3$
                                          $x \equiv 1 \pmod 5$
   b)  $x \equiv 1 \pmod 2$               $x \equiv 6 \pmod 7$
       $x \equiv 2 \pmod 3$
       $x \equiv 3 \pmod 5$        d)  $x \equiv 2 \pmod{11}$
                                          $x \equiv 3 \pmod{12}$
                                          $x \equiv 4 \pmod{13}$
                                          $x \equiv 5 \pmod{17}$
                                          $x \equiv 6 \pmod{19}$

5. A troop of 17 monkeys store their bananas in eleven piles of equal size with a twelfth pile of six left over. When they divide the bananas into 17 equal groups none remain. What is the smallest number of bananas they can have?

6. As an odometer check, a special counter measures the miles a car travels modulo 7. Explain how this counter can be used to determine whether the car has been driven 49335, 149335, or 249335 miles when the odometer reads 49335 and works modulo 100000.

7. Find a multiple of 11 that leaves a remainder of 1 when divided by each of the integers 2,3,5, and 7.

8. Solve the following ancient Indian problem: If eggs are removed from a basket two, three, four, five, and six at a time, there remain, respectively, one, two, three, four, and five eggs. But if the eggs are removed seven at a time no eggs remain. What is the least number of eggs that could have been in the basket?

9. Show that there are arbitrarily long strings of integers each divisible by a perfect square. (*Hint*: Use the Chinese remainder theorem to show that there is a simultaneous solution to the system of congruences $x \equiv 0$ (mod 4), $x \equiv -1$ (mod 9), $x \equiv -2$ (mod 25),..., $x \equiv -k+1$ (mod $p_k^2$), where $p_k$ is the $k$th prime.)

* 10. Show that if $a,b,$ and $c$ are integers with $(a,b) = 1$, then there is an integer $n$ such that $(an+b,c) = 1$.

In Exercises 11-14 we will consider systems of congruences where the moduli of the congruences are not necessarily relatively prime.

11. Show that the system of congruences

$$x \equiv a_1 \ (\text{mod } m_1)$$
$$x \equiv a_2 \ (\text{mod } m_2)$$

has a solution if and only if $(m_1,m_2) \mid (a_1-a_2)$. Show that when there is a solution, it is unique modulo $([m_1,m_2])$. (*Hint*: Write the first congruence as $x = a_1 + km_1$ where $k$ is an integer, and then insert this expression for $x$ into the second congruence.)

12. Using Exercise 11 solve each of the following simultaneous system of congruences,

a) $x \equiv 4 \ (\text{mod } 6)$      b) $x \equiv 7 \ (\text{mod } 10)$
       $x \equiv 13 \ (\text{mod } 15)$          $x \equiv 4 \ (\text{mod } 15)$

13. Show that the system of congruences

$$x \equiv a_1 \ (\text{mod } m_1)$$
$$x \equiv a_2 \ (\text{mod } m_2)$$
$$.$$
$$.$$
$$.$$
$$x \equiv a_r \ (\text{mod } m_r)$$

has a solution if and only if $(m_i,m_j) \mid (a_i - a_j)$ for all pairs of integers $(i,j)$ with $1 \leqslant i < j \leqslant r$. Show that if a solution exists, then it is unique modulo $[m_1, m_2,..., m_r]$. (*Hint*: Use Exercise 11 and mathematical induction.)

14. Using Exercise 13 solve each of the following systems of congruences.

a) $x \equiv 5 \pmod 6$
   $x \equiv 3 \pmod{10}$
   $x \equiv 8 \pmod{15}$

b) $x \equiv 2 \pmod{14}$
   $x \equiv 16 \pmod{21}$
   $x \equiv 10 \pmod{30}$

c) $x \equiv 2 \pmod 9$
   $x \equiv 8 \pmod{15}$
   $x \equiv 10 \pmod{25}$

d) $x \equiv 2 \pmod 6$
   $x \equiv 4 \pmod 8$
   $x \equiv 2 \pmod{14}$
   $x \equiv 14 \pmod{15}$

e) $x \equiv 7 \pmod 9$
   $x \equiv 2 \pmod{10}$
   $x \equiv 3 \pmod{12}$
   $x \equiv 6 \pmod{15}$

15. What is the smallest number of lobsters in a tank if one lobster is left over when they are removed 2,3,5, or 7 at a time, but no lobsters are left over when they are removed 11 at a time?

16. Using the Chinese remainder theorem, explain how to add and how to multiply 784 and 813 on a computer of word size 100.

A positive integer $x \neq 1$ with $n$ base $b$ digits is called an *automorph to the base b* if the last $n$ base $b$ digits of $x^2$ are the same as those of $x$.

* 17. Find the base 10 automorphs with four or fewer digits.

* 18. How many base $b$ automorphs are there with $n$ or fewer base $b$ digits, if $b$ has prime-power factorization $b = p_1^{b_1} \, p_2^{b_2} \, \cdots \, p_k^{b_k}$ ?

According to the theory of *biorhythms*, there are three cycles in your life that start the day you are born. These are the *physical, emotional,* and *intellectual cycles*, of lengths 23,28, and 33 days, respectively. Each cycle follows a sine curve with period equal to the length of that cycle, starting with amplitude zero, climbing to amplitude 1 one quarter of the way through the cycle, dropping back to amplitude zero one half of the way through the cycle, dropping further to amplitude minus one three quarters of the way through the cycle, and climbing back to amplitude zero at the end of the cycle.

Answer the following questions about biorhythms, measuring time in quarter days (so that the units will be integers).

19. For which days of your life will you be at a triple peak, where all of your three cycles are at maximum amplitudes?

20. For which days of your life will you be at a triple nadir, where all three of your cycles have lowest amplitude?

21. When in your life will all three cycles be a neutral position (amplitude 0)?

A set of congruences to distinct moduli greater than one that has the property that every integer satisfies at least one of the congruences is called a *covering set of congruences*.

22. Show the set of congruences $x \equiv 0 \pmod 2$, $x \equiv 0 \pmod 3$, $x \equiv 1 \pmod 4$, $x \equiv 1 \pmod 6$, and $x \equiv 11 \pmod{12}$ is a covering set of congruences.

23. Show that the set of congruences $x \equiv 0 \pmod 2$, $x \equiv 0 \pmod 3$, $x \equiv 0 \pmod 5$, $x \equiv 0 \pmod 7$, $x \equiv 1 \pmod 6$, $x \equiv 1 \pmod{10}$, $x \equiv 1 \pmod{14}$, $x \equiv 2 \pmod{15}$, $x \equiv 2 \pmod{21}$, $x \equiv 23 \pmod{30}$, $x \equiv 4 \pmod{35}$, $x \equiv 5 \pmod{42}$, $x \equiv 59 \pmod{70}$, and $x \equiv 104 \pmod{105}$ is a covering set of congruences.

* 24. Let $m$ be a positive integer with prime-power factorization $m = 2^{a_0} p_1^{a_1} p_2^{a_2} \cdots p_r^{a_r}$. Show that the congruence $x^2 \equiv 1 \pmod m$ has exactly $2^{r+e}$ solutions where $e = 0$ if $a_0 = 0$ or $1$, $e = 1$ if $a_0 = 2$, and $e = 2$ if $a_0 > 2$. (*Hint*: Use Exercise 9 and 10 of Section 3.2.)

25. The three children in a family have feet that are 5 inches, 7 inches, and 9 inches long. When they measure the length of the dining room of their house using their feet, they each find that there are 3 inches left over. How long is the dining room?

## 3.3 Computer Projects

Write programs to do the following:

1. Solve systems of linear congruences of the type found in the Chinese remainder theorem.

2. Solve systems of linear congruences of the type given in Exercises 11-14.

3. Add large integers exceeding the word size of the computer using the Chinese remainder theorem.

4. Multiply large integers exceeding the word size of the computer using the Chinese remainder theorem.

5. Find automorphs to the base $b$, where $b$ is a positive integer greater than one (see preamble to Exercise 17).

6. Plot biorhythm charts and find triple peaks and triple nadirs (see preamble to Exercise 19).

---

## 3.4 Systems of Linear Congruences

We will consider systems of more than one congruence involving the same number of unknowns as congruences, where all congruences have the same modulus. We begin our study with an example.

Suppose we wish to find all integers $x$ and $y$ such that both of the congruences

$$3x + 4y \equiv 5 \pmod{13}$$
$$2x + 5y \equiv 7 \pmod{13}$$

are satisfied. To attempt to find the unknowns $x$ and $y$, we multiply the first congruence by 5 and the second by 4, to obtain

$$15x + 20y \equiv 25 \ (\text{mod } 13)$$
$$8x + 20y \equiv 28 \ (\text{mod } 13).$$

We subtract the first congruence from the second, to find that

$$7x \equiv -3 \ (\text{mod } 13).$$

Since 2 is an inverse of 7 (mod 13), we multiply both sides of the above congruences by 2. This gives

$$2 \cdot 7 \, x \equiv -2 \cdot 3 \ (\text{mod } 13),$$

which tells us that

$$x \equiv 7 \ (\text{mod } 13).$$

Likewise, we can multiply the first congruence by 2 and the second by 3, to see that

$$6x + \ 8y \equiv 10 \ (\text{mod } 13)$$
$$6x + 15y \equiv 21 \ (\text{mod } 13).$$

When we subtract the first congruence from the second, we obtain

$$7y \equiv 11 \ (\text{mod } 13).$$

To solve for $y$, we multiply both sides of this congruence by 2, an inverse of 7 modulo 13. We get

$$2 \cdot 7y \equiv 2 \cdot 11 \ (\text{mod } 13),$$

so that

$$y \equiv 9 \ (\text{mod } 13).$$

What we have shown is that any solution $(x,y)$ must satisfy

$$x \equiv 7 \ (\text{mod } 13), \ y \equiv 9 \ (\text{mod } 13).$$

When we insert these congruences for $x$ and $y$ into the original system, we see that these pairs actually are solutions, since

$$3x + 4y \equiv 3 \cdot 7 + 4 \cdot 9 = 57 \equiv 5 \ (\text{mod } 13)$$
$$2x + 5y \equiv 2 \cdot 7 + 5 \cdot 9 = 59 \equiv 7 \ (\text{mod } 13).$$

Hence, the solutions of this system of congruences are all pairs $(x,y)$ with

$x \equiv 7 \pmod{13}$ and $y \equiv 9 \pmod{13}$.

We now give a general result concerning certain systems of two congruences in two unknowns.

**Theorem 3.14.** Let $a$, $b$, $c$, $d$, $e$, $f$, and $m$ be integers with $m > 0$, such that $(\Delta, m) = 1$, where $\Delta = ad - bc$. Then, the system of congruences

$$ax + by \equiv e \pmod{m}$$
$$cx + dy \equiv f \pmod{m}$$

has a unique solution modulo $m$ given by

$$x \equiv \overline{\Delta}\, (de - bf) \pmod{m}$$
$$y \equiv \overline{\Delta}\, (af - ce) \pmod{m},$$

where $\overline{\Delta}$ is an inverse of $\Delta$ modulo $m$.

*Proof.* We multiply the first congruence of the system by $d$ and the second by $b$, to obtain

$$adx + bdy \equiv de \pmod{m}$$
$$bcx + bdy \equiv bf \pmod{m}.$$

Then, we subtract the second congruence from the first, to find that

$$(ad - bc)\, x \equiv de - bf \pmod{m},$$

or, since $\Delta = ad - bc$,

$$\Delta x \equiv de - bf \pmod{m}.$$

Next, we multiply both sides of this congruence by $\overline{\Delta}$, an inverse of $\Delta$ modulo $m$, to conclude that

$$x \equiv \overline{\Delta}\, (de - bf) \pmod{m}.$$

In a similar way, we multiply the first congruence by $c$ and the second by $a$, to obtain

$$acx + bcy \equiv ce \pmod{m}$$
$$acx + ady \equiv af \pmod{m}.$$

We subtract the first congruence from the second, to find that

$$(ad - bc)y \equiv af - ce \pmod{m}$$

or

$$\Delta y \equiv af - ce \pmod{m}.$$

Finally, we multiply both sides of the above congruence by $\overline{\Delta}$ to see that

$$y \equiv \overline{\Delta}\,(af - ce) \pmod{m}.$$

We have shown that if $(x,y)$ is a solution of the system of congruences, then
$$x \equiv \overline{\Delta}\,(de - bf) \pmod{m}, \quad y \equiv \overline{\Delta}\,(af - ce) \pmod{m}.$$

We can easily check that any such pair $(x,y)$ is a solution. When $x \equiv \overline{\Delta}\,(de - bf) \pmod{m}$ and $y \equiv \overline{\Delta}\,(af - ce) \pmod{m}$, we have

$$\begin{aligned}
ax + by &\equiv a\overline{\Delta}\,(de - bf) + b\overline{\Delta}\,(af - ce)\\
&\equiv \overline{\Delta}\,(ade - abf - abf - bce)\\
&\equiv \overline{\Delta}\,(ad - bc)\,e\\
&\equiv e \pmod{m},
\end{aligned}$$

and

$$\begin{aligned}
cx + dy &\equiv c\overline{\Delta}\,(de - bf) + d\overline{\Delta}\,(af - ce)\\
&\equiv \overline{\Delta}\,(cde - bcf + adf - cde)\\
&\equiv \overline{\Delta}\,(ad - bc)f\\
&\equiv \overline{\Delta}\Delta f\\
&\equiv f \pmod{m}.
\end{aligned}$$

This establishes the theorem. ∎

By similar methods, we may solve systems of $n$ congruences involving $n$ unknowns. However, we will develop the theory of solving such systems, as well as larger systems, by methods taken from linear algebra. Readers unfamiliar with linear algebra may wish to skip the remainder of this section.

Systems of $n$ linear congruences involving $n$ unknowns will arise in our subsequent cryptographic studies. To study these systems when $n$ is large, it is helpful to use the language of matrices. We will use some of the basic notions of matrix arithmetic which are discussed in most linear algebra texts, such as Anton [86].

We need to define congruences of matrices before we proceed.

**Definition**. Let **A** and **B** be $n \times k$ matrices with integer entries, with $(i,j)$th entries $a_{ij}$ and $b_{ij}$, respectively. We say that **A** *is congruent to* **B** *modulo* $m$ if $a_{ij} \equiv b_{ij} \pmod{m}$ for all pairs $(i,j)$ with $1 \leqslant i \leqslant n$ and $1 \leqslant j \leqslant k$. We write $\mathbf{A} \equiv \mathbf{B} \pmod{m}$ if **A** is congruent to **B** modulo $m$.

The matrix congruence $\mathbf{A} \equiv \mathbf{B} \pmod{m}$ provides a succinct way of expressing the $nk$ congruences $a_{ij} \equiv b_{ij} \pmod{m}$ for $1 \leqslant i \leqslant n$ and $1 \leqslant j \leqslant k$.

**Example 3.18**. We easily see that

$$\begin{pmatrix} 15 & 3 \\ 8 & 12 \end{pmatrix} \equiv \begin{pmatrix} 4 & 3 \\ 3 & 1 \end{pmatrix} \pmod{11}.$$

□

The following proposition will be needed.

**Theorem 3.15.** If $\mathbf{A}$ and $\mathbf{B}$ are $n \times k$ matrices with $\mathbf{A} \equiv \mathbf{B} \pmod{m}$, $\mathbf{C}$ is an $k \times p$ matrix and $\mathbf{D}$ is a $p \times n$ matrix, all with integer entries, then $\mathbf{AC} \equiv \mathbf{BC} \pmod{m}$ and $\mathbf{DA} \equiv \mathbf{DB} \pmod{m}$.

*Proof.* Let the entries of $\mathbf{A}$ and $\mathbf{B}$ be $a_{ij}$ and $b_{ij}$, respectively, for $1 \leqslant i \leqslant n$ and $1 \leqslant j \leqslant k$, and let the entries of $\mathbf{C}$ be $c_{ij}$ for $1 \leqslant i \leqslant k$ and $1 \leqslant j \leqslant p$. The $(i,j)$th entries of $\mathbf{AC}$ and $\mathbf{BC}$ are $\sum_{t=1}^{n} a_{it}c_{tj}$ and $\sum_{t=1}^{n} b_{it}c_{tj}$, respectively. Since $\mathbf{A} \equiv \mathbf{B} \pmod{m}$, we know that $a_{it} \equiv b_{it} \pmod{m}$ for all $i$ and $k$. Hence, from Theorem 3.3 see that $\sum_{t=1}^{n} a_{it}c_{tj} \equiv \sum_{t=1}^{n} b_{it}c_{tj} \pmod{m}$. Consequently, $\mathbf{AC} \equiv \mathbf{BC} \pmod{m}$.

The proof that $\mathbf{DA} \equiv \mathbf{DB} \pmod{m}$ is similar and is omitted. ∎

Now let us consider the system of congruences

$$a_{11} x_1 + a_{12} x_2 + \cdots + a_{1n} x_n \equiv b_1 \pmod{m}$$
$$a_{21} x_1 + a_{22} x_2 + \cdots + a_{2n} x_n \equiv b_2 \pmod{m}$$
$$\cdots$$
$$\cdots$$
$$\cdots$$
$$a_{n1} x_1 + a_{n2} x_2 + \cdots + a_{nn} x_n \equiv b_n \pmod{m}.$$

Using matrix notation, we see that this system of $n$ congruences is equivalent to the matrix congruence $\mathbf{AX} \equiv \mathbf{B} \pmod{m}$,

$$\text{where } \mathbf{A} = \begin{pmatrix} a_{11} & a_{12} & \cdots & a_{1n} \\ a_{21} & a_{22} & \cdots & a_{2n} \\ & & \cdot \\ & & \cdot \\ & & \cdot \\ a_{n1} & a_{n2} & \cdots & a_{nn} \end{pmatrix}, \quad \mathbf{X} = \begin{pmatrix} x_1 \\ x_2 \\ \cdot \\ \cdot \\ \cdot \\ x_n \end{pmatrix}, \text{ and } \mathbf{B} = \begin{pmatrix} b_1 \\ b_2 \\ \cdot \\ \cdot \\ \cdot \\ b_n \end{pmatrix}.$$

**Example 3.19.** The system

$$3x + 4y \equiv 5 \pmod{13}$$
$$2x + 5y \equiv 7 \pmod{13}$$

can be written as

$$\begin{pmatrix} 3 & 4 \\ 2 & 5 \end{pmatrix} \begin{pmatrix} x \\ y \end{pmatrix} \equiv \begin{pmatrix} 5 \\ 7 \end{pmatrix} \pmod{13}.$$                                    □

We now develop a method for solving congruences of the form $AX \equiv B \pmod{m}$. This method is based on finding a matrix $\overline{A}$ such that $A\overline{A} \equiv I \pmod{m}$, where $I$ is the identity matrix.

**Definition.** If $A$ and $\overline{A}$ are $n \times n$ matrices of integers and if

$$\overline{A}A \equiv A\overline{A} \equiv I \pmod{m}, \text{ where } I = \begin{pmatrix} 1 & 0 & \cdots & 0 \\ 0 & 1 & \cdots & 0 \\ & & \cdots & \\ 0 & 0 & \cdots & 1 \end{pmatrix} \text{ is the identity matrix of order}$$

$n$, then $\overline{A}$ is said to be an *inverse of* $A$ *modulo* $m$.

If $\overline{A}$ is an inverse of $A$ and $B \equiv \overline{A} \pmod{m}$, then $B$ is also an inverse of $A$. This follows from Theorem 3.15, since $BA \equiv \overline{A}A \equiv I \pmod{m}$.

Conversely, if $B_1$ and $B_2$ are both inverses of $A$, then $B_1 \equiv B_2 \pmod{m}$. To see this, using Theorem 3.15 and the congruence $B_1A \equiv B_2A \equiv I \pmod{m}$, we have $B_1AB_1 \equiv B_2AB_1 \pmod{m}$. Since $AB_1 \equiv I \pmod{m}$, we conclude that $B_1 \equiv B_2 \pmod{m}$.

**Example 3.20.** Since

$$\begin{pmatrix} 1 & 3 \\ 2 & 4 \end{pmatrix} \begin{pmatrix} 3 & 4 \\ 1 & 2 \end{pmatrix} = \begin{pmatrix} 6 & 10 \\ 10 & 16 \end{pmatrix} \equiv \begin{pmatrix} 1 & 0 \\ 0 & 1 \end{pmatrix} \pmod{5}$$

and

$$\begin{pmatrix} 3 & 4 \\ 1 & 2 \end{pmatrix} \begin{pmatrix} 1 & 3 \\ 2 & 4 \end{pmatrix} = \begin{pmatrix} 11 & 25 \\ 5 & 11 \end{pmatrix} \equiv \begin{pmatrix} 1 & 0 \\ 0 & 1 \end{pmatrix} \pmod{5},$$

we see that the matrix $\begin{pmatrix} 3 & 4 \\ 1 & 2 \end{pmatrix}$ is an inverse of $\begin{pmatrix} 1 & 3 \\ 2 & 4 \end{pmatrix}$ modulo 5.                    □

The following proposition gives an easy method for finding inverses for $2 \times 2$ matrices.

**Theorem 3.16.** Let $A = \begin{pmatrix} a & b \\ c & d \end{pmatrix}$ be a matrix of integers, such that $\Delta = \det A = ad - bc$ is relatively prime to the positive integer $m$. Then, the matrix

$$\overline{A} = \overline{\Delta} \begin{pmatrix} d & -b \\ -c & a \end{pmatrix},$$

where $\overline{\Delta}$ is the inverse of $\Delta$ modulo $m$, is an inverse of A modulo $m$.

*Proof.* To verify that the matrix $\overline{A}$ is an inverse of A modulo $m$, we need only verify that $A\overline{A} \equiv \overline{A}A \equiv I \pmod{m}$.

To see this, note that

$$A\overline{A} \equiv \begin{pmatrix} a & b \\ c & d \end{pmatrix} \overline{\Delta} \begin{pmatrix} d & -b \\ -c & a \end{pmatrix} \equiv \overline{\Delta} \begin{pmatrix} ad - bc & 0 \\ 0 & -bc + ad \end{pmatrix}$$

$$\equiv \overline{\Delta} \begin{pmatrix} \Delta & 0 \\ 0 & \Delta \end{pmatrix} \equiv \begin{pmatrix} \overline{\Delta}\Delta & 0 \\ 0 & \overline{\Delta}\Delta \end{pmatrix} \equiv \begin{pmatrix} 1 & 0 \\ 0 & 1 \end{pmatrix} = I \pmod{m}$$

and

$$\overline{A}A \equiv \overline{\Delta} \begin{pmatrix} d & -b \\ -c & a \end{pmatrix} \begin{pmatrix} a & b \\ c & d \end{pmatrix} \equiv \overline{\Delta} \begin{pmatrix} ad - bc & 0 \\ 0 & -bc + ad \end{pmatrix}$$

$$\equiv \overline{\Delta} \begin{pmatrix} \Delta & 0 \\ 0 & \Delta \end{pmatrix} \equiv \begin{pmatrix} \overline{\Delta}\Delta & 0 \\ 0 & \overline{\Delta}\Delta \end{pmatrix} \equiv \begin{pmatrix} 1 & 0 \\ 0 & 1 \end{pmatrix} = I \pmod{m},$$

where $\overline{\Delta}$ is an inverse of $\Delta \pmod{m}$, which exists because $(\Delta, m) = 1$. ∎

**Example 3.21.** Let $A = \begin{pmatrix} 3 & 4 \\ 2 & 5 \end{pmatrix}$. Since 2 is an inverse det $A = 7$ modulo 13, we have

$$\overline{A} \equiv 2 \begin{pmatrix} 5 & -4 \\ -2 & 3 \end{pmatrix} \equiv \begin{pmatrix} 10 & -8 \\ -4 & 6 \end{pmatrix} \equiv \begin{pmatrix} 10 & 5 \\ 9 & 6 \end{pmatrix} \pmod{13}.$$

□

To provide a formula for an inverse of an $n \times n$ matrix where n is a positive integer, we need a result from linear algebra. This result may be found in Anton [86; page 79]. It involves the notion of the adjoint of a matrix, which is defined as follows.

**Definition.** The *adjoint* of an $n \times n$ matrix A is the $n \times n$ matrix with $(i, j)$th entry $C_{ji}$, where $C_{ji}$ is $(-1)^{i+j}$ times the determinant of the matrix obtained by deleting the $i$th row and $j$th column from A. The adjoint of A is denoted by adj(A).

**Theorem 3.17.** If A is an $n \times n$ matrix with det $A \neq 0$, then A (adj A) = (det A)I, where adj A is the adjoint of A.

Using this theorem, the following theorem follows readily.

**Theorem 3.18**. If **A** is an $n \times n$ matrix with integer entries and $m$ is a positive integer such that $(\det \mathbf{A}, m) = 1$, then the matrix $\overline{\mathbf{A}} = \overline{\Delta}$ (adj **A**) is an inverse of **A** modulo $m$, where $\overline{\Delta}$ is an inverse of $\Delta = \det \mathbf{A}$ modulo $m$.

*Proof.* If $(\det \mathbf{A}, m) = 1$, then we know that $\det \mathbf{A} \neq 0$. Hence, by Theorem 3.15 we have

$$\mathbf{A} \text{ adj } \mathbf{A} = (\det \mathbf{A}) \mathbf{I} = \Delta \mathbf{I}.$$

Since $(\det \mathbf{A}, \mathbf{m}) = 1$, there is an inverse $\overline{\Delta}$ of $\Delta = \det \mathbf{A}$ modulo $m$. Hence,

$$\mathbf{A} \, (\overline{\Delta} \text{ adj } \mathbf{A}) \equiv \mathbf{A} \cdot (\text{adj } \mathbf{A})\overline{\Delta} \equiv \Delta\overline{\Delta}\mathbf{I} \equiv \mathbf{I} \pmod{m},$$

and

$$\overline{\Delta} \, (\text{adj } \mathbf{A})\mathbf{A} \equiv \overline{\Delta} \, (\text{adj } \mathbf{A} \cdot \mathbf{A}) \equiv \overline{\Delta}\Delta\mathbf{I} \equiv \mathbf{I} \pmod{m}.$$

This shows that $\overline{\mathbf{A}} = \overline{\Delta} \cdot (\text{adj } \mathbf{A})$ is an inverse of **A** modulo $m$. ∎

**Example 3.22**. Let $\mathbf{A} = \begin{pmatrix} 2 & 5 & 6 \\ 2 & 0 & 2 \\ 1 & 2 & 3 \end{pmatrix}$. Then $\det \mathbf{A} = -5$. Since $(\det \mathbf{A}, 7) = 1$, and an inverse of $\det \mathbf{A} = -5$ is $4 \pmod 7$, we find that

$$\overline{\mathbf{A}} = 4 \, (\text{adj } \mathbf{A}) = 4 \begin{pmatrix} -2 & -3 & 5 \\ -5 & 0 & 10 \\ 4 & 1 & -10 \end{pmatrix} = \begin{pmatrix} -8 & -12 & 20 \\ -20 & 0 & 40 \\ 0 & 4 & -40 \end{pmatrix} \equiv \begin{pmatrix} 6 & 2 & 6 \\ 1 & 0 & 5 \\ 2 & 4 & 2 \end{pmatrix} \pmod 7.$$

□

We can use an inverse of **A** modulo $m$ to solve the system

$$\mathbf{AX} \equiv \mathbf{B} \pmod{m},$$

where $(\det \mathbf{A}, m) = 1$. By Theorem 3.15, when we multiply both sides of this congruence by an inverse $\overline{\mathbf{A}}$ of **A**, we obtain

$$\overline{\mathbf{A}} \, (\mathbf{AX}) \equiv \overline{\mathbf{A}} \, \mathbf{B} \pmod{m}$$
$$(\overline{\mathbf{A}} \mathbf{A})\mathbf{X} \equiv \overline{\mathbf{A}} \, \mathbf{B} \pmod{m}$$
$$\mathbf{X} \equiv \overline{\mathbf{A}} \, \mathbf{B} \pmod{m}.$$

Hence, we find the solution **X** by forming $\overline{\mathbf{A}} \, \mathbf{B} \pmod{m}$.

Note that this method provides another proof of Theorem 3.14. To see this,

let $\mathbf{AX} = \mathbf{B}$, where $\mathbf{A} = \begin{pmatrix} a & b \\ c & d \end{pmatrix}$, $\mathbf{X} = \begin{pmatrix} x \\ y \end{pmatrix}$, and $\mathbf{B} = \begin{pmatrix} e \\ f \end{pmatrix}$. If $\Delta = \det \mathbf{A} = ad - bc$ is relatively prime to $m$, then

$$\begin{pmatrix} x \\ y \end{pmatrix} = \mathbf{X} \equiv \overline{\mathbf{A}}\,\mathbf{B} \equiv \overline{\Delta} \begin{pmatrix} d & -b \\ -c & a \end{pmatrix} \begin{pmatrix} e \\ f \end{pmatrix} = \overline{\Delta} \begin{pmatrix} de - bf \\ af - ce \end{pmatrix} \pmod{m}.$$

This demonstrates that $(x,y)$ is a solution if and only if

$$x \equiv \overline{\Delta}\,(de - bf) \pmod{m}, \quad y \equiv \overline{\Delta}\,(af - ce) \pmod{m}.$$

Next, we give an example of the solution of a system of three congruences in three unknowns using matrices.

**Example 3.23.** We consider the system of three congruences

$$2x_1 + 5x_2 + 6x_3 \equiv 3 \pmod 7$$
$$2x_1 + x_3 \equiv 4 \pmod 7$$
$$x_1 + 2x_2 + 3x_3 \equiv 1 \pmod 7.$$

This is equivalent to the matrix congruence

$$\begin{pmatrix} 2 & 5 & 6 \\ 2 & 0 & 1 \\ 1 & 2 & 3 \end{pmatrix} \begin{pmatrix} x_1 \\ x_2 \\ x_3 \end{pmatrix} \equiv \begin{pmatrix} 3 \\ 4 \\ 1 \end{pmatrix} \pmod 7.$$

We have previously shown that the matrix $\begin{pmatrix} 6 & 2 & 6 \\ 1 & 0 & 5 \\ 2 & 4 & 2 \end{pmatrix}$ is an inverse of $\begin{pmatrix} 2 & 5 & 6 \\ 2 & 0 & 1 \\ 1 & 2 & 3 \end{pmatrix} \pmod 7$. Hence, we have

$$\begin{pmatrix} x_1 \\ x_2 \\ x_3 \end{pmatrix} = \begin{pmatrix} 6 & 2 & 6 \\ 1 & 0 & 5 \\ 2 & 4 & 2 \end{pmatrix} \begin{pmatrix} 3 \\ 4 \\ 1 \end{pmatrix} = \begin{pmatrix} 32 \\ 8 \\ 24 \end{pmatrix} \equiv \begin{pmatrix} 4 \\ 1 \\ 3 \end{pmatrix} \pmod 7. \qquad \square$$

Before leaving this subject, we should mention that many methods used for solving systems of linear equations may be adapted to solve systems of congruences. For instance, Gaussian elimination may be adapted to solve systems of congruences where division is always replaced by multiplication by inverses modulo $m$. Also, there is a method for solving systems of congruences analagous to Cramer's rule. We leave the development of these methods as exercises for those readers familiar with linear algebra.

## 3.4 Exercises

1. Find the solutions of each of the following systems of linear congruences.

    a)  $x + 2y \equiv 1 \pmod 5$
       $2x + y \equiv 1 \pmod 5$

    b)  $x + 3y \equiv 1 \pmod 5$
       $3x + 4y \equiv 2 \pmod 5$

    c)  $4x + y \equiv 2 \pmod 5$
       $2x + 3y \equiv 1 \pmod 5$

2. Find the solutions of each of the following systems of linear congruences.

    a)  $2x + 3y \equiv 5 \pmod 7$
       $x + 5y \equiv 6 \pmod 7$

    b)  $4x + y \equiv 5 \pmod 7$
       $x + 2y \equiv 4 \pmod 7$

\* 3. What are the possibilities for the number of incongruent solutions of the system of linear congruences

$$ax + by \equiv c \pmod p$$
$$dx + ey \equiv f \pmod p,$$

   where $p$ is a prime and $a,b,c,d,e$, and $f$ are positive integers?

4. Find the matrix $\mathbf{C}$ such that

$$\mathbf{C} \equiv \begin{pmatrix} 2 & 1 \\ 4 & 3 \end{pmatrix} \begin{pmatrix} 4 & 0 \\ 2 & 1 \end{pmatrix} \pmod 5$$

   and all entries of $\mathbf{C}$ are nonnegative integers less than 5.

5. Use mathematical induction to prove that if $\mathbf{A}$ and $\mathbf{B}$ are $n \times n$ matrices with integer entries such that $\mathbf{A} \equiv \mathbf{B} \pmod m$, then $\mathbf{A}^k \equiv \mathbf{B}^k \pmod m$ for all positive integers $k$.

   A matrix $\mathbf{A} \neq \mathbf{I}$ is called *involutory modulo m* if $\mathbf{A}^2 \equiv \mathbf{I} \pmod m$.

6. Show that $\begin{bmatrix} 4 & 11 \\ 1 & 22 \end{bmatrix}$ is involutory modulo 26.

7. Prove or disprove that if $\mathbf{A}$ is a $2 \times 2$ involutory matrix modulo $m$, then $\det \mathbf{A} \equiv \pm 1 \pmod m$.

8. Find an inverse modulo 5 of each of the following matrices.

a) $\begin{bmatrix} 0 & 1 \\ 1 & 0 \end{bmatrix}$ b) $\begin{bmatrix} 1 & 2 \\ 3 & 4 \end{bmatrix}$ c) $\begin{bmatrix} 2 & 2 \\ 1 & 2 \end{bmatrix}$

9. Find an inverse modulo 7 of each of the following matrices.

a) $\begin{bmatrix} 1 & 1 & 0 \\ 1 & 0 & 1 \\ 0 & 1 & 1 \end{bmatrix}$ b) $\begin{bmatrix} 1 & 2 & 3 \\ 1 & 2 & 5 \\ 1 & 4 & 6 \end{bmatrix}$ c) $\begin{bmatrix} 1 & 1 & 1 & 0 \\ 1 & 1 & 0 & 1 \\ 1 & 0 & 1 & 1 \\ 0 & 1 & 1 & 1 \end{bmatrix}$

10. Use the results of Exercise 9 to find all solutions of each of the following systems.

a) $x+y \equiv 1 \pmod 7$
$x+z \equiv 2 \pmod 7$
$y+z \equiv 3 \pmod 7$

b) $x+2y+3z \equiv 1 \pmod 7$
$x+2y+5z \equiv 1 \pmod 7$
$x+4y+6z \equiv 1 \pmod 7$

c) $x+y+z \equiv 1 \pmod 7$
$x+y+w \equiv 1 \pmod 7$
$x+z+w \equiv 1 \pmod 7$
$y+z+w \equiv 1 \pmod 7$

11. How many incongruent solutions does each of the following systems of congruences have?

a) $x + y + z \equiv 1 \pmod 5$
$2x + 4y + 3z \equiv 1 \pmod 5$

b) $2x + 3y + z \equiv 3 \pmod 5$
$x + 2y + 3z \equiv 1 \pmod 5$
$2x + z \equiv 1 \pmod 5$

c) $3x + y + 3z \equiv 1 \pmod 5$
$x + 2y + 4z \equiv 2 \pmod 5$
$4x + 3y + 2z \equiv 3 \pmod 5$

d) $2x + y + z \equiv 1 \pmod 5$
$x + 2y + z \equiv 1 \pmod 5$
$x + y + 2z \equiv 1 \pmod 5$

* 12. Develop an analogue of Cramer's rule for solving systems of $n$ linear congruences in $n$ unknowns.

* 13. Develop an analogue of Gaussian elimination to solve systems of $n$ linear congruences in $m$ unknowns (where $m$ and $n$ may be different).

A *magic square* is a square array of integers with the property that the sum of the integers in a row or in a column is always the same. In this exercise, we present a method for producing magic squares.

**\* 14.** Show that the $n^2$ integers $0,1,...,n^2-1$ are put into the $n^2$ positions of an $n \times n$ square, without putting two integers in the same position, if the integer $k$ is placed in the $i$th row and $j$th column, where

$$i \equiv a + ck + e[k/n] \pmod{n},$$
$$j \equiv b + dk + f[k/n] \pmod{n},$$

$1 \leqslant i \leqslant n$, $1 \leqslant j \leqslant n$, and $a,b,c,d,e$, and $f$ are integers with $(cf - de, n) = 1$.

**\* 15.** Show that a magic square is produced in Exercise 14 if $(c,n) = (d,n) = (e,n) = (f,n) = 1$.

**\* 16.** The *positive* and *negative diagonals* of an $n \times n$ square consist of the integers in positions $(i,j)$, where $i + j \equiv k \pmod{n}$ and $i - j \equiv k \pmod{n}$, respectively, where $k$ is a given integer. A square is called *diabolic* if the sum of the integers in a positive or negative diagonal is always the same. Show that a diabolic square is produced using the procedure given in Exercise 14 if $(c+d,n) = (c-d,n) = (e+f,n) = (e-f,n) = 1$.

## 3.4 Computer Projects

Write programs to do the following:

1. Find the solutions of a system of two linear congruences in two unknowns using Theorem 3.14.

2. Find inverses of $2 \times 2$ matrices using Theorem 3.16.

3. Find inverses of $n \times n$ matrices using Theorem 3.18.

4. Solve systems of $n$ linear congruences in $n$ unknowns using inverses of matrices.

5. Solve systems of $n$ linear congruences in $n$ unknowns using an analogue of Cramer's rule (see Exercise 12).

6. Solve system of $n$ linear congruences in $m$ unknowns using an analogue of Gaussian elimination (see Exercise 13).

7. Produce magic squares by the method given in Exercise 14.

# 4

---

# Applications of Congruences

---

## 4.1 Divisibility Tests

Using congruences we can develop divisibility tests for integers based on their expansions with respect to different bases.

We begin with tests which use decimal notation. In the following discussion let $n = (a_k a_{k-1} \cdots a_1 a_0)_{10}$. Then $n = a_k 10^k + a_{k-1} 10^{k-1} + \cdots + a_1 10 + a_0$, with $0 \leqslant a_j \leqslant 9$ for $j = 0, 1, 2, \ldots, k$.

First, we develop tests for divisibility by powers of 2.

Since $10 \equiv 0 \pmod{2}$, it follows that $10^j \equiv 0 \pmod{2^j}$ for all positive integers $j$. Hence,

$$n \equiv (a_0)_{10} \pmod{2},$$
$$n \equiv (a_1 a_0)_{10} \pmod{2^2},$$
$$n \equiv (a_2 a_1 a_0)_{10} \pmod{2^3},$$

$$\cdot$$
$$\cdot$$
$$\cdot$$

$$n \equiv (a_{j-1} a_{j-2} \cdots a_2 a_1 a_0)_{10} \pmod{2^j}.$$

These congruences tell us that to determine whether an integer $n$ is divisible by 2, we only need to examine its last digit for divisibility by 2. Similarly, to determine whether $n$ is divisible by 4, we only need to check the integer made up of the last two digits of $n$ for divisibility by 4. In general, to test $n$ for divisibility by $2^j$, we only need to check the integer made up of the last $j$ digits of $n$ for divisibility by $2^j$.

**Example 4.1.** Let $n = 32688048$. We see that $2 \mid n$ since $2 \mid 8$, $4 \mid n$ since $4 \mid 48$, $8 \mid n$ since $8 \mid 48$, $16 \mid n$ since $16 \mid 8048$, but $32 \nmid n$ since $32 \nmid 88048$.

□

Next, we develop divisibility tests for powers of 5.

To develop tests for divisibility by powers of 5, first note that since $10 \equiv 0 \pmod 5$ we have $10^j \equiv 0 \pmod{5^j}$. Hence, divisibility tests for powers of 5 are analogous to those for powers of 2. We only need to check the integer made up of the last $j$ digits of $n$ to determine whether $n$ is divisible by $5^j$.

**Example 4.2.** Let $n = 15535375$. Since $5 \mid 5$, $5 \mid n$, since $25 \mid 75$, $25 \mid n$, since $125 \mid 375$, $125 \mid n$, but since $625 \nmid 5375$, $625 \nmid n$.

□

Next, we develop tests for divisibility by 3 and by 9.

Note that both the congruences $10 \equiv 1 \pmod 3$ and $10 \equiv 1 \pmod 9$ hold. Hence, $10^k \equiv 1 \pmod 3$ roman and $\pmod 9$. This gives us the useful congruences

$$(a_k a_{k-1}...a_1 a_0) = a_k 10^k + a_{k-1} 10^{k-1} + \cdots + a_1 10 + a_0$$
$$\equiv a_k + a_{k-1} + \cdots + a_1 + a_0 \pmod 3 \text{ and } \pmod 9.$$

Hence, we only need to check whether the sum of the digits of $n$ is divisible by 3, or by 9, to see whether $n$ is divisible by 3, or by 9.

**Example 4.3.** Let $n = 4127835$. Then, the sum of the digits of $n$ is $4 + 1 + 2 + 7 + 8 + 3 + 5 = 30$. Since $3 \mid 30$ but $9 \nmid 30$, $3 \mid n$ but $9 \nmid n$. □

A rather simple test can be found for divisibility by 11.

Since $10 \equiv -1 \pmod{11}$, we have

$$(a_k a_{k-1}...a_1 a_0)_{10} = a_k 10^k + a_{k-1} 10^{k-1} + \cdots + a_1 10 + a_0$$
$$\equiv a_k(-1)^k + a_{k-1}(-1)^{k-1} + \cdots - a_1 + a_0 \pmod{11}.$$

This shows that $(a_k a_{k-1}...a_1 a_0)_{10}$ is divisible by 11 if and only if $a_0 - a_1 + a_2 - \cdots + (-1)^k a_k$, the integer formed by alternately adding and subtracting the digits, is divisible by 11.

**Example 4.4.** We see that 723160823 is divisible by 11, since alternately adding and subtracting its digits yields $3 - 2 + 8 - 0 + 6 - 1 + 3 - 2 + 7 = 22$ which is divisible 11. On the other hand, 33678924 is not divisible by 11, since $4 - 2 + 9 - 8 + 7 - 6 + 3 - 3 = 4$ is not divisible by 11.

□

Next, we develop a test to simultaneously test for divisibility by the primes 7, 11, and 13.

Note that $7 \cdot 11 \cdot 13 = 1001$ and $10^3 = 1000 \equiv -1 \pmod{1001}$. Hence,

$$
\begin{aligned}
(a_k a_{k-1} \ldots a_0)_{10} &= a_k 10^k + a_{k-1} 10^{k-1} + \cdots + a_1 10 + a_0 \\
&\equiv (a_0 + 10a_1 + 100a_2) + 1000(a_3 + 10a_4 + 100a_5) + \\
&\quad (1000)^2 (a_6 + 10a_7 + 100a_8) + \cdots \\
&\equiv (100a_2 + 10a_1 + a_0) - (100a_5 + 10a_4 + a_3) + \\
&\quad (100a_8 + 10a_7 + a_6) - \cdots \\
&\equiv (a_2 a_1 a_0)_{10} - (a_5 a_4 a_3)_{10} + (a_8 a_7 a_6)_{10} - \cdots \pmod{1001}.
\end{aligned}
$$

This congruence tells us that an integer is congruent modulo 1001 to the integer formed by successively adding and subtracting the three-digit integers with decimal expansions formed from successive blocks of three decimal digits of the original number, where digits are grouped starting with the rightmost digit. As a consequence, since 7, 11, and 13 are divisors of 1001, to determine if an integer is divisible by 7, 11, or 13, we only need to check whether this alternating sum and difference of blocks of three digits is divisible by 7, 11, or 13.

**Example 4.5.** Let $n = 59358208$. Since the alternating sum and difference of the integers formed from blocks of three digits, $208 - 358 + 59 = -91$, is divisible by 7 and 13, but not by 11, we see that $n$ is divisible by 7 and 13, but not by 11. ☐

All of the divisibility tests we have developed thus far are based on decimal representations. We now develop divisibility tests using base $b$ representations, where $b$ is a positive integer.

**Theorem 4.1.** If $d \mid b$ and $j$ and $k$ are positive integers with $j < k$, then $(a_k \ldots a_1 a_0)_b$ is divisible by $d^j$ if and only if $(a_{j-1} \ldots a_1 a_0)_b$ is divisible by $d^j$.

*Proof.* Since $b \equiv 0 \pmod{d}$, it follows that $b^j \equiv 0 \pmod{d^j}$. Hence,

$$
\begin{aligned}
(a_k a_{k-1} \ldots a_1 a_0)_b &= a_k b^k + \cdots + a_j b^j + a_{j-1} b^{j-1} + \cdots + a_1 b + a_0 \\
&\equiv a_{j-1} b^{j-1} + \cdots + a_1 b + a_0 \\
&= (a_{j-1} \ldots a_1 a_0)_b \pmod{d^j}.
\end{aligned}
$$

Consequently, $d \mid (a_k a_{k-1} \ldots a_1 a_0)_b$ if and only if $d \mid (a_{j-1} \ldots a_1 a_0)_b$. ☐

**Theorem 4.2.** If $d \mid (b-1)$, then $n = (a_k \ldots a_1 a_0)_b$ is divisible by $d$ if and only if $a_k + \cdots + a_1 + a_0$ is divisible by $d$.

*Proof.* Since $d \mid (b-1)$, we have $b \equiv 1 \pmod{d}$, so that by Theorem 3.7 we know that $b^j \equiv 1 \pmod{d}$ for all positive integers $b$. Hence, $n = (a_k \ldots a_1 a_0)_b = a_k b^k + \cdots + a_1 b + a_0 \equiv a_k + \cdots + a_1 + a_0 \pmod{d}$. This shows that $d \mid n$ if and only if $d \mid (a_k + \cdots + a_1 + a_0)$. ☐

**Theorem 4.3.** If $d \mid (b + 1)$, then $n = (a_k \ldots a_1 a_0)_b$ is divisible by $d$ if and only if $(-1)^k a_k + \cdots -a_1 + a_0$ is divisible by $d$.

*Proof.* Since $d \mid (b + 1)$, we have $b \equiv -1 \pmod{d}$. Hence, $b^j \equiv (-1)^j$ $\pmod{d}$, and consequently, $n = (a_k...a_1a_0)_b \equiv (-1)^k a_k + ... - a_1 + a_0$ $\pmod{d}$. Hence, $d \mid n$ if and only if $d \mid ((-1)^k a_k + ... -a_1 + a_0)$.          □

**Example 4.6.** Let $n = (7F28A6)_{16}$ (in hex notation). Then, since $2 \mid 16$, from Theorem 4.1 we know that $2 \mid n$, since $2 \mid 6$. Likewise, since $4 \mid 16$, we see that $4 \nmid n$, since $4 \nmid 6$. By Theorem 4.2, since $3 \mid (16 - 1)$, $5 \mid (16 -1)$, and $15 \mid (16 - 1)$, and $7 + F + 2 + 8 + A + 6 = (30)_{16}$, we know that $3 \mid n$, since $3 \mid (30)_{16}$, while $5 \nmid n$ and $15 \nmid n$, since $5 \nmid (30)_{16}$ and $15 \nmid (30)_{16}$. Furthermore, by Theorem 4.3, since $17 \mid (16 + 1)$ and $n \equiv 6 - A + 8 - 2 + F - 7 = (A)_{16} \pmod{17}$, we conclude that $17 \nmid n$, since $17 \nmid (A)_{16}$.□

**Example 4.7.** Let $n = (1001001111)_2$. Then, using Theorem 4.3 we see that $3 \mid n$, since $n \equiv 1 - 1 + 1 - 1 + 0 - 0 + 1 - 0 + 0 - 1 \equiv 0 \pmod{3}$ and $3 \mid (2+1)$ .                                                      □

---

## 4.1 Exercises

1.  Determine the highest power of 2 dividing each of the following positive integers.

    a)   201984          c)   89375744
    b)   1423408         d)   41578912246

2.  Determine the highest power of 5 dividing each of the following positive integers.

    a)   112250          c)   235555790
    b)   4860625         d)   48126953125

3.  Which of the following integers are divisible by 3? Of those that are, which are divisible by 9?

    a)   18381           c)   987654321
    b)   65412351        d)   78918239735

4. Which of the following integers are divisible by 11?

   a)   10763732        c)   674310976375
   b)   1086320015      d)   8924310064537

A *repunit* is an integer with decimal expansion containing all 1's.

5. Determine which repunits are divisible by 3; and which are divisible by 9.

6. Determine which repunits are divisible by 11.

7. Determine which repunits are divisible by 1001. Which are divisible by 7? by 13?

8. Determine which repunits with fewer than 10 digits are prime.

A *base b repunit* is an integer with base $b$ expansion containing all 1's.

9. Determine which base $b$ repunits are divisible by factors of $b - 1$.

10. Determine which base $b$ repunits are divisible by factors of $b + 1$.

A *base b palindromic integer* is an integer whose base $b$ representation reads the same forward and backward.

11. Show that every decimal palindromic integer with an even number of digits is divisible by 11.

12. Show that every base 7 palindromic integer with an even number of digits is divisible by 8.

13. Develop a test for divisibility by 37, based on the fact that $10^3 \equiv 1 \pmod{37}$. Use this to check 443692 and 11092785 for divisibility by 37.

14. Devise a divisibility test for integers represented in base $b$ notation for divisibility by $n$ where $n$ in a divisor of $b^2 + 1$. (*Hint*: Split the digits of the base $b$ representation of the integer into blocks of two, starting on the right).

15. Use the test you developed in Exercise 14 to decide whether

    a)   $(101110110)_2$ is divisible by 5.
    b)   $(12100122)_3$ is divisible by 2, and whether it is divisible by 5.
    c)   $(364701244)_8$ is divisible by 5, and whether it is divisible by 13.
    d)   $(5837041320219)_{10}$ is divisible by 101.

16. An old receipt has faded. It reads 88 chickens at a total of $\$x4.2y$ where $x$ and $y$ are unreadable digits. How much did each chicken cost?

17. Use a congruence modulo 9 to find the missing digit, indicated by a question mark:   $89878 \cdot 58965 = 5299\,?\,56270$.

We can check a multiplication $c = ab$ by determining whether the congruence $c \equiv ab \pmod{m}$ is valid, where $m$ is any modulus. If we find that

$c \neq ab$ (mod $m$), then we know an error has been made. When we take $m = 9$ and use the fact that an integer in decimal notation is congruent modulo 9 to the sum of its digits, this check is called *casting out nines*.

**18.** Check each of the following multiplications by casting out nines.

   a)  $875961 \cdot 2753 = 2410520633$
   b)  $14789 \cdot 23567 = 348532367$
   c)  $24789 \cdot 43717 = 1092700713$

**19.** Is a check of a multiplication by casting out nines foolproof?

**20.** What combinations of digits of a decimal expansion of an integer are congruent to this integer modulo 99? Use your answer to devise a check for multiplication based on *casting out ninety nines*. Then use the test to check the multiplications in problem 13.

## 4.1 Computer Projects

Write programs to do the following:

**1.** Determine the highest powers of 2 and of 5 that divide an integer.

**2.** Test an integer for divisibility by 3, 7, 9, 11, and 13. (Use congruences modulo 1001 for divisibility by 7 and 13.)

**3.** Determine the highest power of each factor of $b$ that divides an integer from the base $b$ expansion of the integer.

**4.** Test an integer from its base $b$ expansion, for divisibility by factors of $b - 1$ and of $b + 1$.

## 4.2 The Perpetual Calendar

In this section we derive a formula that gives us the day of the week of any day of any year. Since the days of the week form a cycle of length seven, we use a congruence modulo 7. We denote each day of the week by a number in the set 0, 1, 2, 3, 4, 5, 6, setting

- *Sunday* = 0,
- *Monday* = 1,
- *Tuesday* = 2,
- *Wednesday* = 3,
- *Thursday* = 4,
- *Friday* = 5,
- *Saturday* = 6.

Julius Caesar changed the Egyptian calendar, which was based on a year of exactly 365 days, to a new calendar, called the *Julian calendar*, with a year of average length 365 ¼ days, with leap years every fourth year, to better reflect the true length of the year. However, more recent calculations have shown that the true length of the year is approximately 365.2422 days. As the centuries passed, the discrepancies of 0.0078 days per year added up, so that by the year 1582 approximately 10 extra days had been added unnecessarily in leap years. To remedy this, in 1582 Pope Gregory set up a new calendar. First, 10 days were added to the date, so that October 5, 1582, became October 15, 1582 (and the 6th through the 14th of October were skipped). It was decided that leap years would be precisely the years divisible by 4, except that those exactly divisible by 100, *i.e.*, the years that mark centuries, would be leap years only when divisible by 400. As an example, the years 1700, 1800, 1900, and 2100 are not leap years but 1600 and 2000 are. With this arrangement, the average length of a calendar year is 365.2425 days, rather close to the true year of 365.2422 days. An error of 0.0003 days per year remains, which is 3 days per 10000 years. In the future, this discrepancy will have to be accounted for, and various possibilities have been suggested to correct for this error.

In dealing with calendar dates for various parts of the world, we must also take into account the fact that the Gregorian calendar was not adopted everywhere in 1582. In Britain and what is now the United States, the Gregorian calendar was adopted only in 1752, and by then, it was necessary to add 11 days. In these places September 3, 1752 in the Julian calendar became September 14, 1752 in the Gregorian calendar. Japan changed over 1873, the Soviet Union and nearby countries in 1917, while Greece held out until 1923.

We now set up our procedure for finding the day of the week in the Gregorian calendar for a given date. We first must make some adjustments, because the extra day in a leap year comes at the end of February. We take care of this by renumbering the months, starting each year in March, and considering the months of January and February part of the preceding year. For instance, February 1984, is considered the 12th month of 1983, and May 1984, is considered the 3rd month of 1984. With this convention, for the day of interest, let

- $k$ = day of the month,

- $m$ = month,

  with

  — *January* = 11

- *February* = 12

- *March* = 1

- *April* = 2

- *May* = 3

- *June* = 4

- *July* = 5

- *August* = 6

- *September* = 7

- *October* = 8

- *November* = 9

- *December* = 10

- $N$ = year,

where $N$ is the current year unless the month is January or February in which case $N$ is the previous year with $N = 100C + Y$, where

- $C$ = century,

- $Y$ = particular year of the century.

**Example 4.8**. With April 3, 1951 we have $k = 3$, $m = 2$, $N = 1951$, $C = 19$, and $Y = 51$. $\qquad\qquad\qquad\qquad\qquad\qquad\qquad\qquad\qquad\qquad$ $\square$

We use March 1, of each year as our basis. Let $d_N$ represent the day of the week of March 1, in year $N$. We start with the year 1600 and compute the day of the week March 1, falls on in any given year. Note that between March 1 of year $N - 1$ and March 1 of year $N$, if year $N$ is not a leap year, 365 days have passed, and since $365 = 1 \pmod 7$, we see that $d_N \equiv d_{N-1} + 1 \pmod 7$, while if year $N$ is a leap year, since there is an extra day between the consecutive firsts of March, we see that

$$d_N \equiv d_{N-1} + 2 \pmod 7.$$

Hence, to find $d_N$ from $d_{1600}$, we must first find out how many leap years have occurred between the year 1600 and the year $N$ (not including 1600, but including $N$); let us call this number $x$. To compute $x$, first note that by the division algorithm there are $[(N - 1600)/4]$ years divisible by 4 between 1600 and $N$, there are $[(N-1600)/100]$ years divisible by 100 between 1600 and $N$, and there are $[(N - 1600)/400]$ years divisible by 400 between 1600 and $N$. Hence,

$$x = [(N - 1600)/4] - [(N - 1600)/100] + [(N - 1600)/400]$$
$$= [N/4] - 400 - [N/100] + 16 + [N/400] - 4$$
$$= [N/4] - [N/100] + [N/400] - 388.$$

(We have used Exercise 12 of Section 1.1 to simplify this expression). Now putting this in terms of $C$ and $Y$, we see that

$$x = [25C + (Y/4)] - [C + (Y/100)] + [(C/4) + (Y/400)] - 388$$
$$= 25C + [Y/4] - C + [C/4] - 388$$
$$\equiv 3C + [C/4] + [Y/4] - 3 \pmod 7.$$

Here we have again used Exercise 12 of Section 1.1 the inequality $Y/100 < 1$, and the equation $[(C/4) + (Y/400)] = [C/4]$ (which follows from Exercise 17 of Section 1.5, since $Y/400 < 1/4$).

We can now compute $d_N$ from $d_{1600}$ by shifting $d_{1600}$ by one day for every year that has passed, plus an extra day for each leap year between 1600 and $N$. This gives the following formula:

$$d_N \equiv d_{1600} + N + x$$
$$= d_{1600} + 100C + Y - 1600 + 3C + [C/4] + [Y/4] - 3 \pmod 7.$$

Simplifying, we have

$$d_N \equiv d_{1600} - 2C + Y + [C/4] + [Y/4] \pmod 7.$$

Now that we have a formula relating the day of the week for March 1, of any year, with the day of the week of March 1, 1600, we can use the fact that March 1, 1982, is a Monday to find the day of the week of March 1, 1600. For 1982, since $N = 1982$, we have $C = 19$, and $Y = 82$, and since $d_{1982} = 1$, it follows that

$$1 \equiv d_{1600} - 38 + 82 + [19/4] + [82/4] \equiv d_{1600} - 2 \pmod 7.$$

Hence, $d_{1600} = 3$, so that March 1, 1600, was a Wednesday. When we insert the value of $d_{1600}$, the formula for $d_N$ becomes

$$d_N \equiv 3 - 2C + Y + [C/4] + [Y/4] \pmod 7.$$

We now use this formula to compute the day of the week of the first day of each month of year $N$. To do this, we have to use the number of days of the week that the first of the month of a particular month is shifted from the first of the month of the preceding month. The months with 30 days shift the first of the following month up 2 days, because $30 \equiv 2 \pmod 7$, and those with 31 days shift the first of the following month up 3 days, because $31 \equiv 3 \pmod 7$. Therefore, we must add the following amounts:

| | |
|---|---|
| from March 1, to April 1: | 3 days |
| from April 1, to May 1: | 2 days |

| from May 1, to June 1: | 3 days |
|---|---|
| from June 1, to July 1: | 2 days |
| from July 1, to August 1: | 3 days |
| from August 1, to September 1: | 3 days |
| from September 1, to October 1: | 2 days |
| from October 1, to November 1: | 3 days |
| from November 1, to December 1: | 2 days |
| from December 1, to January 1: | 3 days |
| from January 1, to February 1: | 3 days. |

We need a formula that gives us the same increments. Notice that we have 11 increments totaling 29 days, so that each increment averages 2.6 days. By inspection, we find that the function $[2.6m - 0.2] - 2$ has exactly the same increments as $m$ goes from 1 to 11, and is zero when $m = 1$. (This formula was originally found by a Reverend Zeller; he apparently found it by trial and error.) Hence, the day of the week of the first day of month $m$ of year $N$ is given by by the least nonnegative residue of $d_N + [2.6m - 0.2] - 2$ modulo 7.

To find $W$, the day of the week of day $k$ of month $m$ of year $N$, we simply add $k-1$ to the formula we have devised for the day of the week of the first day of the same month. We obtain the formula:

$$W \equiv k + [2.6m - 0.2] - 2C + Y + [Y/4] + [C/4] \pmod 7.$$

We can use this formula to find the day of the week of any date of any year in the Gregorian calendar.

**Example 4.9.** To find the day of the week of January 1, 1900, we have $C = 18$, $Y = 99$, $m = 11$, and $k = 1$ (since we consider January as the eleventh month of the preceding year). Hence, we have $W \equiv 1 + 28 - 36 + 99 + 4 + 24 \equiv 1 \pmod 7$, so that the first day of the twentieth century was a Monday.  □

---

## 4.2 Exercises

1. Find the day of the week of the day you were born, and of your birthday this year.

2. Find the day of the week of the following important dates in U. S. history (use the Julian calendar before September 3, 1752, and the Gregorian calendar from September 14, 1752 to the present)

   * a) October 12, 1492  (Columbus sights land in the Caribbean)

* b)  May 6, 1692      (Peter Minuit buys Manhattan from the natives)
* c)  June 15, 1752     (Benjamin Franklin invents the lightening rod)
  d)  July 4, 1776       (U. S. Declaration of Independence)
  e)  March 30, 1867    (U. S. buys Alaska from Russia)
  f)  March 17, 1888    (Great blizzard in the Eastern U. S.)
  g)  February 15, 1898 (U. S. Battleship Maine blown up in Havana Harbor)
  h)  July 2, 1925       (Scopes convicted of teaching evolution)
  i)  July 16, 1945      (First atomic bomb exploded)
  j)  July 20, 1969      (First man on the moon)
  k)  August 9, 1974    (Nixon resigns)
  l)  March 28, 1979    (Three Mile Island nuclear mishap).
  m)  June 5, 2013       (First man on Mars)

3. To correct the small discrepancy between the number of days in a year of the Gregorian calendar and an actual year, it has been suggested that the years exactly divisible by 4000 should not be leap years. Adjust the formula for the day of the week of a given date to take this correction into account.

4. Show that days with the same calendar date in two different years of the same century, 28, 56, or 84 years apart, fall on the identical day of the week.

5. Which of your birthdays, until your one hundredth, fall on the same day of the week as the day you were born?

* 6. A new calendar called the *International Fixed Calendar* has been proposed. In this calendar, there are 13 months, including all our present months, plus a new month, called *Sol*, which is placed between June and July. Each month has 28 days, except for the June of leap years which has an extra day (leap years are determined the same way as in the Gregorian calendar). There is an extra day, *Year End Day*, which is not in any month, which we may consider as December 29. Devise a perpetual calendar for the International Fixed Calendar to give day of the week for any calendar date.

## 4.2 Computer Projects

Write programs to do the following:

1. To give the day of the week of any date.

2. To print out a calendar of any year.

3. To print out a calendar for the International Fixed Calendar (See Exercise 6).

## 4.3 Round-Robin Tournaments

Congruences can be used to schedule round-robin tournaments. In this section, we show how to schedule a tournament for $N$ different teams, so that each team plays every other team exactly once. The method we describe was developed by Freund [94].

First note that if $N$ is odd, not all teams can be scheduled in each round, since when teams are paired, the total number of teams playing is even. So, if $N$ is odd, we add a dummy team, and if a team is paired with the dummy team during a particular round, it draws a bye in that round and does not play. Hence, we can assume that we always have an even number of teams, with the addition of a dummy team if necessary.

Now label the $N$ teams with the integers $1, 2, 3,...,N-1, N$. We construct a schedule, pairing teams in the following way. We have team $i$, with $i \neq N$, play team $j$, with $j \neq N$ and $j \neq i$, in the $k$th round if $i + j \equiv k \pmod{N-1}$. This schedules games for all teams in round $k$, except for team $N$ and the one team $i$ for which $2i \equiv k \pmod{N-1}$. There is one such team because Theorem 3.10 tells us that the congruence $2x \equiv k \pmod{N-1}$ has exactly one solution with $1 \leqslant x \leqslant N-1$, since $(2, N-1) = 1$. We match this team $i$ with team $N$ in the $k$th round.

We must now show that each team plays every other team exactly once. We consider the first $N-1$ teams. Note that team $i$, where $1 \leqslant i \leqslant N-1$, plays team $N$ in round $k$ where $2i \equiv k \pmod{N-1}$, and this happens exactly once. In the other rounds, team $i$ does not play the same team twice, for if team $i$ played team $j$ in both rounds $k$ and $k'$, then $i + j \equiv k \pmod{N-1}$, and $i + j \equiv k' \pmod{N-1}$ which is an obvious contradiction because $k \neq k' \pmod{N-1}$. Hence, since each of the first $N-1$ teams plays $N-1$ games, and does not play any team more than once, it plays every team exactly once. Also, team $N$ plays $N-1$ games, and since every other team plays team $N$ exactly once, team $N$ plays every other team exactly once.

**Example 4.10**. To schedule a round-robin tournament with 5 teams, labeled 1, 2, 3, 4, and 5, we include a dummy team labeled 6. In round one, team 1 plays team $j$ where $1 + j \equiv 1 \pmod 5$. This is the team $j = 5$ so that team 1 plays team 5. Team 2 is scheduled in round one with team 4, since the solution of $2 + j \equiv 1 \pmod 5$ is $j = 4$. Since $i = 3$ is the solution of the congruence $2i \equiv 1 \pmod 5$, team 3 is paired with the dummy team 6, and hence, draws a bye in the first round. If we continue this procedure and finish scheduling the other rounds, we end up with the pairings shown in Figure 4.1, where the opponent of team $i$ in round $k$ is given in the $k$th row and $i$th column.                                                                         □

| Round \ Team | 1 | 2 | 3 | 4 | 5 |
|---|---|---|---|---|---|
| 1 | 5 | 4 | bye | 2 | 1 |
| 2 | bye | 5 | 4 | 3 | 2 |
| 3 | 2 | 1 | 5 | bye | 3 |
| 4 | 3 | bye | 1 | 5 | 4 |
| 5 | 4 | 3 | 2 | 1 | bye |

**Figure 4.1.** Round-Robin Schedule for Five Teams.

---

## 4.3 Exercises

1. Set up a round-robin tournament schedule for

   a) 7 teams     c) 9 teams
   b) 8 teams     d) 10 teams.

2. In round-robin tournament scheduling, we wish to assign a *home team* and an *away team* for each game so that each of $n$ teams, where $n$ is odd, plays an equal number of home games and away games. Show that if when $i + j$ is odd, we assign the smaller of $i$ and $j$ as the home team, while if $i + j$ is even, we assign the larger of $i$ and $j$ as the home team, then each team plays an equal number of home and away games.

3. In a round-robin tournament scheduling, use Exercise 2 to determine the home team for each game when there are

   a) 5 teams     b) 7 teams     c) 9 teams.

## 4.3 Computer Projects

Write programs to do the following:

1. Schedule round-robin tournaments.

**2.** Using Exercise 2, schedule round-robin tournaments for an odd number of teams, specifying the home team for each game.

---

## 4.4 Computer File Storage And Hashing Functions

A university wishes to store a file for each of its students in its computer. The identifying number or *key* for each file is the social security number of the student enrolled. The social security number is a nine-digit integer, so it is extremely unfeasible to reserve a memory location for each possible social security number. Instead, a systematic way to arrange the files in memory, using a reasonable amount of memory locations, should be used so that each file can be easily accessed. Systematic methods of arranging files have been developed based on *hashing functions* . A hashing function assigns to the key of each file a particular memory location. Various types of hashing functions have been suggested, but the type most commonly used involves modular arithmetic. We discuss this type of hashing function here. For a general discussion of hashing functions see Knuth [81] or Kronsjö [82].

Let $k$ be the key of the file to be stored; in our example, $k$ is the social security number of a student. Let $m$ be a positive integer. We define the hashing function $h(k)$ by

$$h(k) \equiv k \pmod{m},$$

where $0 \leqslant h(k) < m$, so that $h(k)$ is the least positive residue of $k$ modulo $m$. We wish to pick $m$ intelligently, so that the files are distributed in a reasonable way throughout the $m$ different memory locations $0, 1, 2,..., m-1$.

The first thing to keep in mind is that $m$ should not be a power of the base $b$ which is used to represent the keys. For instance, when using social security numbers as keys, $m$ should not be a power of 10, such as $10^3$, because the value of the hashing function would simply be the last several digits of the key; this may not distribute the keys uniformly throughout the memory locations. For instance, the last three digits of early issued social security numbers may often be between 000 and 099, but seldom between 900 and 999. Likewise, it is unwise to use a number dividing $b^k \pm a$ where $k$ and $a$ are small integers for the modulus $m$. In such a case, $h(k)$ would depend too strongly on the particular digits of the key, and different keys with similar, but rearranged, digits may be sent to the same memory location, For instance, if $m = 111$, then, since $111 \mid (10^3 - 1) = 999$, we have $10^3 \equiv 1 \pmod{111}$, so that the social security numbers 064 212 848 and 064 848 212 are sent to the same memory location, since

$$h(064\ 212\ 848) \equiv 064\ 212\ 848 \equiv 064 + 212 + 848 \equiv 1124 \equiv 14 \pmod{111},$$

and

$h\,(064\ 848\ 212) \equiv 064\ 848\ 212 \equiv 064 + 848 + 212 \equiv 1124 \equiv 14\ (\mathrm{mod}\ 111).$

To avoid such difficulties, $m$ should be a prime approximating the number of available memory locations devoted to file storage. For instance, if there are 5000 memory locations available for storage of 2000 student files we could pick $m$ to be equal to the prime 4969.

We have avoided mentioning the problem that arises when the hashing function assigns the same memory location to two different files. When this occurs, we say the there is a *collision*. We need a method to resolve collisions, so that files are assigned to different memory locations. There are two kinds of collision resolution policies. In the first kind, when a collision occurs, extra memory locations are linked together to the first memory location. When one wishes to access a file where this collision resolution policy has been used, it is necessary to first evaluate the hashing function for the particular key involved. Then the list linked to this memory location is searched.

The second kind of collision resolution policy is to look for an open memory location when an occupied location is assigned to a file. Various suggestions, such as the following technique have been made for accomplishing this.

Starting with our original hashing function $h_0(k) = h(k)$, we define a sequence of memory locations $h_1(k), h_2(k),\ldots$ . We first attempt to place the file with key $k$ at location $h_0(k)$. If this location is occupied, we move to location $h_1(k)$. If this is occupied, we move to location $h_2(k)$, *etc.*

We can choose the sequence of functions $h_j(k)$ in various ways. The simplest way is to let

$$h_j(k) \equiv h(k) + j\ (\mathrm{mod}\ m),\ 0 \leqslant h_j(k) < m.$$

This places the file with key $k$ as near as possible past location $h(k)$. Note that with this choice of $h_j(k)$, all memory locations are checked, so if there is an open location, it will be found. Unfortunately, this simple choice of $h_j(k)$ leads to difficulties; files tend to *cluster*. We see that if $k_1 \neq k_2$ and $h_i(k_1) = h_j(k_2)$ for nonnegative integers $i$ and $j$, then $h_{i+k}(k_1) = h_{j+k}(k_2)$ for $k = 1, 2, 3, \ldots$, so that exactly the same sequence of locations are traced out once there is a collision. This lowers the efficiency of the search for files in the table. We would like to avoid this problem of clustering, so we choose the function $h_j(k)$ in a different way.

To avoid clustering, we use a technique called *double hashing*. We choose, as before,

$$h(k) \equiv k\ (\mathrm{mod}\ m),$$

with $0 \leqslant h(k) < m$, where $m$ is prime, as the hashing function. We take a second hashing function

$$g(k) \equiv k + 1 \pmod{m-2},$$

where $0 < g(k) \leqslant m - 1$, so that $(g(k), m) = 1$. We take as a *probing sequence*

$$h_j(k) \equiv h(k) + j \, g(k) \pmod{m},$$

where $0 \leqslant h_j(k) < m$. Since $(g(k), m) = 1$, as $j$ runs through the integers 0, 1, 2, ..., $m - 1$, all memory locations are traced out. The ideal situation would be for $m-2$ to also be prime, so that the values $g(k)$ are distributed in a reasonable way. Hence, we would like $m-2$ and $m$ to be twin primes.

**Example 4.11**. In our example using social security numbers, both $m = 4969$, and $m-2 = 4967$ are prime. Our probing sequence is

$$h_j(k) \equiv h(k) + j \, g(k) \pmod{4969},$$

where $0 \leqslant h_j(k) < 4969$, $h(k) \equiv k \pmod{4969}$, and $g(k) \equiv k + 1 \pmod{4967}$.

Suppose we wish to assign memory locations to files for students with social security numbers:

$$
\begin{array}{ll}
k_1 = 344\ 401\ 659 & k_6 = 372\ 500\ 191 \\
k_2 = 325\ 510\ 778 & k_7 = 034\ 367\ 980 \\
k_3 = 212\ 228\ 844 & k_8 = 546\ 332\ 190 \\
k_4 = 329\ 938\ 157 & k_9 = 509\ 496\ 993 \\
k_5 = 047\ 900\ 151 & k_{10} = 132\ 489\ 973.
\end{array}
$$

Since $k_1 \equiv 269$, $k_2 \equiv 1526$, and $k_3 \equiv 2854 \pmod{4969}$, we assign the first three files to locations 269, 1526, and 2854, respectively. Since $k_4 \equiv 1526 \pmod{4969}$, but location 1526 is taken, we compute $h_1(k_4) \equiv h(k_4) + g(k_4) = 1526 + 216 = 1742 \pmod{4969}$; this follows since $g(k_4) \equiv 1 + k_4 \equiv 216 \pmod{4967}$. Since location 1742 is free, we assign the fourth file to this location. The fifth, six, seventh, and eighth files go into the available locations 3960, 4075, 2376, and 578, respectively, because $k_5 \equiv 3960$, $k_6 \equiv 4075$, $k_7 \equiv 2376$, and $k_8 \equiv 578 \pmod{4969}$. We find that $k_9 \equiv 578 \pmod{4969}$; because location 578 is occupied, we compute $h_1(k_9) + g(k_9) = 578 + 2002 = 2580 \pmod{4969}$, where $g(k_9) = 1 + k_9 \equiv 2002 \pmod{4967}$. Hence, we assign the ninth file to the free location 2580. Finally, we find that $k_{10} \equiv 1526 \pmod{4967}$, but location 1526 is taken. We compute $h_1(k_{10}) \equiv h(k_{10}) + g(k_{10}) = 1526 + 216 = 1742 \pmod{4969}$, because $g(k_{10}) = k_{10} \equiv 216 \pmod{4967}$, but location 1742 is taken. Hence, we continue by finding $h_2(k_{10}) \equiv h(k_{10}) + 2g(k_{10}) \equiv 1958 \pmod{4969}$ and in this available location, we place the tenth file.

Table 4.1 lists the assignments for the files of students by their social security numbers. In the table, the file locations are shown in boldface.   □

| Social Security Number | $h(k)$ | $h_1(k)$ | $h_2(k)$ |
|---|---|---|---|
| 344 401 659 | **269** | | |
| 325 510 778 | **1526** | | |
| 212 228 844 | **2854** | | |
| 329 938 157 | 1526 | **1742** | |
| 047 900 151 | **3960** | | |
| 372 500 191 | **4075** | | |
| 034 367 980 | **2376** | | |
| 546 332 190 | **578** | | |
| 509 496 993 | 578 | **2580** | |
| 132 489 973 | 1526 | 1742 | **1958** |

Table 4.1.  Hashing Function for Student Files.

We wish to find conditions where double hashing leads to clustering. Hence, we find conditions when

(4.1)                     $$h_i(k_1) = h_j(k_2)$$

and

(4.2)                     $$h_{i+1}(k_1) = h_{j+1}(k_2),$$

so that the two consecutive terms of two probe sequences agree. If both (4.1) and (4.2) occur, then

$$h(k_1) + ig(k_1) \equiv h(k_2) + jg(k_2) \pmod{m}$$

and

$$h(k_1) + (i+1)g(k_1) \equiv h(k_2) + (j+1)g(k_2) \pmod{m}.$$

Subtracting the first of these two congruences from the second, we obtain

$$g(k_1) \equiv g(k_2) \pmod{m},$$

so that

$$k_1 \equiv k_2 \pmod{m-2}.$$

Since $g(k_1) = g(k_2)$, we can substitute this into the first congruence to obtain

$$h(k_1) \equiv h(k_2) \ (\text{mod } m),$$

which shows that

$$k_1 \equiv k_2 \ (\text{mod } m).$$

Consequently, since $(m-2, m) = 1$, Theorem 3.8 tells us that

$$k_1 \equiv k_2 \ (\text{mod } m(m-2)).$$

Therefore, the only way that two probing sequences can agree for two consecutive terms is if the two keys involved, $k_1$ and $k_2$, are congruent modulo $m(m-2)$. Hence, clustering is extremely rare. Indeed, if $m(m-2) > k$ for all keys $k$, clustering will never occur.

---

## 4.4 Exercises

1. A parking lot has 101 parking places. A total of 500 parking stickers are sold and only 50-75 vehicles are expected to be parked at a time. Set up a hashing function and collision resolution policy for assigning parking places based on license plates displaying six-digit numbers.

2. Assign memory locations for students in your class, using as keys the day of the month of birthdays of students with hashing function $h(K) \equiv K$ (mod 19),

   a) with probing sequence $h_j(K) \equiv h(K) + j$ (mod 19).

   b) with probing sequence $h_j(K) \equiv h(K) + j \cdot g(K), 0 \leqslant j \leqslant 16$, where $g(K) \equiv 1 + K(\text{mod } 17)$.

* 3. Let the hashing function be $h(K) \equiv K(\text{mod } m)$, with $0 \leqslant h(K) < m$, and let the probing sequence for collision resolution be $h_j(K) \equiv h(K) + jq$ (mod $m$), $0 \leqslant h_j(K) < m$, for $j = 1, 2,..., m-1$. Show that all memory locations are probed

   a) if $m$ is prime and $1 \leqslant q \leqslant m-1$.

   b) if $m = 2^r$ and $q$ is odd.

* 4. A probing sequence for resolving collisions where the hashing function is $h(K) \equiv K(\text{mod } m)$, $0 \leqslant h(K) < m$, is given by $h_j(K) \equiv h(K) + j(2h(K) + 1)$ (mod $m$), $0 \leqslant h_j(K) < m$.

   a) Show that if $m$ is prime, then all memory sequences are probed.

   b) Determine conditions for clustering to occur, i.e., when $h_j(K_1) = h_j(K_2)$ and $h_{j+r}(K_1) = h_{j+r}(K_2)$ for $r = 1, 2,...$ .

5. Using the hashing function and probing sequence of the example in the text, find open memory locations for the files of students with social security numbers: $k_{11} = 137612044, k_{12} = 505576452, k_{13} = 157170996, k_{14} = 131220418$. (Add these to the ten files already stored.)

## 4.4 Computer Projects

Write programs to assign memory locations to student files, using the hashing function $h(k) \equiv k \pmod{1021}$, $0 \leqslant h(k) < 1021$, where the keys are the social security numbers of students.

**1.** Linking files together when collisions occur.

**2.** Using $h_j(k) \equiv h(k) + j \pmod{1021}$, $j = 0, 1, 2,...$ as the probing sequence.

**3.** Using $h_j(k) \equiv h(k) + j \cdot g(k)$, $j = 0, 1, 2,...$ where $g(k) \equiv 1 + k \pmod{1019}$ as the probing sequence.

# 5

# Some Special Congruences

## 5.1 Wilson's Theorem and Fermat's Little Theorem

In this section we discuss two important congruences that are often useful in number theory. We first discuss a congruence for factorials called Wilson's theorem.

**Theorem 5.1. Wilson's Theorem.** If $p$ is prime, then $(p-1)! \equiv -1 \pmod{p}$.

The first proof of Wilson's Theorem was given by the French mathematician Joseph Lagrange in 1770*. The mathematician after whom the theorem is named, John Wilson, conjectured, but did not prove it. Before proving Wilson's theorem, we use an example to illustrate the idea behind the proof.

---

* **JOSEPH LOUIS LAGRANGE** (1736-1813) was born in Italy and studied physics and mathematics at the University of Turin. Although he originally planned to pursue a career in physics, Lagrange growing interest in mathematics led him to change course. At the age he was appointed as a mathematics professor at the Royal Artillery School in Turin. In 1766 he filled the post Euler vacated at the Royal Academy of Berlin when Frederick the Great sought him out. Lagrange directed the mathematics section of the Royal Academy for 20 years. In 1787, when his patron Frederick the Great died, Lagrange moved to France at the invitation of Louis XVI, to join the French Academy. In France he had a distinguished career in teaching and writing. He was a favorite of Marie Antoinette, but managed to win the favor of the new regime that came into power after the French Revolution. Langrange's contributions to mathematics include unifying the mathematical theory of mechanics. He made fundamental discoveries in group theory and helped put calculus on a rigorous foundation. His contributions to number theory include the first proof of Wilson's theorem and the result that every positive integer can be written as the sum of four squares.

**Example 5.1.** Let $p=7$. We have $(7-1)! = 6! = 1 \cdot 2 \cdot 3 \cdot 4 \cdot 5 \cdot 6$. We will rearrange the factors in the product, grouping together pairs of inverses modulo 7. We note that $2 \cdot 4 \equiv 1 \pmod 7$ and $3 \cdot 5 \equiv 1 \pmod 7$. Hence, $6! \equiv 1 \cdot (2 \cdot 4) \cdot (3 \cdot 5) \cdot 6 \equiv 1 \cdot 6 \equiv -1 \pmod 7$. Thus, we have verified a special case of Wilson's theorem. □

We now use the technique illustrated in the example to prove Wilson's theorem.

*Proof.* When $p=2$ we have $(p-1)! \equiv 1 \equiv -1 \pmod 2$. Hence, the theorem is true for $p=2$. Now, let $p$ be a prime greater than 2. Using Theorem 3.10, for each integer $a$ with $1 \leqslant a \leqslant p-1$, there is an inverse $\bar{a}$, $1 \leqslant \bar{a} \leqslant p-1$, with $a\bar{a} \equiv 1 \pmod p$. By Theorem 3.11 the only positive integers less than $p$ that are their own inverses are 1 and $p-1$. Therefore, we can group the integers from 2 to $p-2$ into $(p-3)/2$ pairs of integers, with the product of each pair congruent to 1 modulo $p$. Hence, we have

$$2 \cdot 3 \ \cdots \ (p-3) \cdot (p-2) \equiv 1 \pmod p.$$

We conclude the proof by multiplying both sides of the above congruence by 1 and $p-1$ to obtain

$$(p-1)! = 1 \cdot 2 \cdot 3 \cdots (p-3)(p-2)(p-1) \equiv 1 \cdot (p-1) \equiv -1 \pmod p. \ \square$$

An interesting observation is that the converse of Wilson's theorem is also true, as the following theorem shows.

**Theorem 5.2.** If $n$ is a positive integer such that $(n-1)! \equiv -1 \pmod n$, then $n$ is prime.

*Proof.* Assume that $n$ is a composite integer and that $(n-1)! \equiv -1 \pmod n$. Since $n$ is composite, we have $n=ab$, where $1 < a < n$ and $1 < b < n$. Since $a < n$, we know that $a \mid (n-1)!$, because $a$ is one of the $n-1$ numbers multiplied together to form $(n-1)!$. Since $(n-1)! \equiv -1 \pmod n$ it follows that $n \mid [(n-1)! + 1]$. This means, by Theorem 1.7, that $a$ also divides $(n-1)! + 1$. By Theorem 1.8 since $a \mid (n-1)!$ and $a \mid [(n-1)! + 1]$, we conclude that $a \mid [(n-1)! + 1] - (n-1)! = 1$. This is an obvious contradiction, since $a > 1$. □

We illustrate the use of this result with an example.

**Example 5.2.** Since $(6-1)! = 5! = 120 \equiv 0 \pmod 6$, Theorem 5.2 verifies the obvious fact that 6 is not prime. □

As we can see, the converse of Wilson's theorem gives us a primality test. To decide whether an integer $n$ is prime, we determine whether $(n-1)! \equiv -1 \pmod n$. Unfortunately, this is an impractical test because $n - 1$ multiplications modulo $n$ are needed to find $(n-1)!$ requiring $O(n(\log_2 n)^2)$ bit operations.

When working with congruences involving exponents, the following theorem* is of great importance.

**Theorem 5.3. Fermat's Little Theorem.**  If $p$ is prime and $a$ is a positive integer with $p \nmid a$, then $a^{p-1} \equiv 1 \pmod{p}$.

*Proof.* Consider the $p-1$ integers $a$, $2a$, ..., $(p-1)a$. None of these integers are divisible by $p$, for if $p \mid ja$, then by Lemma 2.3, $p \mid j$, since $p \nmid a$. This is impossible because $1 \leqslant j \leqslant p-1$. Furthermore, no two of the integers $a$, $2a$, ..., $(p-1)a$ are congruent modulo $p$. To see this, assume that $ja \equiv ka \pmod{p}$. Then, by Corollary 3.4.1, since $(a,p) = 1$, we have $j \equiv k \pmod{p}$. This is impossible, since $j$ and $k$ are positive integers less than $p-1$.

Since the integers $a$, $2a$, ..., $(p-1)a$ are a set of $p-1$ integers all incongruent to zero, and no two congruent modulo $p$, we know that the least positive residues of $a$, $2a$, ..., $(p-1)a$, taken in some order, must be the integers $1$, $2$, ..., $p-1$. As a consequence, the product of the integers $a$, $2a$, ..., $(p-1)a$ is congruent modulo $p$ to the product of the first $p-1$ positive integers. Hence,

$$a \cdot 2a \ \cdots \ (p-1)a \ \equiv 1 \cdot 2 \ \cdots \ (p-1) \pmod{p}.$$

Therefore,

$$a^{p-1}(p-1)! \equiv (p-1)! \pmod{p}.$$

Since $((p-1)!, p) = 1$, using Corollary 3.4.1, we cancel $(p-1)!$ to obtain

$$a^{p-1} \equiv 1 \pmod{p}. \qquad \square$$

We illustrate the ideas of the proof with an example.

**Example 5.3.**  Let $p=7$ and $a=3$. Then, $1 \cdot 3 \equiv 3 \pmod{7}$, $2 \cdot 3 \equiv 6 \pmod{7}$, $3 \cdot 3 \equiv 2 \pmod{7}$, $4 \cdot 3 \equiv 5 \pmod{7}$, $5 \cdot 3 \equiv 1 \pmod{7}$, and $6 \cdot 3 \equiv 4 \pmod{7}$. Consequently,

$$(1 \cdot 3) \cdot (2 \cdot 3) \cdot (3 \cdot 3) \cdot (4 \cdot 3) \cdot (5 \cdot 3) \cdot (6 \cdot 3) \equiv 3 \cdot 6 \cdot 2 \cdot 5 \cdot 1 \cdot 4 \pmod{7},$$

so that $3^6 \cdot 1 \cdot 2 \cdot 3 \cdot 4 \cdot 5 \cdot 6 \equiv 3 \cdot 6 \cdot 2 \cdot 5 \cdot 1 \cdot 4 \pmod{7}$. Hence, $3^6 \cdot 6! \equiv 6! \pmod{7}$, and therefore $3^6 \equiv 1 \pmod{7}$. $\qquad \square$

---

\* This is called Fermat's Little Theorem to distinguish it from the famous conjecture known as Fermat's Last Theorem which we discuss in Chapter 11.

On occasion, we would like to have a congruence like Fermat's little theorem that holds for all integers $a$, given the prime $p$. This is supplied by the following result.

**Theorem 5.4.** If $p$ is prime and $a$ is a positive integer, then $a^p \equiv a \pmod{p}$.

*Proof.* If $p \nmid a$, by Fermat's little theorem we know that $a^{p-1} \equiv 1 \pmod{p}$. Multiplying both sides of this congruence by $a$, we find that $a^p \equiv a \pmod{p}$. If $p \mid a$, then $p \mid a^p$ as well, so that $a^p \equiv a \equiv 0 \pmod{p}$. This finishes the proof, since $a^p \equiv a \pmod{p}$ if $p \nmid a$ and if $p \mid a$. □

Fermat's little theorem is useful in finding the least positive residues of powers.

**Example 5.4.** We can find the least positive residue of $3^{201}$ modulo 11 with the help of Fermat's little theorem. We know that $3^{10} \equiv 1 \pmod{11}$. Hence, $3^{201} = (3^{10})^{20} \cdot 3 \equiv 3 \pmod{11}$ . □

A useful application of Fermat's little theorem is provided by the following result.

**Theorem 5.5.** If $p$ is prime and $a$ is an integer with $p \nmid a$, then $a^{p-2}$ is an inverse of $a$ modulo $p$.

*Proof.* If $p \nmid a$, then Fermat's little theorem tells us that $a \cdot a^{p-2} = a^{p-1} \equiv 1 \pmod{p}$. Hence, $a^{p-2}$ is an inverse of $a$ modulo $p$. □

**Example 5.5.** By Theorem 5.5 we know that $2^9 = 512 \equiv 6 \pmod{11}$ is an inverse of 2 modulo 11. □

Theorem 5.5 gives us another way to solve linear congruences with respect to prime moduli.

**Corollary 5.5.1.** If $a$ and $b$ are positive integers and $p$ is prime with $p \nmid a$, then the solutions of the linear congruence $ax \equiv b \pmod{p}$ are the integers $x$ such that $x \equiv a^{p-2}b \pmod{p}$.

*Proof.* Suppose that $ax \equiv b \pmod{p}$. Since $p \nmid a$, we know from Theorem 5.5 that $a^{p-2}$ is an inverse of $a \pmod{p}$. Multiplying both sides of the original congruence by $a^{p-2}$, we have

$$a^{p-2}ax \equiv a^{p-2}b \pmod{p}.$$

Hence,

$$x \equiv a^{p-2}b \pmod{p}. \qquad \blacksquare$$

## 5.1  Exercises

1. Using Wilson's theorem, find the least positive residue of $8 \cdot 9 \cdot 10 \cdot 11 \cdot 12 \cdot 13$ modulo 7.

2. Using Fermat's little theorem, find the least positive residue of $2^{1000000}$ modulo 17.

3. Show that $3^{10} \equiv 1 \pmod{11^2}$.

4. Using Fermat's little theorem, find the last digit of the base 7 expansion of $3^{100}$.

5. Using Fermat's little theorem, find the solutions of the linear congruences

   a)   $7x \equiv 12 \pmod{17}$   b)   $4x \equiv 11 \pmod{19}$.

6. Show that if $n$ is a composite integer with $n \neq 4$, then $(n-1)! \equiv 0 \pmod{n}$.

7. Show that if $p$ is an odd prime, then $2(p-3)! \equiv -1 \pmod{p}$.

8. Show that if $n$ is odd and $3 \nmid n$, then $n^2 \equiv 1 \pmod{24}$.

9. Show that $42 \mid (n^7 - n)$ for all positive integers n.

10. Show that if $p$ and $q$ are distinct primes, then $p^{q-1} + q^{p-1} \equiv 1 \pmod{pq}$.

11. Show that $p$ is prime and $a$ and $b$ are integers such that $a^p \equiv b^p \pmod{p}$, then $a^p \equiv b^p \pmod{p^2}$.

12. Show that if $p$ is an odd prime, then $1^2 3^2 \cdots (p-4)^2 (p-2)^2 \equiv (-1)^{(p+1)/2} \pmod{p}$.

13. Show that if $p$ is prime and $p \equiv 3 \pmod{4}$, then $\{(p-1)/2\}! \equiv \pm 1 \pmod{p}$.

14. a) Let $p$ be prime and suppose that $r$ is a positive integer less then $p$ such that $(-1)^r r! \equiv -1 \pmod{p}$. Show that $(p-r+1)! \equiv -1 \pmod{p}$.

    b) Using part (a), show that $61! \equiv 63! \equiv -1 \pmod{71}$.

15. Using Wilson's theorem, show that if $p$ is a prime and $p \equiv 1 \pmod{4}$, then the congruence $x^2 \equiv -1 \pmod{p}$ has two incongruent solutions given by $x \equiv \pm [(p-1)/2]! \pmod{p}$.

16. Show that if $p$ is a prime and $0 < k < p$, then $(p-k)!(k-1)! \equiv (-1)^k \pmod{p}$.

17. Show that if $p$ is prime and $a$ is an integer, then $p \mid [a^p + (p-1)! \, a]$.

\* 18. For which positive integers $n$ is $n^4 + 4^n$ prime?

19. Show that the pair of positive integers $n$ and $n+2$ are twin primes if and only if $4[(n-1)! + 1] + n \equiv 0 \pmod{n(n+2)}$, where $n \neq 1$.

20. Show that the positive integers $n$ and $n + k$, where $n > k$ and $k$ is an even positive integer, are both prime if and only if $(k!)^2[(n-1)! + 1] + n(k! - 1)(k - 1)! \equiv 0 \pmod{n(n + k)}$.

21. Show that if $p$ is prime, then $\binom{2p}{p} \equiv 2 \pmod{p}$.

22. In Exercise 21 of Section 1.9, we showed that the binomial coefficient $\binom{p}{k}$, where $1 \leqslant k \leqslant p - 1$, is divisible by $p$ when $p$ is prime. Use this fact and the binomial theorem to show that if $a$ and $b$ are integers, then $(a + b)^p \equiv a^p + b^p \pmod{p}$.

23. Use Exercise 22 to prove Fermat's little theorem by mathematical induction. (*Hint:* In the induction step, use part (a) to obtain a congruence for $(a + 1)^p$.)

\* 24. Using Exercise 24 of Section 3.3, prove *Gauss' generalization of Wilson's theorem*, namely that the product of all the positive integers less than $m$ that are relatively prime to $m$ is congruent to 1 (mod $m$), unless $m = 4, p^t$, or $2p^t$ where $p$ is an odd prime and $t$ is a positive integer, in which case, it is congruent to $-1$ (mod $m$).

25. A deck of cards is shuffled by cutting the deck into two piles of 26 cards. Then, the new deck is formed by alternating cards from the two piles, starting with the bottom pile.

    a) Show that if a card begins in the $c$th position in the deck, it will be in the $b$th position in the new deck where $b \equiv 2c \pmod{53}$ and $1 \leqslant b \leqslant 52$.

    b) Determine the number of shuffles of the type described above that are needed to return the deck of cards to its original order.

26. Let $p$ be prime and let $a$ be a positive integer not divisible by $p$. We define the *Fermat quotient* $q_p(a)$ by $q_p(a) = (a^{p-1} - 1)/p$. Show that if $a$ and $b$ are positive integers not divisible by the prime $p$, then $q_p(ab) \equiv q_p(a) + q_p(b) \pmod{p}$.

27. Let $p$ be prime and let $a_1, a_2, ..., a_p$ and $b_1, b_2, ..., b_p$ be complete systems of residues modulo $p$. Show that $a_1b_1, a_2b_2, ..., a_pb_p$ is not a complete system of residues modulo $p$.

\* 28. Show that if $n$ is a positive integer with $n \geqslant 2$ then $n$ does not divide $2^n - 1$.

## 5.1 Computer Projects

Write programs to do the following:

1. Find all Wilson primes less than 10000. A *Wilson prime* is a prime $p$ for which $(p - 1)! \equiv -1 \pmod{p^2}$.

2. Find the primes $p$ less than 10000 for which $2^{p-1} \equiv 1 \pmod{p^2}$.

**3.** Solve linear congruences with prime moduli via Fermat's little theorem.

---

## 5.2 Pseudoprimes

Fermat's little theorem tells us that if $n$ is prime and $b$ is any integer, then $b^n \equiv b \pmod{n}$. Consequently, if we can find an integer $b$ such that $b^n \not\equiv b \pmod{n}$, then we know that $n$ is composite.

**Example 5.6.** We can show 63 is not prime by observing that

$$2^{63} = 2^{60} \cdot 2^3 = (2^6)^{10} \cdot 2^3 = 64^{10} 2^3 \equiv 2^3 \equiv 8 \not\equiv 2 \pmod{63}. \qquad \square$$

Using Fermat's little theorem, we can show that an integer is composite. It would be even more useful if it also provided a way to show that an integer is prime. The ancient Chinese believed that if $2^n \equiv 2 \pmod{n}$, then $n$ must be prime. Unfortunately, the converse of Fermat's little theorem is not true, as the following example shows.

**Example 5.7.** Let $n = 341 = 11 \cdot 31$. By Fermat's little theorem we see that $2^{10} \equiv 1 \pmod{11}$, so that $2^{340} = (2^{10})^{34} \equiv 1 \pmod{11}$. Also $2^{340} = (2^5)^{68} \equiv (32)^{68} \equiv 1 \pmod{31}$. Hence, by Theorem 3.1, we have $2^{340} \equiv 1 \pmod{341}$. By multiplying both sides of this congruence by 2, we have $2^{341} \equiv 2 \pmod{341}$, even though 341 is not prime. $\qquad \square$

Examples such as this lead to the following definition.

**Definition.** Let $b$ be a positive integer. If $n$ is a composite positive integer and $b^n \equiv b \pmod{n}$, then $n$ is called a *pseudoprime to the base $b$*.

Note that if $(b,n) = 1$, then the congruence $b^n \equiv b \pmod{n}$ is equivalent to the congruence $b^{n-1} \equiv 1 \pmod{n}$. To see this, note that by Corollary 3.4.1 we can divide both sides of the first congruence by $b$, since $(b,n) = 1$, to obtain the second congruence. By part (iii) of Theorem 3.3 we can multiply both sides of the second congruence by $b$ to obtain the first. We will often use this equivalent condition.

**Example 5.8.** The integers $341 = 11 \cdot 31$, $561 = 3 \cdot 11 \cdot 17$ and $645 = 3 \cdot 5 \cdot 43$ are pseudoprimes to the base 2, since it is easily verified that $2^{340} \equiv 1 \pmod{341}$, $2^{560} \equiv 1 \pmod{561}$, and $2^{644} \equiv 1 \pmod{645}$. $\qquad \square$

If there are relatively few pseudoprimes to the base $b$, then checking to see whether the congruence $b^n \equiv b \pmod{n}$ holds is an effective test; only a small fraction of composite numbers pass this test. In fact, the pseudoprimes to the base $b$ have been shown to be much rarer than prime numbers. In particular, there are 455052512 primes, but only 14884 pseudoprimes to the base 2, less than $10^{10}$. Although pseudoprimes to any given base are rare, there are, nevertheless, infinitely many pseudoprimes to any given base. We will prove this for the base 2. The following lemma is useful in the proof.

**Lemma 5.1.** If $d$ and $n$ are positive integers such that $d$ divides $n$, then $2^d - 1$ divides $2^n - 1$.

*Proof.* Since $d \mid n$, there is a positive integer $t$ with $dt = n$. By setting $x = 2^d$ in the identity $x^t - 1 = (x - 1) \ (x^{t-1} + x^{t-2} + \cdots + 1)$, we find that $2^n - 1 = (2^d - 1) \ (2^{d(t-1)} + 2^{d(t-2)} + \cdots + 2^d + 1)$. Consequently, we see that $(2^d - 1) \mid (2^n - 1)$. ∎

We can now prove that there are infinitely many pseudoprimes to the base 2.

**Theorem 5.6.** There are infinitely many pseudoprimes to the base 2.

*Proof.* We will show that if $n$ is an odd pseudoprime to the base 2, then $m = 2^n - 1$ is also an odd pseudoprime to the base 2. Since we have at least one odd pseudoprime to the base 2, namely $n_0 = 341$, we will be able to construct infinitely many odd pseudoprimes to the base 2 by taking $n_0 = 341$ and $n_{k+1} = 2^{n_k} - 1$ for $k = 0, 1, 2, 3, \dots$. These odd integers are all different, since $n_0 < n_1 < n_2 < \cdots < n_k < n_{k+1} < \cdots$.

To continue the proof, let $n$ be an odd pseudoprime, so that $n$ is composite and $2^{n-1} \equiv 1 \pmod{n}$. Since $n$ is composite, we have $n = dt$ with $1 < d < n$ and $1 < t < n$. We will show that $m = 2^n - 1$ is also pseudoprime by first showing that it is composite, and then by showing that $2^{m-1} \equiv 1 \pmod{m}$.

To see that $m$ is composite we use Lemma 5.1 to note that $(2^d - 1) \mid (2^n - 1) = m$. To show that $2^{m-1} \equiv 1 \pmod{m}$ since $2^n \equiv 2 \pmod{n}$, there is an integer $k$ with $2^n - 2 = kn$. Hence, $2^{m-1} = 2^{2^n-2} = 2^{kn}$. By Lemma 5.1 we have $m = (2^n - 1) \mid (2^{kn} - 1) = 2^{m-1} - 1$. Hence, $2^{m-1} - 1 \equiv 0 \pmod{m}$, so that $2^{m-1} \equiv 1 \pmod{m}$. We conclude that $m$ is also a pseudoprime to the base 2. ∎

If we want to know whether an integer $n$ is prime, and we find that $2^{n-1} \equiv 1 \pmod{n}$, we know that $n$ is either prime or $n$ is a pseudoprime to the base 2. One follow-up approach is to test $n$ with other bases. That is, we check to see whether $b^{n-1} \equiv 1 \pmod{n}$ for various positive integers $b$. If we find any values of $b$ with $(b,n) = 1$ and $b^{n-1} \not\equiv 1 \pmod{n}$, then we know that $n$ is composite.

**Example 5.9.** We have seen that 341 is a pseudoprime to the base 2. Since

$$7^3 = 343 \equiv 2 \pmod{341}$$

and

$$2^{10} = 1024 \equiv 1 \pmod{341} ,$$

we have

$$7^{340} = (7^3)^{113}7 \equiv 2^{113}7 = (2^{10})^{11}\cdot2^3\cdot7$$
$$\equiv 8\cdot7 \equiv 56 \not\equiv 1 \pmod{341}.$$

Hence, we see that 341 is composite, since $7^{340} \not\equiv 1 \pmod{341}$.     □

Unfortunately, there are composite integers $n$ that cannot be shown to be composite using the above approach, because there are integers which are pseudoprimes to every base, that is, there are composite integers $n$ such that $b^{n-1} \equiv 1 \pmod{n}$, for all $b$ with $(b,n) = 1$. This leads to the following definition.

**Definition.** A composite integer which satisfies $b^{n-1} \equiv 1 \pmod{n}$ for all positive integers $b$ with $(b,n) = 1$ is called a *Carmichael number*.

**Example 5.10.** The integer $561 = 3\cdot11\cdot17$ is a Carmichael number. To see this, note that if $(b, 561) = 1$, then $(b, 3) = (b, 11) = (b, 17) = 1$. Hence, from Fermat's little theorem, we have $b^2 \equiv 1 \pmod 3$, $b^{10} \equiv 1 \pmod{11}$, and $b^{16} \equiv 1 \pmod{17}$. Consequently, $b^{560} = (b^2)^{280} \equiv 1 \pmod 3, b^{560} = (b^{10})^{56} \equiv 1 \pmod{11}$, and $b^{560} = (b^{16})^{35} \equiv 1 \pmod{17}$. Therefore, by Theorem 3.8 $b^{560} \equiv 1 \pmod{561}$ for all $b$ with $(b,n) = 1$.     □

It has been conjectured that there are infinitely many Carmichael numbers, but so far this has not been demonstrated. We can prove the following theorem, which provides conditions which produce Carmichael numbers.

**Theorem 5.7.** If $n = q_1 q_2 \cdots q_k$, where the $q_j$'s are distinct primes that satisfy $(q_j - 1) \mid (n - 1)$ for all $j$, then $n$ is a Carmichael number.

*Proof.* Let $b$ be a positive integer with $(b,n) = 1$. Then $(b,q_j) = 1$ for $j = 1, 2, ..., k$, and hence, by Fermat's little theorem, $b^{q_j-1} \equiv 1 \pmod{q_j}$ for $j = 1, 2, ..., k$. Since $(q_j - 1) \mid (n - 1)$ for each integer $j = 1, 2, ..., k$, there are integers $t_j$ with $t_j(q_j - 1) = n - 1$. Hence, for each $j$, we know that $b^{n-1} = b^{(q_j-1)t_j} \equiv 1 \pmod{q_j}$. Therefore, by Corollary 3.8.1 we see that $b^{n-1} \equiv 1 \pmod{n}$, and we conclude that $n$ is a Carmichael number.     ■

**Example 5.11.** Theorem 5.7 shows that $6601 = 7\cdot23\cdot41$ is a Carmichael number, because 7, 23, and 41 are all prime, $6 = (7 - 1) \mid 6600$, $22 = (23 - 1) \mid 6600$, and $40 = (41 - 1) \mid 6600$.

The converse of Theorem 5.7 is also true, that is, all Carmichael numbers are of the form $q_1q_2 \cdots q_k$ where the $q_j$'s are distinct primes and $(q_j-1) \mid (n-1)$ for all $j$. We will prove this fact in Chapter 8.

Once the congruence $b^{n-1} \equiv 1 \pmod{n}$ has been verified, another possible approach is to consider the least positive residue of $b^{(n-1)/2}$ modulo $n$. We note that if $x = b^{(n-1)/2}$, then $x^2 = b^{n-1} \equiv 1 \pmod{n}$. If $n$ is prime, by Theorem 3.11 we know that either $x \equiv 1$ or $x \equiv -1 \pmod{n}$. Consequently, once we have found that $b^{n-1} \equiv 1 \pmod{n}$, we can check to see whether $b^{(n-1)/2} \equiv \pm 1 \pmod{n}$. If this congruence does not hold, then we know that $n$ is composite.

**Example 5.12.** Let $b = 5$ and let $n = 561$, the smallest Carmichael number. We find that $5^{(561-1)/2} = 5^{280} \equiv 67 \pmod{561}$. Hence, 561 is composite.   $\square$

We continue developing primality tests with the following definitions.

**Definition.** Let $n$ be a positive integer with $n-1 = 2^s t$, where $s$ is a nonnegative integer and $t$ is an odd positive integer. We say that $n$ passes *Miller's test* for *the base $b$* if either $b^t \equiv 1 \pmod{n}$ or $b^{2^j t} \equiv -1 \pmod{n}$ for some $j$ with $0 \leqslant j \leqslant s -1$.

We now show that if $n$ is prime, then $n$ passes Miller's test for all bases $b$ with $n \nmid b$.

**Theorem 5.8.** If $n$ is prime and $b$ is a positive integer with $n \nmid b$, then $n$ passes Miller's test for the base $b$.

*Proof.* Let $n-1 = 2^s t$, where $s$ is a nonnegative integer and $t$ is an odd positive integer. Let $x_k = b^{(n-1)/2^k} = b^{2^{s-k} t}$, for $k = 0, 1, 2, ..., s$. Since $n$ is prime, Fermat's little theorem tells us that $x_0 = b^{n-1} \equiv 1 \pmod{n}$. By Theorem 3.11, since $x_1^2 = (b^{(n-1)/2})^2 = x_0 \equiv 1 \pmod{n}$, either $x_1 \equiv -1 \pmod{n}$ or $x_1 \equiv 1 \pmod{n}$. If $x_1 \equiv 1 \pmod{n}$, since $x_2^2 = x_1 \equiv 1 \pmod{n}$, either $x_2 \equiv -1 \pmod{n}$ or $x_2 \equiv 1 \pmod{n}$. In general, if we have found that $x_0 \equiv x_1 \equiv x_2 \equiv \cdots \equiv x_k \equiv 1 \pmod{n}$, with $k < s$, then, since $x_{k+1}^2 = x_k \equiv 1 \pmod{n}$, we know that either $x_{k+1} \equiv -1 \pmod{n}$ or $x_{k+1} \equiv 1 \pmod{n}$.

Continuing this procedure for $k = 1, 2,..., s$, we find that either $x_k \equiv 1 \pmod{n}$, for $k = 0, 1, ..., s$, or $x_k \equiv -1 \pmod{n}$ for some integer $k$. Hence, $n$ passes Miller's test for the base $b$.   ∎

If the positive integer $n$ passes Miller's test for the base $b$, then either $b^t \equiv 1 \pmod{n}$ or $b^{2^j t} \equiv -1 \pmod{n}$ for some $j$ with $0 \leqslant j \leqslant s -1$, where $n - 1 = 2^s t$ and $t$ is odd.

In either case, we have $b^{n-1} \equiv 1 \pmod{n}$, since $b^{n-1} = (b^{2^j t})^{2^{s-j}}$ for $j = 0, 1, 2, ..., s$, so that an integer $n$ that passes Miller's test for the base $b$ is automatically a pseudoprime to the base $b$. With this observation, we are led to the following definition.

**Definition**. If $n$ is composite and passes Miller's test for the base $b$, then we say $n$ is a *strong pseudoprime to the base b*.

**Example 5.13**. Let $n = 2047 = 23 \cdot 89$. Then $2^{2046} = (2^{11})^{186} = (2048)^{186} \equiv 1$ (mod 2047), so that 2047 is a pseudoprime to the base 2. Since $2^{2046/2} = 2^{1023} = (2^{11})^{93} = (2048)^{93} \equiv 1$ (mod 2047), 2047 passes Miller's test for the base 2. Hence, 2047 is a strong pseudoprime to the base 2.                                    $\square$

Although strong pseudoprimes are exceedingly rare, there are still infinitely many of them. We demonstrate this for the base 2 with the following theorem.

**Theorem 5.9**. There are infinitely many strong pseudoprimes to the base 2.

*Proof.* We shall show that if $n$ is a pseudoprime to the base 2, then $N = 2^n - 1$ is a *strong* pseudoprime to the base 2.

Let $n$ be an odd integer which is a pseudoprime to the base 2. Hence, $n$ is composite, and $2^{n-1} \equiv 1$ (mod $n$). From this congruence, we see that $2^{n-1} - 1 = nk$ for some integer $k$; furthermore, $k$ must be odd. We have

$$N - 1 = 2^n - 2 = 2(2^{n-1} - 1) = 2^1 nk;$$

this is the factorization of $N-1$ into an odd integer and a power of 2.

We now note that

$$2^{(N-1)/2} = 2^{nk} = (2^n)^k \equiv 1 \pmod{N}$$

because $2^n = (2^n - 1) + 1 = N + 1 \equiv 1$ (mod $N$). This demonstrates that $N$ passes Miller's test.

In the proof of Lemma 5.1 we showed that if $n$ is composite, then $N = 2^n - 1$ also is composite. Hence, $N$ passes Miller's Test and is composite, so that $N$ is a strong pseudoprime to the base 2. Since every pseudoprime $n$ to the base 2 yields a strong pseudoprime $2^n - 1$ to the base 2 and since there are infinitely many pseudoprimes to the base 2, we conclude that there are infinitely many strong pseudoprimes to the base 2.                                    ∎

The following observations are useful in combination with Miller's test for checking the primality of relatively small integers. The smallest odd strong pseudoprime to the base 2 is 2047, so that if $n < 2047$, $n$ is odd, and $n$ passes Miller's test to the base 2, then $n$ is prime. Likewise, 1373653 is the smallest odd strong pseudoprime to both the bases 2 and 3, giving us a primality test for integers less than 1373653. The smallest odd strong pseudoprime to the bases 2, 3, and 5 is 25326001, and the smallest odd strong pseudoprime to all the bases 2, 3, 5, and 7 is 3215031751. Also, less than $25 \cdot 10^9$ the only odd integer which is a pseudoprime to all the bases 2, 3, 5, and 7 is 3251031751. This leads us to a primality test for integers less than $25 \cdot 10^9$. An odd integer $n$ is prime if $n < 25 \cdot 10^9$, $n$ passes Miller's test for the bases 2, 3, 5, and 7, and $n \neq 3215031751$.

There is no analogy of a Carmichael number for strong pseudoprimes. This is a consequence of the following theorem.

**Theorem 5.10.** If $n$ is an odd composite positive integer, then $n$ passes Miller's test for at most $(n-1)/4$ bases $b$ with $1 \leqslant b \leqslant n - 1$.

We prove Theorem 5.10 in Chapter 8. Note that Theorem 5.10 tells us that if $n$ passes Miller's tests for more than $(n-1)/4$ bases less than $n$, then $n$ must be prime. However, this is a rather lengthy way, worse than performing trial divisions, to show that a positive integer $n$ is prime. Miller's test does give an interesting and quick way of showing an integer $n$ is "probably prime". To see this, take at random an integer $b$ with $1 \leqslant b \leqslant n - 1$ (we will see how to make this "random" choice in Chapter 8). From Theorem 5.10, we see that if $n$ is composite the probability that $n$ passes Miller's test for the base $b$ is less than $1/4$. If we pick $k$ different bases less than $n$ and perform Miller's tests for each of these bases we are led to the following result.

**Theorem 5.11. Rabin's Probabilistic Primality Test.** Let $n$ be a positive integer. Pick $k$ different positive integers less than $n$ and perform Miller's test on $n$ for each of these bases. If $n$ is composite the probability that $n$ passes all $k$ tests is less than $(1/4)^k$.

Let $n$ be a composite positive integer. Using Rabin's probabilistic primality test, if we pick 100 different integers at random between 1 and $n$ and perform Miller's test for each of these 100 bases, then the probability than $n$ passes all the tests is less than $10^{-60}$, an extremely small number. In fact, it may be more likely that a computer error was made than that a composite integer passes all the 100 tests. Using Rabin's primality test does not definitely prove that an integer $n$ that passes all 100 tests is prime, but does give extremely strong, indeed almost overwhelming, evidence that the integer is prime.

There is a famous conjecture in analytic number theory called the *generalized Riemann hypothesis*. A consequence of this hypothesis is the following conjecture.

**Conjecture 5.1.** For every composite positive integer $n$, there is a base $b$ with $b < 70 \, (\log_2 n)^2$, such that $n$ fails Miller's test for the base $b$.

If this conjecture is true, as many number theorists believe, the following result provides a rapid primality test.

**Theorem 5.12.** If the generalized Riemann hypothesis is valid, then there is an algorithm to determine whether a positive integer $n$ is prime using $O((\log_2 n)^5)$ bit operations.

*Proof.* Let $b$ be a positive integer less than $n$. To perform Miller's test for the base $b$ on $n$ takes $O((\log_2 n)^3)$ bit operations, because this test requires that we perform no more than $\log_2 n$ modular exponentiations, each using $O((\log_2 b)^2)$ bit operations. Assume that the generalized Riemann hypothesis is true. If $n$ is composite, then by Conjective 5.1, there is a base $b$ with $1 < b < 70 (\log_2 n)^2$ such that $n$ fails Miller's test for $b$. To discover this $b$ requires less than $O((\log_2 n)^3) \cdot O((\log_2 n)^2) = O((\log_2 n)^5)$ bit operations. Hence, $O((\log_2 n)^5)$ bit operations we can determine whether $n$ is composite or prime.          ∎

The important point about Rabin's probabilistic primality test and Theorem 5.12 is that both results indicate that it is possible to check an integer $n$ for primality using only $O((\log_2 n)^k)$ bit operations, where $k$ is a positive integer. This contrasts strongly with the problem of factoring. We have seen that the best algorithm known for factoring an integer requires a number of bit operations exponential in the square root of the logarithm of the number of bits in the integer being factored, while primality testing seems to require only a number of bit operations less than a polynomial in the number bits of the integer tested. We capitalize on this difference by presenting a recently invented cipher system in Chapter 7.

---

## 5.2 Exercises

1.  Show that 91 is a pseudoprime to the base 3.

2.  Show that 45 is a pseudoprime to the bases 17 and 19.

3.  Show that the even integer $n = 161038 = 2 \cdot 73 \cdot 1103$ satisfies the congruence $2^n \equiv 2 \pmod{n}$. The integer 161038 is the smallest even pseudoprime to the base 2.

4.  Show that every odd composite integer is a pseudoprime to both the base 1 and the base $-1$.

5.  Show that if $n$ is an odd composite integer and $n$ is a pseudoprime to the base $a$, then $n$ is a pseudoprime to the base $n - a$.

\* 6.  Show that if $n = (a^{2p} - 1)/(a^2 - 1)$, where $a$ is an integer, $a > 1$, and $p$ is an odd prime not dividing $a(a^2 - 1)$, then $n$ is a pseudoprime to the base $a$. Conclude that there are infinitely many pseudoprimes to any base $a$. (*Hint*: To establish that $a^{n-1} \equiv 1 \pmod{n}$, show that $2p \mid (n - 1)$, and demonstrate that $a^{2p} \equiv 2 \pmod{n}$.)

7.  Show that every composite Fermat number $F_m = 2^{2^m} + 1$ is a pseudoprime to the base 2.

8. Show that if $p$ is prime and the Mersenne number $M_p = 2^p - 1$ is composite, then $M_p$ is a pseudoprime to the base 2.

9. Show that if $n$ is a pseudoprime to the bases $a$ and $b$, then $n$ is also a pseudoprime to the base $ab$.

10. Show that if $n$ is a pseudoprime to the base $a$, then $n$ is a pseudoprime to the base $\bar{a}$, where $\bar{a}$ is an inverse of $a$ modulo $n$.

11. a) Show that if $n$ is a pseudoprime to the base $a$, but not a pseudoprime to the base $b$, then $n$ is not a pseudoprime to the base $ab$.

    b) Show that if there is an integer $b$ with $(b,n) = 1$ such that $n$ is not a pseudoprime to the base $b$, then $n$ is a pseudoprime to less than or equal to $\phi(n)$ different bases $a$ with $1 \leqslant a < n$. (*Hint*: Show that the sets $a_1, a_2, ..., a_r$ and $ba_1, ba_2, ..., ba_r$ have no common elements, where $a_1, a_2, ..., a_r$ are the bases less than $n$ to which $n$ is a pseudoprime.)

12. Show that 25 is a strong pseudoprime to the base 7.

13. Show that 1387 is a pseudoprime, but not a strong pseudoprime to the base 2.

14. Show that 1373653 is a strong pseudoprime to both bases 2 and 3.

15. Show that 25326001 is a strong pseudoprime to bases 2, 3, and 5.

16. Show that the following integers are Carmichael numbers

    a) $2821 = 7 \cdot 13 \cdot 31$
    b) $10585 = 5 \cdot 29 \cdot 73$
    c) $29341 = 13 \cdot 37 \cdot 61$
    d) $314821 = 13 \cdot 61 \cdot 397$
    e) $278545 = 5 \cdot 17 \cdot 29 \cdot 113$
    f) $172081 = 7 \cdot 13 \cdot 31 \cdot 61$
    g) $564651361 = 43 \cdot 3361 \cdot 3907$.

17. Find a Carmichael number of the form $7 \cdot 23 \cdot q$ where $q$ is an odd prime.

18. a) Show that every integer of the form $(6m+1)(12m+1)(18m+1)$, where $m$ is a positive integer such that $6m+1$, $12m+1$, and $18m+1$ are all primes, is a Carmichael number.

    b) Conclude from part (a) that $1729 = 7 \cdot 13 \cdot 19$, $294409 = 37 \cdot 73 \cdot 109$, $56052361 = 211 \cdot 421 \cdot 631$, $118901521 = 271 \cdot 541 \cdot 811$, and $172947529 = 307 \cdot 613 \cdot 919$ are Carmichael numbers.

19. Show that if $n$ is a positive integer with $n \equiv 3 \pmod 4$, then Miller's test takes $O((\log_2 n)^3)$ bit operations.

## 5.2 Computer Projects

Write programs to do the following:

**1.** Given a positive integer $n$, determine whether $n$ satisfies the congruence $b^{n-1} \equiv 1 \pmod{n}$ where $b$ is a positive integer less than $n$; if it does, then $n$ is either a prime or a pseudoprime to the base $b$.

**2.** Given a positive integer integer $n$, determine whether n passes Miller's test to the base $b$; if it does then $n$ is either prime or a strong pseudoprime to the base $b$.

**3.** Perform a primality test for integers less than $25 \cdot 10^9$ based on Miller's tests for the bases 2, 3, 5, and 7. (Use the remarks that follow Theorem 5.9.)

**4.** Perform Rabin's probabilistic primality test.

**5.** Find Carmichael numbers.

---

## 5.3 Euler's Theorem

Fermat's little theorem tells us how to work with certain congruences involving exponents when the modulus is a prime. How do we work with the corresponding congruences modulo a composite integer? For this purpose, we first define a special counting function.

**Definition.** Let $n$ be a positive integer. The *Euler $\phi$-function* $\phi(n)$ is defined to be the number of positive integers not exceeding $n$ which are relatively prime to $n$.

This function is named after the great Swiss mathematician Leonhard Euler* who introduced this function and proved the results of this section.

---

* **LEONHARD EULER** (1707-1783) was the son of minister from the vicinity of Basel, Switzerland, who, besides theology, had also studied mathematics. At 13 Euler entered the University of Basel with the aim of pursuing a career in theology, as his father wished. At the university, Euler was tutored in mathematics by Johann Bernoulli, of the famous Bernoulli family of mathematicians, and became friends with Johann's sons Nicklaus and Daniel. His interest in mathematics led him to abandon his plans to follow in his father's footsteps. Euler obtained his master's degree in philosophy at the age of 16. In 1727, Peter the Great invited Euler to join the Imperial Academy in St. Petersburg (now Leningrad), at the insistance of the Nicklaus and Daniel Bernoulli who entered the academy in 1725 when it was founded. Euler spent the years 1727-1741 and 1766-1783 at the Imperial Academy. He spent the interval 1741-1766 at the Royal academy of Berlin. Euler was incredibly prolific. He wrote over 700 books and papers, and he left so much unpublished work that the Imperial Academy did not finish publication of Euler's work for 47 years after his death. During his life, his papers accumulated so rapidly that he kept a pile of papers to be published for the academy. They published the top papers in the pile first, so that later results were published before results they superceded or depended on. Euler was blind for the last 17 years of his life, but had a

In Table 5.1 we display the values of $\phi(n)$ for $1 \leqslant n \leqslant 12$. The values of $\phi(n)$ for $1 \leqslant n \leqslant 100$ are given in Table 2 of the Appendix.

| $n$ | 1 | 2 | 3 | 4 | 5 | 6 | 7 | 8 | 9 | 10 | 11 | 12 |
|---|---|---|---|---|---|---|---|---|---|---|---|---|
| $\phi(n)$ | 1 | 1 | 2 | 2 | 4 | 2 | 6 | 4 | 6 | 4 | 10 | 4 |

**Table 5.1.** The Values of Euler's Phi-function for $1 \leqslant n \leqslant 12$.

In Chapter 6, we study the Euler phi-function further. In this section, we use the phi-function to give an analogue of Fermat's little theorem for composite moduli. To do this, we need to lay some groundwork.

**Definition.**  A *reduced residue system modulo n* is a set of $\phi(n)$ integers such that each element of the set is relatively prime to $n$, and no two different elements of the set are congruent modulo $n$.

**Example 5.14.**  The set 1, 3, 5, 7 is a reduced residue system modulo 8.  The set $-3, -1, 1, 3$ is also such a set.                                    □

We will need the following theorem about reduced residue systems.

**Theorem 5.13.**  If $r_1, r_2, ..., r_{\phi(n)}$ is a reduced residue system modulo $n$, and if $a$ is a positive integer with $(a,n) = 1$, then the set $ar_1, ar_2, ..., ar_{\phi(n)}$ is also a reduced residue system modulo $n$.

*Proof.*  To show that each integer $ar_j$ is relatively prime to $n$, we assume that $(ar_j,n) > 1$. Then, there is a prime divisor $p$ of $(ar_j,n)$. Hence, either $p \mid a$ or $p \mid r_j$. Thus, we either have $p \mid a$ and $p \mid n$, or $p \mid r_j$ and $p \mid n$. However, we cannot have both $p \mid r_j$ and $p \mid n$, since $r_j$ is a member of a reduced residue modulo $n$, and both $p \mid a$ and $p \mid n$ cannot hold since $(a,n) = 1$. Hence, we can conclude that $ar_j$ and $n$ are relatively prime for $j = 1, 2, ..., \phi(n)$.

To demonstrate that no two $ar_j$'s are congruent modulo $n$, we assume that $ar_j \equiv ar_k \pmod{n}$, where $j$ and $k$ are distinct positive integers with $1 \leqslant j \leqslant \phi(n)$ and $1 \leqslant k \leqslant \phi(n)$. Since $(a,n) = 1$, by Corollary 3.4.1 we see

---

fantastic memory, so that his blindness did not deter his mathematical output. He also had 13 children, and was able to continue his research while a child or two bounced on his knees. The publication of the collected works of Euler by the Swiss Society of Natural Science will require more than 75 large volumes.

that $r_j \equiv r_k \pmod{n}$. This is a contradiction, since $r_j$ and $r_k$ come from the original set of reduced residues modulo $n$, so that $r_j \not\equiv r_k \pmod{n}$.   ∎

We illustrate the use of Theorem 5.13 by the following example.

**Example 5.15.** The set 1, 3, 5, 7 is a reduced residue system modulo 8. Since $(3,8) = 1$, from Theorem 5.13, the set $3 \cdot 1 = 3$, $3 \cdot 3 = 9$, $3 \cdot 5 = 15$, $3 \cdot 7 = 21$ is also a reduced residue system modulo 8.   ∎

We now state Euler's theorem.

**Theorem 5.14. Euler's Theorem.** If $m$ is a positive integer and $a$ is an integer with $(a,m) = 1$, then $a^{\phi(m)} \equiv 1 \pmod{m}$.   ∎

Before we prove Euler's theorem, we illustrate the idea behind the proof with an example.

**Example 5.16.** We know that both the sets 1, 3, 5, 7 and $3 \cdot 1$, $3 \cdot 3$, $3 \cdot 5$, $3 \cdot 7$ are reduced residue systems modulo 8. Hence, they have the same least positive residues modulo 8. Therefore,

$$(3 \cdot 1) \cdot (3 \cdot 3) \cdot (3 \cdot 5) \cdot (3 \cdot 7) \equiv 1 \cdot 3 \cdot 5 \cdot 7 \pmod{8},$$

and

$$3^4 \cdot 1 \cdot 3 \cdot 5 \cdot 7 \equiv 1 \cdot 3 \cdot 5 \cdot 7 \pmod{8}.$$

Since $(1 \cdot 3 \cdot 5 \cdot 7, 8) = 1$, we conclude that

$$3^4 = 3^{\phi(8)} \equiv 1 \pmod{8}.$$   □

We now use the ideas illustrated by this example to prove Euler's theorem.

*Proof.* Let $r_1, r_2, ..., r_{\phi(m)}$ denote the reduced residue system made up of the positive integers not exceeding $m$ that are relatively prime to $m$. By Theorem 5.13, since $(a,m) = 1$, the set $ar_1, ar_2, ..., ar_{\phi(m)}$ is also a reduced residue system modulo $m$. Hence, the least positive residues of $ar_1, ar_2, ..., ar_{\phi(m)}$ must be the integers $r_1, r_2, ..., r_{\phi(m)}$ in some order. Consequently, if we multiply together all terms in each of these reduced residue systems, we obtain

$$ar_1 ar_2 \cdots ar_{\phi(m)} \equiv r_1 r_2 \cdots r_{\phi(m)} \pmod{m} .$$

Thus,

$$a^{\phi(m)} r_1 r_2 \cdots r_{\phi(m)} \equiv r_1 r_2 \cdots r_{\phi(m)} \pmod{m}.$$

Since $(r_1 r_2 \cdots r_{\phi(m)}, m) = 1$, from Corollary 3.4.1, we can conclude that $a^{\phi(m)} \equiv 1 \pmod{m}$.   ∎

We can use Euler's Theorem to find inverses modulo $m$. If $a$ and $m$ are relatively prime, we know that

$$a \cdot a^{\phi(m)-1} = a^{\phi(m)} \equiv 1 \pmod{m}.$$

Hence, $a^{\phi(m)-1}$ is an inverse of $a$ modulo $m$.

**Example 5.17.** We know that $2^{\phi(9)-1} = 2^{6-1} = 2^5 = 32 \equiv 5 \pmod{9}$ is an inverse of 2 modulo 9.

We can solve linear congruences using this observation. To solve $ax \equiv b \pmod{m}$, where $(a,m) = 1$, we multiply both sides of this congruence by $a^{\phi(m)-1}$ to obtain

$$a^{\phi(m)-1} ax \equiv a^{\phi(m)-1} b \pmod{m}.$$

Therefore, the solutions are those integers $x$ such that $x \equiv a^{\phi(m)-1} b \pmod{m}$. □

**Example 5.18.** The solutions of $3x \equiv 7 \pmod{10}$ are given by $x \equiv 3^{\phi(10)-1} \cdot 7 \equiv 3^3 \cdot 7 \equiv 9 \pmod{10}$, since $\phi(10) = 4$. □

## 5.3 Exercises

1. Find a reduced residue system modulo

   a)  6      d)  14
   b)  9      e)  16
   c)  10     f)  17.

2. Find a reduced residue system modulo $2^m$, where $m$ is a positive integer.

3. Show that if $c_1, c_2, ..., c_{\phi(m)}$ is a reduced residue system modulo $m$, where $m$ is a positive integer with $m \neq 2$, then $c_1 + c_2 + \cdots + c_{\phi(m)} \equiv 0 \pmod{m}$.

4. Show that if $m$ is a positive integer and $a$ is an integer relatively prime to $m$ and $a - 1$ is also relatively prime to $m$, then $1 + a + a^2 + \cdots + a^{\phi(m)-1} \equiv 0 \pmod{m}$.

5. Use Euler's theorem to find the least positive residue of $3^{100000}$ modulo 35.

6. Show that if a is an integer such that $a$ is not divisble by 3 or such that $a$ is divisible by 9, then $a^7 \equiv a \pmod{63}$.

7. Show that if a is an integer relatively prime to 32760, then $a^{12} \equiv 1 \pmod{32760}$.

8. Show that $a^{\phi(b)} + b^{\phi(a)} \equiv 1 \pmod{ab}$, if $a$ and $b$ are relatively prime positive integers.

9. Solve each of the following linear congruences using Euler's theorem.
   a)  $5x \equiv 3 \pmod{14}$
   b)  $4x \equiv 7 \pmod{15}$
   c)  $3x \equiv 5 \pmod{16}$

10. Show that the solutions to the simultaneous system of congruences

$$x \equiv a_1 \pmod{m_1}$$
$$x \equiv a_2 \pmod{m_2}$$
$$\vdots$$
$$x \equiv a_r \pmod{m_r},$$

where the $m_j$ are pairwise relatively prime, are given by

$$x \equiv a_1 M_1^{\phi(m_1)} + a_2 M_2^{\phi(m_2)} + \cdots + a_r M_r^{\phi(m_r)} \pmod{M},$$

where $M = m_1 m_2 \cdots m_r$ and $M_j = M/m_j$ for $j = 1, 2, ..., r$.

11. Use Euler's theorem to find the last digit in the decimal expansion of $7^{1000}$

12. Use Euler's theorem to find the last digit in the hexadecimal expansion of $5^{1000000}$.

13. Find $\phi(n)$ for the integers $n$ with $13 \leq n \leq 20$.

14. Show every positive integer relatively prime to 10 divides infinitely many repunits (see the preamble to Exercise 5 of Section 4.1). (*Hint*: Note that the $n$-digit repunit $111 \cdots 11 = (10^n-1)/9$.)

15. Show every positive integer relatively prime to $b$ divides infinitely many base $b$ repunits (see the preamble to Exercise 9 of Section 4.1).

* 16. Show that if $m$ is a positive integer, $m > 1$, then $a^m \equiv a^{m-\phi(m)} \pmod{m}$ for all positive integers $a$.

## 5.3 Computer Projects

Write programs to do the following:

1. Solve linear congruences using Euler's theorem.

2. Find the solutions of a system of linear congruences using Euler's theorem and the Chinese remainder theorem (see Exercise 10).

# 6

# Multiplicative Functions

## 6.1 The Euler Phi-function

The Euler-phi function has the property that its value at an integer $n$ is the product of the values of the Euler-phi function at the prime powers that occur in the factorization of $n$. Functions with this property are called multiplicative; such functions arise throughout number theory. In this chapter we will show that the Euler phi-function is multiplicative. From this fact we will derive a formula for its values based on prime factorizations. Later in this chapter we will study other multiplicative functions, including the number of divisors function and the sum of divisors function.

We first present some definitions.

**Definition.** An *arithmetic function* is a function that is defined for all positive integers.

Throughout this chapter, we are interested in arithmetic functions that have a special property.

**Definition.** An arithmetic function $f$ is called *multiplicative* if $f(mn) = f(m)f(n)$ whenever $m$ and $n$ are relatively prime positive integers. It is called *completely multiplicative* if $f(mn) = f(m)f(n)$ for all positive integers $m$ and $n$.

**Example 6.1.** The function $f(n) = 1$ for all $n$ is completely multiplicative, and hence also multiplicative, because $f(mn) = 1$, $f(m) = 1$, and $f(n) = 1$, so that $f(mn) = f(m)f(n)$. Similarly, the function $g(n) = n$ is completely multiplicative, and hence multiplicative, since $g(mn) = mn = g(m)g(n)$. $\square$

If $f$ is a multiplicative function, then we can find a simple formula for $f(n)$ given the prime-power factorization of $n$.

**Theorem 6.1.** If $f$ is a multiplicative function and if $n = p_1^{a_1} p_2^{a_2} \cdots p_s^{a_s}$ is the prime-power factorization of the positive integer $n$, then $f(n) = f(p_1^{a_1}) f(p_2^{a_2}) \cdots f(p_s^{a_s})$.

*Proof.* We will use the fact that $f$ is multiplicative to successively split off prime powers from the factorization of $n$, obtaining factors $f(p_i^{a_i})$ in the evaluation of $f(n)$. First, since $f$ is multiplicative and $(p_1^{a_1}, p_2^{a_2} \cdots p_s^{a_s}) = 1$, we see that $f(n) = f(p_1^{a_1} p_2^{a_2} \cdots p_s^{a_s}) = f(p_1^{a_1} \cdot (p_2^{a_2} \cdots p_s^{a_s})) = f(p_1^{a_1}) f(p_2^{a_2} p_3^{a_3} \cdots p_s^{a_s})$. Since $(p_2^{a_2}, p_3^{a_3} \cdots p_s^{a_s}) = 1$, we know that $f(p_2^{a_2} p_3^{a_3} \cdots p_s^{a_s}) = f(p_2^{a_2}) f(p_3^{a_3} \cdots p_s^{a_s})$, so that $f(n) = f(p_1^{a_1}) f(p_2^{a_2}) f(p_3^{a_3} \cdots p_s^{a_s})$. Continuing in this way, we find that $f(n) = f(p_1^{a_1}) f(p_2^{a_2}) f(p_3^{a_3}) \cdots f(p_s^{a_s})$. ∎

We now return to the Euler phi-function. We first consider its values at primes and then at prime powers.

**Theorem 6.2.** If $p$ is prime, then $\phi(p) = p - 1$. Conversely, if $p$ is a positive integer with $\phi(p) = p - 1$, then $p$ is prime.

*Proof.* If $p$ is prime then every positive integer less than $p$ is relatively prime to $p$. Since there are $p - 1$ such integers, we have $\phi(p) = p - 1$.

Conversely, if $p$ is composite, then $p$ has a divisor $d$ with $1 < d < p$, and, of course, $p$ and $d$ are not relatively prime. Since we know that at least one of the $p - 1$ integers $1, 2, ..., p - 1$, namely $d$, is not relatively prime to $p$, $\phi(p) \leqslant p - 2$. Hence, if $\phi(p) = p - 1$, then $p$ must be prime. ∎

We now find the values of the phi-function at prime powers.

**Theorem 6.3.** Let $p$ be a prime and $a$ a positive integer. Then $\phi(p^a) = p^a - p^{a-1}$.

*Proof.* The positive integers less than $p^a$ that are not relatively prime to $p$ are those integers not exceeding $p^a$ that are divisible by $p$. These are the integers $kp$ where $1 \leqslant k \leqslant p^{a-1}$. Since there are exactly $p^{a-1}$ such integers, there are $p^a - p^{a-1}$ integers less than $p^a$ that are relatively prime to $p^a$. Hence, $\phi(p^a) = p^a - p^{a-1}$. ∎

**Example 6.2.** Using Theorem 6.3, we find that $\phi(5^3) = 5^3 - 5^2 = 100$, $\phi(2^{10}) = 2^{10} - 2^9 = 512$, and $\phi(11^2) = 11^2 - 11 = 110$. □

To find a formula for $\phi(n)$, given the prime factorization of $n$, it suffices to show that $\phi$ is multiplicative. We illustrate the idea behind the proof with the following example.

**Example 6.3.** Let $m = 4$ and $n = 9$, so that $mn = 36$. We list the integers from 1 to 36 in a rectangular chart, as shown in Figure 6.1.

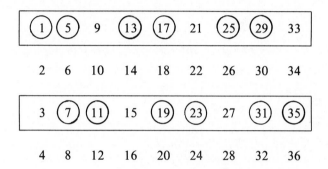

**Figure 6.1.** Demonstating that $\phi(36) = \phi(4)\phi(9)$.

Neither the second nor fourth row contains integers relatively prime to 36, since each element in these rows is not relatively prime to 4, and hence not relatively prime to 36. We enclose the other two rows; each element of these rows is relatively prime to 4. Within each of these rows, there are 6 integers relatively prime to 9. We circle these; they are the 12 integers in the list relatively prime to 36. Hence $\phi(36) = 2\cdot6 = \phi(4)\phi(9)$.                              □

We now state and prove the theorem that shows that $\phi$ is multiplicative.

**Theorem 6.4.** Let $m$ and $n$ be relatively prime positive integers. Then $\phi(mn) = \phi(m)\phi(n)$.

*Proof.* We display the positive integers not exceeding $mn$ in the following way.

$$
\begin{array}{ccccc}
1 & m+1 & 2m+1 & \ldots & (n-1)m+1 \\[8pt]
2 & m+2 & 2m+2 & \ldots & (n-1)m+2 \\[8pt]
3 & m+3 & 2m+3 & \ldots & (n-1)m+3 \\[8pt]
\cdot & \cdot & \cdot & & \cdot \\
\cdot & \cdot & \cdot & & \cdot \\
\cdot & \cdot & \cdot & & \cdot \\[8pt]
r & m+r & 2m+r & \ldots & (n\text{-}1)m+r \\[8pt]
\cdot & \cdot & \cdot & & \cdot \\
\cdot & \cdot & \cdot & & \cdot \\
\cdot & \cdot & \cdot & & \cdot \\[8pt]
m & 2m & 3m & \ldots & mn
\end{array}
$$

Now suppose $r$ is a positive integer not exceeding $m$. Suppose $(m,r) = d > 1$. Then no number in the $r$th row is relatively prime to $mn$, since any element of this row is of the form $km + r$, where $k$ is an integer with $1 \leqslant k \leqslant n - 1$, and $d \mid (km+r)$, since $d \mid m$ and $d \mid r$.

Consequently, to find those integers in the display that are relatively prime to $mn$, we need to look at the $r$th row only if $(m,r) = 1$. If $(m,r) = 1$ and $1 \leqslant r \leqslant m$, we must determine how many integers in this row are relatively prime to $mn$. The elements in this row are $r$, $m + r$, $2m + r$, ..., $(n-1)m + r$. Since $(r,m) = 1$, each of these integers is relatively prime to $m$. By Theorem 3.6 the $n$ integers in the $r$th row form a complete system of residues modulo $n$. Hence, exactly $\phi(n)$ of these integers are relatively prime to $n$. Since these $\phi(n)$ integers are also relatively prime to $m$, they are relatively prime to $mn$.

Since there are $\phi(m)$ rows, each containing $\phi(n)$ integers relatively prime to $mn$, we can conclude that $\phi(mn) = \phi(m)\phi(n)$. ∎

Combining Theorems 6.3 and 6.4, we derive the following formula for $\phi(n)$.

**Theorem 6.5.** Let $n = p_1^{a_1} p_2^{a_2} \cdots p_k^{a_k}$ be the prime-power factorization of the positive integer $n$. Then

$$
\phi(n) = n(1 - \frac{1}{p_1})(1 - \frac{1}{p_2}) \cdots (1 - \frac{1}{p_k}).
$$

*Proof.* Since $\phi$ is multiplicative, Theorem 6.1 tells us that if the prime-power factorization of $n$ is $n = p_1^{a_1} p_2^{a_2} \cdots p_k^{a_k}$, then

$$\phi(n) = \phi(p_1^{a_1})\phi(p_2^{a_2}) \cdots \phi(p_k^{a_k}).$$

In addition, by Theorem 6.3 we know that

$$\phi(p_j^{a_j}) = p_j^{a_j} - p_j^{a_j-1} = p_j^{a_j}(1 - \frac{1}{p_j})$$

for $j = 1,2,...,k$. Hence,

$$\phi(n) = p_1^{a_1}(1 - \frac{1}{p_1})p_2^{a_2}(1 - \frac{1}{p_2}) \cdots p_k^{a_k}(1 - \frac{1}{p_k})$$

$$= p_1^{a_1} p_2^{a_2} \cdots p_k^{a_k}(1 - \frac{1}{p_1})(1 - \frac{1}{p_2}) \cdots (1 - \frac{1}{p_k})$$

$$= n(1 - \frac{1}{p_1})(1 - \frac{1}{p_2}) \cdots (1 - \frac{1}{p_k}).$$

This is the desired formula for $\phi(n)$. ■

We illustrate the use of Theorem 6.5 with the following example.

**Example 6.4**. Using Theorem 6.5 we note that

$$\phi(100) = \phi(2^2 5^2) = 100(1 - \frac{1}{2})(1 - \frac{1}{5}) = 40.$$

and

$$\phi(720) = \phi(2^4 3^2 5) = 720(1 - \frac{1}{2})(1 - \frac{1}{3})(1 - \frac{1}{5}) = 192. \quad \square$$

We now introduce a type of summation notation which is useful in working with multiplicative functions.

Let $f$ be an arithmetic function. Then

$$\sum_{d|n} f(d)$$

represents the sum of the values of $f$ at all the positive divisors of $n$.

**Example 6.5**. If $f$ is an arithmetic function, then

$$\sum_{d|12} f(d) = f(1) + f(2) + f(3) + f(4) + f(6) + f(12).$$

For instance,

$$\sum_{d|12} d^2 = 1^2 + 2^2 + 3^2 + 4^2 + 6^2 + 12^2$$

$$= 1 + 4 + 9 + 16 + 36 + 144 = 210. \quad \square$$

The following result, which states that $n$ is the sum of the values of the phi-function at all the positive divisors of $n$, will also be useful in the sequel.

**Theorem 6.6.** Let $n$ be a positive integer. Then

$$\sum_{d|n} \phi(d) = n.$$

*Proof.* We split the set of integers from 1 to $n$ into classes. Put the integer $m$ into the class $C_d$ if the greatest common divisor of $m$ and $n$ is $d$. We see that $m$ is in $C_d$, *i.e.* $(m,n) = d$, if and only if $(m/d, n/d) = 1$. Hence, the number of integers in $C_d$ is the number of positive integers not exceeding $n/d$ that are relatively prime to the integer $n/d$. From this observation, we see that there are $\phi(n/d)$ integers in $C_d$. Since we divided the integers 1 to $n$ into disjoint classes and each integer is in exactly one class, $n$ is the sum of the numbers of elements in the different classes. Consequently, we see that

$$n = \sum_{d|n} \phi(n/d).$$

As $d$ runs through the positive integers that divide $n$, $n/d$ also runs through these divisors, so that

$$n = \sum_{d|n} \phi(n/d) = \sum_{d|n} \phi(d).$$

This proves the theorem.                                                                                     ∎

**Example 6.6.** We illustrate the proof of Theorem 6.6 when $n = 18$. The integers from 1 to 18 can be split into classes $C_d$ where $d \mid 18$ such that the class $C_d$ contains those integers $m$ with $(m, 18) = d$. We have

$$\begin{aligned}
C_1 &= \{1, 5, 7, 11, 13, 17\} & C_6 &= \{6, 12\} \\
C_2 &= \{2, 4, 8, 10, 14, 16\} & C_9 &= \{9\} \\
C_3 &= \{3, 15\} & C_{18} &= \{18\}.
\end{aligned}$$

We see that the class $C_d$ contains $\phi(18/d)$ integers, as the six classes contain $\phi(18) = 6$, $\phi(9) = 6$, $\phi(6) = 2$, $\phi(3) = 2$, $\phi(2) = 1$, and $\phi(1) = 1$ integers, respectively. We note that $18 = \phi(18) + \phi(9) + \phi(6) + \phi(3) + \phi(2) + \phi(1) = \sum_{d|18} \phi(d)$.                                                                  □

---

## 6.1 Exercises

  1. Determine whether each of the following arithmetic functions is completely multiplicative. Prove your answers.

a) $f(n) = 0$     d) $f(n) = \log n$   g) $f(n) = n+1$
b) $f(n) = 2$     e) $f(n) = n^2$      h) $f(n) = n^n$
c) $f(n) = n/2$   f) $f(n) = n!$       i) $f(n) = \sqrt{n}$

2.  Find the value of the Euler phi-function at each of the following integers.

    a) 100        d) $2 \cdot 3 \cdot 5 \cdot 7 \cdot 11 \cdot 13$
    b) 256        e) 10!
    c) 1001       f) 20!

3.  Show that $\phi(5186) = \phi(5187) = \phi(5188)$.

* 4.  Find all positive integers $n$ such that $\phi(n)$ has the value

    a) 1          d) 6
    b) 2          e) 14
    c) 3          f) 24.

5.  For which positive integers $n$ is $\phi(n)$ odd?

6.  For which positive integers $n$ is $\phi(n)$ divisible by 4?

7.  For which positive integers $n$ is $\phi(n)$ equal to $n/2$?

8.  For which positive integers $n$ does $\phi(n)|n$?

9.  Show that if $n$ is a positive integer, then

$$\phi(2n) = \begin{cases} \phi(n) & \text{if } n \text{ is odd} \\ 2\phi(n) & \text{if } n \text{ is even.} \end{cases}$$

10. Show that if $n$ is a positive integer having $k$ distinct odd prime divisors, then $\phi(n)$ is divisible by $2^k$.

11. For which positive integers $n$ is $\phi(n)$ a power of 2?

12. Show that if $m$ and $k$ are positive integers, then $\phi(m^k) = m^{k-1}\phi(m)$.

13. Show that if $a$ and $b$ are positive integers, then

$$\phi(ab) = (a,b)\phi(a)\phi(b)/\phi((a,b)).$$

14. Show that if $m$ and $n$ are positive integers with $m \mid n$, then $\phi(m) \mid \phi(n)$.

* 15. Prove Theorem 6.5, using the principle of inclusion-exclusion (see Exercise 18 of Section 1.4).

16. Show that a positive integer $n$ is composite if and only if $\phi(n) \leqslant n - \sqrt{n}$.

17. Let $n$ be a positive integer. Define the sequence of positive integers $n_1, n_2, n_3, \ldots$ recursively by $n_1 = \phi(n)$ and $n_{k+1} = \phi(n_k)$ for $k = 1, 2, 3, \ldots$ . Show that there is a positive integer $r$ such that $n_r = 1$.

A multiplicative function is called *strongly multiplicative* if and only if $f(p^k) = f(p)$ for every prime $p$ and every positive integer $k$.

**18.** Show that $f(n) = \phi(n)/n$ is a strongly multiplicative function.

Two arithmetic functions $f$ and $g$ may be multiplied using the *Dirichlet product* which is defined by

$$(f*g)(n) = \sum_{d|n} f(d)g(n/d).$$

**19.** Show that $f*g = g*f$.

**20.** Show that $(f*g)*h = f*(g*h)$ .

**21.** Show that if $\iota$ is the multiplicative function defined by

$$\iota(n) = \begin{cases} 1 & \text{if } n = 1 \\ 0 & \text{if } n > 1, \end{cases}$$

then $\iota*f = f*\iota = f$ for all arithmetic functions $f$.

**22.** The arithmetic function $g$ is said to be the *inverse* of the arithmetic function $f$ if $f*g = g*f = \iota$. Show that the arithmetic function $f$ has an *inverse* if and only if $f(1) \neq 0$. Show that if $f$ has an inverse it is unique. (*Hint*: When $f(1) \neq 0$, find the inverse $f^{-1}$ of $f$ by calculating $f(n)$ recursively, using the fact that $\iota(n) = \sum_{d|n} f(d)f^{-1}(n/d)$.)

**23.** Show that if $f$ and $g$ are multiplicative functions, then the Dirichlet product $f*g$ is also multiplicative.

**24.** Show that the *Möbius function* defined by

$$\mu(n) = \begin{cases} 1 & \text{if } n = 1 \\ (-1)^s & \text{if } n \text{ is square--free with prime factorization} \\ & \qquad n = p_1 p_2 \cdots p_s \\ 0 & \text{if } n \text{ has square factor larger than } 1 \end{cases}$$

is multiplicative.

**25.** Show that if $n$ is a positive integer greater than one, then $\sum_{d|n} \mu(d) = 0$.

**\* 26.** Let $f$ be an arithmetic function. Show that if $F$ is the arithmetic function defined by

$$F(n) = \sum_{d|n} f(d),$$

then

$$f(n) = \sum_{d|n} \mu(d)F(n/d).$$

This result is called the *Möbius inversion formula*.

* **27.** Use the Möbius inversion formula to show that if $f$ is an arithmetic function and $F$ is the arithmetic function defined by

$$F(n) = \sum_{d|n} f(d),$$

then if $F$ is multiplicative, so is $f$.

**28.** Using the Möbius inversion formula and the fact that $n = \sum_{d|n} \phi(n/d)$, prove that

a) $\phi(p^t) = p^t - p^{t-1}$, where $p$ is a prime and $t$ is a positive integer.

b) $\phi(n)$ is multiplicative.

**29.** Show that the function $f(n) = n^k$ is completely multiplicative for every real number $k$.

We define *Liouville's function* $\lambda(n)$ by $\lambda(1) = 1$ and for $n > 1$ by $\lambda(n) = (-1)^{a_1+a_2+\dots+a_m}$, if the prime-power factorization of $n$ is $n = p_1^{a_1} p_2^{a_2} \cdots p_m^{a_m}$.

**30.** Show that $\lambda(n)$ is completely multiplicative.

**31.** Show that if $n$ is a positive integer then $\sum_{d|n} \lambda(n)$ equals 0 if $n$ is not a perfect square, and equals 1 if $n$ is a perfect square.

**32.** Show that if $f$ and $g$ are multiplicative functions then $fg$ is also multiplicative.

**33.** Show that if $f$ and $g$ are completely multiplicative functions then $fg$ is also completely multiplicative.

**34.** Show that if $f$ is completely multiplicative, then $f(n) = f(p_1)^{a_1} f(p_2)^{a_2} \cdots f(p_m)^{a_m}$ when the prime-power factorization of $n$ is $n = p_1^{a_1} p_2^{a_2} \cdots p_m^{a_m}$.

A function $f$ that satisfies the equation $f(mn) = f(m) + f(n)$ for all relatively prime positive integers $m$ and $n$ is called *additive*, and if the above equation holds for all positive integers $m$ and $n$, $f$ is called *completely additive*.

**35.** Show that the function $f(n) = \log n$ is completely additive.

**36.** Show that if $\omega(n)$ is the function that denotes the number of distinct prime factors of $n$, then $\omega$ is additive, but not completely additive.

**37.** Show that if $f$ is an additive function and if $g(n) = 2^{f(n)}$, then $g$ is multiplicative.

## 6.1 Computer Projects

Write programs to do the following:

    **1.**   Find values of the Euler phi-function.

    **2.**   Find the integer $r$ in Exercise 17.

---

## 6.2 The Sum and Number of Divisors

As we mentioned in Section 6.1, the number of divisors function and the sum of divisors function are both multiplicative functions. We will show these functions are multiplicative and derive formulae for their values at a positive integer $n$ from the prime factorization of $n$.

**Definition.** The sum of the divisors function, denoted by $\sigma$, is defined by setting $\sigma(n)$ equal to the sum of all the positive divisors of $n$.

In Table 6.1 we give $\sigma(n)$ for $1 \leqslant n \leqslant 12$. The values of $\sigma(n)$ for $1 \leqslant n \leqslant 100$ are given in Table 2 of the Appendix.

| $n$ | 1 | 2 | 3 | 4 | 5 | 6 | 7 | 8 | 9 | 10 | 11 | 12 |
|---|---|---|---|---|---|---|---|---|---|---|---|---|
| $\sigma(n)$ | 1 | 3 | 4 | 7 | 6 | 12 | 8 | 15 | 13 | 18 | 12 | 28 |

**Table 6.1.** The Sum of the Divisors for $1 \leqslant n \leqslant 12$.

The other function which we will study is the number of divisors.

**Definition.** The number of divisors function, denoted by $\tau$, is defined by setting $\tau(n)$ equal to the number of positive divisors of $n$.

In Table 6.2 we give $\tau(n)$ for $1 \leqslant n \leqslant 12$. The values of $\tau(n)$ for $1 \leqslant n \leqslant 100$ are given in Table 2 of the Appendix.

| $n$ | 1 | 2 | 3 | 4 | 5 | 6 | 7 | 8 | 9 | 10 | 11 | 12 |
|---|---|---|---|---|---|---|---|---|---|---|---|---|
| $\tau(n)$ | 1 | 2 | 2 | 3 | 2 | 4 | 2 | 4 | 3 | 4 | 2 | 6 |

**Table 6.2.** The Number of Divisors for $1 \leqslant n \leqslant 12$.

Note that we can express $\sigma(n)$ and $\tau(n)$ in terms of summation notation. It is simple to see that

$$\sigma(n) = \sum_{d\mid n} d$$

and

$$\tau(n) = \sum_{d\mid n} 1.$$

To prove that $\sigma$ and $\tau$ are multiplicative, we use the following theorem.

**Theorem 6.7.** If $f$ is a multiplicative function, then the arithmetic function $F(n) = \sum_{d\mid n} f(d)$ is also multiplicative.

Before we prove the theorem, we illustrate the idea behind its proof with the following example. Let $f$ be a multiplicative function, and let $F(n) = \sum_{d\mid n} f(d)$. Let $m = 4$ and $n = 15$. We will show that $F(60) = F(4)F(15)$. Each of the divisors of 60 may be written as the product of a divisor of 4 and a divisor of 15 in the following way: $1 = 1\cdot1$, $2 = 2\cdot1$, $3 = 1\cdot3$, $4 = 4\cdot1$, $5 = 1\cdot5$, $6 = 2\cdot3$, $10 = 2\cdot5$, $12 = 4\cdot3$, $15 = 1\cdot15$, $20 = 4\cdot5$, $30 = 2\cdot15$, $60 = 4\cdot15$ (in each product, the first factor is the divisor of 4 , and the second is the divisor of 15). Hence,

$$
\begin{aligned}
F(60) &= f(1) + f(2) + f(3) + f(4) + f(5) + f(6) + f(10) + f(12) \\
&\quad + f(15) + f(20) + f(30) + f(60) \\
&= f(1\cdot1) + f(2\cdot1) + f(1\cdot3) + f(4\cdot1) + f(1\cdot5) + f(2\cdot3) \\
&\quad + f(2\cdot5) + f(4\cdot3) + f(1\cdot15) + f(4\cdot5) + f(2\cdot15) + f(4\cdot15) \\
&= f(1)f(1) + f(2)f(1) + f(1)f(3) + f(4)f(1) + f(1)f(5) \\
&\quad + f(2)f(3) + f(2)f(5) + f(4)f(3) + f(1)f(15) + f(4)f(5) \\
&\quad + f(2)f(15) + f(4)f(15) \\
&= (f(1) + f(2) + f(4))(f(1) + f(3) + f(5) + f(15)) \\
&= F(4)F(15).
\end{aligned}
$$

We now prove Theorem 6.7 using the idea illustrated by the example.

*Proof.* To show that $F$ is a multiplicative function, we must show that if $m$ and $n$ are relatively prime positive integers, then $F(mn) = F(m)F(n)$. So let us assume that $(m,n) = 1$. We have

$$F(mn) = \sum_{d\mid mn} f(d).$$

By Lemma 2.5, since $(m,n) = 1$ , each divisor of $mn$ can be written uniquely as the product of relatively prime divisors $d_1$ of $m$ and $d_2$ of $n$, and each pair of divisors $d_1$ of $m$ and $d_2$ of $n$ corresponds to a divisor $d = d_1 d_2$ of $mn$. Hence, we can write

$$F(mn) = \sum_{\substack{d_1 \mid m \\ d_2 \mid n}} f(d_1 d_2).$$

Since $f$ is multiplicative and since $(d_1, d_2) = 1$, we see that

$$F(mn) = \sum_{\substack{d_1 \mid n \\ d_2 \mid n}} f(d_1) f(d_2)$$

$$= \sum_{d_1 \mid m} f(d_1) \sum_{d_2 \mid n} f(d_2)$$

$$= F(m) F(n). \qquad \blacksquare$$

Now that we know $\sigma$ and $\tau$ are multiplicative, we can derive formulae for their values based on prime factorizations. First, we find formulae for $\sigma(n)$ and $\tau(n)$ when $n$ is the power of a prime.

**Lemma 6.1.** Let $p$ be prime and $a$ a positive integer. Then

$$\sigma(p^a) = (1+p+p^2+ \cdots +p^a) = \frac{p^{a+1}-1}{p-1}$$

and

$$\tau(p^a) = a + 1.$$

*Proof.* The divisors of $p^a$ are $1, p, p^2, ..., p^{a-1}, p^a$. Consequently, $p^a$ has exactly $a + 1$ divisors, so that $\tau(p^a) = a + 1$. Also, we note that $\sigma(p^a) = 1 + p + p_2 + \cdots + p^{a-1} + p^a = \frac{p^{a+1}-1}{p-1}$, using the formula in Example 1.10 for the sum of terms of a geometric progression. $\qquad \blacksquare$

**Example 6.7.** When we apply Lemma 6.1 with $p = 5$ and $a = 3$ we find that $\sigma(5^3) = 1 + 5 + 5^2 + 5^3 = \frac{5^4-1}{5-1} = 156$ and $\tau(5^3) = 1 + 3 = 4$. $\qquad \square$

The above lemma and the fact that $\sigma$ and $\tau$ are multiplicative lead to the following formulae.

**Theorem 6.8.** Let the positive integer $n$ have prime factorization $n = p_1^{a_1} p_2^{a_2} \cdots p_s^{a_s}$. Then

$$\sigma(n) = \frac{p_1^{a_1+1}-1}{p_1-1} \cdot \frac{p_2^{a_2+1}-1}{p_2-1} \cdot \cdots \cdot \frac{p_s^{a_s+1}-1}{p_s-1} = \prod_{j=1}^{s} \frac{p_j^{a_j+1}-1}{p_j-1}$$

and

$$\tau(n) = (a_1+1)(a_2+1) \cdots (a_s+1) = \prod_{j=1}^{s} (a_j+1).$$

*Proof.* Since both $\sigma$ and $\tau$ are multiplicative, we see that $\sigma(n) = \sigma(p_1^{a_1} p_2^{a_2} \cdots p_s^{a_s}) = \sigma(p_1^{a_1})\sigma(p_2^{a_2}) \cdots \sigma(p_s^{a_s})$ and $\tau(n) = \tau(p_1^{a_1} p_2^{a_2} \cdots p_s^{a_s}) = \tau(p_1^{a_1})\tau(p_2^{a_2}) \cdots \tau(p_s^{a_s})$. Inserting the values for $\sigma(p_i^{a_i})$ and $\tau(p_i^{a_i})$ found in Lemma 6.1, we obtain the desired formulae. ∎

We illustrate how to use Theorem 6.8 with the following example.

**Example 6.8.** Using Theorem 6.8 we find that

$$\sigma(200) = \sigma(2^3 5^2) = \frac{2^4-1}{2-1} \cdot \frac{5^3-1}{5-1} = 15 \cdot 31 = 465$$

and

$$\tau(200) = \tau(2^3 5^2) = (3+1)(2+1) = 12.$$

Also

$$\sigma(720) = \sigma(2^4 \cdot 3^2 \cdot 5) = \frac{2^5-1}{2-1} \cdot \frac{3^2-1}{3-1} \cdot \frac{5^2-1}{5-1} = 31 \cdot 13 \cdot 6 = 2418$$

and

$$\tau(2^4 \cdot 3^2 \cdot 5) = (4+1)(2+1)(1+1) = 30.$$ □

---

## 6.2 Exercises

1.  Find the sum of the positive integer divisors of each of the following integers.

    a)  35
    b)  196
    c)  1000
    d)  $2^{100}$

    e)  $2 \cdot 3 \cdot 5 \cdot 7 \cdot 11$
    f)  $2^5 3^4 5^3 7^2 11$
    g)  10!
    h)  20!

2.  Find the number of positive integer divisors of each of the following integers.

    a)  36
    b)  99
    c)  144

    d)  $2 \cdot 3 \cdot 5 \cdot 7 \cdot 11 \cdot 13 \cdot 17 \cdot 19$
    e)  $2 \cdot 3^2 \cdot 5^3 \cdot 7^4 \cdot 11^5 \cdot 13^4 \cdot 17^5 \cdot 19^5$
    f)  20!

3.  Which positive integers have an odd number of positive divisors?

4.  For which positive integers $n$ is the sum of divisors of $n$ odd?

* 5. Find all positive integers $n$ with $\sigma(n)$ equal to each of the following integers.

   a) 12                    d) 48
   b) 18                    e) 52
   c) 24                    f) 84

* 6. Find the smallest positive integer $n$ with $\tau(n)$ equal to each of the following integers.

   a) 1                     d) 6
   b) 2                     e) 14
   c) 3                     f) 100

7. Show that if $k > 1$ is an integer, then the equation $\tau(n) = k$ has infinitely many solutions.

8. Which positive integers have exactly two positive divisors?

9. Which positive integers have exactly three positive divisors?

10. Which positive integers have exactly four positive divisors?

11. What is the product of the positive divisors of a positive integer $n$?

Let $\sigma_k(n)$ denote the sum of the $k$th powers of the divisors of $n$, so that $\sigma_k(n) = \sum_{d|n} d^k$. Note that $\sigma_1(n) = \sigma(n)$.

12. Find $\sigma_3(4)$, $\sigma_3(6)$ and $\sigma_3(12)$.

13. Give a formula for $\sigma_k(p)$, where $p$ is prime.

14. Give a formula for $\sigma_k(p^a)$, where $p$ is prime, and $a$ is a positive integer.

15. Show that the function $\sigma_k$ is multiplicative.

16. Using Exercises 14 and 15, find a formula for $\sigma_k(n)$, where $n$ has prime-power factorization $n = p_1^{a_1} p_2^{a_2} \cdots p_m^{a_m}$.

* 17. Find all positive integers $n$ such that $\phi(n) + \sigma(n) = 2n$.

* 18. Show that no two positive integers have the same product of divisors.

19. Show that the number of pairs of positive integers with least common multiple equal to the positive integer $n$ is $\tau(n^2)$.

20. Let $n$ be a positive integer. Define the sequence of integers $n_1, n_2, n_3, \ldots$ by $n_1 = \tau(n)$ and $n_{k+1} = \tau(n_k)$ for $k = 1, 2, 3, \ldots$. Show that there is a positive integer $r$ such that $2 = n_r = n_{r+1} = n_{r+2} = \cdots$.

21. Show that a positive integer $n$ is composite if and only if $\sigma(n) > n + \sqrt{n}$.

22. Let $n$ be a positive integer. Show that $\tau(2^n - 1) \geq \tau(n)$.

* 23. Show that $\sum_{j=1}^{n} \tau(j) = [\sqrt{n}\,]^2$ whenever $n$ is a positive integer.

* 24. Let $a$ and $b$ be positive integers with $b > 1$. Show that $\sigma(a)/a \leq \sigma(ab)/ab \leq \sigma(a)\sigma(b)/ab$.

* **25.** Show that if $a$ and $b$ and positive integers then
$$\sigma(a)\sigma(b) \;=\; \sum_{d\mid (a,b)} d\sigma(ab/d^2).$$

* **26.** Show that if $n$ is a positive integer then $\left[\sum_{d\mid n}\tau(n)\right]^2 = \sum_{d\mid n}\tau(d)^3$.

* **27.** Find the determinant of the $n\times n$ matrix with $(i,j)$th entry equal to $(i,j)$.

* **28.** Let $n$ be a positive integer such that $24\mid (n+1)$. Show that $\sigma(n)$ is divisible by 24.

## 6.2  Computer Projects

Write programs to do the following:

**1.** Find the number of divisors of a positive integer.

**2.** Find the sum of the divisors of a positive integer.

**3.** Find the integer $r$ defined in Exercise 20.

---

# 6.3  Perfect Numbers and Mersenne Primes

Because of certain mystical beliefs, the ancient Greeks were interested in those integers that are equal to the sum of all their proper positive divisors. These integers are called *perfect numbers*.

**Definition.** If $n$ is a positive integer and $\sigma(n) = 2n$, then $n$ is called a *perfect number*.

**Example 6.9.** Since $\sigma(6) = 1 + 2 + 3 + 6 = 12$, we see that 6 is perfect. We also note that $\sigma(28) = 1 + 2 + 4 + 7 + 14 + 28 = 56$, so that 28 is another perfect number.                                                    □

The ancient Greeks knew how to find all even perfect numbers. The following theorem tells us which even positive integers are perfect.

**Theorem 6.9.** The positive integer $n$ is an even perfect number if and only if
$$n = 2^{m-1}(2^m-1)$$
where $m$ is an integer such that $m \geqslant 2$ and $2^m-1$ is prime.

*Proof.* First, we show that if $n = 2^{m-1}(2^m-1)$ where $2^m-1$ is prime, then $n$ is perfect. We note that since $2^m-1$ is odd, we have $(2^{m-1}, 2^m-1) = 1$. Since $\sigma$ is a multiplicative function, we see that
$$\sigma(n) = \sigma(2^{m-1})\sigma(2^m-1).$$

Lemma 6.1 tells us that $\sigma(2^{m-1}) = 2^m-1$ and $\sigma(2^m-1) = 2^m$, since we are assuming that $2^m-1$ is prime. Consequently,

$$\sigma(n) = (2^m-1)2^m = 2n,$$

demonstrating that $n$ is a perfect number.

To show that the converse is true, let $n$ be an even perfect number. Write $n = 2^s t$ where $s$ and $t$ are positive integers and $t$ is odd. Since $(2^s, t) = 1$, we see from Lemma 6.1 that

$$(6.1) \qquad \sigma(n) = \sigma(2^s t) = \sigma(2^s)\sigma(t) = (2^{s+1}-1)\sigma(t).$$

Since n is perfect, we have

$$(6.2) \qquad \sigma(n) = 2n = 2^{s+1}t.$$

Combining (6.1) and (6.2) shows that

$$(6.3) \qquad (2^{s+1}-1)\sigma(t) = 2^{s+1}t.$$

Since $(2^{s+1}, 2^{s+1}-1) = 1$, from Lemma 2.3 we see that $2^{s+1} \mid \sigma(t)$. Therefore, there is an integer $q$ such that $\sigma(t) = 2^{s+1}q$. Inserting this expression for $\sigma(t)$ into (6.3) tells us that

$$(2^{s+1}-1)2^{s+1}q = 2^{s+1}t,$$

and, therefore,

$$(6.4) \qquad (2^{s+1}-1)q = t.$$

Hence, $q \mid t$ and $q \neq t$.

When we replace $t$ by the expression on the left-hand side of (6.4), we find that

$$(6.5) \qquad t + q = (2^{s+1}-1)q + q = 2^{s+1}q = \sigma(t).$$

We will show that $q = 1$. Note that if $q \neq 1$, then there are at least three distinct positive divisors of $t$, namely 1, $q$, and $t$. This implies that $\sigma(t) \geqslant t + q + 1$, which contradicts (6.5). Hence, $q = 1$ and, from (6.4), we conclude that $t = 2^{s+1}-1$. Also, from (6.5), we see that $\sigma(t) = t + 1$, so that $t$ must be prime, since its only positive divisors are 1 and $t$. Therefore, $n = 2^s(2^{s+1}-1)$, where $2^{s+1}-1$ is prime. ∎

By Theorem 6.9 we see that to find even perfect numbers, we must find primes of the form $2^m-1$. In our search for primes of this form, we first show that the exponent $m$ must be prime.

**Theorem 6.10.** If $m$ is a positive integer and $2^m-1$ is prime, then $m$ must be prime.

*Proof.* Assume that $m$ is not prime, so that $m = ab$ where $1 < a < m$ and

$1 < b < m$. Then

$$2^m - 1 = 2^{ab} - 1 = (2^a - 1)(2^{a(b-1)} + 2^{a(b-2)} + \ldots + 2^a + 1).$$

Since both factors on the right side of the equation are greater than 1, we see that $2^m - 1$ is composite if m is not prime. Therefore, if $2^m - 1$ is prime, then $m$ must also be prime. ∎

From Theorem 6.10 we see that to search for primes of the form $2^m - 1$, we need to consider only integers $m$ that are prime. Integers of the form $2^m - 1$ have been studied in great depth; these integers are named after a French monk of the seventeenth century, Mersenne*, who studied these integers.

**Definition.** If $m$ is a positive integer, then $M_m = 2^m - 1$ is called the $m$th *Mersenne number*, and, if $p$ is prime and $M_p = 2^p - 1$ is also prime, then $M_p$ is called a *Mersenne prime*.

**Example 6.10.** The Mersenne number $M_7 = 2^7 - 1$ is prime, whereas the Mersenne number $M_{11} = 2^{11} - 1 = 2047 = 23 \cdot 89$ is composite. □

It is possible to prove various theorems that help decide whether Mersenne numbers are prime. One such theorem will now be given. Related results are found in the exercises in Chapter 9.

**Theorem 6.11.** If $p$ is an odd prime, then any divisor of the Mersenne number $M_p = 2^p - 1$ is of the form $2kp + 1$ where $k$ is a positive integer.

*Proof.* Let $q$ be a prime dividing $M_p = 2^p - 1$. From Fermat's little theorem, we know that $q \mid (2^{q-1} - 1)$. Also, from Lemma 3.2 we know that

$$(6.6) \qquad (2^p - 1, 2^{q-1} - 1) = 2^{(p, q-1)} - 1.$$

Since $q$ is a common divisor of $2^p - 1$ and $2^{q-1} - 1$, we know that $(2^p - 1, 2^{q-1} - 1) > 1$. Hence, $(p, q-1) = p$, since the only other possibility, namely $(p, q-1) = 1$, would imply from (6.6) that $(2^p - 1, 2^{q-1} - 1) = 1$.

---

* **MARIN MERSENNE** (1588-1648) was a Franciscan friar who lived most of his life in Parisian cloisters. He was the author of *Cognitata Physico-Mathematica* which stated without proof that $M_p$ is prime for $p = 2, 3, 5, 7, 13, 17, 19, 31, 67, 127,$ and $257$ and for no other primes $p$ with $p < 257$. It took over 300 years to totally settle this claim made by Mersenne. The final work done in 1947 showed that Mersenne made five errors in his work (namely that $M_{61}$ is prime, $M_{67}$ is composite, $M_{89}$ is prime, $M_{107}$ is prime, and $M_{257}$ is composite). Besides his famous statement about primes of the form $M_p$, Mersenne contributed to the development of number theory through his extensive correspondence with many mathematicians, including Fermat. Mersenne effectively served as as a clearing house and a disseminator of new mathematical ideas in the 17th century.

Hence $p \mid (q-1)$, and, therefore, there is a positive integer $m$ with $q - 1 = mp$. Since $q$ is odd we see that $m$ must be even, so that $m = 2k$, where $k$ is a positive integer. Hence, $q = mp + 1 = 2kp + 1$. ∎

We can use Theorem 6.11 to help decide whether Mersenne numbers are prime. We illustrate this with the following examples.

**Example 6.11.** To decide whether $M_{13} = 2^{13} - 1 = 8191$ is prime, we only need look for a prime factor not exceeding $\sqrt{8191} = 90.504...$. Furthermore, by Theorem 6.11, any such prime divisor must be of the form $26k + 1$. The only candidates for primes dividing $M_{13}$ less than or equal to $\sqrt{M_{13}}$ are 53 and 79. Trial division easily rules out these cases, so that $M_{13}$ is prime. □

**Example 6.12.** To decide whether $M_{23} = 2^{23} - 1 = 8388607$ is prime, we only need to determine whether $M_{23}$ is divisible by a prime less than or equal to $\sqrt{M_{23}} = 2896.309...$ of the form $46k + 1$. The first prime of this form is 47. A trial division shows that $8388607 = 47 \cdot 178481$, so that $M_{23}$ is composite. □

Because there are special primality tests for Mersenne numbers, it has been possible to determine whether extremely large Mersenne numbers are prime. Following is one such primality test. This test has been used to find the largest known Mersenne primes, which are the largest known primes. The proof of this test may be found in Lenstra [101] and Sierpiński [39].

**Theorem 6.12. The Lucas-Lehmer Test.** Let $p$ be a prime and let $M_p = 2^p - 1$ denote the $p$th Mersenne number. Define a sequence of integers recursively by setting $r_1 = 4$, and for $k \geqslant 2$,

$$r_k \equiv r_{k-1}^2 - 2 \pmod{M_p}, \quad 0 \leqslant r_k < M_p.$$

Then $M_p$ is prime if and only if $r_{p-1} \equiv 0 \pmod{M_p}$.

We use an example to illustrate an application of the Lucas-Lehmer test.

**Example 6.13.** Consider the Mersenne number $M_5 = 2^5 - 1 = 31$. Then $r_1 = 4$, $r_2 \equiv 4^2 - 2 = 14 \pmod{31}$, $r_3 \equiv 14^2 - 2 \equiv 8 \pmod{31}$, and $r_4 \equiv 8^2 - 2 \equiv 0 \pmod{31}$. Since $r_4 \equiv 0 \pmod{31}$, we conclude that $M_5 = 31$ is prime. □

The Lucas-Lehmer test can be performed quite rapidly as the following corollary states.

**Corollary 6.1.** Let $p$ be prime and let $M_p = 2^p - 1$ denote the $p$th Mersenne number. It is possible to determine whether $M_p$ is prime using $O(p^3)$ bit operations.

*Proof.* To determine whether $M_p$ is prime using the Lucas-Lehmer test requires $p - 1$ squarings modulo $M_p$, each requiring $O((\log M_p)^2) = O(p^2)$ bit operations. Hence, the Lucas-Lehmer test requires $O(p^3)$ bit operations. ∎

Much activity has been directed toward the discovery of Mersenne primes, especially since each new Mersenne prime discovered has become the largest prime known, and for each new Mersenne prime, there is a new perfect number. At the present time, a total of 30 Mersenne primes are known; these include all Mersenne primes $M_p$ with $p \leqslant 132049$. The latest Mersenne prime found is $M_{216091}$ which was shown to be prime in 1985 by Slowinski using three hours of time on a supercomputer. Unfortunately the range $132049 < p < 216091$ has not been searched, so that there may be additional Mersenne primes with $p$ in this range. The known Mersenne primes, the number of decimal digits in these primes, and their dates of discovery are displayed in Table 6.3.

Computers were used to find the 18 largest Mersenne primes known. The discovery by high school students of the 25th and 26th Mersenne prime received much publicity, including coverage on the nightly news of a major television network. An interesting account of the search for the 27th Mersenne prime and related historical and computational information may be found in [109]. A report of the discovery of the 28th Mersenne prime is given in [93]. It has been conjectured but has not been proved, that there are infinitely many Mersenne primes.

We have reduced the study of even perfect numbers to the study of Mersenne primes. We may ask whether there are odd perfect numbers. The answer is still unknown. It is possible to demonstrate that if they exist, odd perfect numbers must have certain properties (see Exercises 26-30, for example). Furthermore, it is known that there are no odd perfect numbers less than $10^{200}$, and it has been shown that any odd perfect number must have at least eight different prime factors. A discussion of odd perfect numbers may be found in Guy [19], and information concerning recent results about odd perfect numbers is given by Hagis [97].

| $p$ | Number of decimal digits in $M_p$ | Date of Discovery |
|---|---|---|
| 2 | 1 | ancient times |
| 3 | 1 | ancient times |
| 5 | 2 | ancient times |
| 7 | 3 | ancient times |
| 13 | 4 | Mid 15th century |
| 17 | 6 | 1603 |
| 19 | 6 | 1603 |
| 31 | 10 | 1772 |
| 61 | 19 | 1883 |
| 89 | 27 | 1911 |
| 107 | 33 | 1914 |
| 127 | 39 | 1876 |
| 521 | 157 | 1952 |
| 607 | 183 | 1952 |
| 1279 | 386 | 1952 |
| 2203 | 664 | 1952 |
| 2281 | 687 | 1952 |
| 3217 | 969 | 1957 |
| 4253 | 1281 | 1961 |
| 4423 | 1332 | 1961 |
| 9689 | 2917 | 1963 |
| 9941 | 2993 | 1963 |
| 11213 | 3376 | 1963 |
| 19937 | 6002 | 1971 |
| 21701 | 6533 | 1978 |
| 23209 | 6987 | 1979 |
| 44497 | 13395 | 1979 |
| 86243 | 25962 | 1983 |
| 132049 | 39751 | 1983 |
| 216091 | 65050 | 1985 |

**Table 6.3.** The Known Mersenne Primes.

## 6.3 Exercises

**1.** Find the six smallest even perfect numbers.

**2.** Show that if $n$ is a positive integer greater than 1, then the Mersenne number $M_n$ cannot be the power of a positive integer.

If $n$ is a positive integer, then we say that $n$ is *deficient* if $\sigma(n) < 2n$ , and we say that $n$ is *abundant* if $\sigma(n) > 2n$. Every integer is either deficient, perfect, or abundant.

**3.** Find the six smallest abundant positive integers.

**\* 4.** Find the smallest odd abundant positive integer.

**5.** Show that every prime power is deficient.

**6.** Show that any divisor of a deficient or perfect number is deficient.

**7.** Show that any multiple of an abundant or perfect number is abundant.

**8.** Show that if $n = 2^{m-1}(2^m-1)$ , where $m$ is a positive integer such that $2^m-1$ is composite, then $n$ is abundant.

Two positive integers $m$ and $n$ are called an *amicable pair* if $\sigma(m) = \sigma(n) = m + n$.

**9.** Show that each of the following pairs of integers are amicable pairs.

a) 220, 284
b) 1184, 1210
c) 79750, 88730

**10.** a) Show that if $n$ is a positive integer with $n \geqslant 2$, such that $3 \cdot 2^{n-1}-1$, $3 \cdot 2^n-1$, and $3^2 \cdot 2^{2n-1}-1$ are all prime, then $2^n(3 \cdot 2^{n-1}-1)(3 \cdot 2^n-1)$ and $2^n(3^2 \cdot 2^{2n-1}-1)$ form an amicable pair.

b) Find three amicable pairs using part (a).

An integer $n$ is called *k-perfect* if $\sigma(n) = kn$. Note that a perfect number is 2-perfect.

**11.** Show that $120 = 2^3 \cdot 3 \cdot 5$ is 3-perfect.

**12.** Show that $30240 = 2^5 \cdot 3^2 \cdot 5 \cdot 7$ is 4-perfect.

**13.** Show that $14182439040 = 2^7 \cdot 3^4 \cdot 5 \cdot 7 \cdot 11^2 \cdot 17 \cdot 19$ is 5-perfect.

**14.** Find all 3-perfect numbers of the form $n = 2^k \cdot 3 \cdot p$, where $p$ is an odd prime.

**15.** Show that if $n$ is 3-perfect and $3 \nmid n$, then $3n$ is 4-perfect.

An integer $n$ is *k-abundant* if $\sigma(n) > (k+1)n$.

**16.** Find a 3-abundant integer.

17. Find a 4-abundant integer.

* 18. Show that for each positive integer $k$ there are an infinite number of $k$-abundant integers.

A positive integer $n$ is called *superperfect* if $\sigma(\sigma(n)) = 2n$.

19. Show that 16 is superperfect.

20. Show that if $n = 2^q$ where $2^{q+1}-1$ is prime, then $n$ is superperfect.

* 21. Show that every even superperfect number is of the form $n = 2^q$ where $2^{q+1}-1$ is prime.

22. Show that if $n = p^2$ where $p$ is an odd prime, then $n$ is not superperfect.

23. Use Theorem 6.11 to determine whether each of the following Mersenne numbers is prime.

    a) $M_7$                   c) $M_{17}$
    b) $M_{11}$                d) $M_{29}$

24. Use the Lucas-Lehmer test, Theorem 6.12, to determine whether each of the following Mersenne numbers is prime.

    a) $M_3$                   c) $M_{11}$
    b) $M_7$                  d) $M_{13}$

* 25. a) Show that if $n$ is a positive integer and $2n + 1$ is prime, then either $(2n+1) \mid M_n$ or $(2n+1) \mid (M_n+2)$. (*Hint:* Use Fermat's little theorem to show that $M_n(M_n+2) \equiv 0 \pmod{2n+1}$.)

    b) Use part (a) to show that $M_{11}$ and $M_{23}$ are composite.

* 26. a) Show that if $n$ is an odd perfect number, then $n = p^a m^2$ where $p$ is an odd prime and $p \equiv a \equiv 1 \pmod 4$.

    b) Use part (a) to show that if $n$ is an odd perfect number, then $n \equiv 1 \pmod 4$.

* 27. Show that if $n = p^a m^2$ is an odd perfect number where $p$ is prime, then $n \equiv p \pmod 8$.

* 28. Show that if $n$ is an odd perfect number, then 3, 5, and 7 are not all divisors of $n$.

* 29. Show that if $n$ is an odd perfect number then $n$ has at least three different prime divisors.

** 30. Show that if $n$ is an odd perfect number then $n$ has at least four different prime divisors.

31. Find all positive integers $n$ such that the product of all divisors of $n$ other than $n$ is exactly $n^2$. (These integers are multiplicative analogues of perfect numbers.)

**32.** Let $n$ be a positive integer. Define the sequence $n_1, n_2, n_3, \ldots$, recursively by $n_1 = \sigma(n) - n$ and $n_{k+1} = \sigma(n_k) - n_k$ for $k = 1, 2, 3, \ldots$.

   a) Show that if $n$ is perfect, then $n = n_1 = n_2 = n_3 = \cdots$.

   b) Show that if $n$ and $m$ are an amicable pair, then $n_1 = m$, $n_2 = n$, $n_3 = m$, $n_4 = n, \ldots$ and so on, *i.e.*, the sequence $n_1, n_2, n_3, \ldots$ is periodic with period 2.

   c) Find the sequence of integers generated if $n = 12496 = 2^4 \cdot 11 \cdot 71$.

It has been conjectured that for all integers $n$, the sequence of integers $n_1, n_2, n_3, \ldots$ is periodic.

## 6.3 Computer Projects

Write programs to do the following:

**1.** Classify positive integers according to whether they are deficient, perfect, or abundant (see the preamble to Exercise 3).

**2.** Use Theorem 6.11 to look for factors of Mersenne numbers.

**3.** Determine whether Mersenne numbers are prime using the Lucas-Lehmer test.

**4.** Given a positive integer $n$, determine if the sequence defined in Exercise 32 is periodic.

**5.** Find amicable pairs.

# 7

---

# Cryptology

---

## 7.1 Character Ciphers

From ancient times to the present, secret messages have been sent. Classically, the need for secret communication has occurred in diplomacy and in military affairs. Now, with electronic communication coming into widespread use, secrecy has become an important issue. Just recently, with the advent of electronic banking, secrecy has become necessary even for financial transactions. Hence, there is a great deal of interest in the techniques of making messages unintelligible to everyone except the intended receiver.

Before discussing specific secrecy systems, we present some terminology. The discipline devoted to secrecy systems is called *cryptology*. *Cryptography* is the part of cryptology that deals with the design and implementation of secrecy systems, while *cryptanalysis* is aimed at breaking these systems. A message that is to be altered into a secret form is called *plaintext*. A *cipher* is a method for altering a plaintext message into *ciphertext* by changing the letters of the plaintext using a transformation. The *key* determines the particular transformation from a set of possible transformations that is to be used. The process of changing plaintext into ciphertext is called *encryption* or *enciphering*, while the reverse process of changing the ciphertext back to the plaintext by the intended receiver, possessing knowledge of the method for doing this, is called *decryption* or *deciphering*. This, of course, is different from the process someone other than the intended receiver uses to make the message intelligible through cryptanalysis.

In this chapter, we present secrecy systems based on modular arithmetic. The first of these had its origin with Julius Ceasar. The newest secrecy system we will discuss was invented in the late 1970's. In all these systems we start by translating letters into numbers. We take as our standard alphabet the letters of English and translate them into the integers from 0 to 25, as shown in Table 7.1.

| letter | A | B | C | D | E | F | G | H | I | J | K | L | M | N | O | P | Q | R | S | T | U | V | W | X | Y | Z |
|--------|---|---|---|---|---|---|---|---|---|---|---|---|---|---|---|---|---|---|---|---|---|---|---|---|---|---|
| numerical equivalent | 0 | 1 | 2 | 3 | 4 | 5 | 6 | 7 | 8 | 9 | 10 | 11 | 12 | 13 | 14 | 15 | 16 | 17 | 18 | 19 | 20 | 21 | 22 | 23 | 24 | 25 |

**Table 7.1.** The Numerical Equivalents of Letters.

Of course, if we were sending messages in Russian, Greek, Hebrew or any other language we would use the appropriate alphabet range of integers. Also, we may want to include punctuation marks, a symbol to indicate blanks, and perhaps the digits for representing numbers as part of the message. However, for the sake of simplicity, we restrict ourselves to the letters of the English alphabet.

First, we discuss secrecy systems based on transforming each letter of the plaintext message into a different letter to produce the ciphertext. Such ciphers are called *character* or *monographic ciphers*, since each letter is changed individually to another letter by a *substitution*. Altogether there are 26! possible ways to produce a monographic transformation. We will discuss a set that is based on modular arithmetic.

A cipher, that was used by Julius Caesar, is based on the substitution in which each letter is replaced by the letter three further down the alphabet, with the last three letters shifted to the first three letters of the alphabet. To describe this cipher using modular arithmetic, let $P$ be the numerical equivalent of a letter in the plaintext and $C$ the numerical equivalent of the corresponding ciphertext letter. Then

$$C \equiv P+3 \pmod{26}, \quad 0 \leqslant C \leqslant 25.$$

The correspondence between plaintext and ciphertext is given in Table 7.2.

| plaintext | A | B | C | D | E | F | G | H | I | J | K | L | M | N | O | P | Q | R | S | T | U | V | W | X | Y | Z |
|---|---|---|---|---|---|---|---|---|---|---|---|---|---|---|---|---|---|---|---|---|---|---|---|---|---|---|
|  | 0 | 1 | 2 | 3 | 4 | 5 | 6 | 7 | 8 | 9 | 10 | 11 | 12 | 13 | 14 | 15 | 16 | 17 | 18 | 19 | 20 | 21 | 22 | 23 | 24 | 25 |
|  | 3 | 4 | 5 | 6 | 7 | 8 | 9 | 10 | 11 | 12 | 13 | 14 | 15 | 16 | 17 | 18 | 19 | 20 | 21 | 22 | 23 | 24 | 25 | 0 | 1 | 2 |
| ciphertext | D | E | F | G | H | I | J | K | L | M | N | O | P | Q | R | S | T | U | V | W | X | Y | Z | A | B | C |

**Table 7.2.** The Correspondence of Letters for the Caesar Cipher.

To encipher a message using this transformation, we first change it to its numerical equivalent, grouping letters in blocks of five. Then we transform each number. The grouping of letters into blocks helps to prevent successful cryptanalysis based on recognizing particular words. We illustrate this procedure by enciphering the message

THIS MESSAGE IS TOP SECRET.

Broken into groups of five letters, the message is

THISM ESSAG EISTO PSECR ET.

Converting the letters into their numerical equivalents, we obtain

    19   7   8   18   12     4   18    18   0   6     4   8   18   19   14
    15   18   4   3   17     4   19 .

Using the Caesar transformation $C \equiv P+3 \pmod{26}$, this becomes

    22   10   11   21   15     7   21    21   3   9     7   11   21   22   17
    18   21   7   6   20     7   22 .

Translating back to letters, we have

WKLVP  HVVDJ  HLVWR  SVHGU  HW.

This is the message we send.

The receiver deciphers it in the following manner. First, the letters are converted to numbers. Then, the relationship $P \equiv C-3 \pmod{26}$, $0 \leqslant P \leqslant 25$, is used to change the ciphertext back to the numerical version of the plaintext, and finally the message is converted to letters.

We illustrate the deciphering procedure with the following message enciphered by the Caesar cipher:

### WKLVL  VKRZZ  HGHFL  SKHU.

First, we change these letters into their numerical equivalents, to obtain

22 10 11 21 11   21 10 17 25 25   7 6 7 5 11   18 10 7 20.

Next, we perform the transformation $P \equiv C-3 \pmod{26}$ to change this to plaintext, and we obtain

19 7 8 18 8   18 7 14 22 22   4 3 4 2 8   15 7 4 17.

We translate this back to letters and recover the plaintext message

### THISI  SHOWW  EDECI  PHER.

By combining the appropriate letters into words, we find that the message reads

### THIS IS HOW WE DECIPHER.

The Caesar cipher is one of a family of similar ciphers described by a *shift transformation*

$$C \equiv P+k \pmod{26}, \quad 0 \leqslant C \leqslant 25,$$

where $k$ is the key representing the size of the shift of letters in the alphabet. There are 26 different transformations of this type, including the case of $k \equiv 0 \pmod{26}$, where letters are not altered, since in this case $C \equiv P \pmod{26}$.

More generally, we will consider transformations of the type

(7.1)            $$C \equiv aP+b \pmod{26}, \quad 0 \leqslant C \leqslant 25,$$

where $a$ and $b$ are integers with $(a, 26) = 1$. These are called *affine transformations*. Shift transformations are affine transformations with $a=1$. We require that $(a, 26) = 1$, so that as P runs through a complete system of residues modulo 26, $C$ also does. There are $\phi(26) = 12$ choices for $a$, and 26 choices for $b$, giving a total of $12 \cdot 26 = 312$ transformations of this type (one of these is $C \equiv P \pmod{26}$ obtained when $a=1$ and $b=0$). If the relationship between plaintext and ciphertext is described by (7.1), then the inverse relationship is given by

$$P \equiv \bar{a}(C-b) \pmod{26}, \quad 0 \leqslant P \leqslant 25,$$

where $\bar{a}$ is an inverse of $a$ (mod 26).

As an example of such a cipher, let $a=7$ and $b=10$, so that $C \equiv 7P + 10 \pmod{26}$. Hence, $P \equiv 15(C-10) \equiv 15C+6 \pmod{26}$, since 15 is an inverse of 7 modulo 26. The correspondence between letters is given in Table 7.3.

| plaintext | A | B | C | D | E | F | G | H | I | J | K | L | M | N | O | P | Q | R | S | T | U | V | W | X | Y | Z |
|---|---|---|---|---|---|---|---|---|---|---|---|---|---|---|---|---|---|---|---|---|---|---|---|---|---|---|
| | 0 | 1 | 2 | 3 | 4 | 5 | 6 | 7 | 8 | 9 | 10 | 11 | 12 | 13 | 14 | 15 | 16 | 17 | 18 | 19 | 20 | 21 | 22 | 23 | 24 | 25 |
| ciphertext | 10 | 17 | 24 | 5 | 12 | 19 | 0 | 7 | 14 | 21 | 2 | 9 | 16 | 23 | 4 | 11 | 18 | 25 | 6 | 13 | 20 | 1 | 8 | 15 | 22 | 3 |
| | K | R | Y | F | M | T | A | H | O | V | C | J | Q | X | E | L | S | Z | G | N | V | B | I | P | W | D |

**Table 7.3.** The Correspondence of Letters for the Cipher with $C \equiv 7P+10 \pmod{26}$.

To illustrate how we obtained this correspondence, note that the plaintext letter L with numerical equivalent 11 corresponds to the ciphertext letter J, since $7 \cdot 11 + 10 = 87 \equiv 9 \pmod{26}$ and 9 is the numerical equivalent of J.

To illustrate how to encipher, note that

PLEASE SEND MONEY

is transformed to

LJMKG MGXFQ EXMW.

Also note that the ciphertext

FEXEN XMBMK JNHMG MYZMN

corresponds to the plaintext

DONOT REVEA LTHES ECRET,

or combining the appropriate letters

DO NOT REVEAL THE SECRET.

We now discuss some of the techniques directed at the cryptanalysis of ciphers based on affine transformations. In attempting to break a monographic cipher, the frequency of letters in the ciphertext is compared with the frequency of letters in ordinary text. This gives information concerning the correspondence between letters. In various frequency counts of English text, one finds the percentages listed in Table 7.4 for the occurrence of the 26 letters of the alphabet. Counts of letter frequencies in other languages may be found in Friedman [69] and Kullback [73].

| letter | A | B | C | D | E | F | G | H | I | J | K | L | M | N | O | P | Q | R | S | T | U | V | W | X | Y | Z |
|---|---|---|---|---|---|---|---|---|---|---|---|---|---|---|---|---|---|---|---|---|---|---|---|---|---|---|
| frequency (in %) | 7 | 1 | 3 | 4 | 13 | 3 | 2 | 3 | 8 | <1 | <1 | 4 | 3 | 8 | 7 | 3 | <1 | 8 | 6 | 9 | 3 | 1 | 1 | <1 | 2 | <1 |

**Table 7.4.** The Frequencies of Occurrence of the Letters of the Alphabet.

From this information, we see that the most frequently occurring letters are E, T, N, R, I, O and A, in that order. We can use this information to determine which cipher based on an affine transformation has been used to encipher a message.

First, suppose that we know in advance that a shift cipher has been employed to encipher a message; each letter of the message has been transformed by a correspondence $C \equiv P + k \pmod{26}$, $0 \leqslant C \leqslant 25$. To cryptanalyze the ciphertext

$$\text{YFXMP} \quad \text{CESPZ} \quad \text{CJTDF} \quad \text{DPQFW} \quad \text{QZCPY}$$
$$\text{NTASP} \quad \text{CTYRX} \quad \text{PDDLR} \quad \text{PD}$$

we first count the number of occurrences of each letter in the ciphertext. This is displayed in Table 7.5.

| letter | A | B | C | D | E | F | G | H | I | J | K | L | M | N | O | P | Q | R | S | T | U | V | W | X | Y | Z |
|---|---|---|---|---|---|---|---|---|---|---|---|---|---|---|---|---|---|---|---|---|---|---|---|---|---|---|
| number of occurrences | 1 | 0 | 4 | 5 | 1 | 3 | 0 | 0 | 0 | 1 | 0 | 1 | 1 | 1 | 0 | 7 | 2 | 2 | 1 | 3 | 0 | 0 | 1 | 1 | 3 | 2 |

**Table 7.5.** The Number of Occurrences of Letters in a Ciphertext.

We notice that the most frequently occurring letter in the ciphertext is P with the letters C,D,F,T, and Y occurring with relatively high frequency. Our initial guess would be that P represents E, since E is the most frequently occurring letter in English text. If this is so, then $15 \equiv 4 + k \pmod{26}$, so that $k \equiv 11 \pmod{26}$. Consequently, we would have $C \equiv P+11 \pmod{26}$ and $P \equiv C-11 \pmod{26}$. This correspondence is given in Table 7.6.

| ciphertext | A | B | C | D | E | F | G | H | I | J | K | L | M | N | O | P | Q | R | S | T | U | V | W | X | Y | Z |
|---|---|---|---|---|---|---|---|---|---|---|---|---|---|---|---|---|---|---|---|---|---|---|---|---|---|---|---|
| | 0 | 1 | 2 | 3 | 4 | 5 | 6 | 7 | 8 | 9 | 10 | 11 | 12 | 13 | 14 | 15 | 16 | 17 | 18 | 19 | 20 | 21 | 22 | 23 | 24 | 25 |
| plaintext | 15 | 16 | 17 | 18 | 19 | 20 | 21 | 22 | 23 | 24 | 25 | 0 | 1 | 2 | 3 | 4 | 5 | 6 | 7 | 8 | 9 | 10 | 11 | 12 | 13 | 14 |
| | P | Q | R | S | T | U | V | W | Z | Y | Z | A | B | C | D | E | F | G | H | I | J | K | L | M | N | O |

**Table 7.6.** Correspondence of Letters for the Sample Ciphertext.

Using this correspondence, we attempt to decipher the message. We obtain

NUMBE RTHEO RYISU SEFUL FOREN
CIPHE RINGM ESSAG ES

This can easily be read as

NUMBER THEORY IS USEFUL FOR
ENCIPHERING MESSAGES.

Consequently, we made the correct guess. If we had tried this transformation, and instead of the plaintext, it has produced garbled text, we would have tried another likely transformation based on the frequency count of letters in the ciphertext.

Now, suppose we know that an affine transformation of the form $C \equiv aP+b \pmod{26}$, $0 \leqslant C \leqslant 25$, has been used for enciphering. For instance, suppose we wish to cryptanalyze the enciphered message

USLEL JUTCC YRTPS URKLT YGGFV
ELYUS LRYXD JURTU ULVCU URJRK
QLLQL YXSRV LBRYZ CYREK LVEXB
RYZDG HRGUS LJLLM LYPDJ LJTJU
FALGU PTGVT JULYU SLDAL TJRWU
SLJFE OLPU

The first thing to do is to count the occurrences of each letter; this count is displayed in Table 7.7.

| letter | A | B | C | D | E | F | G | H | I | J | K | L | M | N | O | P | Q | R | S | T | U | V | W | X | Y | Z |
|---|---|---|---|---|---|---|---|---|---|---|---|---|---|---|---|---|---|---|---|---|---|---|---|---|---|---|
| number of occurrences | 2 | 2 | 4 | 4 | 5 | 3 | 6 | 1 | 0 | 10 | 3 | 22 | 1 | 0 | 1 | 4 | 2 | 12 | 5 | 8 | 16 | 3 | 1 | 3 | 10 | 2 |

**Table 7.7.** The Number of Occurrences of Letters in a Ciphertext.

With this information, we guess that the letter L, which is the most frequently occurring letter in the ciphertext, corresponds to E, while the letter U, which occurs with the second highest frequency, corresponds to T. This implies, if the transformation is of the form $C \equiv aP + b \pmod{26}$, the pair of congruences

$$4a + b \equiv 11 \pmod{26}$$
$$19a + b \equiv 20 \pmod{26}.$$

By Theorem 3.14 we see that the solution of this system is $a \equiv 11 \pmod{26}$ and $b \equiv 19 \pmod{26}$.

If this is the correct enciphering transformation, then using the fact that 19 is an inverse of 11 modulo 26, the deciphering transformation is

$$P \equiv 19 \ (C-19) \equiv 19C - 361 \equiv 19C + 3 \pmod{26}, \ 0 \leqslant P \leqslant 25.$$

This gives the correspondence found in Table 7.8.

| ciphertext | A | B | C | D | E | F | G | H | I | J | K | L | M | N | O | P | Q | R | S | T | U | V | W | X | Y | Z |
|---|---|---|---|---|---|---|---|---|---|---|---|---|---|---|---|---|---|---|---|---|---|---|---|---|---|---|
| | 0 | 1 | 2 | 3 | 4 | 5 | 6 | 7 | 8 | 9 | 10 | 11 | 12 | 13 | 14 | 15 | 16 | 17 | 18 | 19 | 20 | 21 | 22 | 23 | 24 | 25 |
| plaintext | 3 | 22 | 15 | 8 | 1 | 20 | 13 | 6 | 25 | 19 | 11 | 4 | 23 | 16 | 9 | 2 | 21 | 14 | 7 | 0 | 19 | 12 | 5 | 24 | 17 | 10 |
| | D | W | P | I | B | U | N | G | Z | S | L | E | X | Q | J | C | V | O | H | A | T | M | P | Y | R | K |

**Table 7.8.** The Correspondence of Letters for the Sample Ciphertext.

With this correspondence, we try to read the ciphertext. The ciphertext becomes

```
THEBE STAPP ROACH TOLEA RNNUM
BERTH EORYI STOAT TEMPT TOSOL
VEEVE RYHOM EWORK PROBL EMBYW
ORKIN GONTH ESEEX ERCIS ESAST
UDENT CANMA STERT HEIDE ASOFT
HESUB JECT
```

We leave it to the reader to combine the appropriate letters into words to see that the message is intelligible.

---

## 7.1 Exercises

1. Using the Caesar cipher, encipher the message ATTACK AT DAWN.

2. Decipher the ciphertext message LFDPH LVDZL FRQTX HUHG that has been enciphered using the Caesar cipher.

3. Encipher the message SURRENDER IMMEDIATELY using the affine transformation $C \equiv 11P + 18 \pmod{26}$.

4. Decipher the message RTOLK TOIK, which was enciphered using the affine transformation $C \equiv 3P + 24 \pmod{26}$.

5. If the most common letter in a long ciphertext, enciphered by a shift transformation $C \equiv P + k \pmod{26}$ is Q, then what is the most likely value of $k$?

6. The message KYVMR CLVFW KYVBV PZJJV MVEKV VE was enciphered using a shift transformation $C \equiv P + k \pmod{26}$. Use frequencies of letters to determine the value of $k$. What is the plaintext message?

7. If the two most common letters in a long ciphertext, enciphered by an affine transformation $C \equiv aP + b \pmod{26}$ are W and B, respectively, then what are the most likely values for $a$ and $b$?

8. The message PJXFJ SWJNX JMRTJ FVSUJ OOJWF OVAJR WHEOF JRWJO DJFFZ BJF was enciphered using an affine transformation $C \equiv aP + b \pmod{26}$. Use frequencies of letters to determine the values of $a$ and $b$. What is the plaintext message?

Given two ciphers, plaintext may be enciphered by using one of the ciphers, and by then using the other cipher. This procedure produces a *product cipher*.

9. Find the product cipher obtained by using the transformation $C \equiv 5P + 13 \pmod{26}$ followed by the transformation $C \equiv 17P + 3 \pmod{26}$.

10. Find the product cipher obtained by using the transformation $C \equiv aP + b$ (mod 26) followed by the transformation $C \equiv cP + d$ (mod 26), where $(a, 26) = (c, 26) = 1$.

A *Vigenère* cipher operates in the following way. A sequence of letters $l_1, l_2, ..., l_n$, with numerical equivalents $k_1, k_2, ..., k_n$, serves as the key. Plaintext messages are split into blocks of length $n$. To encipher a plaintext block of letters with numerical equivalents $p_1, p_2, ..., p_n$ to obtain a ciphertext block of letters with numerical equivalents $c_1, c_2, ..., c_n$, we use a sequence of shift ciphers with

$$c_i \equiv p_i + k_i \text{ (mod 26)}, \ 0 \leqslant c_i \leqslant 25,$$

for $i = 1, 2, ..., n$. In Exercise 11-13 we use the word SECRET as the key for a Vigenère cipher.

11. Using this Vigenère cipher, encipher the message

    DO NOT OPEN THIS ENVELOPE.

12. Decipher the following message which was enciphered using this Vigenère cipher:

    WBRCSL AZGJMG KMFV.

* 13. Describe how cryptanalysis of ciphertext, which was enciphered using a Vigenère cipher, can be carried out.

## 7.1 Computer Projects

Write programs to do the following:

1. Encipher messages using the Caesar cipher.

2. Encipher messages using the transformation $C \equiv P + k$ (mod 26), where $k$ is a given integer.

3. Encipher messages using the transformation $C \equiv aP + b$ (mod 26), where $a$ and $b$ are integers with $(a, 26) = 1$.

4. Decipher messages that have been enciphered using the Caesar cipher.

5. Decipher messages that have been enciphered using the transformation $C \equiv P + k$ (mod 26), where $k$ is a given integer.

6. Decipher messages that have been enciphered using the transformation $C \equiv aP + b$ (mod 26), where $a$ and $b$ are integers with $(a, 26) = 1$.

* 7. Cryptanalyze, using frequency counts, ciphertext that was enciphered using a transformation of the form $C \equiv P + k$ (mod 26) where $k$ is an unknown integer.

* 8. Cryptanalyze, using frequency counts, ciphertext that was enciphered using a transformation of the form $C \equiv aP + b$ (mod 26) where $a$ and $b$ are unknown integers with $(a, 26) = 1$.

9. Encipher messages using Vigenère ciphers (see the preamble to Exercise 11).

10. Decipher messages that have been enciphered using Vigenère ciphers.

---

## 7.2 Block Ciphers

We have seen that monographic ciphers based on substitution are vulnerable to cryptanalysis based on the frequency of occurrence of letters in the ciphertext. To avoid this weakness, cipher systems were developed that substitute for each block of plaintext letters of a specified length, a block of ciphertext letters of the same length. Ciphers of this sort are called *block* or *polygraphic ciphers*. In this section, we will discuss some polygraphic ciphers based on modular arithmetic; these were developed by Hill [122] around 1930.

First, we consider *digraphic ciphers*; in these ciphers each block of two letters of plaintext is replaced by a block of two letters of ciphertext. We illustrate this process with an example.

The first step is to split the message into blocks of two letters (adding a dummy letter, say X, at the end of the message, if necessary, so that the final block has two letters). For instance, the message

THE GOLD IS BURIED IN ORONO

is split up as

TH EG OL DI SB UR IE DI NO RO NO.

Next, these letters are translated into their numerical equivalents (as previously done) to obtain

19 7     4 6     14 11     3 8     18 1     20 17     8 4     3 8
13 14     17 14     13 14.

Each block of two plaintext numbers $P_1P_2$ is converted into a block of two ciphertext numbers $C_1C_2$:

$$C_1 \equiv 5P_1 + 17P_2 \pmod{26}$$
$$C_2 \equiv 4P_1 + 15P_2 \pmod{26}.$$

For instance, the first block 19 7 is converted to 6 25, because

$$C_1 \equiv 5{\cdot}19 + 17{\cdot}7 \equiv 6 \pmod{26}$$
$$C_2 \equiv 4{\cdot}19 + 15{\cdot}7 \equiv 25 \pmod{26}.$$

After performing this operation on the entire message, the following ciphertext is obtained:

   6 25  18 2  23 13  21 2  3 9  25 23  4 14  21 2  17 2  11 18  17 2.

When these blocks are translated into letters, we have the ciphertext message

   GZ  SC  XN  VC  DJ  ZX  EO  VC  RC  LS  RC.

The deciphering procedure for this cipher system is obtained by using Theorem 3.14. To find the plaintext block $P_1P_2$ corresponding to the ciphertext block $C_1C_2$, we use the relationship

$$P_1 \equiv 17C_1 + 5C_2 \pmod{26}$$
$$P_2 \equiv 18C_1 + 23C_2 \pmod{26}.$$

The digraphic cipher system we have presented here is conveniently described using matrices. For this cipher system, we have

$$\begin{pmatrix} C_1 \\ C_2 \end{pmatrix} \equiv \begin{pmatrix} 5 & 17 \\ 4 & 15 \end{pmatrix} \begin{pmatrix} P_1 \\ P_2 \end{pmatrix} \pmod{26}.$$

By Theorem 3.16, we see that the matrix $\begin{pmatrix} 17 & 5 \\ 18 & 23 \end{pmatrix}$ is an inverse of $\begin{pmatrix} 5 & 17 \\ 4 & 15 \end{pmatrix}$ modulo 26. Hence, Theorem 3.15 tells us that deciphering can be done using the relationship

$$\begin{pmatrix} P_1 \\ P_2 \end{pmatrix} \equiv \begin{pmatrix} 17 & 5 \\ 18 & 23 \end{pmatrix} \begin{pmatrix} C_1 \\ C_2 \end{pmatrix} \pmod{26}.$$

In general, a Hill cipher system may be obtained by splitting plaintext into blocks of $n$ letters, translating the letters into their numerical equivalents, and forming ciphertext using the relationship

$$C \equiv AP \pmod{26},$$

where $A$ is an $n \times n$ matrix, $(\det A, 26) = 1$, $C = \begin{pmatrix} C_1 \\ C_2 \\ \cdot \\ \cdot \\ \cdot \\ C_n \end{pmatrix}$ and $P = \begin{pmatrix} P_1 \\ P_2 \\ \cdot \\ \cdot \\ \cdot \\ P_n \end{pmatrix}$ and

where $C_1C_2...C_n$ is the ciphertext block that corresponds to the plaintext block $P_1P_2...P_n$. Finally, the ciphertext numbers are translated back to letters. For deciphering, we use the matrix $\overline{A}$, an inverse of $A$ modulo 26, which may be obtained using Theorem 3.18. Since $\overline{A}A \equiv I \pmod{26}$, we have

$$\overline{A}C \equiv \overline{A}(AP) \equiv (\overline{A}A)P \equiv P \pmod{26}.$$

Hence, to obtain plaintext from ciphertext, we use the relationship

$$P \equiv \overline{A}C \pmod{26}.$$

We illustrate this procedure using $n = 3$ and the enciphering matrix

$$A = \begin{pmatrix} 11 & 2 & 19 \\ 5 & 23 & 25 \\ 20 & 7 & 1 \end{pmatrix}.$$

Since det $A \equiv 5 \pmod{26}$, we have (det $A, 26) = 1$. To encipher a plaintext block of length three, we use the relationship

$$\begin{pmatrix} C_1 \\ C_2 \\ C_3 \end{pmatrix} \equiv A \begin{pmatrix} P_1 \\ P_2 \\ P_3 \end{pmatrix} \pmod{26}.$$

To encipher the message STOP PAYMENT, we first split the message into blocks of three letters, adding a final dummy letter X to fill out the last block. We have plaintext blocks

$$\text{STO PPA YME NTX.}$$

We translate these letters into their numerical equivalents

$$18\ 19\ 14 \quad 15\ 15\ 0 \quad 24\ 12\ 4 \quad 13\ 19\ 23.$$

We obtain the first block of ciphertext in the following way:

$$\begin{pmatrix} C_1 \\ C_2 \\ C_3 \end{pmatrix} = \begin{pmatrix} 11 & 2 & 19 \\ 5 & 23 & 25 \\ 20 & 7 & 1 \end{pmatrix} \begin{pmatrix} 18 \\ 19 \\ 14 \end{pmatrix} = \begin{pmatrix} 8 \\ 19 \\ 13 \end{pmatrix} \pmod{26}.$$

Enciphering the entire plaintext message in the same manner, we obtain the ciphertext message

$$8\ 19\ 13 \quad 13\ 4\ 15 \quad 0\ 2\ 22 \quad 20\ 11\ 0.$$

Translating this message into letters, we have our ciphertext message

$$\text{ITN NEP ACW ULA.}$$

The deciphering process for this polygraphic cipher system takes a ciphertext block and obtains a plaintext block using the transformation

$$\begin{pmatrix} P_1 \\ P_2 \\ P_3 \end{pmatrix} \equiv \overline{A} \begin{pmatrix} C_1 \\ C_2 \\ C_3 \end{pmatrix} \pmod{26}$$

where

$$\overline{A} = \begin{pmatrix} 6 & -5 & 11 \\ -5 & -1 & -10 \\ -7 & 3 & 7 \end{pmatrix}$$

is an inverse of $A$ modulo 26, which may be obtained using Theorem 3.18.

Because polygraphic ciphers operate with blocks, rather than with individual letters, they are not vulnerable to cryptanalysis based on letter frequency. However, polygraphic ciphers operating with blocks of size $n$ are vulnerable to cryptanalysis based on frequencies of blocks of size $n$. For instance, with a digraphic cipher system, there are $26^2 = 676$ digraphs, blocks of length two. Studies have been done to compile the relative frequencies of digraphs in typical English text. By comparing the frequencies of digraphs in the ciphertext with the average frequencies of digraphs, it is often possible to successfully attack digraphic ciphers. For example, according to some counts, the most common digraph in English is TH, followed closely by HE. If a Hill digraphic cipher system has been employed and the most common digraph is KX, followed by VZ, we may guess that the ciphertext digraphs KX and VZ correspond to TH and HE, respectively. This would mean that the blocks 19 7 and 7 4 are sent to 10 23 and 21 25, respectively. If $A$ is the enciphering matrix, this implies that

$$A \begin{pmatrix} 19 & 7 \\ 7 & 4 \end{pmatrix} \equiv \begin{pmatrix} 10 & 21 \\ 23 & 25 \end{pmatrix} \pmod{26}.$$

Since $\begin{pmatrix} 4 & 19 \\ 19 & 19 \end{pmatrix}$ is an inverse of $\begin{pmatrix} 19 & 7 \\ 7 & 4 \end{pmatrix} \pmod{26}$, we find that

$$A \equiv \begin{pmatrix} 10 & 21 \\ 23 & 25 \end{pmatrix} \begin{pmatrix} 4 & 19 \\ 19 & 19 \end{pmatrix} \equiv \begin{pmatrix} 23 & 17 \\ 21 & 2 \end{pmatrix} \pmod{26},$$

which gives a possible key. After attempting to decipher the ciphertext using $\overline{A} = \begin{pmatrix} 2 & 9 \\ 5 & 23 \end{pmatrix}$ to transform the ciphertext, we would know if our guess was correct.

In general, if we know $n$ correspondences between plaintext blocks of size $n$ and ciphertext blocks of size $n$, for instance if we know that the ciphertext blocks $C_{1j}C_{2j}...C_{nj}, j = 1,2,...,n$, correspond to the plaintext blocks $P_{1j}P_{2j}...P_{nj}, j = 1,2,...,n$, respectively, then we have

$$
\mathbf{A} \begin{pmatrix} P_{1j} \\ . \\ . \\ . \\ P_{nj} \end{pmatrix} \equiv \begin{pmatrix} C_{1j} \\ . \\ . \\ . \\ C_{nj} \end{pmatrix} \quad (\text{mod } 26),
$$

for $j = 1, 2, \ldots, n$.

These $n$ congruences can be succinctly expressed using the matrix congruence

$$
\mathbf{AP} \equiv \mathbf{C} \quad (\text{mod } 26),
$$

where $\mathbf{P}$ and $\mathbf{C}$ are $n \times n$ matrices with $ij$th entries $P_{ij}$ and $C_{ij}$, respectively. If $(\det \mathbf{P}, 26) = 1$, then we can find the enciphering matrix $\mathbf{A}$ via

$$
\mathbf{A} \equiv \mathbf{C}\overline{\mathbf{P}} \quad (\text{mod } 26),
$$

where $\overline{\mathbf{P}}$ is an inverse of $\mathbf{P}$ modulo 26.

Cryptanalysis using frequencies of polygraphs is only worthwhile for small values of $n$, where $n$ is the size of the polygraphs. When $n = 10$, for example, there are $26^{10}$, which is approximately $1.4 \times 10^{14}$, polygraphs of this length. Any analysis of the relative frequencies of these polygraphs is extremely infeasible.

---

### 7.2 Exercises

1. Using the digraphic cipher that sends the plaintext block $P_1 P_2$ to the ciphertext block $C_1 C_2$ with

   $$
   \begin{aligned}
   C_1 &\equiv 3P_1 + 10P_2 \quad (\text{mod } 26) \\
   C_2 &\equiv 9P_1 + 7P_2 \quad (\text{mod } 26),
   \end{aligned}
   $$

   encipher the message BEWARE OF THE MESSENGER.

2. Decipher the ciphertext message UW DM NK QB EK, which was enciphered using the digraphic cipher which sends the plaintext block $P_1 P_2$ into the ciphertext block $C_1 C_2$ with

   $$
   \begin{aligned}
   C_1 &\equiv 23P_1 + 3P_2 \quad (\text{mod } 26) \\
   C_2 &\equiv 10P_1 + 25P_2 \quad (\text{mod } 26).
   \end{aligned}
   $$

3. A cryptanalyst has determined that the two most common digraphs in a ciphertext message are RH and NI and guesses that these ciphertext digraphs correspond to the two most common diagraphs in English text, TH and HE. If the plaintext was enciphered using a Hill digraphic cipher described by

$$C_1 \equiv aP_1 + bP_2 \pmod{26}$$
$$C_2 \equiv cP_1 + dP_2 \pmod{26},$$

what are $a, b, c,$ and $d$?

4. How many pairs of letters remain unchanged when encryption is performed using each of the following digraphic ciphers?

a) $C_1 \equiv 4P_1 + 5P_2 \pmod{26}$
   $C_2 \equiv 3P_1 + P_2 \pmod{26}$

b) $C_1 \equiv 7P_1 + 17P_2 \pmod{26}$
   $C_2 \equiv P_1 + 6P_2 \pmod{26}$

c) $C_1 \equiv 3P_1 + 5P_2 \pmod{26}$
   $C_2 \equiv 6P_1 + 3P_2 \pmod{26}$

5. Show that if the enciphering matrix **A** in the Hill cipher system is involutory modulo 26, *i.e.*, $\mathbf{A}^2 \equiv \mathbf{I} \pmod{26}$, then **A** also serves as a deciphering matrix for this cipher system.

6. A cryptanalyst has determined that the three most common trigraphs (blocks of length three) in a ciphertext are, LME, WRI and ZYC and guesses that these ciphertext trigraphs correspond to the three most common trigraphs in English text, THE, AND, and THA. If the plaintext was enciphered using a Hill trigraphic cipher described by $\mathbf{C} \equiv \mathbf{AP} \pmod{26}$, what are the entries of the 3×3 enciphering matrix **A**?

7. Find the product cipher obtained by using the digraphic Hill cipher with enciphering matrix $\begin{pmatrix} 2 & 3 \\ 1 & 17 \end{pmatrix}$ followed by using the digraphic Hill cipher with enciphering matrix $\begin{pmatrix} 5 & 1 \\ 25 & 4 \end{pmatrix}$.

8. Show that the product cipher obtained from two digraphic Hill ciphers is again a digraphic Hill cipher.

* 9. Show that the product cipher obtained by enciphering first using a Hill cipher with blocks of size $m$ and then using a Hill cipher with blocks of size $n$ is again a Hill cipher using blocks of size $[m,n]$.

10. Find the 6×6 enciphering matrix corresponding to the product cipher obtained by first using the Hill cipher with enciphering matrix $\begin{pmatrix} 3 & 1 \\ 2 & 1 \end{pmatrix}$, followed by using the Hill cipher with enciphering matrix $\begin{pmatrix} 1 & 1 & 0 \\ 1 & 0 & 1 \\ 0 & 1 & 1 \end{pmatrix}$.

* 11.  A *transposition cipher* is a cipher where blocks of a specified size are enciphered by permuting their characters in a specified manner. For instance, plaintext blocks of length five, $P_1P_2P_3P_4P_5$, may be sent to ciphertext blocks $C_1C_2C_3C_4C_5 = P_4P_5P_2P_1P_3$. Show that every such transposition cipher is a Hill cipher with an enciphering matrix that contains only 0's and 1's as entries with the property that each row and each column contains exactly one 1.

## 7.2 Computer Projects

Write programs to do the following:

   1.  Encipher messages using a Hill cipher.

   2.  Decipher messages that were enciphered using a Hill cipher.

* 3.  Cryptanalyze messages that were enciphered using a digraphic Hill cipher, by analyzing the frequency of digraphs in the ciphertext.

---

## 7.3 Exponentiation Ciphers

In this section, we discuss a cipher, based on modular exponentiation, that was invented in 1978 by Pohlig and Hellman [126]. We will see that ciphers produced by this system are resistant to cryptanalysis.

Let $p$ be an odd prime and let $e$, the enciphering key, be a positive integer with $(e,p-1) = 1$. To encipher a message, we first translate the letters of the message into numerical equivalents (retaining initial zeros in the two-digit numerical equivalents of letters). We use the same relationship we have used before, as shown in Table 7.9.

| letter | A | B | C | D | E | F | G | H | I | J | K | L | M | N | O | P | Q | R | S | T | U | V | W | X | Y | Z |
|---|---|---|---|---|---|---|---|---|---|---|---|---|---|---|---|---|---|---|---|---|---|---|---|---|---|---|
| numerical equivalent | 00 | 01 | 02 | 03 | 04 | 05 | 06 | 07 | 08 | 09 | 10 | 11 | 12 | 13 | 14 | 15 | 16 | 17 | 18 | 19 | 20 | 21 | 22 | 23 | 24 | 25 |

**Table 7.9.** Two-digit Numerical Equivalents of Letters.

Next, we group the resulting numbers into blocks of $2m$ decimal digits, where $2m$ is the largest positive even integer such that all blocks of numerical equivalents corresponding to $m$ letters (viewed as a single integer with $2m$ decimal digits) are less than $p$, e.g. if $2525 < p < 252525$, then $m = 2$.

For each plaintext block $P$, which is an integer with $2m$ decimal digits, we form a ciphertext block $C$ using the relationship

$$C \equiv P^e \pmod{p}, \ 0 \leqslant C < p.$$

The ciphertext message consists of these ciphertext blocks which are integers less than $p$. Notice that different values of $e$ determine different ciphers and hence $e$ is aptly called the enciphering key. We illustrate the enciphering technique with the following example.

**Example 7.1.** Let the prime to be used as the modulus in the enciphering procedure be $p = 2633$ and let the enciphering key to be used as the exponent in the modular exponentiation be $e = 29$, so that $(e, p-1) = (29, 2632) = 1$. To encipher the plaintext message,

THIS IS AN EXAMPLE OF AN EXPONENTIATION CIPHER,

we first convert the letters of the message into their numerical equivalents, and then form blocks of length four from these digits, to obtain

| | | | | |
|---|---|---|---|---|
| 1907 | 0818 | 0818 | 0013 | 0423 |
| 0012 | 1511 | 0414 | 0500 | 1304 |
| 2315 | 1413 | 0413 | 1908 | 0019 |
| 0814 | 1302 | 0815 | 0704 | 1723. |

Note that we have added the two digits 23, corresponding to the letter X, at the end of the message to fill out the final block of four digits.

We next translate each plaintext block $P$ into a ciphertext block $C$ using the relationship

$$C \equiv P^{29} \pmod{2633}, \ 0 < C < 2633.$$

For instance, to obtain the first ciphertext block from the first plaintext block we compute

$$C \equiv 1907^{29} \equiv 2199 \pmod{2633}.$$

To efficiently carry out the modular exponentiation, we use the algorithm given in Section 3.1. When we encipher the blocks in this way, we find that the ciphertext message is

| | | | | |
|---|---|---|---|---|
| 2199 | 1745 | 1745 | 1206 | 2437 |
| 2425 | 1729 | 1619 | 0935 | 0960 |
| 1072 | 1541 | 1701 | 1553 | 0735 |
| 2064 | 1351 | 1704 | 1841 | 1459.     □ |

To decipher a ciphertext block $C$, we need to know a deciphering key, namely an integer $d$ such that $de \equiv 1 \pmod{p-1}$, so that $d$ is an inverse of $e \pmod{p-1}$, which exists since $(e, p-1) = 1$. If we raise the ciphertext block $C$ to the $d$th power modulo $p$, we recover our plaintext block $P$, since

$$C^d \equiv (P^e)^d = P^{ed} \equiv P^{k(p-1)+1} \equiv (P^{p-1})^k P \equiv P \pmod{p},$$

where $de = k(p-1) + 1$, for some integer $k$, since $de \equiv 1 \pmod{p-1}$. (Note that we have used Fermat's little theorem to see that $P^{p-1} \equiv 1 \pmod{p}$.)

**Example 7.2**. To decipher the ciphertext blocks generated using the prime modulus $p = 2633$ and the enciphering key $e = 29$, we need an inverse of $e$ modulo $p-1 = 2632$. An easy computation, as done in Section 3.2, shows that $d = 2269$ is such an inverse. To decipher the ciphertext block $C$ in order to find the corresponding plaintext block $P$, we use the relationship

$$P \equiv C^{2269} \pmod{2633}.$$

For instance, to decipher the ciphertext block 2199, we have

$$P \equiv 2199^{2269} \equiv 1907 \pmod{2633}.$$

Again, the modular exponentiation is carried out using the algorithm given in Section 3.2. □

For each plaintext block $P$ that we encipher by computing $P^e \pmod{p}$, we use only $O((\log_2 p)^3)$ bit operations, as Theorem 3.9 demonstrates. Before we decipher we need to find an inverse $d$ of $e$ modulo $p-1$. This can be done using $O(\log^3 p)$ bit operations (see Exercise 11 of Section 3.2), and this needs to be done only once. Then, to recover the plaintext block $P$ from a ciphertext block $C$, we simply need to compute the least positive residue of $C^d$ modulo $p$; we can do this using $O((\log_2 p)^3)$ bit operations. Consequently, the processes of enciphering and deciphering using modular exponentiation can be done rapidly.

On the other hand, cryptanalysis of messages enciphered using modular exponentiation generally cannot be done rapidly. To see this, suppose we know the prime $p$ used as the modulus, and moreover, suppose we know the plaintext block $P$ corresponding to a ciphertext block $C$, so that

(7.2) $$C \equiv P^e \pmod{p}.$$

For successful cryptanalysis, we need to find the enciphering key $e$. When the relationship (7.2) holds, we say that $e$ is the *logarithm of $C$ to the base $P$ modulo $p$*. There are various algorithms for finding logarithms to a given base modulo a prime. The fastest such algorithm requires approximately $\exp(\sqrt{\log p \log \log p})$ bit operations (see [116]). To find logarithms modulo a prime with $n$ decimal digits using the fastest known algorithm requires approximately the same number of bit operations as factoring integers with the same number of decimal digits, when the fastest known factoring algorithm is used. Consulting Table 2.1, we see that finding logarithms modulo a prime $p$ requires an extremely long time. For instance, when $p$ has 100 decimal digits, finding logarithms modulo $p$ requires approximately

74 years, whereas when $p$ has 200 decimal digits, approximately $3.8 \times 10^9$ years are required.

We should mention that for primes $p$ where $p-1$ has only small prime factors, it is possible to use special techniques to find logarithms modulo $p$ using $O(\log^2 p)$ bit operations. Clearly, this sort of prime should not be used as a modulus in this cipher system. Taking a prime $p = 2q + 1$, where $q$ is also prime, obviates this difficulty.

Modular exponentiation is useful for establishing *common keys* to be used by two or more individuals. These common keys may, for instance, be used as keys in a cipher system for sessions of data communication, and should be constructed so that unauthorized individuals cannot discover them in a feasible amount of computer time.

Let $p$ be a large prime and let $a$ be an integer relatively prime to $p$. Each individual in the network picks a key $k$ that is an integer relatively prime to $p-1$. When two individuals with keys $k_1$ and $k_2$ wish to exchange a key, the first individual sends the second the integer $y_1$, where

$$y_1 \equiv a^{k_1} (\text{mod } p), \quad 0 < y_1 < p,$$

and the second individual finds the common key $K$ by computing

$$K \equiv y_1^{k_2} \equiv a^{k_1 k_2} (\text{mod } p), \quad 0 < K < p.$$

Similarly, the second individual sends the first the integer $y_2$ where

$$y_2 \equiv a^{k_2} (\text{mod } p), \quad 0 < y_2 < p,$$

and the first individual finds the common key $K$ by computing

$$K \equiv y_2^{k_1} \equiv a^{k_1 k_2} (\text{mod } p), \quad 0 < K < p.$$

We note that other individuals in the network cannot find this common key $K$ in a feasible amount of computer time, since they must compute logarithms modulo $p$ to find $K$.

In a similar manner, a common key can be shared by any group of $n$ individuals. If these individuals have keys $k_1, k_2, ..., k_n$, they can share the common key

$$K = a^{k_1 k_2 \cdots k_n} (\text{mod } p).$$

We leave an explicit description of a method used to produce this common key K as a problem for the reader.

An amusing application of exponentiation ciphers has been described by Shamir, Rivest, and Adleman [131]. They show that by using exponentiation ciphers, a fair game of poker may be played by two players communicating via computers. Suppose Alex and Betty wish to play poker. First, they jointly choose a large prime $p$. Next, they individually choose secret keys $e_1$ and $e_2$, to be used as exponents in modular exponentiation. Let $E_{e_1}$ and $E_{e_2}$ represent the corresponding enciphering transformations, so that

$$E_{e_1}(M) \equiv M^{e_1} \pmod{p}$$
$$E_{e_2}(M) \equiv M^{e_2} \pmod{p},$$

where $M$ is a plaintext message. Let $d_1$ and $d_2$ be the inverses of $e_1$ and $e_2$ modulo $p$ respectively, and let $D_{e_1}$ and $D_{e_2}$ be the corresponding deciphering transformations, so that

$$D_{e_1}(C) \equiv C^{d_1} \pmod{p}$$
$$D_{e_2}(C) \equiv C^{d_2} \pmod{p},$$

where $C$ is a ciphertext message.

Note that enciphering transformations commute, that is

$$E_{e_1}(E_{e_2}(M)) = E_{e_2}(E_{e_1}(M)),$$

since

$$(M^{e_2})^{e_1} \equiv (M^{e_1})^{e_2} \pmod{p}.$$

To play electronic poker, the deck of cards is represented by the 52 messages

$$M_1 = \text{"TWO OF CLUBS"}$$
$$M_2 = \text{"THREE OF CLUBS"}$$
$$\cdot$$
$$\cdot$$
$$\cdot$$
$$M_{52} = \text{"ACE OF SPADES."}$$

When Alex and Betty wish to play poker electronically, they use the following sequence of steps. We suppose Betty is the dealer.

i. Betty uses her enciphering transformation to encipher the 52 messages for the cards. She obtains $E_{e_2}(M_1)$, $E_{e_2}(M_2)$,...,$E_{e_2}(M_{52})$. Betty shuffles the deck, by randomly reordering the enciphered messages. Then she sends the 52 shuffled enciphered messages to Alex.

ii. Alex selects, at random, five of the enciphered messages that Betty has sent him. He returns these five messages to Betty and she deciphers them to find her hand, using her deciphering transformation $D_{e_2}$, since $D_{e_2}(E_{e_2}(M)) = M$ for all messages $M$. Alex cannot determine which cards Betty has, since he cannot decipher the enciphered messages $E_{e_2}(M_j)$, $j = 1,2,...,52$.

iii. Alex selects five other enciphered messages at random. Let these messages be $C_1$, $C_2$, $C_3$, $C_4$, and $C_5$, where

$$C_j = E_{e_2}(M_{i_j}),$$

$j = 1,2,3,4,5$. Alex enciphers these five previously enciphered messages using his enciphering transformation. He obtains the five messages

$$C_j^* = E_{e_1}(C_j) = E_{e_1}(E_{e_2}(M_{i_j}))$$

$j = 1,2,3,4,5$. Alex sends these five messages that have been enciphered twice (first by Betty and afterwards by Alex) to Betty.

iv. Betty uses her deciphering transformation $D_{e_2}$ to find

$$\begin{aligned} D_{e_2}(C_j^*) &= D_{e_2}(E_{e_1}(E_{e_2}(M_{i_j}))) \\ &= D_{e_2}(E_{e_2}(E_{e_1}(M_{i_j}))) \\ &= E_{e_1}(M_{i_j}), \end{aligned}$$

since $E_{e_1}(E_{e_2}(M)) = E_{e_2}(E_{e_1}(M))$ and $D_{e_2}(E_{e_2}(M)) = M$ for all messages $M$. Betty sends the fives message $E_{e_1}(M_{i_j})$ back to Alex.

v. Alex uses his deciphering transformation $D_{e_1}$ to obtain his hand, since

$$D_{e_1}(E_{e_1}(M_{i_j})) = M_{i_j}.$$

When a game is played where it is necessary to deal additional cards, such as draw poker, the same steps are followed to deal additional cards from the remaining deck. Note that using the procedure we have described, neither player knows the cards in the hand of the other player, and all hands are equally likely for each player. To guarantee that no cheating has occurred, at the end of the game both players reveal their keys, so that each player can verify that the other player was actually dealt the cards claimed.

A description of a possible weakness in this scheme, and how it may be overcome, may be found in the exercise set of Section 9.1.

---

### 7.3 Exercises

1. Using the prime $p = 101$ and enciphering key $e = 3$, encipher the message GOOD MORNING using modular exponentiation.

2. What is the plaintext message that corresponds to the ciphertext 1213 0902 0539 1208 1234 1103 1374 that is produced using modular exponentiation with modulus $p = 2591$ and enciphering key $e = 13$?

3. Show that the enciphering and deciphering procedures are identical when enciphering is done using modular exponentiation with modulus $p = 31$ and enciphering key $e = 11$.

**4.** With modulus $p = 29$ and unknown enciphering key $e$, modular exponentiation produces the ciphertext 04 19 19 11 04 24 09 15 15. Cryptanalyze the above cipher, if it is also known that the ciphertext block 24 corresponds to the plaintext letter U (with numerical equivalent 20). (*Hint*: First find the logarithm of 24 to the base 20 modulo 29 using some guesswork.)

**5.** Using the method described in the text for exchanging common keys, what is the common key that can be used by individuals with keys $k_1 = 27$ and $k_2 = 31$ when the modulus is $p = 101$ and the base is $a = 5$?

**6.** What is the group key $K$ that can be shared by four individuals with keys $k_1 = 11$, $k_2 = 12$, $k_3 = 17$, $k_4 = 19$ using the modulus $p = 1009$ and base $a = 3$?

**\* 7.** Describe a procedure to allow $n$ individuals to share the common key described in the text.

## 7.3 Computer Projects

Write programs to do the following:

**1.** Encipher messages using modular exponentiation.

**2.** Decipher messages that have been enciphered using modular exponentiation.

**3.** Cryptanalyze ciphertext that has been enciphered using modular exponentiation when a correspondence between a plaintext block $P$ and a ciphertext block $C$ is known.

**4.** Produce common keys for individuals in a network.

**\* 5.** Play electronic poker using encryption via modular exponentiation.

## 7.4 Public-Key Cryptography

If one of the cipher systems previously described in this chapter is used to establish secure communications within a network, then each pair of communicants must employ an enciphering key that is kept secret from the other individuals in the network, since once the enciphering key in one of those cipher systems is known, the deciphering key can be found using a small amount of computer time. Consequently, to maintain secrecy the enciphering keys must themselves be transmitted over a channel of secure communications.

To avoid assigning a key to each pair of individuals that must be kept secret from the rest of the network, a new type of cipher system, called a *public-key cipher system*, has been recently introduced. In this type of cipher system, enciphering keys can be made public, since an unrealistically large amount of computer time is required to find a deciphering transformation from an enciphering transformation. To use a public-key cipher system to establish secret communications in a network of $n$ individuals, each individual produces a key of the type specified by the cipher system, retaining certain private information that went into the construction of the enciphering transformation $E(k)$, obtained from the key $k$ according to a specified rule. Then a directory of the $n$ keys $k_1, k_2, ..., k_n$ is published. When individual $i$ wishes to send a message to individual $j$, the letters of the message are translated into their numerical equivalents and combined into blocks of specified size. Then, for each plaintext block $P$ a corresponding ciphertext block $C = E_{k_j}(P)$ is computed using the enciphering transformation $E_{k_j}$. To decipher the message, individual $j$ applies the deciphering transformation $D_{k_j}$ to each ciphertext block $C$ to find $P$, i.e.

$$D_{k_j}(C) = D_{k_j}(E_{k_j}(P)) = P.$$

Since the deciphering transformation $D_{k_j}$ cannot be found in a realistic amount of time by anyone other than individual $j$, no unauthorized individuals can decipher the message, even though they know the key $k_j$. Furthermore, cryptanalysis of the ciphertext message, even with knowledge of $k_j$, is extremely infeasible due to the large amount of computer time needed.

The *RSA cipher system*, recently invented by Rivest, Shamir, and Adleman [128], is a public-key cipher system based on modular exponentiation where the keys are pairs $(e,n)$, consisting of an exponent $e$ and a modulus $n$ that is the product of two large primes, i.e. $n = pq$, where $p$ and $q$ are large primes, so that $(e, \phi(n)) = 1$. To encipher a message, we first translate the letters into their numerical equivalents and then form blocks of the largest possible size (with an even number of digits). To encipher a plaintext block $P$, we form a ciphertext block $C$ by

$$E(P) = C \equiv P^e \pmod{n}, \quad 0 < C < n.$$

The deciphering procedure requires knowledge of an inverse $d$ of $e$ modulo $\phi(n)$, which exists since $(e, \phi(n)) = 1$. To decipher the ciphertext block $C$, we find

$$D(C) \equiv C^d = (P^e)^d = P^{ed} = P^{k\phi(n)+1}$$
$$\equiv (P^{\phi(n)})^k P \equiv P \pmod{n},$$

where $ed = k\phi(n) + 1$ for some integer $k$, since $ed \equiv 1 \pmod{\phi(n)}$, and by Euler's theorem, we have $P^{\phi(n)} \equiv 1 \pmod{n}$, when $(P, n) = 1$ (the probability that $P$ and $n$ are not relatively prime is extremely small; see Exercise 4 at the end of this section). The pair $(d,n)$ is a *deciphering key*.

To illustrate how the RSA cipher system works, we present an example where the enciphering modulus is the product of the two primes 43 and 59 (which are smaller than the large primes that would actually be used). We have $n = 43 \cdot 59 = 2537$ as the modulus and $e = 13$ as the exponent for the RSA cipher. Note that we have $(e, \phi(n)) = (13, 42 \cdot 58) = 1$. To encipher the message

### PUBLIC KEY CRYPTOGRAPHY,

we first translate the letters into their numerical equivalents, and then group these numbers together into blocks of four. We obtain

$$
\begin{array}{cccc}
1520 & 0111 & 0802 & 1004 \\
2402 & 1724 & 1519 & 1406 \\
1700 & 1507 & 2423,
\end{array}
$$

where we have added the dummy letter $X = 23$ at the end of the passage to fill out the final block.

We encipher each plaintext block into a ciphertext block, using the relationship

$$C \equiv P^{13} \pmod{2537}.$$

For instance, when we encipher the first plaintext block 1520, we obtain the ciphertext block

$$C \equiv (1520)^{13} \equiv 95 \pmod{2537}.$$

Enciphering all the plaintext blocks, we obtain the ciphertext message

$$
\begin{array}{cccc}
0095 & 1648 & 1410 & 1299 \\
0811 & 2333 & 2132 & 0370 \\
1185 & 1457 & 1084.
\end{array}
$$

In order to decipher messages that were enciphered using the RSA cipher, we must find an inverse of $e = 13$ modulo $\phi(2537) = \phi(43 \cdot 59) = 42 \cdot 58 = 2436$. A short computation using the Euclidean algorithm, as done in Section 3.2, shows that $d = 937$ is an inverse of 13 modulo 2436. Consequently, to decipher the cipher text block $C$, we use the relationship

$$P \equiv C^{937} \pmod{2537}, \ 0 \leqslant P \leqslant 2537,$$

which is valid because

$$C^{937} \equiv (P^{13})^{937} \equiv (P^{2436})^5 P \equiv P \pmod{2537};$$

note that we have used Euler's theorem to see that

$$P^{\phi(2537)} = P^{2436} \equiv 1 \pmod{2537},$$

when $(P, 2537) = 1$ (which is true for all of the plaintext blocks in our example).

To understand how the RSA cipher system fulfills the requirements of a public-key cipher system, first note that each individual can find two large primes $p$ and $q$, with 100 decimal digits, in just a few minutes of computer time. These primes can be found by picking odd integers with 100 digits at random; by the prime number theorem, the probability that such an integer is prime is approximately $2/\log 10^{100}$. Hence, we expect to find a prime after examining an average of $1/(2/\log 10^{100})$, or approximately 115, such integers. To test these randomly chosen odd integers for primality, we use Rabin's probabilistic primality test discussed in Section 5.2. For each of these 100-digit odd integers we perform Miller's test for 100 bases less than the integer; the probability that a composite integer passes all these tests is less than $10^{-60}$. The procedure we have just outlined requires only a few minutes of computer time to find a 100-digit prime, and each individual need do it only twice.

Once the primes $p$ and $q$ have been found, an enciphering exponent $e$ should be chosen with $(e, \phi(pq)) = 1$. One suggestion for choosing $e$ is to take any prime greater than both $p$ and $q$. No matter how $e$ is found, it should be true that $2^e > n = pq$, so that it is impossible to recover the plaintext block $P$, $P \neq 0$ or 1, just by taking the $e$th root of the integer $C$ with $C \equiv P^e \pmod{n}$, $0 < C < n$. As long as $2^e > n$, every message other than $P = 0$ and 1, is enciphered by exponentiation followed by a reduction modulo $n$.

We note that the modular exponentiation needed for enciphering messages using the RSA cipher system can be done using only a few seconds of computer time when the modulus, exponent, and base in the modular exponentiation have as many as 200 decimal digits. Also, using the Euclidean algorithm, we can rapidly find an inverse $d$ of the enciphering exponent $e$ modulo $\phi(n)$ when the primes $p$ and $q$ are known, so that $\phi(n) = \phi(pq) = (p-1)(q-1)$ is known.

To see why knowledge of the enciphering key $(e, n)$ does not easily lead to the deciphering key $(d, n)$, note that to find $d$, an inverse of $e$ modulo $\phi(n)$, requires that we first find $\phi(n) = \phi(pq) = (p-1)(q-1)$. Note that finding $\phi(n)$ is not easier than factoring the integer $n$. To see why, note that $p + q = n - \phi(n) + 1$ and $p - q = \sqrt{(p+q)^2 - 4pq} = \sqrt{(p+q)^2 - 4n}$, so that $p = \frac{1}{2}[(p+q) + (p-q)]$ and $q = \frac{1}{2}[(p+q) + (p-q)]$, and consequently $p$ and $q$ can easily be found when $n = pq$ and $\phi(n) = (p-1)(q-1)$ are known. Note that when $p$ and $q$ both have around 100 decimal digits, $n = pq$ has around 200 decimal digits. From Table 2.1, we see that using the fastest factorization algorithm known, $3.8 \times 10^9$ years of computer time are required to factor an integer of this size. Also, if the integer $d$ is known, but $\phi(n)$ is not, then $n$ may also be factored easily, since $ed - 1$ is a multiple of $\phi(n)$ and there are special algorithms for factoring an integer $n$ using any multiple of

$\phi(n)$ (see Miller [102]). It has not been proven that it is impossible to decipher messages enciphered using the RSA cipher system without factoring $n$, but so far no such method has been discovered. As yet, all deciphering methods suggested that work in general are equivalent to factoring $n$, and as we have remarked, factoring large integers seems to be an intractable problem, requiring tremendous amounts of computer time.

A few extra precautions should be taken in choosing the primes $p$ and $q$ to be used in the RSA cipher system to prevent the use of special rapid techniques to factor $n = pq$. For example, both $p - 1$ and $q - 1$ should have large prime factors, $(p - 1, q - 1)$ should be small, and $p$ and $q$ should have decimal expansions differing in length by a few digits.

For the RSA cipher system, once the modulus $n$ has been factored, it is easy to find the deciphering transformation from the enciphering transformation. It may be possible to somehow find the deciphering transformation from the enciphering transformation without factoring $n$, although this seems unlikely. Rabin [127] has discovered a variant of the RSA cipher system for which factorization of the modulus $n$ has almost the same computational complexity as obtaining the deciphering transformation from the enciphering transformation. To describe Rabin's cipher system, let $n = pq$, where $p$ and $q$ are odd primes, and let $b$ be an integer with $0 \leqslant b < n$. To encipher the plaintext message $P$, we form

$$C \equiv P(P + b) \pmod{n}.$$

We will not discuss the deciphering procedure for Rabin ciphers here, because it relies on some concepts we have not yet developed (see Exercise 43 in Section 9.1). However, we remark that there are four possible values of $P$ for each ciphertext $C$ such that $C \equiv P(P + b) \pmod{n}$, an ambiguity which complicates the deciphering process. When $p$ and $q$ are known, the deciphering procedure for a Rabin cipher can be carried out rapidly since $O(\log n)$ bit operations are needed.

Rabin has shown that if there is an algorithm for deciphering in this cipher system, without knowledge of the primes $p$ and $q$, that requires $f(n)$ bit operations, then there is an algorithm for the factorization of $n$ requiring only $2(f(n) + \log n)$ bit operations. Hence the process of deciphering messages enciphered with a Rabin cipher without knowledge of $p$ and $q$ is a problem of computational complexity similar to that of factorization.

Public-key cipher systems can also be used to send signed messages. When signatures are used, the recipient of a message is sure that the message came from the sender, and can convince an impartial judge that only the sender could be the source of the message. This authentication is needed for electronic mail, electronic banking, and electronic stock market transactions. To see how the RSA cipher system can be used to send signed messages, suppose that individual $i$ wishes to send a signed message to individual $j$. The first thing that individual $i$ does to a plaintext block $P$ is to compute

$$S = D_{k_i}(P) \equiv P^{d_i} \pmod{n_i},$$

where $(d_i, n_i)$ is the deciphering key for individual $i$, which only individual $i$ knows. Then, if $n_j > n_i$, where $(e_j, n_j)$ is the enciphering key for individual $j$, individual $i$ enciphers $S$ by forming

$$C = E_{k_j}(S) \equiv S^{e_j} \pmod{n_j}, \quad 0 < C < n_j.$$

When $n_j < n_i$ individual $i$ splits $S$ into blocks of size less than $n_j$ and enciphers each block using the enciphering transformation $E_{k_j}$.

For deciphering, individual $j$ first uses the private deciphering transformation $D_{k_j}$ to recover $S$, since

$$D_{k_j}(C) = D_{k_j}(E_{k_j}(S)) = S.$$

To find the plaintext message $P$, supposedly sent by individual $i$, individual $j$ next uses the public enciphering transformation $E_{k_i}$, since

$$E_{k_i}(S) = E_{k_i}(D_{k_i}(P)) = P.$$

Here, we have used the identity $E_{k_i}(D_{k_i}(P)) = P$, which follows from the fact that

$$E_{k_i}(D_{k_i}(P)) \equiv (P^{d_i})^{e_i} \equiv P^{d_i e_i} \equiv P \pmod{n_i},$$

since

$$d_i e_i \equiv 1 \pmod{\phi(n_i)}.$$

The combination of the plaintext block $P$ and the signed version $S$ convinces individual $j$ that the message actually came from individual $i$. Also, individual $i$ cannot deny sending the message, since no one other than individual $i$ could have produced the signed message $S$ from the original message $P$.

The RSA cipher system relies on the difference in the computer time needed to find primes and the computer time needed to factor. In Chapter 9, we will use this same difference to develop a technique to "flip coins" electronically.

---

## 7.4 Exercises

1. Find the primes $p$ and $q$ if $n = pq = 14647$ and $\phi(n) = 14400$.

2. Find the primes $p$ and $q$ if $n = pq = 4386607$ and $\phi(n) = 4382136$.

3. Suppose a cryptanalyst discovers a message $P$ that is not relatively prime to the enciphering modulus $n = pq$ used in a RSA cipher. Show that the

cryptanalyst can factor $n$.

**4.** Show that it is extremely unlikely that a message such as that described in Exercise 3 can be discovered. Do this by demonstrating that the probability that a message $P$ is not relatively prime to $n$ is $\frac{1}{p} + \frac{1}{q} - \frac{1}{pq}$, and if $p$ and $q$ are both larger than $10^{100}$, this probability is less than $10^{-99}$.

**5.** What is the ciphertext that is produced when the RSA cipher with key $(e,n) = (3,2669)$ is used to encipher the message BEST WISHES?

**6.** If the ciphertext message produced by the RSA cipher with key $(e,n) = (5,2881)$ is 0504 1874 0347 0515 2088 2356 0736 0468, what is the plaintext message?

**7.** Harold and Audrey have as their RSA keys $(3,23\cdot47)$ and $(7,31\cdot59)$, respectively.

    a) Using the method in the text, what is the signed ciphertext sent by Harold to Audrey, when the plaintext message is CHEERS HAROLD?

    b) Using the method in the text, what is the signed ciphertext sent by Audrey to Harold when the plaintext message is SINCERELY AUDREY?

In Exercises 8 and 9, we present two methods for sending signed messages using the RSA cipher system, avoiding possible changes in block sizes.

**\* 8.** Let $H$ be a fixed integer. Let each individual have two pairs of enciphering keys: $k = (e,n)$ and $k^* = (e,n^*)$ with $n < H < n^*$, where $n$ and $n^*$ are both the product of two primes. Using the RSA cipher system, individual $i$ can send a signed message $P$ to individual $j$ by sending $E_{k_j^*}(D_{k_i}(P))$.

    a) Show that it is not necessary to change block sizes when the transformation $E_{k_j^*}$ is applied after $D_{k_i}$ has been applied.

    b) Explain how individual $j$ can recover the plaintext message $P$, and why no one other than individual $i$ could have sent the message.

    c) Let individual $i$ have enciphering keys $(3,11\cdot71)$ and $(3,29\cdot41)$ so that $781 = 11\cdot71 < 1000 < 1189 = 29\cdot41$, and let individual $j$ have enciphering keys $(7,19\cdot47)$ and $(7,31\cdot37)$, so that $893 = 19\cdot47 < 1000 < 1147 = 31\cdot37$. What ciphertext message does individual $i$ send to individual $j$ using the method given in this problem when the signed plaintext message is HELLO ADAM? What ciphertext message does individual $j$ send to individual $i$ when the signed plaintext message is GOODBYE ALICE?

**\* 9.** a) Show that if individuals $i$ and $j$ have enciphering keys $k_i = (e_i,n_i)$ and $k_j = (e_j,n_j)$, respectively, where both $n_i$ and $n_j$ are products of two distinct primes, then individual $i$ can send a signed message $P$ to individual $j$ without needing to change the size of blocks by sending

$$E_{k_j}(D_{k_i}(P)) \text{ if } n_i < n_j$$
$$D_{k_i}(E_{k_j}(P)) \text{ if } n_i > n_j .$$

b) How can individual $j$ recover $P$?

c) How can individual $j$ guarantee that a message came from individual $i$?

d) Let $k_i = (11,47·61)$ and $k_j = (13,43·59)$. Using the method described in part (a), what does individual $i$ send to individual $j$ if the message is REGARDS FRED, and what does individual $j$ send to individual $i$ if the message is REGARDS ZELDA?

10. Encipher the message SELL NOW using the Rabin cipher $C \equiv P(P+5) \pmod{2573}$.

## 7.4 Computer Projects

Write programs to do the following:

1. Encipher messages with an RSA cipher.

2. Decipher messages that were enciphered using an RSA cipher.

3. Send signed messages using an RSA cipher and the method described in the text.

4. Send signed messages using an RSA cipher and the method in Exercise 8.

5. Send signal messages using an RSA cipher and the method in Exercise 9.

6. Encipher messages using a Rabin cipher.

## 7.5 Knapsack Ciphers

In this section, we discuss cipher systems based on the knapsack problem. Given a set of positive integers $a_1, a_2,..., a_n$ and an integer $S$, the *knapsack problem* asks which of these integers, if any, add together to give $S$. Another way to phrase the knapsack problem is to ask for values of $x_1, x_2,..., x_n$, each either 0 or 1, such that

(7.3) $$S = a_1 x_1 + a_2 x_2 + \cdots + a_n x_n.$$

We use an example to illustrate the knapsack problem.

**Example 7.3.** Let $(a_1, a_2, a_3, a_4, a_5) = (2,7,8,11,12)$. By inspection, we see that there are two subsets of these five integers that add together to give 21, namely $21 = 2+8+11 = 2+7+12$. Equivalently, there are exactly two solutions to the equation $2x_1 + 7x_2 + 8x_3 + 11x_4 + 12x_5 = 21$, with $x_i = 0$

or 1 for $i = 1,2,3,4,5$, namely $x_1 = x_3 = x_4 = 1$, $x_2 = x_5 = 0$, and $x_1 = x_2 = x_5 = 1$, $x_3 = x_4 = 0$. $\qquad\square$

To verify that equation (7.3) holds, where each $x_i$ is either 0 or 1, requires that we perform at most $n$ additions. On the other hand, to search by trial and error for solutions of (7.3), may require that we check all $2^n$ possibilities for $(x_1, x_2,..., x_n)$. The best method known for finding a solution of the knapsack problem requires $O(2^{n/2})$ bit operations, which makes a computer solution of a general knapsack problem extremely infeasible even when $n = 100$.

Certain values of the integers $a_1, a_2,..., a_n$ make the solution of the knapsack problem much easier than the solution in the general case. For instance, if $a_j = 2^{j-1}$, to find the solution of $S =$ $a_1 x_1 + a_2 x_2 + \cdots + a_n x_n$, where $x_i = 0$ or 1 for $i = 1,2,...,n$, simply requires that we find the binary expansion of $S$. We can also produce easy knapsack problems by choosing the integers $a_1, a_2,..., a_n$ so that the sum of the first $j-1$ of these integers is always less than the $j$th integer, i.e. so that

$$\sum_{i=1}^{j-1} a_i < a_j, \qquad j = 2,3, ...,n.$$

If a sequence of integers $a_1, a_2,..., a_n$ satisfies this inequality, we call the sequence *super-increasing*.

**Example 7.4.** The sequence 2, 3, 7, 14, 27 is super-increasing because $3 > 2$, $7 > 3+2$, $14 > 7+3+2$, and $27 > 14+7+3+2$. $\qquad\square$

To see that knapsack problems involving super-increasing sequences are easy to solve, we first consider an example.

**Example 7.5.** Let us find the integers from the set 2, 3, 7, 14, 27 that have 37 as their sum. First, we note that since $2 + 3 + 7 + 14 < 27$, a sum of integers from this set can only be greater than 27 if the sum contains the integer 27. Hence, if $2x_1 + 3x_2 + 7x_3 + 14x_4 + 27x_5 = 37$ with each $x_i = 0$ or 1, we must have $x_5 = 1$ and $2x_1 + 3x_2 + 7x_3 + 14x_4 = 10$. Since $14 > 10$, $x_4$ must be 0 and we have $2x_1 + 3x_2 + 7x_3 = 10$. Since $2 + 3 < 7$, we must have $x_3 = 1$ and therefore $2x_1 + 3x_2 = 3$. Obviously, we have $x_2 = 1$ and $x_1 = 0$. The solution is $37 = 3 + 7 + 27$. $\qquad\square$

In general, to solve knapsack problems for a super-increasing sequence $a_1$, $a_2,..., a_n$, i.e. to find the values of $x_1, x_2, ..., x_n$ with $S = a_1 x_1 + a_2 x_2 + \cdots + a_n x_n$ and $x_i = 0$ or 1 for $i = 1, 2,..., n$ when $S$ is given, we use the following algorithm. First, we find $x_n$ by noting that

$$x_n = \begin{cases} 1 \text{ if } & S \geqslant a_n \\ 0 \text{ if } & S < a_n. \end{cases}$$

Then, we find $x_{n-1}, x_{n-2}, ..., x_1$, in succession, using the equations

$$x_j = \begin{cases} 1 \text{ if } & S - \sum_{i=j+1}^{n} x_i a_i \geqslant a_j \\ 0 \text{ if } & S - \sum_{i=j+1}^{n} x_i a_i < a_j, \end{cases}$$

for $j = n-1, n-2, ..., 1$.

To see that this algorithm works, first note that if $x_n = 0$ when $S \geqslant a_n$, then $\sum_{i=1}^{n} a_i x_i \leqslant \sum_{i=1}^{n-1} a_i < a_n \leqslant S$, contradicting the condition $\sum_{j=1}^{n} a_j x_j = S$. Similarly, if $x_j = 0$ when $S - \sum_{i=j+1}^{n} x_i a_i \geqslant a_j$, then $\sum_{i=1}^{n} a_i x_i \leqslant \sum_{i=1}^{j-1} x_i + \sum_{i=j+1}^{n} x_i a_i < a_j + \sum_{i=j+1}^{n} x_i a_i \leqslant S$, which is again a contradiction.

Using this algorithm, knapsack problems based on super-increasing sequences can be solved extremely quickly. We now discuss a cipher system based on this observation. This cipher system was invented by Merkle and Hellman [125], and was considered a good choice for a public-key cipher system until recently. We will comment more about this later.

The ciphers that we describe here are based on transformed super-increasing sequences. To be specific, let $a_1, a_2, ..., a_n$ be super-increasing and let $m$ be a positive integer with $m > 2a_n$. Let $w$ be an integer relatively prime to $m$ with inverse $\overline{w}$ modulo $m$. We form the sequence $b_1, b_2, ..., b_n$ where $b_j \equiv wa_j \pmod{m}$ and $0 < b_j < m$. We cannot use a special technique to solve a knapsack problem of the type $S = \sum_{i=1}^{n} b_i x_i$ where $S$ is a positive integer, since the sequence $b_1, b_2, ..., b_n$ is not super-increasing. However, when $\overline{w}$ is known, we can find

(7.4)  $$\overline{w}S = \sum_{i=1}^{n} \overline{w} b_i x_i \equiv \sum_{i=1}^{n} a_i x_i \pmod{m},$$

since $\overline{w} b_j \equiv a_j \pmod{m}$. From (7.4) we see that

$$S_0 = \sum_{i=1}^{n} a_i x_i$$

where $S_0$ is the least positive residue of $\overline{w}S$ modulo $m$. We can easily solve the equation

$$S_0 = \sum_{i=1}^{n} a_i x_i,$$

since $a_1, a_2,..., a_n$ is super-increasing. This solves the knapsack problem

$$S = \sum_{i=1}^{n} b_i x_i,$$

since $b_j \equiv wa_j$ (mod $m$) and $0 \leqslant b_j < m$. We illustrate this procedure with an example.

**Example 7.6.** The super-increasing sequence $(a_1,a_2,a_3,a_4,a_5) = (3,5,9,20,44)$ can be transformed into the sequence $(b_1,b_2,b_3,b_4,b_5) = (23,68,69,5,11)$ by taking $b_j \equiv 67a_j$ (mod 89), for $j = 1,2,3,4,5$. To solve the knapsack problem $23x_1 + 68x_2 + 69x_3 + 5x_4 + 11x_5 = 84$, we can multiply both sides of this equation by 4, an inverse of 67 modulo 89, and then reduce modulo 89, to obtain the congruence $3x_1 + 5x_2 + 9x_3 + 20x_4 + 44x_5 \equiv 336 \equiv 69$ (mod 89). Since $89 > 3 + 5 + 9 + 20 + 44$, we can conclude that $3x_1 + 5x_2 + 9x_3 + 20x_4 + 44x_5 = 69$. The solution of this easy knapsack problem is $x_5 = x_4 = x_2 = 1$ and $x_3 = x_1 = 0$. Hence, the original knapsack problem has as its solution $68 + 5 + 11 = 84$.

The cipher system based on the knapsack problem works as follows. Each individual chooses a super-increasing sequence of positive integers of a specified length, say $N$, e.g. $a_1, a_2,..., a_N$, as well as a modulus $m$ with $m > 2a_N$ and a multiplier $w$ with $(m,w) = 1$. The transformed sequence $b_1, b_2,..., b_N$, where $b_j \equiv wa_j$ (mod $m$), $0 < b_j < m$, for $j = 1,2, ...,N$, is made public. When someone wishes to send a message $P$ to this individual, the message is first translated into a string of 0's and 1's using the binary equivalents of letters, as shown in Table 7.10. This string of zeros and ones is next split into segments of length $N$ (for simplicity we suppose that the length of the string is divisible by $N$; if not, we can simply fill out the last block with all 1's). For each block, a sum is computed using the sequence $b_1, b_2,..., b_N$; for instance, the block $x_1x_2...x_N$ gives $S = b_1x_1 + b_2x_2 + \cdots + b_Nx_N$. Finally, the sums generated by each block form the ciphertext message.

We note that to decipher ciphertext generated by the knapsack cipher, without knowledge of $m$ and $w$, requires that a group of hard knapsack problems of the form

(7.5)                     $S = b_1x_1 + b_2x_2 + \cdots + b_Nx_N$

be solved. On the other hand, when $m$ and $w$ are known, the knapsack problem (7.5) can be transformed into an easy knapsack problem, since

$$\overline{w}S = \overline{w}b_1x_1 + \overline{w}b_2x_2 + \cdots + \overline{w}b_Nx_N$$
$$\equiv a_1x_1 + a_2x_2 + \cdots + a_Nx_N \pmod{m},$$

where $\overline{w}b_j = a_j \pmod{m}$, where $\overline{w}$ is an inverse of w modulo m, so that

$$(7.6) \qquad\qquad S_0 = a_1x_1 + a_2x_2 + \cdots + a_Nx_N,$$

where $S_0$ is the least positive residue of $\overline{w}S$ modulo $m$. We have equality in (7.6), since both sides of the equation are positive integers less than $m$ which are congruent modulo $m$.

We illustrate the enciphering and deciphering procedures of the knapsack cipher with an example. We start with the super-increasing sequence $(a_1,a_2,a_3,a_4,a_5,a_6,a_7,a_8,a_9,a_{10}) = (2,11,14,29,58,119,241,480,959,1917)$. We take $m = 3837$ as the enciphering modulus, so that $m > 2a_{10}$, and $w = 1001$ as the multiplier, so that $(m,w) = 1$, to transform the super-increasing sequence into the sequence $(2002,3337,2503,2170, 503,$ $172,3347,855,709,417)$.

To encipher the message

### REPLY IMMEDIATELY,

we first translate the letters of the message into their five digit binary equivalents, as shown in Table 7.10, and then group these digits into blocks of ten, to obtain

| | | |
|---|---|---|
| 1000100100 | 0111101011 | 1100001000 |
| 0110001100 | 0010000011 | 0100000000 |
| 1001100100 | 0101111000. | |

For each block of ten binary digits, we form a sum by adding together the appropriate terms of the sequence (2002, 3337, 2503, 2170, 503, 172, 3347, 855, 709, 417) in the slots corresponding to positions of the block containing a digit equal to 1. This gives us

$$3360 \quad 12986 \quad 8686 \quad 10042 \quad 3629 \quad 3337 \quad 5530 \quad 9729.$$

For instance, we compute the first sum, 3360, by adding 2002, 503, and 855.

To decipher, we find the least positive residue modulo 3837 of 23 times each sum, since 23 is an inverse of 1001 modulo 3837, and then we solve the corresponding easy knapsack problem with respect to the original super-increasing sequence $(2,11,14,29,58,119,241,480,959,1917)$. For example, to decipher the first block, we find that $3360\cdot23 \equiv 540 \pmod{3837}$, and then note that $540 = 480 + 58 + 2$. This tells us that the first block of plaintext binary digits is 1000100100.

| letter | binary equivalent | letter | binary equivalent |
|--------|-------------------|--------|-------------------|
| A | 00000 | N | 01101 |
| B | 00001 | O | 01110 |
| C | 00010 | P | 01111 |
| D | 00011 | Q | 10000 |
| E | 00100 | R | 10001 |
| F | 00101 | S | 10010 |
| G | 00110 | T | 10011 |
| H | 00111 | U | 10100 |
| I | 01000 | V | 10101 |
| J | 01001 | W | 10110 |
| K | 01010 | X | 10111 |
| L | 01011 | Y | 11000 |
| M | 01100 | Z | 11001 |

**Table 7.10.** The Binary Equivalents of Letters.

Recently, Shamir [129] has shown that knapsack ciphers are not satisfactory for public-key cryptography. The reason is that there is an efficient algorithm for solving knapsack problems involving sequences $b_1, b_2,..., b_n$ with $b_j \equiv wa_j \pmod{m}$, where $w$ and $m$ are relatively prime positive integers and $a_1, a_2,..., a_n$ is a super-increasing sequence. The algorithm found by Shamir can solve these knapsack problems using only $O(P(n))$ bit operations, where $P$ is a polynomial, instead of requiring exponential time, as is required for known algorithms for general knapsack problems, involving sequences of a general nature.

There are several possibilities for altering this cipher system to avoid the weakness found by Shamir. One such possibility is to choose a sequence of pairs of relatively prime integers $(w_1, m_1)$, $(w_2, m_2)$ ,..., $(w_r, m_r)$, and then form the series of sequences

$$b_j^{(1)} \equiv w_1 a_j \pmod{m_1}$$
$$b_j^{(2)} \equiv w_2 b_j^{(1)} \pmod{m_2}$$

.

.

.

$$b_j^{(r)} \equiv w_r b_j^{(r-1)} \pmod{m_r},$$

for $j = 1,2,...,n$. We then use the final sequence $b_1^{(r)}, b_2^{(r)},..., b_n^{(r)}$ as the enciphering sequence. Unfortunately, efficient algorithms have been found for solving knapsack problems involving sequences obtained by iterating modular multiplications with different moduli.

---

## 7.5 Exercises

1. Decide whether each of the following sequences is super-increasing

   a)  (3,5,9,19,40)        c)  (3,7,17,30,59)
   b)  (2,6,10,15,36)       d)  (11,21,41,81,151).

2. Show that if $a_1$, $a_2$,..., $a_n$ is a super-increasing sequence, then $a_j \geqslant 2^{j-1}$ for $j = 1,2, ..., n$.

3. Show that the sequence $a_1$, $a_2$,..., $a_n$ is super-increasing if $a_{j+1} > 2a_j$ for $j = 1,2, ..., n-1$.

4. Find all subsets of the integers 2, 3, 4, 7, 11, 13, 16 that have 18 as their sum.

5. Find the sequence obtained from the super-increasing sequence $(1,3,5,10,20,41,80)$ when modular multiplication is applied with multiplier $w = 17$ and modulus $m = 162$.

6. Encipher the message BUY NOW using the knapsack cipher based on the sequence obtained from the super-increasing sequence $(17,19,37,81,160)$, by performing modular multiplication with multiplier $w = 29$ and modulus $m = 331$.

7. Decipher the ciphertext 402 105 150 325 that was enciphered by the knapsack cipher based on the sequence $(306,374,233,19,259)$. This sequence is obtained by using modular multiplication with multiplier $w = 17$ and modulus $m = 464$, to transform the super-increasing sequence $(18,22,41,83,179)$.

8. Find the sequence obtained by applying successively the modular multiplications with multipliers and moduli $(7,92)$, $(11,95)$, and $(6,101)$, respectively, on the super-increasing sequence $(3,4,8,17,33,67)$.

9. What process can be employed to decipher messages that have been enciphered using knapsack ciphers that involve sequences arising from iterating modular multiplications with different moduli?

A *multiplicative knapsack problem* is a problem of the following type: Given positive integers $a_1,a_2,...,a_n$ and a positive integer $P$, find the subset, or subsets, of these integers with product $P$, or equivalently, find all solutions of

$$P = a_1^{x_1} a_2^{x_2} \cdots a_n^{x_n}$$

where $x_j = 0$ or 1 for $j = 1,2,...,n$.

10. Find all products of subsets of the integers 2,3,5,6, and 10 equal to 60.

11. Find all products of subsets of the integers $8,13,17,21,95,121$ equal to $15960$.

12. Show that if the integers $a_1,a_2,...,a_n$ are pairwise relatively prime, then the multiplicative knapsack problem $P = a_1^{x_1} a_2^{x_2} \cdots a_n^{x_n}$, $x_j = 0$ or $1$ for $j = 1,2,...,n$, is easily solved from the prime factorizations of the integers $P,a_1,a_2,...,a_n$, and show that if there is a solution, then it is unique.

13. Show that by taking logarithms to the base $b$ modulo $m$, where $(b,m) = 1$ and $0 < b < m$, the multiplicative knapsack problem

    $$P = a_1^{x_1} a_2^{x_2} \cdots a_n^{x_n}$$

    is converted into an additive knapsack problem

    $$S = \alpha_1 x_1 + \alpha_2 x_2 + \cdots + \alpha_n x_n$$

    where $S$, $\alpha_1$, $\alpha_2$,..., $\alpha_n$ are the logarithms of $P$, $a_1$, $a_2$,..., $a_n$ to the base $b$ modulo $m$, respectively.

14. Explain how Exercises 12 and 13 can be used to produce ciphers where messages are easily deciphered when the mutually relatively prime integers $a_1$, $a_2$,..., $a_n$ are known, but cannot be deciphered quickly when the integers $\alpha_1$, $\alpha_2$,..., $\alpha_n$ are known.

## 7.5 Computer Projects

Write programs to do the following:

1. Solve knapsack problems by trial and error.

2. Solve knapsack problems involving super-increasing sequences.

3. Encipher messages using knapsack ciphers.

4. Decipher messages that were enciphered using knapsack ciphers.

5. Encipher and decipher messages using knapsack ciphers involving sequences arising from iterating modular multiplications with different moduli.

6. Solve multiplicative knapsack problems involving sequences of mutually relatively prime integers (see Exercise 14).

## 7.6 Some Applications to Computer Science

In this section we describe two applications of cryptography to computer science. The Chinese remainder theorem is used in both applications.

The first application involves the enciphering of a database. A *database* is a collection of computer files or records. Here we will show how to encipher an entire database so that individual files may be deciphered without jeopardizing the security of other files in the database.

Suppose that a database $B$ contains the $n$ files $F_1$, $F_2$, ..., $F_n$. Since each file is a string of 0's and 1's, we can consider each file to be a binary integer. We first choose $n$ distinct primes $m_1$, $m_2$, ..., $m_n$ with $m_j > F_j$ for $j = 1, 2, ..., n$. As the ciphertext we use an integer $C$ that is congruent to $F_j$ modulo $m_j$ for $j = 1, 2, ..., n$; the existence of such an integer is guaranteed by the Chinese remainder theorem. We let $M = m_1 m_2 \cdots m_n$ and $M_j = M/m_j$ for $j = 1, 2, ..., n$. Furthermore, let $e_j = M_j \cdot y_j$ where $y_j$ is an inverse of $M_j$ modulo $m_j$. For the ciphertext, we take the integer $C$ with

$$C \equiv \sum_{j=1}^{n} e_j F_j \pmod{M}, \quad 0 \leqslant C < M.$$

The integers $e_1$, $e_2$, ..., $e_n$ serve as the *write subkeys* of the cipher.

To retrieve the $j$th file $F_j$ from the ciphertext $C$, we simply note that

$$F_j \equiv C \pmod{m_j}, \quad 0 \leqslant F_j < m_j.$$

We call the moduli $m_1$, $m_2$, ..., $m_n$ the *read subkeys* of the cipher. Note that knowledge of $m_j$ permits access only to file $j$; for access to the other files, it is necessary to know the moduli other than $m_j$.

We illustrate the enciphering and deciphering procedures for databases with the following example.

**Example 7.7.** Suppose our database contains four files $F_1$, $F_2$, $F_3$, and $F_4$, represented by the binary integers $(0111)_2$, $(1001)_2$, $(1100)_2$, and $(1111)_2$, or in decimal notation $F_1 = 7$, $F_2 = 9$, $F_3 = 12$ and $F_4 = 15$. We pick four primes, $m_1 = 11$, $m_2 = 13$, $m_3 = 17$, and $m_4 = 19$, greater than the corresponding integers representing the files. To encipher this database, we use the Chinese remainder theorem to find the ciphertext $C$ which is the positive integer with $C \equiv 7 \pmod{11}$, $C \equiv 9 \pmod{13}$, $C \equiv 12 \pmod{17}$, and $C \equiv 15 \pmod{19}$, less than $M = 11 \cdot 13 \cdot 17 \cdot 19 = 46189$. To compute $C$ we first find $M_1 = 13 \cdot 17 \cdot 19 = 4199$, $M_2 = 11 \cdot 17 \cdot 19 = 3553$, $M_3 = 11 \cdot 13 \cdot 19 = 2717$, and $M_4 = 11 \cdot 13 \cdot 17 = 2431$. We easily find that $y_1 = 7$, $y_2 = 10$, $y_3 = 11$ and $y_4 = 18$ are inverses of $M_j$ modulo $m_j$ for $j = 1, 2, 3, 4$. Hence, the write subkeys are $e_1 = 4199 \cdot 7 = 29393$, $e_2 = 3553 \cdot 10 = 35530$, $e_3 = 2717 \cdot 11 = 29887$, and $e_4 = 2431 \cdot 18 = 43758$. To construct the ciphertext, we note that

$$\begin{aligned} C &\equiv e_1 F_1 + e_2 F_2 + e_3 F_3 + e_4 F_4 \\ &\equiv 29393 \cdot 7 + 35530 \cdot 9 + 29887 \cdot 12 + 43758 \cdot 15 \\ &\equiv 1540535 \\ &\equiv 16298 \pmod{46189}, \end{aligned}$$

so that $C = 16298$. The read subkeys are the integers $m_j$, $j = 1, 2, 3, 4$. To recover the file $F_j$ from $C$, we simply find the least positive residue of $C$ modulo $m_j$. For instance, we find $F_1$ by noting that

$$F_1 \equiv 16298 \equiv 7 \pmod{11}. \qquad \square$$

We now discuss another application of cryptography, namely a method for sharing secrets. Suppose that in a communications network, there is some vital, but extremely sensitive information. If this information is distributed to several individuals, it becomes much more vulnerable to exposure; on the other hand, if this information is lost, there are serious consequences. An example of such information is the *master key K* used for access to the password file in a computer system.

In order to protect this master key $K$ from both loss and exposure, we construct *shadows* $k_1, k_2, ..., k_r$ which are given to $r$ different individuals. We will show that the key $K$ can be produced easily from any $s$ of these shadows, where $s$ is a positive integer less than $r$, whereas the knowledge of less than $s$ of these shadows does not permit the key $K$ to be found. Because at least $s$ different individuals are needed to find $K$, the key is not vulnerable to exposure. In addition, the key $K$ is not vulnerable to loss, since any $s$ individuals from the $r$ individuals with shadows can produce $K$. Schemes with the properties we have just described are called $(s,r)$ *threshold schemes*.

To develop a system that can be used to generate shadows with these properties, we use the Chinese remainder theorem. We choose a prime $p$ greater than the key $K$ and a sequence of pairwise relatively prime integers $m_1, m_2, ..., m_r$ that are not divisible by $p$, such that

$$m_1 < m_2 < \cdots < m_r,$$

and

(7.7) $$m_1 m_2 \cdots m_s > p m_r m_{r-1} \cdots m_{r-s+2}.$$

Note that the inequality (7.7) states that the product of the $s$ smallest of the integers $m_j$ is greater than the product of $p$ and the $s-1$ largest of the integers $m_j$. From (7.7), we see that if $M = m_1 m_2 \cdots m_s$, then $M/p$ is greater than the product of any set of $s-1$ of the integers $m_j$.

Now let $t$ be a nonnegative integer less than $M/p$ that is chosen at random. Let

$$K_0 = K + tp,$$

so that $0 \leqslant K_0 \leqslant M-1$ (since $0 \leqslant K_0 = K + tp < p + tp = (t+1)p \leqslant (M/p)p = M$).

To produce the shadows $k_1, k_2, ..., k_r$, we let $k_j$ be the integer with

$$k_j \equiv K_0 \pmod{m_j}, \quad 0 \leqslant k_j < m_j,$$

for $j = 1, 2, ..., r$. To see that the master key $K$ can be found by any $s$

individuals possessing shadows, from the total of $r$ individuals with shadows, suppose that the $s$ shadows $k_{j_1}, k_{j_2}, \ldots, k_{j_s}$ are available. Using the Chinese remainder theorem, we can easily find the least positive residue of $K_0$ modulo $M_i$ where $M_j = m_{j_1} m_{j_2} \cdots m_{j_s}$. Since we know that $0 \leqslant K_0 < M \leqslant M_j$, we can determine $K_0$, and then find $K = K_0 - tp$.

On the other hand, suppose that we know only the $s-1$ shadows $k_{i_1}, k_{i_2}, \ldots, k_{i_{s-1}}$. By the Chinese remainder theorem, we can determine the least positive residue $a$ of $K_0$ modulo $M_i$ where $M_i = m_{i_1} m_{i_2} \cdots m_{i_{s-1}}$. With these shadows, the only information we have about $K_0$ is that $a$ is the least positive residue of $K_0$ modulo $M_j$ and $0 \leqslant K_0 < M$. Consequently, we only know that

$$K_0 = a + xM_i ,$$

where $0 \leqslant x < M/M_i$. From (7.7), we can conclude that $M/M_i > p$, so that as $x$ ranges through the positive integers less than $M/M_i$, $x$ takes every value in a full set of residues modulo $p$. Since $(m_j, p) = 1$ for $j = 1, 2, \ldots, s$, we know that $(M_i, p) = 1$, and consequently, $a + xM_i$ runs through a full set of residues modulo $p$ as $x$ does. Hence, we see that the knowledge of $s-1$ shadows is insufficient to determine $K_0$, as $K_0$ could be in any of the $p$ congruence classes modulo $p$.

We use an example to illustrate this threshold scheme.

**Example 7.8.** Let $K = 4$ be the master key. We will use a $(2,3)$ threshold scheme of the kind just described with $p = 7$, $m_1 = 11$, $m_2 = 12$, and $m_3 = 17$, so that $M = m_1 m_2 = 132 > pm_3 = 119$. We pick $t = 14$ randomly from among the positive integers less than $M/p = 132/7$. This gives us

$$K_0 = K + tp = 4 + 14 \cdot 7 = 102.$$

The three shadows $k_1, k_2$, and $k_3$ are the least positive residues of $K_0$ modulo $m_1$, $m_2$, and $m_3$, *i.e.*

$$k_1 \equiv 102 \equiv 3 \pmod{11}$$
$$k_2 \equiv 102 \equiv 6 \pmod{12}$$
$$k_3 \equiv 102 \equiv 0 \pmod{17},$$

so that the three shadows are $k_1 = 3$, $k_2 = 6$, and $k_3 = 0$.

We can recover the master key $K$ from any two of the three shadows. Suppose we know that $k_1 = 3$ and $k_3 = 0$. Using the Chinese remainder theorem, we can determine $K_0$ modulo $m_1 m_3 = 11 \cdot 17 = 187$, *i.e.* since $K_0 \equiv 3 \pmod{11}$ and $K_0 \equiv 0 \pmod{17}$ we have $k_0 \equiv 102 \pmod{187}$. Since $0 \leqslant K_0 < M = 132 < 187$, we know that $K_0 = 102$, and consequently the master key is $K = K_0 - tp = 102 - 14 \cdot 7 = 4$.

□

We will develop another threshold scheme in Exercise 12 of Section 8.2. The interested reader should also consult Denning [68] for related topics in cryptography.

---

## 7.6 Exercises

1. Suppose that the database $B$ contains four files, $F_1 = 4$, $F_2 = 6$, $F_3 = 10$, and $F_4 = 13$. Let $m_1 = 5$, $m_2 = 7$, $m_3 = 11$, and $m_4 = 16$ be the read subkeys of the cipher used to encipher the database.

   a) What are the write subkeys of the cipher?

   b) What is the ciphertext $C$ corresponding to the database?

2. When the database $B$ with three files $F_1$, $F_2$, and $F_3$ is enciphered using the method described in the text, with read subkeys $m_1 = 14$, $m_2 = 15$, and $m_3 = 19$, the corresponding ciphertext is $C = 619$. If file $F_3$ is changed from $F_3 = 11$ to $F_3 = 12$, what is the updated value of the ciphertext $C$?

3. Decompose the master key $K = 3$ into three shadows using a $(2,3)$ threshold scheme of the type described in the text with $p = 5$, $m_1 = 8$, $m_2 = 9$, $m_3 = 11$ and with $t = 13$.

4. Show how to recover the master key $K$ from each of the three pairs of shadows found in Exercise 3.

## 7.6 Computer Projects

Write programs to do the following:

1. Using the system described in the text, encipher databases and recover files from the ciphertext version of databases.

2. Update files in the ciphertext version of databases (see Exercise 2).

3. Find the shadows in a threshold scheme of the type described in the text.

4. Recover the master key from a set of shadows.

---

# 8

# Primitive Roots

## 8.1 The Order of an Integer and Primitive Roots

By Euler's theorem if $m$ is a positive integer and if $a$ is an integer relatively prime to $m$, then $a^{\phi(m)} \equiv 1 \pmod{m}$. Therefore, at least one positive integer $x$ satisfies the congruence $a^x \equiv 1 \pmod{m}$. Consequently, by the well-ordering property, there is a least positive integer $x$ satisfying this congruence.

**Definition.** Let $a$ and $m$ be relatively prime positive integers. Then, the least positive integer $x$ such that $a^x \equiv 1 \pmod{m}$ is called the *order of $a$ modulo $m$*.

We denote the order of $a$ modulo $m$ by $\text{ord}_m a$.

**Example 8.1.** To find the order of 2 modulo 7, we compute the least positive residues modulo 7 of powers of 2. We find that

$$2^1 \equiv 2 \pmod 7, \ 2^2 \equiv 4 \pmod 7, \ 2^3 \equiv 1 \pmod 7.$$

Therefore, $\text{ord}_7 2 = 3$.

Similarly, to find the order of 3 modulo 7 we compute

$$3^1 \equiv 3 \pmod 7, \ 3^2 \equiv 2 \pmod 7, \ 3^3 \equiv 6 \pmod 7$$
$$3^4 \equiv 4 \pmod 7, \ 3^5 \equiv 5 \pmod 7, \ 3^6 \equiv 1 \pmod 7.$$

$\square$

We see that $\text{ord}_7 3 = 6$.

In order to find all solutions of the congruence $a^x \equiv 1 \pmod{m}$, we need the following theorem.

**Theorem 8.1.** If $a$ and $n$ are relatively prime integers with $n > 0$, then the positive integer $x$ is a solution of the congruence $a^x \equiv 1 \pmod{n}$ if and only if

$\text{ord}_n a \mid x$.

*Proof.* If $\text{ord}_n a \mid x$, then $x = k \cdot \text{ord}_n a$ where $k$ is a positive integer. Hence,

$$a^x = a^{k \cdot \text{ord}_n a} = (a^{\text{ord}_n a})^k \equiv 1 \ (\text{mod } n).$$

Conversely, if $a^x \equiv 1 \ (\text{mod } n)$, we first use the division algorithm to write

$$x = q \cdot \text{ord}_n a + r, \quad 0 \leqslant r < \text{ord}_n a.$$

From this equation, we see that

$$a^x = a^{q \cdot \text{ord}_n a + r} = (a^{\text{ord}_n a})^q a^r \equiv a^r \ (\text{mod } n).$$

Since $a^x \equiv 1 \ (\text{mod } n)$, we know that $a^r \equiv 1 \ (\text{mod } n)$. From the inequality $0 \leqslant r < \text{ord}_n a$, we conclude that $r = 0$, since, by definition, $y = \text{ord}_n a$ is the least positive integer such that $a^y \equiv 1 \ (\text{mod } n)$. Because $r = 0$, we have $x = q \cdot \text{ord}_n a$. Therefore, $\text{ord}_n a \mid x$. ∎

This theorem leads to the following corollary.

**Corollary 8.1.1.** If $a$ and $n$ are relatively prime integers with $n > 0$, then $\text{ord}_n a \mid \phi(n)$.

*Proof.* Since $(a, n) = 1$, Euler's theorem tells us that

$$a^{\phi(n)} \equiv 1 \ (\text{mod } n).$$

Using Theorem 8.1 we conclude that $\text{ord}_n a \mid \phi(n)$. ∎

We can use Corollary 8.1.1 as a shortcut when we compute orders. The following example illustrates the procedure.

**Example 8.2.** To find the order of 5 modulo 17, we first note that $\phi(17) = 16$. Since the only positive divisors of 16 are 1, 2, 4, 8, and 16, by Corollary 8.1 these are the only possible values of $\text{ord}_{17} 5$. Since

$$5^1 \equiv 5 \ (\text{mod } 17), \ 5^2 \equiv 8 \ (\text{mod } 17), \ 5^4 \equiv 13 \ (\text{mod } 17),$$
$$5^8 \equiv 16 \ (\text{mod } 17), \ 5^{16} \equiv 1 \ (\text{mod } 17),$$

we conclude that $\text{ord}_{17} 5 = 16$. □

The following theorem will be useful in our subsequent discussions.

**Theorem 8.2.** If $a$ and $n$ are relatively prime integers with $n > 0$, then $a^i \equiv a^j$, $(\text{mod } n)$ where $i$ and $j$ are nonnegative integers, if and only if $i \equiv j \ (\text{mod } \text{ord}_n a)$.

*Proof.* Suppose that $i \equiv j \ (\text{mod } \text{ord}_n a)$, and $0 \leqslant j \leqslant i$. Then we have $i = j + k \cdot \text{ord}_n a$, where $k$ is a positive integer. Hence

$$a^i = a^{j+k \cdot \text{ord}_n a} = a^j (a^{\text{ord}_n a})^k \equiv a^j \pmod{n},$$

since $a^{\text{ord}_n a} \equiv 1 \pmod{n}$.

Conversely, assume that $a^i \equiv a^j \pmod{n}$ with $i \geq j$. Since $(a,n) = 1$, we know that $(a^j, n) = 1$. Hence, using Corollary 3.4.1, the congruence

$$a^i \equiv a^j a^{i-j} \equiv a^j \pmod{n}$$

implies, by cancellation of $a^j$, that

$$a^{i-j} \equiv 1 \pmod{n}.$$

By Theorem 8.1 it follows that $\text{ord}_n a$ divides $i - j$, or equivalently, $i \equiv j \pmod{\text{ord}_n a}$.  ∎

Given an integer $n$, we are interested in integers $a$ with order modulo $n$ equal to $\phi(n)$. This is the largest possible order modulo $n$.

**Definition.** If $r$ and $n$ are relatively prime integers with $n > 0$ and if $\text{ord}_n r = \phi(n)$, then $r$ is called a *primitive root modulo n.*

**Example 8.3.** We have previously shown that $\text{ord}_7 3 = 6 = \phi(7)$. Consequently, 3 is a primitive root modulo 7. Likewise, since $\text{ord}_7 5 = 6$, as can easily be verified, 5 is also a primitive root modulo 7.  □

Not all integers have primitive roots. For instance, there are no primitive roots modulo 8. To see this, note that only integers less than 8 and relatively prime to 8 are 1, 3, 5, and 7, and $\text{ord}_8 1 = 1$, while $\text{ord}_8 3 = \text{ord}_8 5 = \text{ord}_8 7 = 2$. Since $\phi(8) = 4$, there are no primitive roots modulo 8. In our subsequent discussions, we will find all integers possessing primitive roots.

To indicate one way in which primitive roots are useful, we give the following theorem.

**Theorem 8.3.** If $r$ and $n$ are relatively prime positive integers with $n > 0$ and if $r$ is a primitive root modulo $n$, then the integers

$$r^1, r^2, ..., r^{\phi(n)}$$

form a reduced residue set modulo $n$.

*Proof.* To demonstrate that the first $\phi(n)$ powers of the primitive root $r$ form a reduced residue set modulo $n$, we only need to show that they are all relatively prime to $n$, and that no two are congruent modulo $n$.

Since $(r,n) = 1$, it follows from Exercise 14 of Section 2.1 that $(r^k, n) = 1$ for any positive integer $k$. Hence, these powers are all relatively prime to $n$.

To show that no two of these powers are congruent modulo $n$, assume that

$$r^i \equiv r^j \pmod{n}.$$

By Theorem 8.2 we see that $i \equiv j \pmod{\phi(n)}$. However, for $1 \leqslant i \leqslant \phi(n)$ and $1 \leqslant j \leqslant \phi(n)$, the congruence $i \equiv j \pmod{\phi(n)}$ implies that $i = j$. Hence, no two of these powers are congruent modulo $n$. This shows that we do have a reduced residue system modulo $n$.                                                       ∎

**Example 8.4**. Note that 2 is a primitive root modulo 9, since $2^2 \equiv 4$, $2^3 \equiv 8$, and $2^6 \equiv 1 \pmod 9$. By Theorem 8.3 we see that the first $\phi(9) = 6$ powers of 2 form a reduced residue system modulo 9. These are $2^1 \equiv 2 \pmod 9$, $2^2 \equiv 4 \pmod 9$, $2^3 \equiv 8 \pmod 9$, $2^4 \equiv 7 \pmod 9$, $2^5 \equiv 5 \pmod 9$, and $2^6 \equiv 1 \pmod 9$.                                                                    □

When an integer possesses a primitive root, it usually has many primitive roots. To demonstrate this, we first prove the following theorem.

**Theorem 8.4**. If $\mathrm{ord}_m a = t$ and if $u$ is a positive integer, then

$$\mathrm{ord}_m(a^u) = t/(t,u).$$

*Proof.* Let $s = \mathrm{ord}_m(a^u)$, $v = (t,u)$, $t = t_1 v$, and $u = u_1 v$. By Theorem 2.1 we know that $(t_1, u_1) = 1$.

Note that

$$(a^u)^{t_1} = (a^{u_1 v})^{(t/v)} = (a^t)^{u_1} \equiv 1 \pmod m,$$

since $\mathrm{ord}_m a = t$. Hence Theorem 8.1 tells us that $s \mid t_1$.

On the other hand, since

$$(a^u)^s = a^{us} \equiv 1 \pmod m,$$

we know that $t \mid us$. Hence, $t_1 v \mid u_1 vs$, and consequently, $t_1 \mid u_1 s$. Since $(t_1, u_1) = 1$, using Lemma 2.3, we see that $t_1 \mid s$.

Now, since $s \mid t_1$ and $t_1 \mid s$, we conclude that $s = t_1 = t/v = t/(t,u)$. This proves the result.                                                                    ∎

We have the following corollary of Theorem 8.4.

**Corollary 8.4.1**. Let $r$ be a primitive root modulo $m$ where $m$ is an integer, $m > 1$. Then $r^u$ is a primitive root modulo $m$ if and only if $(u, \phi(m)) = 1$.

*Proof.* By Theorem 8.4 we know that

$$\begin{aligned} \mathrm{ord}_m r^u &= \mathrm{ord}_m r/(u, \mathrm{ord}_m r) \\ &= \phi(m)/(u, \phi(m)). \end{aligned}$$

Consequently, $\mathrm{ord}_m r^u = \phi(m)$, and $r^u$ is a primitive root modulo $m$, if and only if $(u, \phi(m)) = 1$.                                                                    ∎

This leads immediately to the following theorem.

**Theorem 8.5.** If the positive integer $m$ has a primitive root, then it has a total of $\phi(\phi(m))$ incongruent primitive roots.

*Proof.* Let $r$ be a primitive root modulo $m$. Then Theorem 8.3 tells us that the integers $r, r^2, ..., r^{\phi(m)}$ form a reduced residue system modulo $m$. By Corollary 8.4.1 we know that $r^u$ is a primitive root modulo $m$ if and only if $(u, \phi(m)) = 1$. Since there are exactly $\phi(\phi(m))$ such integers $u$, there are exactly $\phi(\phi(m))$ primitive roots modulo $m$. ∎

**Example 8.5.** Let $m = 11$. A little computation tells us that 2 is a primitive root modulo 11. Since 11 has a primitive root, we know that 11 has $\phi(\phi(11)) = 4$ incongruent primitive roots. It is easily seen that 2, 6, 7, and 8 are four incongruent primitive roots modulo 11. □

---

## 8.1 Exercises

1. Determine the

   a) order of 2 modulo 5      c) order of 10 modulo 13
   b) order of 3 modulo 10     d) order of 7 modulo 19.

2. Find a primitive root modulo each of the following integers.

   a) 4        d) 13
   b) 5        e) 14
   c) 10       f) 18

3. Show that the integer 12 has no primitive roots.

4. How many incongruent primitive roots does 13 have? Find a set of this many incongruent primitive roots modulo 13.

5. Show that if $\bar{a}$ is an inverse of $a$ modulo $n$, then $\text{ord}_n a = \text{ord}_n \bar{a}$.

6. Show that if $n$ is a positive integer and $a$ and $b$ are integers relatively prime to $n$ such that $(\text{ord}_n a, \text{ord}_n b) = 1$, then $\text{ord}_n (ab) = \text{ord}_n a \cdot \text{ord}_n b$.

7. What can be said about $\text{ord}_n (ab)$ if $a$ and $b$ are integers relatively prime to $n$ when $\text{ord}_n a$ and $\text{ord}_n b$ are not necessarily relatively prime?

8. Decide whether it is true that if $n$ is a positive integer and $d$ is a divisor of $\phi(n)$, then there is an integer $a$ with $\text{ord}_n a = d$. Give reasons for your answers.

9. Show that if $a$ is an integer relatively prime to the positive integer $m$ and $\text{ord}_m a = st$, then $\text{ord}_m a^t = s$.

10. Show that if $m$ is a positive integer and $a$ is an integer relatively prime to $m$ such that $\text{ord}_m a = m - 1$, then $m$ is prime.

11. Show that $r$ is a primitive root modulo the odd prime $p$ if and only if

$$r^{(p-1)/q} \not\equiv 1 \ (\text{mod } p)$$

for all prime divisors $q$ of $p-1$.

12. Show that if $r$ is a primitive root modulo the positive integer $m$, then $\bar{r}$ is also a primitive root modulo $m$, if $\bar{r}$ is an inverse of $r$ modulo $m$.

13. Show that $\text{ord}_{F_n} 2 \leqslant 2^{n+1}$, where $F_n = 2^{2^n} + 1$ is the $n$ th Fermat number.

* 14. Let $p$ be a prime divisor of the Fermat number $F_n = 2^{2^n} + 1$.

   a) Show that $\text{ord}_p 2 = 2^{n+1}$.

   b) From part (a), conclude that $2^{n+1} \mid (p-1)$, so that $p$ must be of the form $2^{n+1}k + 1$.

15. Let $m = a^n - 1$, where $a$ and $n$ are positive integers. Show that $\text{ord}_m a = n$ and conclude that $n \mid \phi(m)$.

* 16. a) Show that if $p$ and $q$ are distinct odd primes, then $pq$ is a pseudoprime to the base 2 if and only if $\text{ord}_q 2 \mid (p-1)$ and $\text{ord}_p 2 \mid (q-1)$.

   b) Use part (a) to decide which of the following integers are pseudoprimes to the base 2: $13 \cdot 67$, $19 \cdot 73$, $23 \cdot 89$, $29 \cdot 97$.

* 17. Show that if $p$ and $q$ are distinct odd primes, then $pq$ is a pseudoprime to the base 2 if and only if $M_p M_q = (2^p - 1)(2^q - 1)$ is a pseudoprime to the base 2.

There is a method for deciphering messages that were enciphered by an RSA cipher, without knowledge of the deciphering key. This method is based on iteration. Suppose that the public key $(e,n)$ used for enciphering is known, but the deciphering key $(d,n)$ is not. To decipher a ciphertext block $C$, we form a sequence $C_1$, $C_2$, $C_3$,... setting $C_1 \equiv C^e \ (\text{mod } n)$, $0 < C_1 < n$     and $C_{j+1} \equiv C_j^e \ (\text{mod } n)$, $0 < C_{j+1} < n$ for $j = 1,2,3,...$ .

18. Show that $C_j \equiv C^{e^j} \ (\text{mod } n)$, $0 < C_j < n$.

* 19. Show that there is an index $j$ such that $C_j = C$ and $C_{j-1} = P$, where $P$ is the original plaintext message. Show that this index $j$ is a divisor of $\text{ord}_{\phi(n)} e$.

20. Let $n = 47 \cdot 59$ and $e = 17$. Using iteration, find the plaintext corresponding to the ciphertext 1504.

   (*Note*: This iterative method for attacking RSA ciphers is seldom successful in a reasonable amount of time. Moreover, the primes $p$ and $q$ may be chosen so that this attack is almost always futile. See Exercise 13 of Section 8.2.)

## 8.1 Computer Projects

Write projects to do the following:

1. Find the order of $a$ modulo $m$, when $a$ and $m$ are relatively prime positive integers.

2. Find primitive roots when they exist.

3. Attempt to decipher RSA ciphers by iteration (see the preamble to Exercise 18).

---

# 8.2 Primitive Roots for Primes

In this section and in the one following, our objective is to determine which integers have primitive roots. In this section, we show that every prime has a primitive root. To do this, we first need to study polynomial congruences.

Let $f(x)$ be a polynomial with integer coefficients. We say that an integer $c$ is a *root of $f(x)$ modulo m* if $f(c) \equiv 0 \pmod{m}$. It is easy to see that if $c$ is a root of $f(x)$ modulo $m$, then every integer congruent to $c$ modulo $m$ is also a root.

**Example 8.6.** The polynomial $f(x) = x^2 + x + 1$ has exactly two incongruent roots modulo 7, namely $x \equiv 2 \pmod 7$ and $x \equiv 4 \pmod 7$. $\quad\square$

**Example 8.7.** The polynomial $g(x) = x^2 + 2$ has no roots modulo 5. $\quad\square$

**Example 8.8.** Fermat's little theorem tells us that if $p$ is prime, then the polynomial $h(x) = x^{p-1} - 1$ has exactly $p-1$ incongruent roots modulo $p$, namely $x \equiv 1, 2, 3, ..., p-1 \pmod p$. $\quad\square$

We will need the following important theorem concerning roots of polynomials modulo $p$ where $p$ is a prime.

**Theorem 8.6. Lagrange's Theorem.** Let $f(x) = a_n x^n + a_{n-1} x^{n-1} + ... + a_1 x + a_0$ be a polynomial of degree $n$, $n \geqslant 1$, with integer coefficients and with leading coefficient $a_n$ not divisible by $p$. Then $f(x)$ has at most $n$ incongruent roots modulo $p$.

*Proof.* We use mathematical induction to prove the theorem. When $n = 1$, we have $f(x) = a_1 x + a_0$ with $p \nmid a_1$. A root of $f(x)$ modulo $p$ is a solution of the linear congruence $a_1 x \equiv -a_0 \pmod p$. By Theorem 3.10 since $(a_1, p) = 1$, this linear congruence has exactly one solution, so that there is exactly one root modulo $p$ of $f(x)$. Clearly, the theorem is true for $n = 1$.

Now suppose that the theorem is true for polynomials of degree $n - 1$, and let $f(x)$ be a polynomial of degree $n$ with leading coefficient not divisible by $p$. Assume that the polynomial $f(x)$ has $n + 1$ incongruent roots modulo $p$, say $c_0, c_1, ..., c_n$, so that $f(c_k) \equiv 0 \pmod{p}$ for $k = 0, 1, ..., n$. We have

$$\begin{aligned}
f(x) - f(c_0) &= a_n(x^n - c_0^n) + a_{n-1}(x^{n-1} - c_0^{n-1}) + \cdots + a_1(x - c_0) \\
&= a_n(x - c_0)(x^{n-1} + x^{n-2}c_0 + \cdots + xc_0^{n-2} + c_0^{n-1}) \\
&\quad + a_{n-1}(x - c_0)(x^{n-2} + x^{n-3}c_0 + \cdots + xc_0^{n-3} + c_0^{n-2}) \\
&\quad + \cdots + a_1(x - c_0) \\
&= (x - c_0)g(x),
\end{aligned}$$

where $g(x)$ is a polynomial of degree $n - 1$ with leading coefficient $a_n$. We now show that $c_1, c_2, ..., c_n$ are all roots of $g(x)$ modulo $p$. Let $k$ be an integer, $1 \leqslant k \leqslant n$. Since $f(c_k) \equiv f(c_0) \equiv 0 \pmod{p}$, we have

$$f(c_k) - f(c_0) = (c_k - c_0)g(c_k) \equiv 0 \pmod{p}.$$

It follows that $g(c_k) \equiv 0 \pmod{p}$, since $c_k - c_0 \not\equiv 0 \pmod{p}$. Hence, $c_k$ is a root of $g(x)$ modulo $p$. This shows that the polynomial $g(x)$, which is of degree $n - 1$ and has a leading coefficient not divisible by $p$, has $n$ incongruent roots modulo $p$. This contradicts the induction hypothesis. Hence, $f(x)$ must have no more than $n$ incongruent roots modulo $p$. The induction argument is complete.  ∎

We use Lagrange's theorem to prove the following result.

**Theorem 8.7.** Let $p$ be prime and let $d$ be a divisor of $p - 1$. Then the polynomial $x^d - 1$ has exactly $d$ incongruent roots modulo $p$.

*Proof.* Let $p - 1 = de$. Then

$$\begin{aligned}
x^{p-1} - 1 &= (x^d - 1)(x^{d(e-1)} + x^{d(e-2)} + \cdots + x^d + 1) \\
&= (x^d - 1)g(x).
\end{aligned}$$

From Fermat's little theorem, we see that $x^{p-1} - 1$ has $p - 1$ incongruent roots modulo $p$. Furthermore, any root of $x^{p-1} - 1$ modulo $p$ is either a root of $x^d - 1$ modulo $p$ or a root of $g(x)$ modulo $p$.

Lagrange's theorem tells us that $g(x)$ has at most $d(e-1) = p - d - 1$ roots modulo $p$. Since every root of $x^{p-1} - 1$ modulo $p$ that is not a root of $g(x)$ modulo $p$ must be a root of $x^d - 1$ modulo $p$, we know that the polynomial $x^d - 1$ has at least $(p-1) - (p-d-1) = d$ incongruent roots modulo $p$. On the other hand, Lagrange's theorem tells us that it has at most $d$ incongruent roots modulo $p$. Consequently, $x^d - 1$ has precisely $d$ incongruent roots modulo $p$.  ∎

Theorem 8.7 can be used to prove the following result which tells us how many incongruent integers have a given order modulo $p$.

**Theorem 8.8.** Let $p$ be a prime and let $d$ be a positive divisor of $p - 1$. Then the number of incongruent integers of order $d$ modulo $p$ is equal to $\phi(d)$.

*Proof.* For each positive integer $d$ dividing $p-1$, let $F(d)$ denote the number of positive integers of order $d$ modulo $p$ that are less than $p$. Since the order modulo $p$ of an integer not divisible by $p$ divides $p-1$, it follows that

$$p-1 = \sum_{d \mid p-1} F(d).$$

By Theorem 6.6 we know that

$$p-1 = \sum_{d \mid p-1} \phi(d).$$

We will show that $F(d) \leqslant \phi(d)$ when $d \mid (p-1)$. This inequality, together with the equality

$$\sum_{d \mid p-1} F(d) = \sum_{d \mid p-1} \phi(d),$$

implies that $F(d) = \phi(d)$ for each positive divisor $d$ of $p-1$.

Let $d \mid (p-1)$. If $F(d) = 0$, it is clear that $F(d) \leqslant \phi(d)$. Otherwise, there is an integer $a$ of order $d$ modulo $p$. Since $\mathrm{ord}_p a = d$, the integers

$$a, \ a^2, \ ..., \ a^d$$

are incongruent modulo $p$. Furthermore, each of these powers of $a$ is a root of $x^d - 1$ modulo $p$, since $(a^k)^d \equiv (a^d)^k \equiv 1 \pmod{p}$ for all positive integers $k$. By Theorem 8.7 we know that $x^d - 1$ has exactly $d$ incongruent roots modulo $p$, so every root modulo $p$ is congruent to one of these powers of $a$. However, by Theorem 8.4 we know that the powers of $a$ with order $d$ are those of the form $a^k$ with $(k,d) = 1$. There are exactly $\phi(d)$ such integers $k$ with $1 \leqslant k \leqslant d$, and consequently, if there is one element of order $d$ modulo $p$, there must be exactly $\phi(d)$ such positive integers less than $d$. Hence, $F(d) \leqslant \phi(d)$.

Therefore, we can conclude that $F(d) = \phi(d)$, which tells us that there are precisely $\phi(d)$ incongruent integers of order $d$ modulo $p$.  ∎

The following corollary is derived immediately from Theorem 8.8.

**Corollary 8.8.1**. Every prime has a primitive root.

*Proof.* Let $p$ be a prime. By Theorem 8.7 we know that there are $\phi(p-1)$ incongruent integers of order $p-1$ modulo $p$. Since each of these is, by definition, a primitive root, $p$ has $\phi(p-1)$ primitive roots.  ∎

The smallest positive primitive root of each prime less than 1000 is given in Table 3 of the Appendix.

## 8.2 Exercises

1.  Find the number of primitive roots of each of the following primes.

    a)  7                               d)  19
    b)  13                              e)  29
    c)  17                              f)  47

2.  Let $r$ be a primitive root of the prime $p$ with $p \equiv 1 \pmod 4$. Show that $-r$ is also a primitive root.

3.  Show that if $p$ is a prime and $p \equiv 1 \pmod 4$, there is an integer $x$ such that $x^2 \equiv -1 \pmod p$. (*Hint*: Use Theorem 8.8 to show that there is an integer $x$ of order 4 modulo $p$.)

4.  a)  Find the number of incongruent roots modulo 6 of the polynomial $x^2 - x$.

    b)  Explain why the answer to part (a) does not contradict Lagrange's theorem.

5.  a)  Use Lagrange's theorem to show that if $p$ is a prime and $f(x)$ is a polynomial of degree $n$ with integer coefficients and more than $n$ roots modulo $p$, then $p$ divides every coefficient of $f(x)$.

    b)  Let $p$ be prime. Using part (a), show that every coefficient of the polynomial $f(x) = (x-1)(x-2) \cdots (x-p+1) - x^{p-1} + 1$ is divisible by $p$.

    c)  Using part (b), give a proof of Wilson's theorem. (*Hint*: Consider the constant term of $f(x)$.)

6.  Find the least positive residue of the product of a set of $\phi(p-1)$ incongruent primitive roots modulo a prime $p$.

\* 7.  A systematic method for constructing a primitive root modulo a prime $p$ is outlined in this problem. Let the prime factorization of $\phi(p) = p-1$ be $p-1 = q_1^{t_1} q_2^{t_2} \cdots q_r^{t_r}$ where $q_1, q_2, \ldots, q_r$ are prime.

    a)  Use Theorem 8.8 to show that there are integers $a_1, a_2, \ldots, a_r$ such that $\mathrm{ord}_p a_1 = q_1^{t_1}$, $\mathrm{ord}_p a_2 = q_2^{t_2}$, $\ldots$, $\mathrm{ord}_p a_r = q_r^{t_r}$.

    b)  Use Exercise 6 of Section 8.1 to show that $a = a_1 a_2 \cdots a_r$ is a primitive root modulo $p$.

    c)  Follow the procedure outlined in parts (a) and (b) to find a primitive root modulo 29.

\* 8.  Let the positive integer $n$ have prime-power factorization $n = p_1^{a_1} p_2^{a_2} \cdots p_r^{a_r}$. Show that the number of incongruent bases modulo $n$ for which $n$ is a pseudoprime to that base is $\prod_{j=1}^{r} (n-1, p_j-1)$.

9.  Use Exercise 8 to show that every odd composite integer that is not a power of 3 is a pseudoprime to at least two bases other than $\pm 1$.

10. Show that if $p$ is prime and $p = 2q + 1$, where $q$ is prime and $a$ is a positive integer with $1 < a < p-1$, then $p-a^2$ is a primitive root modulo $p$.

* 11. a) Suppose that $f(x)$ is a polynomial with integer coefficients of degree $n-1$. Let $x_1, x_2, \ldots, x_n$ be $n$ incongruent integers modulo $p$. Show that for all integers $x$, the congruence

$$f(x) \equiv \sum_{j=1}^{n} f(x_j) \prod_{\substack{i=1 \\ i \neq j}}^{n} (x-x_i) \overline{(x_j-x_i)} \pmod{p}$$

holds, where $\overline{x_j-x_i}$ is an inverse of $x_j-x_i \pmod{n}$. This technique for finding $f(x)$ modulo $p$ is called *Lagrange interpolation*.

b) Find the least positive residue of $f(5)$ modulo 11 if $f(x)$ is a polynomial of degree 3 with $f(1) \equiv 8$, $f(2) \equiv 2$, and $f(3) \equiv 4 \pmod{11}$.

12. In this exercise we develop a threshold scheme for protection of master keys in a computer system, different than the scheme discussed in Section 7.6. Let $f(x)$ be a randomly chosen polynomial of degree $r-1$, with the condition that $K$, the master key, is the constant term of the polynomial. Let $p$ be a prime, such that $p > K$ and $p > s$. The $s$ shadows $k_1, k_2, \ldots, k_s$ are computed by finding the least positive residue of $f(x_j)$ modulo $p$ for $j = 1, 2, \ldots, s$ where $x_1, x_2, \ldots, x_s$ are randomly chosen integers incongruent modulo $p$, i.e.,

$$k_j \equiv f(x_j) \pmod{p}, \ 0 \leqslant k_j < p,$$

for $j = 1, 2, \ldots, s$.

a) Use Lagrange interpolation, described in Exercise 11, to show that the master key $K$ can be determined from any $r$ shadows.

b) Show that the master key $K$ cannot be determined from less than $r$ shadows.

c) Let $K = 33$, $p = 47$, $r = 4$, and $s = 7$. Let $f(x) = 4x^3 + x^2 + 31x + 33$. Find the seven shadows corresponding to the values of $f(x)$ at 1,2,3,4,5,6, and 7.

d) Show how to find the master key from the four shadows $f(1)$, $f(2)$, $f(3)$, and $f(4)$.

13. Show that an RSA cipher with enciphering modulus $n = pq$ is resistant to attack by iteration (see the preamble to Exercise 18 of Section 8.1) if $p = 2p' + 1$ and $q = 2q' + 1$, where $p'$ and $q'$ are primes.

## 8.2 Computer Projects

Write programs to do the following:

1. Find a primitive root of a prime using Exercise 7.

2. Implement the threshold scheme given in Exercise 12.

---

## 8.3 The Existence of Primitive Roots

In the previous section, we showed that every prime has a primitive root. In this section, we will find all positive integers having primitive roots. First, we will show that every power of an odd prime possesses a primitive root. We begin by considering squares of primes.

**Theorem 8.9.** If $p$ is an odd prime with primitive root $r$, then either $r$ or $r + p$ is a primitive root modulo $p^2$.

*Proof.* Since $r$ is a primitive root modulo $p$, we know that

$$\text{ord}_p r = \phi\ (p) = p-1.$$

Let $n = \text{ord}_{p^2} r$, so that

$$r^n \equiv 1 \pmod{p^2}.$$

Since a congruence modulo $p^2$ obviously holds modulo $p$, we have

$$r^n \equiv 1 \pmod{p}.$$

By Theorem 8.1 it follows that

$$p - 1 = \text{ord}_p r \mid n.$$

On the other hand, Corollary 8.1.1 tells us that

$$n \mid \phi(p^2) = p\ (p-1).$$

Since $n \mid p\ (p-1)$ and $p-1 \mid n$, either $n = p-1$ or $n = p\ (p-1)$. If $n = p\ (p-1)$, then $r$ is a primitive root modulo $p^2$, since $\text{ord}_{p^2} r = \phi(p^2)$. Otherwise, we have $n = p-1$, so that

(8.1)                                $$r^{p-1} \equiv 1 \pmod{p^2}.$$

Let $s = r + p$. Then, since $s \equiv r \pmod{p}$, $s$ is also a primitive root modulo $p$. Hence, $\text{ord}_{p^2} s$ equals either $p-1$ or $p\ (p-1)$. We will show that $\text{ord}_{p^2} s \neq p-1$. The binomial theorem tells us that

$$s^{p-1} = (r+p)^{p-1} = r^{p-1} + (p-1)r^{p-2}p + \binom{p-1}{2}r^{p-3}p^2 + \cdots + p^{p-1}$$
$$\equiv r^{p-1} + (p-1)p \cdot r^{p-2} \pmod{p^2}.$$

Hence, using (8.1), we see that

$$s^{p-1} \equiv 1 + (p-1)p \cdot r^{p-2} \equiv 1 - pr^{p-2} \pmod{p^2}.$$

From this last congruence, we can conclude that

$$s^{p-1} \not\equiv 1 \pmod{p^2}.$$

To see this, note that if $s^{p-1} \equiv 1 \pmod{p^2}$, then $pr^{p-2} \equiv 0 \pmod{p^2}$. This last congruence implies that $r^{p-2} \equiv 0 \pmod{p}$, which is impossible, since $p \nmid r$ (remember $r$ is a primitive root of $p$). Hence, $\text{ord}_{p^2}s = p(p-1) = \phi(p^2)$. Consequently, $s = r+p$ is a primitive root of $p^2$. ∎

**Example 8.9.** The prime $p = 7$ has $r = 3$ as a primitive root. From the proof of Theorem 8.9, we see that $r = 3$ is also a primitive root modulo $p^2 = 49$, since

$$r^{p-1} = 3^6 \not\equiv 1 \pmod{49}.$$

We note that it is extremely rare for the congruence

$$r^{p-1} \equiv 1 \pmod{p^2}$$

to hold when $r$ is a primitive root modulo the prime $p$. Consequently, it is very seldom that a primitive root $r$ modulo the prime $p$ is not also a primitive root modulo $p^2$. An example is provided by $p = 487$. For the primitive root 10 modulo 487, we have

$$10^{486} \equiv 1 \pmod{487^2}.$$

Hence, 10 is not a primitive root modulo $487^2$, but by Theorem 8.9, we know that $497 = 10 + 487$ is a primitive root modulo $487^2$. □

We now turn our attention to arbitrary powers of primes.

**Theorem 8.10.** Let $p$ be an odd prime, then $p^k$ has a primitive root for all positive integers $k$. Moreover, if $r$ is a primitive root modulo $p^2$, then $r$ is a primitive root modulo $p^k$, for all positive integers $k$.

*Proof.* By Theorem 8.9 we know that $p$ has a primitive root $r$ that is also a primitive root modulo $p^2$, so that

(8.2) $$r^{p-1} \not\equiv 1 \pmod{p^2}.$$

Using mathematical induction, we will prove that for this primitive root r,

(8.3) $$r^{p^{k-2}(p-1)} \not\equiv 1 \pmod{p^k}$$

for all positive integers k. Once we have established this congruence, we can show that $r$ is also a primitive root modulo $p^k$ by the following reasoning. Let

$$n = \text{ord}_{p^k}r.$$

By Theorem 6.8 we know that $n \mid \phi(p^k) = p^{k-1}(p-1)$. On the other hand, since

$$r^n \equiv 1 \pmod{p^k},$$

we also know that

$$r^n \equiv 1 \pmod{p}.$$

By Theorem 8.1 we see that $p-1 = \phi(p) \mid n$. Because $(p-1) \mid n$, and $n \mid p^{k-1}(p-1)$, we know that $n = p^t(p-1)$, where $t$ is an integer such that $0 \leqslant t \leqslant k-1$. If $n = p^t(p-1)$ with $t \leqslant k-2$, then

$$r^{p^{k-2}(p-1)} = (r^{p^t(p-1)})^{p^{k-2-t}} \equiv 1 \pmod{p^k},$$

which would contradict (8.3). Hence, $\operatorname{ord}_{p^k} r = p^{k-1}(p-1) = \phi(p^k)$. Consequently, $r$ is also a primitive root modulo $p^k$.

All that remains is to prove (8.3) using mathematical induction. The case of $k=2$ follows from (8.2). Let us assume the assertion is true for the positive integer $k \geqslant 2$. Then

$$r^{p^{k-2}(p-1)} \not\equiv 1 \pmod{p^k}.$$

Since $(r,p) = 1$, we know that $(r,p^{k-1}) = 1$. Consequently, from Euler's theorem, we know that

$$r^{p^{k-2}(p-1)} \equiv r^{\phi(p^{k-1})}.$$

Therefore, there is an integer $d$ such that

$$r^{p^{k-2}(p-1)} = 1 + dp^{k-1},$$

where $p \nmid d$, since by hypothesis $r^{p^{k-2}(p-1)} \not\equiv 1 \pmod{p^k}$. We take the $p$th power of both sides of the above equation, to obtain, via the binomial theorem,

$$
\begin{aligned}
r^{p^{k-1}(p-1)} &= (1 + dp^{k-1})^p \\
&= 1 + p(dp^{k-1}) + \binom{p}{2}p^2(dp^{k-1})^2 + \cdots + (dp^{k-1})^p \\
&\equiv 1 + dp^k \pmod{p^{k+1}}.
\end{aligned}
$$

Since $p \nmid d$, we can conclude that

$$r^{p^{k-1}(p-1)} \not\equiv 1 \pmod{p^{k+1}}.$$

This completes the proof by induction. ■

**Example 8.10.** By Example 8.9 we know that $r = 3$ is a primitive root modulo 7 and $7^2$. Hence Theorem 8.10 tells us that $r = 3$ is also a primitive root modulo $7^k$ for all positive integers $k$. □

It is now time to discuss whether there are primitive roots modulo powers of 2. We first note that both 2 and $2^2 = 4$ have primitive roots, namely 1 and 3, respectively. For higher powers of 2, the situation is different, as the following theorem shows; there are no primitive roots modulo these powers of 2.

**Theorem 8.11.** If $a$ is an odd integer, and if $k$ is an integer, $k \geqslant 3$, then

$$a^{\phi(2^k)/2} = a^{2^{k-2}} \equiv 1 \pmod{2^k}.$$

*Proof.* We prove this result using mathematical induction. If $a$ is an odd integer, then $a = 2b + 1$, where $b$ is an integer. Hence,

$$a^2 = (2b + 1)^2 = 4b^2 + 4b + 1 = 4b(b + 1) + 1.$$

Since either $b$ or $b + 1$ is even, we see that $8 \mid 4b(b + 1)$, so that

$$a^2 \equiv 1 \pmod 8.$$

This is the congruence of interest when $k = 3$.

Now to complete the induction argument, let us assume that

$$a^{2^{k-2}} \equiv 1 \pmod{2^k}.$$

Then there is an integer d such that

$$a^{2^{k-2}} = 1 + d \cdot 2^k.$$

Squaring both sides of the above equality, we obtain

$$a^{2^{k-1}} = 1 + d2^{k+1} + d^2 2^{2k}.$$

This yields

$$a^{2^{k-1}} \equiv 1 \pmod{2^{k+1}},$$

which completes the induction argument. ∎

Theorem 8.11 tells us that no power of 2, other than 2 and 4, has a primitive root, since when $a$ is an odd integer, $\mathrm{ord}_{2^k} a \neq \phi(2^k)$, since $a^{\phi(2^k)/2} \equiv 1 \pmod{2^k}$.

Even though there are no primitive roots modulo $2^k$ for $k \geq 3$, there always is an element of largest possible order, namely $\phi(2^k)/2$, as the following theorem shows.

**Theorem 8.12.** Let $k \geq 3$ be an integer. Then

$$\mathrm{ord}_{2^k} 5 = \phi(2^k)/2 = 2^{k-2}.$$

*Proof.* Theorem 8.11 tells us that

$$5^{2^{k-2}} \equiv 1 \pmod{2^k},$$

for $k \geq 3$. By Theorem 8.1 we see that $\mathrm{ord}_{2^k} 5 \mid 2^{k-2}$. Therefore, if we show that $\mathrm{ord}_{2^k} 5 \nmid 2^{k-3}$, we can conclude that

$$\mathrm{ord}_{2^k} 5 = 2^{k-2}.$$

To show that $\mathrm{ord}_{2^k} 5 \nmid 2^{k-3}$, we will prove by mathematical induction that for $k \geq 3$,

$$5^{2^{k-3}} \equiv 1 + 2^{k-1} \not\equiv 1 \pmod{2^k}.$$

For $k = 3$, we have

$$5 = 1 + 4 \ (\text{mod } 8).$$

Now assume that

$$5^{2^{k-3}} \equiv 1 + 2^{k-1} \ (\text{mod } 2^k).$$

This means that there is a positive integer $d$ such that

$$5^{2^{k-3}} = (1 + 2^{k-1}) + d \, 2^k.$$

Squaring both sides, we find that

$$5^{2^{k-2}} = (1 + 2^{k-1})^2 + 2(1 + 2^{k-1})d \, 2^k + (d \, 2^k)^2 \ ,$$

so that

$$5^{2^{k-2}} \equiv (1 + 2^{k-1})^2 = 1 + 2^k + 2^{2k-2} \equiv 1 + 2^k \ (\text{mod } 2^{k+1}).$$

This completes the induction argument and shows that

$$\text{ord}_{2^k} 5 = \phi(2^k)/2. \qquad \blacksquare$$

We have now demonstrated that all powers of odd primes possess primitive roots, while the only powers of 2 having primitive roots are 2 and 4. Next, we determine which integers not powers of primes, *i.e.* those integers divisible by two or more primes, have primitive roots. We will demonstrate that the only positive integers not powers of primes possessing primitive roots are twice powers of odd primes.

We first narrow down the set of positive integers we need consider with the following result.

**Theorem 8.13**. If $n$ is a positive integer that is not a prime power or twice a prime power, then $n$ does not have a primitive root.

*Proof.* Let $n$ be a positive integer with prime-power factorization

$$n = p_1^{t_1} p_2^{t_2} \cdots p_m^{t_m}.$$

Let us assume that the integer $n$ has a primitive root $r$. This means that $(r,n) = 1$ and $\text{ord}_n r = \phi(n)$. Since $(r,n) = 1$, we know that $(r,p^t) = 1$, whenever $p^t$ is one of the prime powers occurring in the factorization of $n$. By Euler's theorem, we know that

$$r^{\phi(p^t)} \equiv 1 \ (\text{mod } p^t).$$

Now let $U$ be the least common multiple of $\phi(p_1^{t_1})$, $\phi(p_2^{t_2}),...,\phi(p_m^{t_m})$, *i.e.*

$$U = [\phi(p_1^{t_1}), \ \phi(p_2^{t_2}),...,\phi(p_m^{t_m})].$$

Since $\phi(p_i^{t_i}) \mid U$, we know that

$$r^U \equiv 1 \ (\text{mod } p_i^{t_i})$$

for $i = 1, 2,..., m$. From this last congruence, we see that

$$\operatorname{ord}_n r = \phi(n) \leqslant U.$$

By Theorem 6.4, since $\phi$ is multiplicative, we have

$$\phi(n) = \phi(p_1^{t_1} p_2^{t_2} \cdots p_m^{t_m}) = \phi(p_1^{t_1})\phi(p_2^{t_2}) \cdots \phi(p_m^{t_m}).$$

This formula for $\phi(n)$ and the inequality $\phi(n) \leqslant U$ imply that

$$\phi(p_1^{t_1})\,\phi(p_2^{t_2}) \cdots \phi(p_m^{t_m}) \leqslant [\phi(p_1^{t_1}),\phi(p_2^{t_2}),\ldots,\phi(p_m^{t_m})].$$

Since the product of a set of integers is less than or equal to their least common multiple only if the integers are pairwise relatively prime (and then the less than or equal to relation is really just an equality), the integers $\phi(p_1^{t_1}),\phi(p_2^{t_2}),\ldots,\phi(p_m^{t_m})$ must be pairwise relatively prime.

We note that $\phi(p^t) = p^{t-1}(p-1)$, so that $\phi(p^t)$ is even if $p$ is odd, or if $p = 2$ and $t \geqslant 2$. Hence, the numbers $\phi(p_1^{t_1})$, $\phi(p_2^{t_2}),\ldots,\phi(p_m^{t_m})$ are not pairwise relatively prime unless $m = 1$ and $n$ is a prime power or $m = 2$ and $n$ is $n = 2p^t$, where $p$ is an odd prime and $t$ is a positive integer. ∎

We have now limited consideration to integers of the form $n = 2p^t$, where $p$ is an odd prime and $t$ is a positive integer. We now show that all such integers have primitive roots.

**Theorem 8.14.** If $p$ is an odd prime and $t$ is a positive integer, then $2p^t$ possesses a primitive root. In fact, if $r$ is a primitive root modulo $p^t$, then if $r$ is odd it is also a primitive root modulo $2p^t$, while if $r$ is even, $r + p^t$ is a primitive root modulo $2p^t$.

*Proof.* If $r$ is a primitive root modulo $p^t$, then

$$r^{\phi(p^t)} \equiv 1 \pmod{p^t},$$

and no positive exponent smaller than $\phi(p^t)$ has this property. By Theorem 6.4 we note that $\phi(2p^t) = \phi(2)\,\phi(p^t) = \phi(p^t)$, so that $r^{\phi(2p^t)} \equiv 1 \pmod{p^t}$.

If $r$ is odd, then

$$r^{\phi(2p^t)} \equiv 1 \pmod{2}.$$

Thus, by Corollary 3.8.1 we see that $r^{\phi(2p^t)} \equiv 1 \pmod{2p^t}$. Since no smaller power of $r$ is congruent to 1 modulo $2p^t$, we conclude that $r$ is a primitive root modulo $2p^t$.

On the other hand, if $r$ is even, then $r + p^t$ is odd. Hence,

$$(r + p^t)^{\phi(2p^t)} \equiv 1 \pmod{2}.$$

Since $r + p^t \equiv r \pmod{p^t}$, we see that

$$(r + p^t)^{\phi(2p^t)} \equiv 1 \pmod{p^t}.$$

Therefore, $(r + p^t)^{\phi(2p^t)} \equiv 1 \pmod{2p^t}$, and as no smaller power of $r + p^t$ is congruent to 1 modulo $2p^t$, we conclude that $r + p^t$ is a primitive root modulo $2p^t$.                                                                        ∎

**Example 8.11.** Earlier in this section we showed that 3 is a primitive root modulo $7^t$ for all positive integers $t$. Hence, since 3 is odd, Theorem 8.14 tells us that 3 is also a primitive root modulo $2 \cdot 7^t$ for all positive integers $t$. For instance, 3 is a primitive root modulo 14.

Similarly, we know that 2 is a primitive root modulo $5^t$ for all positive integers $t$. Hence, since $2 + 5^t$ is odd, Theorem 8.14 tells us that $2 + 5^t$ is a primitive root modulo $2 \cdot 5^t$ for all positive integers $t$. For instance, 27 is a primitive root modulo 50.                                                □

Combining Corollary 8.8.1 and Theorems 8.10, 8.13, and 8.14, we can now describe which positive integers have a primitive root.

**Theorem 8.15.** The positive integer $n$ possesses a primitive root if and only if

$$n = 2, 4, p^t, \text{ or } 2p^t,$$

where $p$ is an odd prime and $t$ is a positive integer.

---

## 8.3 Exercises

1. Which of the integers $4, 10, 16, 22$ and $28$ have a primitive root?

2. Find a primitive root modulo

   a)    $11^2$    c)    $17^2$
   b)    $13^2$    d)    $19^2$.

3. Find a primitive root, for all positive integers $k$, modulo

   a)    $3^k$    c)    $13^k$
   b)    $11^k$    d)    $17^k$.

4. Find a primitive root modulo

   a)    6    c)    26
   b)    18    e)    338.

5. Find all the primitive roots modulo 22.

6. Show that there are the same number of primitive roots modulo $2p^t$ as there are of $p^t$, where $p$ is an odd prime and $t$ is a positive integer.

☞ 7. Show that if $m$ has a primitive root, then the only solutions of the congruence $x^2 \equiv 1 \pmod{m}$ are $x \equiv \pm 1 \pmod{m}$.

* 8. Let $n$ be a positive integer possessing a primitive root. Using this primitive root, prove that the product of all positive integers less than $n$ and relatively prime to $n$ is congruent to $-1$ modulo $n$. (When $n$ is prime, this result is Wilson's Theorem.)

* 9. Show that although there are no primitive roots modulo $2^k$ where $k$ is an integer, $k \geqslant 3$, every odd integer is congruent to exactly one of the integers $(-1)^\alpha 5^\beta$, where $\alpha = 0$ or $1$ and $\beta$ is an integer satisfying $0 \leqslant \beta \leqslant 2^{k-2}-1$.

## 8.3 Computer Projects

Write computer programs to do the following:

1. Find primitive roots modulo powers of odd primes.

2. Find primitive roots modulo twice powers of odd primes.

---

## 8.4 Index Arithmetic

In this section we demonstrate how primitive roots may be used to do modular arithmetic. Let $r$ be a primitive root modulo the positive integer $m$ (so that $m$ is of the form described in Theorem 8.15). By Theorem 8.3 we know that the integers

$$r, r^2, r^3, ..., r^{\phi(m)}$$

form a reduced system of residues modulo $m$. From this fact, we see that if $a$ is an integer relatively prime to $m$, then there is a unique integer $x$ with $1 \leqslant x \leqslant \phi(m)$ such that

$$r^x \equiv a \pmod{m}.$$

This leads to the following definition.

**Definition.** Let $m$ be a positive integer with primitive root $r$. If $a$ is a positive integer with $(a,m) = 1$, then the unique integer $x$ with $1 \leqslant x \leqslant \phi(m)$ and $r^x \equiv a \pmod{m}$ is called the *index of a to the base r modulo m*. With this definition, we have $a \equiv r^{\text{ind}_r a} \pmod{m}$.

If $x$ is the index of $a$ to the base $r$ modulo $m$, then we write $x = \text{ind}_r a$, where we do not indicate the modulus $m$ in the notation, since it is assumed to be fixed. From the definition, we know that if $a$ and $b$ are integers relatively prime to $m$ and $a \equiv b \pmod{m}$, then $\text{ind}_r a = \text{ind}_r b$.

**Example 8.12.** Let $m = 7$. We have seen that 3 is a primitive root modulo 7 and that $3^1 \equiv 3 \pmod 7$, $3^2 \equiv 2 \pmod 7$, $3^3 \equiv 6 \pmod 7$, $3^4 \equiv 4 \pmod 7$, $3^5 \equiv 5 \pmod 5$, and $3^6 \equiv 1 \pmod 7$.

Hence, modulo 7 we have

$$\text{ind}_3 1 = 6, \ \text{ind}_3 2 = 2, \ \text{ind}_3 3 = 1,$$
$$\text{ind}_3 4 = 4, \ \text{ind}_3 5 = 5, \ \text{ind}_3 6 = 3.$$

With a different primitive root modulo 7, we obtain a different set of indices. For instance, calculations show that with respect to the primitive root 5,

$$\text{ind}_5 1 = 6, \ \text{ind}_5 2 = 4, \ \text{ind}_5 3 = 5,$$
$$\text{ind}_5 4 = 2, \ \text{ind}_5 5 = 1, \ \text{ind}_5 6 = 3. \qquad \square$$

We now develop some properties of indices. These properties are somewhat similar to those of logarithms, but instead of equalities, we have congruences modulo $\phi(m)$.

**Theorem 8.16.** Let $m$ be a positive integer with primitive root $r$, and let $a$ and $b$ be integers relatively prime to $m$. Then

  (i)   $\text{ind}_r 1 \equiv 0 \pmod{\phi(m)}$.
  (ii)  $\text{ind}_r(ab) \equiv \text{ind}_r a + \text{ind}_r b \pmod{\phi(m)}$
  (iii) $\text{ind}_r a^k \equiv k \cdot \text{ind}_r a \pmod{\phi(m)}$ if $k$ is a positive integer.

*Proof of* (i). From Euler's theorem, we know that $r^{\phi(m)} \equiv 1 \pmod m$. Since $r$ is a primitive root modulo $m$, no smaller positive power of $r$ is congruent to 1 modulo $m$. Hence, $\text{ind}_r 1 = \phi(m) \equiv 0 \pmod{\phi(m)}$.

*Proof of* (ii). To prove this congruence, note that from the definition of indices,

$$r^{\text{ind}_r(ab)} \equiv ab \pmod m$$

and

$$r^{\text{ind}_r a + \text{ind}_r b} \equiv r^{\text{ind}_r a} \cdot r^{\text{ind}_r b} \equiv ab \pmod m.$$

Hence,

$$r^{\text{ind}_r(ab)} \equiv r^{\text{ind}_r a + \text{ind}_r b} \pmod m.$$

Using Theorem 8.2, we conclude that

$$\text{ind}_r(ab) \equiv \text{ind}_r a + \text{ind}_r b \pmod{\phi(m)}.$$

*Proof of* (iii). To prove the congruence of interest, first note that, by definition, we have

$$r^{\text{ind}_r a^k} \equiv a^k \pmod m$$

and

$$r^{k \cdot \operatorname{ind}_r a} \equiv (r^{\operatorname{ind}_r a})^k \equiv a^k \pmod{m}.$$

Hence,

$$r^{\operatorname{ind}_r a^k} \equiv r^{k \cdot \operatorname{ind}_r a} \pmod{m}.$$

Using Theorem 8.2, this leads us immediately to the congruence we want, namely

$$\operatorname{ind}_r a^k \equiv k \cdot \operatorname{ind}_r a \pmod{\phi(m)}. \qquad \blacksquare$$

**Example 8.13.** From the previous examples, we see that modulo 7, $\operatorname{ind}_5 2 = 4$ and $\operatorname{ind}_5 3 = 5$. Since $\phi(7) = 6$, part (ii) of Theorem 8.16 tells us that

$$\operatorname{ind}_5 6 = \operatorname{ind}_5 2 \cdot 3 = \operatorname{ind}_5 2 + \operatorname{ind}_5 3 = 4 + 5 = 9 \equiv 3 \pmod{6}.$$

Note that this agrees with the value previously found for $\operatorname{ind}_5 6$.

From part (iii) of Theorem 8.16, we see that

$$\operatorname{ind}_5 3^4 \equiv 4 \cdot \operatorname{ind}_5 3 \equiv 4 \cdot 5 = 20 \equiv 2 \pmod{6}.$$

Note that direct computation gives the same result, since

$$\operatorname{ind}_5 3^4 = \operatorname{ind}_5 81 = \operatorname{ind}_5 4 = 2. \qquad \square$$

Indices are helpful in the solution of certain types of congruences. Consider the following examples.

**Example 8.14.** We will use indices to solve the congruence $6x^{12} \equiv 11 \pmod{17}$. We find that 3 is a primitive root of 17 (since $3^8 \equiv -1 \pmod{17}$). The indices of integers to the base 3 modulo 17 are given in Table 8.1.

| a | 1 | 2 | 3 | 4 | 5 | 6 | 7 | 8 | 9 | 10 | 11 | 12 | 13 | 14 | 15 | 16 |
|---|---|---|---|---|---|---|---|---|---|----|----|----|----|----|----|----|
| $\operatorname{ind}_3 a$ | 16 | 14 | 1 | 12 | 5 | 15 | 11 | 10 | 2 | 3 | 7 | 13 | 4 | 9 | 6 | 8 |

**Table 8.1.** Indices to the Base 3 Modulo 17.

Taking the index of each side of the congruence to the base 3 modulo 17, we obtain a congruence modulo $\phi(17) = 16$, namely

$$\operatorname{ind}_3(6x^{12}) \equiv \operatorname{ind}_3 11 = 7 \pmod{16}.$$

Using (ii) and (iii) of Theorem 8.16, we obtain

$$\operatorname{ind}_3(6x^{12}) \equiv \operatorname{ind}_3 6 + \operatorname{ind}_3(x^{12}) \equiv 15 + 12 \cdot \operatorname{ind}_3 x \pmod{16}.$$

Hence,

$$15 + 12 \cdot \text{ind}_3 x \equiv 7 \pmod{16}$$

or

$$12 \cdot \text{ind}_3 x \equiv 8 \pmod{16}.$$

From this congruence it follows (as the reader should show) that

$$\text{ind}_3 x \equiv 2 \pmod{4}.$$

Hence,

$$\text{ind}_3 x \equiv 2, \ 6, \ 10, \ \text{or} \ 14 \pmod{16}.$$

Consequently, from the definition of indices, we find that

$$x \equiv 3^2, \ 3^6, \ 3^{10} \ \text{or} \ 3^{14} \pmod{17},$$

(note that this congruence holds modulo 17). Since $3^2 \equiv 9$, $3^6 \equiv 15$, $3^{10} \equiv 8$, and $3^{14} \equiv 2 \pmod{17}$, we conclude that

$$x \equiv 9, \ 15, \ 8, \ \text{or} \ 2 \pmod{17}.$$

Since each step in the computations is reversible, there are four incongruent solutions of the original congruence modulo 17.                                □

**Example 8.15.** We wish to find all solutions of the congruence $7^x \equiv 6 \pmod{17}$. When we take indices to the base 3 modulo 17 of both sides of this congruence, we find that

$$\text{ind}_3 (7^x) \equiv \text{ind}_3 6 = 15 \pmod{16}.$$

By part (iii) of Theorem 8.16 we obtain

$$\text{ind}_3 (7^x) \equiv x \cdot \text{ind}_3 7 \equiv 11x \pmod{16}.$$

Hence,

$$11x \equiv 15 \pmod{16}.$$

Since 3 is an inverse of 11 modulo 16, we multiply both sides of the linear congruence above by 3, to find that

$$x \equiv 3 \cdot 15 = 45 \equiv 13 \pmod{16}.$$

All steps in this computation are reversible. Therefore, the solutions of

$$7^x \equiv 6 \pmod{17}$$

are given by

$$x \equiv 13 \pmod{16}.$$                                □

Next, we discuss congruences of the form $x^k \equiv a \pmod{m}$, where $m$ is a positive integer with a primitive root and $(a,m) = 1$. First, we present a definition.

**Definition.** If $m$ and $k$ are positive integers and $a$ is an integer relatively prime to $m$, then we say that $a$ is a *kth power residue of m* if the congruence $x^k \equiv a \pmod{m}$ has a solution.

When $m$ is an integer possessing a primitive root, the following theorem gives a useful criterion for an integer $a$ relatively prime to $m$ to be a $k$th power residue of $m$.

**Theorem 8.17.** Let $m$ be a positive integer with a primitive root. If $k$ is a positive integer and $a$ is an integer relatively prime to $m$, then the congruence $x^k \equiv a \pmod{m}$ has a solution if and only if

$$a^{\phi(m)/d} \equiv 1 \pmod{m}$$

where $d = (k, \phi(m))$. Furthermore, if there are solutions of $x^k \equiv a \pmod{m}$, then there are exactly $d$ incongruent solutions modulo $m$.

*Proof.* Let $r$ be a primitive root modulo the positive integer $m$. We note that the congruence

$$x^k \equiv a \pmod{m}$$

holds if and only if

(8.1) $$k \cdot \text{ind}_r x \equiv \text{ind}_r a \pmod{\phi(m)}.$$

Now let $d = (k, \phi(m))$ and $y = \text{ind}_r x$, so that $x \equiv r^y \pmod{m}$. By Theorem 3.10, we note that if $d \nmid \text{ind}_r a$, then the linear congruence

(8.2) $$ky \equiv \text{ind}_r a \pmod{\phi(m)}$$

has no solutions, and hence, there are no integers $x$ satisfying (8.1). If $d \mid \text{ind}_r a$, then there are exactly $d$ integers $y$ incongruent modulo $\phi(m)$ such that (8.2) holds, and hence, exactly $d$ integers $x$ incongruent modulo $m$ such that (8.1) holds. Since $d \mid \text{ind}_r a$ if and only if

$$(\phi(m)/d)\text{ind}_r a \equiv 0 \pmod{\phi(m)},$$

and this congruence holds if and only if

$$a^{\phi(m)/d} \equiv 1 \pmod{m},$$

the theorem is true. ∎

We note that Theorem 8.17 tells us that if $p$ is a prime, $k$ is a positive integer, and $a$ is an integer relatively prime to $p$, then $a$ is a $k$th power residue of $p$ if and only if

$$a^{(p-1)/d} \equiv 1 \pmod{p},$$

where $d = (k, p-1)$. We illustrate this observation with an example.

**Example 8.16.** To determine whether 5 is a sixth power residue of 17, *i.e.*

whether the congruence

$$x^6 \equiv 5 \pmod{17}$$

has a solution, we determine that

$$5^{16/(6,16)} = 5^8 \equiv -1 \pmod{17}. \qquad \square$$

Hence, 5 is not a sixth power residue of 17.

A table of indices with respect to the least primitive root modulo each prime less than 100 is given in Table 4 of the Appendix.

We now present the proof of Theorem 5.10. We state this theorem again for convenience.

**Theorem 5.10**. If $n$ is an odd composite positive integer, then $n$ passes Miller's test for at most $(n-1)/4$ bases $b$ with $1 \leqslant b < n-1$.

We need the following lemma in the proof.

**Lemma 8.1**. Let $p$ be an odd prime and let $e$ and $q$ be positive integers. Then the number of incongruent solutions of the congruence $x^{q-1} \equiv 1 \pmod{p^e}$ is $(q, p^{e-1}(p-1))$.

*Proof.* Let $r$ be a primitive root of $p^e$. By taking indices with respect to $r$, we see that $x^q \equiv 1 \pmod{p^e}$ if and only if $qy \equiv 0 \pmod{\phi(p^e)}$ where $y = \text{ind}_r x$. Using Theorem 3.10, we see that there are exactly $(q, \phi(p^e))$ incongruent solutions of $qy \equiv 0 \pmod{\phi(p^e)}$. Consequently, there are $(q, \phi(p^e)) = (q, p^{e-1}(p-1))$ incongruent solutions of $x^q \equiv 1 \pmod{p^e}$. ∎

We now proceed with a proof of Theorem 5.10.

*Proof.* Let $n-1 = 2^s t$, where $s$ is a positive integer and $t$ is an odd positive integer. For $n$ to be a strong pseudoprime to the base $b$, either

$$b^t \equiv 1 \pmod{n}$$

or

$$b^{2^j t} \equiv -1 \pmod{n}$$

for some integer $j$ with $0 \leqslant j \leqslant s - 1$. In either case, we have

$$b^{n-1} \equiv 1 \pmod{n}.$$

Let the prime-power factorization of $n$ be $n = p_1^{e_1} p_2^{e_2} \cdots p_r^{e_r}$. By Lemma 8.1 we know that there are $(n-1, p_j^{e_j}(p_j-1)) = (n-1, p_j-1)$ incongruent solutions of $x^{n-1} \equiv 1 \pmod{p_j^{e_j}}$, $j = 1,2,...,r$. Consequently, the Chinese remainder theorem tells us that there are exactly $\prod_{j=1}^{r} (n-1, p_j-1)$ incongruent solutions of $x^{n-1} \equiv 1 \pmod{n}$.

To prove the theorem, we first consider the case where the prime-power factorization of n contains a prime power $p_k^{e_k}$ with exponent $e_k \geqslant 2$. Since

$$(p_k - 1)/p_k^{e_k} = 1/p_k^{e_k - 1} - 1/p_k^{e_k} \leqslant 2/9$$

(the largest possible value occurs when $p_j = 3$ and $e_j = 2$), we see that

$$\prod_{j=1}^{r} (n-1, p_j - 1) \leqslant \prod_{j=1}^{r} (p_j - 1)$$

$$\leqslant \left( \prod_{\substack{j=1 \\ j \neq k}}^{r} p_j \right) \left( \frac{2}{9} p_k^{e_k} \right)$$

$$\leqslant \frac{2}{9} n.$$

Since $\frac{2}{9} n \leqslant \frac{1}{4}(n-1)$ for $n \geqslant 9$, it follows that

$$\prod_{j=1}^{r} (n-1, p_j - 1) \leqslant (n-1)/4.$$

Consequently, there are at most $(n-1)/4$ integers $b$, $1 \leqslant b \leqslant n$, for which $n$ is a strong pseudoprime to the base $b$.

The other case to consider is when $n = p_1 p_2 \cdots p_r$, where $p_1, p_2, ..., p_r$ are distinct odd primes. Let

$$p_i - 1 = 2^{s_i} t_{i}, \quad i = 1, 2, ..., r,$$

where $s_i$ is a positive integer and $t_i$ is an odd positive integer. We reorder the primes $p_1, p_2, ..., p_r$ (if necessary) so that $s_1 \leqslant s_2 \leqslant \cdots \leqslant s_r$. We note that

$$(n-1, p_i - 1) = 2^{\min(s, s_i)} (t, t_i).$$

The number of incongruent solutions of $x^t \equiv 1 \pmod{p_i}$ is $T = (t, t_i)$. From Exercise 22 at the end of this section, there are $2^j t_i$ incongruent solutions of $x^{2^j t} \equiv -1 \pmod{p_i}$ when $0 \leqslant j \leqslant s_i - 1$, and no solutions otherwise. Hence, using the Chinese remainder theorem, there are $T_1 T_2 \cdots T_r$ incongruent solutions of $x^t \equiv 1 \pmod{n}$, and $2^{jr} T_1 T_2 \cdots T_r$ incongruent solutions of $x^{2^j t} \equiv -1 \pmod{n}$ when $0 \leqslant j \leqslant s_1 - 1$. Therefore, there are a total of

$$T_1 T_2 \cdots T_r \left[ 1 + \sum_{j=0}^{s_1 - 1} 2^{jr} \right] = T_1 T_2 \cdots T_r \left[ 1 + \frac{2^{rs_1} - 1}{2^r - 1} \right]$$

integers $b$ with $1 \leqslant b \leqslant n-1$, for which $n$ is a strong pseudoprime to the base $b$.

Now note that

$$\phi(n) = (p_1-1)(p_2-1) \cdots (p_r-1) = t_1 t_2 \cdots t_r 2^{s_1+s_2+\cdots+s_r}.$$

We will show that

$$T_1 T_2 \cdots T_r \left(1+\frac{2^{rs_1}-1}{2^r-1}\right) \leqslant \phi(n)/4,$$

which proves the desired result. Because $T_1 T_2 \cdots T_r \leqslant t_1 t_2 \cdots t_r$, we can achieve our goal by showing that

(8.3)
$$\left(1+\frac{2^{rs_1}-1}{2^r-1}\right)/2^{s_1+s_2+\cdots+s_r} \leqslant 1/4.$$

Since $s_1 \leqslant s_2 \leqslant \cdots \leqslant s_r$, we see that

$$\begin{aligned}
\left(1+\frac{2^{rs_1}-1}{2^r-1}\right)/2^{s_1+s_2+\cdots+s_r} &\leqslant \left(1+\frac{2^{rs_1}-1}{2^r-1}\right)/2^{rs_1}\\
&= \frac{1}{2^{rs_1}} + \frac{2^{rs_1}-1}{2^{rs_1}(2^r-1)}\\
&= \frac{1}{2^{rs_1}} + \frac{1}{2^r-1} - \frac{1}{2^{rs_1}(2^r-1)}\\
&= \frac{1}{2^r-1} + \frac{2^r-2}{2^{rs_1}(2^r-1)}\\
&\leqslant \frac{1}{2^{r-1}}.
\end{aligned}$$

From this inequality, we conclude that (8.3) is valid when $r \leqslant 3$.

When $r = 2$, we have $n = p_1 p_2$ with $p_1-1 = 2^{s_1} t_1$ and $p_2-1 = 2^{s_2} t_2$, with $s_1 \leqslant s_2$. If $s_1 < s_2$, then (8.3) is again valid, since

$$\begin{aligned}
\left(1+\frac{2^{2s_1}-1}{3}\right)/2^{s_1+s_2} &= \left(1+\frac{2^{2s_1}-1}{3}\right)/\left(2^{2s_1}\cdot 2^{s_2-s_1}\right)\\
&= \left(\frac{1}{3} + \frac{1}{3\cdot 2^{2s_1-1}}\right)/2^{s_2-s_1}\\
&\leqslant \frac{1}{4}.
\end{aligned}$$

When $s_1 = s_2$, we have $(n-1, p_1-1) = 2^s T_1$ and $(n-1, p_2-1) = 2^s T_2$. Let us assume that $p_1 > p_2$. Note that $T_1 \neq t_1$, for if $T_1 = t_1$, then $(p_1-1) \mid (n-1)$, so that

$$n = p_1 p_2 \equiv p_2 \equiv 1 \pmod{p_1 - 1},$$

which implies that $p_2 > p_1$, a contradiction. Since $T_1 \neq t_1$, we know that $T_1 \leqslant t_1 / 3$. Similarly, if $p_1 < p_2$ then $T_2 \neq t_2$, so that $T_2 \leqslant t_2/3$. Hence, $T_1 T_2 \leqslant t_1 t_2 /3$, and since $\left[ 1 + \dfrac{2^{2s_1} - 1}{3} \right] /2^{2s_1} \leqslant \dfrac{1}{2}$, we have

$$T_1 T_2 \left[ 1 + \frac{2^{2s_1} - 1}{3} \right] \leqslant t_1 t_2 \, 2^{2s_1} /6 = \phi(n)/6,$$

which proves the theorem for this final case, since $\phi(n)/6 \leqslant (n-1)/6 < (n-1)/4$. ∎

By analyzing the inequalities in the proof of Theorem 5.10, we can see that the probability that $n$ is a strong pseudoprime to the randomly chosen base $b$, $1 \leqslant b \leqslant n-1$, is close to $1/4$ only for integers $n$ with prime factorizations of the form $n = p_1 p_2$ with $p_1 = 1 + 2q_1$ and $p_2 = 1 + 4q_2$, where $q_1$ and $q_2$ are odd primes, or $n = q_1 q_2 q_3$ with $p_1 = 1 + 2q_1$, $p_2 = 1 + 2q_2$, and $p_3 = 1 + 2q_3$, where $q_1, q_2$, and $q_3$ are distinct odd primes (see Exercise 23).

---

## 8.4 Exercises

1. Write out a table of indices modulo 23 with respect to the primitive root 5.

2. Find all the solutions of the congruences

   a) $3x^5 \equiv 1 \pmod{23}$     b) $3x^{14} \equiv 2 \pmod{23}$.

3. Find all the solutions of the congruences

   a) $3^x \equiv 2 \pmod{23}$     b) $13^x \equiv 5 \pmod{23}$.

4. For which positive integers $a$ is the congruence $ax^4 \equiv 2 \pmod{13}$ solvable?

5. For which positive integers $b$ is the congruence $8x^7 \equiv b \pmod{29}$ solvable?

6. Find the solutions of $2^x \equiv x \pmod{13}$, using indices to the base 2 modulo 13.

7. Find all the solutions of $x^x \equiv x \pmod{23}$.

8. Show that if $p$ is an odd prime and $r$ is a primitive root of $p$, then $\text{ind}_r(p-1) = (p-1)/2$.

9. Let $p$ be an odd prime. Show that the congruence $x^4 \equiv -1 \pmod{p}$ has a solution if and only if $p$ is of the form $8k + 1$.

**10.** Prove that there are infinitely many primes of the form $8k+1$. (*Hint*: Assume that $p_1,p_2,...,p_n$ are the only primes of this form. Let $Q = (p_1 p_2 \cdots p_n)^4 + 1$. Show that $Q$ must have an odd prime factor different than $p_1,p_2,...,p_n$ and by Exercise 9, necessarily of the form $8k+1$.)

From Exercise 9 of Section 8.3, we know that if $a$ is a positive integer, then there are unique integers $\alpha$ and $\beta$ with $\alpha = 0$ or 1 and $0 \leqslant \beta \leqslant 2^{k-2}-1$ such that $a \equiv (-1)^\alpha 5^\beta \pmod{2^k}$. Define the *index system of a modulo $2^k$* to be equal to the pair $(\alpha,\beta)$.

**11.** Find the index systems of 7 and 9 modulo 16.

**12.** Develop rules for the index systems modulo $2^k$ of products and powers analogous to the rules for indices.

**13.** Use the index system modulo 32 to find all solutions of $7x^9 \equiv 11 \pmod{32}$ and $3^x \equiv 17 \pmod{32}$.

Let $n = 2^{t_0} p_1^{t_1} p_2^{t_2} \cdots p_m^{t_m}$ be the prime-power factorization of $n$. Let $a$ be an integer relatively prime to $n$. Let $r_1, r_2, ..., r_m$ be primitive roots of $p_1^{t_1}, p_2^{t_2}, ..., p_m^{t_m}$, respectively, and let $\gamma_1 = \mathrm{ind}_{r_1} a \pmod{p_1^{t_1}}$, $\gamma_2 = \mathrm{ind}_{r_2} a \pmod{p_2^{t_2}}$, $..., \gamma_m = \mathrm{ind}_{r_m} a \pmod{p_m^{t_m}}$. If $t_0 \leqslant 2$, let $r_0$ be a primitive root of $2^{t_0}$, and let $\gamma_0 = \mathrm{ind}_{r_0} a \pmod{2^t}$. If $t_0 \geqslant 3$, let $(\alpha,\beta)$ be the index system of $a$ modulo $2^k$, so that $a \equiv (-1)^\alpha 5^\beta \pmod{2^k}$. Define the *index system of a modulo n* to be $(\gamma_0, \gamma_1, \gamma_2, ..., \gamma_m)$ if $t_0 \leqslant 2$ and $(\alpha,\beta,\gamma_1,\gamma_2,...,\gamma_m)$ if $t_0 \geqslant 3$.

**14.** Show that if $n$ is a positive integer, then every integer has a unique index system modulo $n$.

**15.** Find the index systems of 17 and 41 (mod 120) (in your computations, use 2 as a primitive root of the prime factor 5 of 120).

**16.** Develop rules for the index systems modulo $n$ of products and powers analogous to those for indices.

**17.** Use an index system modulo 60 to find the solutions of $11x^7 \equiv 43 \pmod{60}$.

**18.** Let $p$ be a prime, $p > 3$. Show that if $p \equiv 2 \pmod 3$ then every integer not divisible by 3 is a third-power, or *cubic*, residue of $p$, while if $p \equiv 1 \pmod 3$, an integer $a$ is a cubic residue of $p$ if and only if $a^{(p-1)/3} \equiv 1 \pmod p$.

**19.** Let $e$ be a positive integer with $e \geqslant 2$. Show that if $k$ is a positive integer, then every odd integer $a$ is a $k$th power residue of $2^e$.

**\* 20.** Let $e$ be a positive integer with $e \geqslant 2$. Show that if $k$ is even, then an integer $a$ is a $k$th power residue of $2^e$ if and only if $a \equiv 1 \pmod{(4k, 2^e)}$.

**\* 21.** Let $e$ be a positive integer with $e \geqslant 2$. Show that if $k$ is a positive integer, then the number of incongruent $k$th power residues of $2^e$ is

$$\frac{2^{e-1}}{(n, 2)(n, 2^{e-2})}.$$

☛ **22.** Let $N = 2^j u$ be a positive integer with $j$ a nonnegative integer and $u$ an odd positive integer and let $p-1 = 2^s t$, where $s$ and $t$ are positive integers with $t$ odd. Show that there are $2^j(t,u)$ incongruent solutions of $x^N \equiv -1 \pmod{p}$ if $0 \leqslant j \leqslant s-1$, and no solutions otherwise.

\* **23.**  a)  Show that the probability that $n$ is a strong pseudoprime for a base $b$ randomly chosen with $1 \leqslant b \leqslant n-1$ is near $(n-1)/4$ only when $n$ has a prime factorization of the form $n = p_1 p_2$ where $p_1 = 1 + 2q_1$ and $p_2 = 1 + 4q_2$ with $q_1$ and $q_2$ prime or $n = p_1 p_2 p_3$ where $p_1 = 1 + 2q_1$, $p_2 = 1 + 2q_2$, $p_3 = 1 + 2q_3$ with $q_1, q_2, q_3$ distinct odd primes.

   b)  Find the probability that $n = 49939 \cdot 99877$ is a strong pseudoprime to the base $b$ randomly chosen with $1 \leqslant b \leqslant n - 1$.

## 8.4  Computer Projects

Write programs to do the following:

1.  Construct a table of indices modulo a particular primitive root of an integer.

2.  Using indices, solve congruences of the form $ax^b \equiv c \pmod{m}$ where $a, b, c,$ and $m$ are integers with $c > 0$, $m > 0$, and where $m$ has a primitive root.

3.  Find $k$th power residues of a positive integer $m$ having a primitive root, where $k$ is a positive integer.

4.  Find index systems modulo powers of 2 (see preamble to Exercise 11).

5.  Find index systems modulo arbitrary positive integers (see preamble to Exercise 14).

## 8.5  Primality Tests Using Primitive Roots

From the concepts of orders of integers and primitive roots, we can produce useful primality tests. The following theorem presents such a test.

**Theorem 8.18.** If $n$ is a positive integer and if an integer $x$ exists such that

$$x^{n-1} \equiv 1 \pmod{n}$$

and

$$x^{(n-1)/q} \not\equiv 1 \pmod{n}$$

for all prime divisors $q$ of $n - 1$, then $n$ is prime.

*Proof.* Since $x^{n-1} \equiv 1 \pmod{n}$, Theorem 8.1 tells us that $\mathrm{ord}_n x \mid (n-1)$.

We will show that $\mathrm{ord}_n x = n - 1$. Suppose that $\mathrm{ord}_n x \neq n - 1$. Since $\mathrm{ord}_n x \mid (n-1)$, there is an integer $k$ with $n - 1 = k \cdot \mathrm{ord}_n x$ and since $\mathrm{ord}_n x \neq n - 1$, we know that $k > 1$. Let $q$ be a prime divisor of $k$. Then

$$x^{(n-1)/q} = x^{k/q \, \mathrm{ord}_n x} = (x^{\mathrm{ord}_n x})^{(k/q)} \equiv 1 \pmod{n}.$$

However, this contradicts the hypotheses of the theorem, so we must have $\mathrm{ord}_n x = n - 1$. Now, since $\mathrm{ord}_n x \leqslant \phi(n)$ and $\phi(n) \leqslant n - 1$, it follows that $\phi(n) = n - 1$. By Theorem 6.2 we know that $n$ must be prime. ∎

Note that Theorem 8.18 is equivalent to the fact that if there is an integer with order modulo $n$ equal to $n-1$, then $n$ must be prime. We illustrate the use of Theorem 8.18 with an example.

**Example 8.17.** Let $n = 1009$. Then $11^{1008} \equiv 1 \pmod{1009}$. The prime divisors of 1008 are 2,3, and 7. We see that $11^{1008/2} = 11^{504} \equiv -1 \pmod{1009}$, $11^{1008/3} = 11^{336} \equiv 374 \pmod{1009}$, and $11^{1008/7} = 11^{144} \equiv 935 \pmod{1009}$. Hence, by Theorem 8.18 we know that 1009 is prime. □

The following corollary of Theorem 8.18 gives a slightly more efficient primality test.

**Corollary 8.18.1.** If $n$ is an odd positive integer and if $x$ is a positive integer such that

$$x^{(n-1)/2} \equiv -1 \pmod{n}$$

and

$$x^{(n-1)/q} \not\equiv 1 \pmod{n}$$

for all odd prime divisors $q$ of $n - 1$, then $n$ is prime.

*Proof.* Since $x^{(n-1)/2} \equiv -1 \pmod{n}$, we see that

$$x^{n-1} = (x^{(n-1)/2})^2 \equiv (-1)^2 \equiv 1 \pmod{n}.$$

Since the hypotheses of Theorem 8.18 are met, we know that $n$ is prime. ∎

**Example 8.18.** Let $n = 2003$. The odd prime divisors of $n-1 = 2002$ are 7, 11, and 13. Since $5^{2002/2} = 5^{1001} \equiv -1 \pmod{2003}$, $5^{2002/7} = 5^{286} \equiv 874 \pmod{2003}$, $5^{2002/11} = 5^{183} \equiv 886 \pmod{2003}$, and $5^{2002/13} = 5^{154} \equiv 633 \pmod{2003}$, we see from Corollary 8.18.1 that 2003 is prime. □

To determine whether an integer $n$ is prime using either Theorem 8.18 or Corollary 8.18.1 it is necessary to know the prime factorization of $n - 1$. As we have remarked before, finding the prime factorization of an integer is a time-consuming process. Only when we have some *a priori* information about the factorization of $n - 1$ are the primality tests given by these results practical. Indeed, with such information these tests can be useful. Such a situation occurs with the Fermat numbers; in Chapter 9 we give a primality test for these numbers based on the ideas of this section.

It is of interest to ask how quickly a computer can verify primality or compositeness. We answer these questions as follows.

**Theorem 8.19.** If $n$ is composite, this can be proved with $O((\log_2 n)^2)$ bit operations.

*Proof.* If $n$ is composite, there are integers $a$ and $b$ with $1 < a < n$, $1 < b < n$, and $n = ab$. Hence, given the two integers $a$ and $b$, we multiply $a$ and $b$ and verify that $n = ab$. This takes $O((\log_2 n)^2)$ bit operations and proves that $n$ is composite. ∎

We can use Theorem 8.19 to estimate the number of bit operations needed to prove primality when the appropriate information is known.

**Theorem 8.20.** If $n$ is prime, this can be proven using $O((\log_2 n)^4)$ bit operations.

*Proof.* We use the second principle of mathematical induction. The induction hypothesis is an estimate for $f(n)$, where $f(n)$ is the total number of multiplications and modular exponentiations needed to verify that the integer $n$ is prime.

We demonstrate that

$$f(n) \leqslant 3 (\log n/\log 2) - 2.$$

First, we note that $f(2) = 1$. We assume that for all primes $q$, with $q < n$, the inequality

$$f(q) \leqslant 3 (\log q/\log 2) - 2$$

holds.

To prove that $n$ is prime, we use Corollary 8.18.1. Once we have the numbers $2^a$, $q_1,..., q_t$, and $x$ that supposedly satisfy

    (i)   $n - 1 = 2^a q_1 q_2 \cdots q_t$,

    (ii)  $q_i$ is prime for $i = 1, 2,..., t$,

    (iii) $x^{(n-1)/2} \equiv -1 \pmod{n}$,

and

    (iv) $x^{(n-1)/q_j} \equiv 1 \pmod{n}$, for $i = 1, 2,... t$,

we need to do $t$ multiplications to check (i), $t + 1$ modular exponentiations to check (iii) and (iv), and $f(q_i)$ multiplications and modular exponentiations to check (ii), that $q_i$ is prime for $i = 1, 2,..., t$. Hence,

$$f(n) = t + (t+1) + \sum_{i=1}^{t} f(q_i)$$

$$\leqslant 2t + 1 + \sum_{i=1}^{t} ((3 \log q_i / \log 2) - 2)$$

Now each multiplication requires $O((\log_2 n)^2)$ bit operations and each modular exponentiation requires $O((\log_2 n)^3)$ bit operations. Since the total number of multiplications and modular exponentiations needed is $f(n) = O(\log_2 n)$, the total number of bit operations needed is $O((\log_2 n)(\log_2 n)^3) = O((\log_2 n)^4)$. ∎

Theorem 8.20 was discovered by Pratt. He interpreted the result as showing that every prime has a "succinct certification of primality." It should be noted that Theorem 8.20 cannot be used to find this short proof of primality, for the factorization of $n - 1$ and the primitive root $x$ of $n$ are required. More information on this subject may be found in Lenstra [101].

Recently, an extremely efficient primality test has been developed by Adleman, Pomerance, and Rumely. We will not describe the test here because it relies on concepts not developed in this book. We note, that to determine whether an integer is prime using this test requires less than $(\log_2 n)^{c \log_2 \log_2 \log_2 n}$ bit operations, where $c$ is a constant. For instance, to determine whether a 100-digit integer is prime requires just 40 seconds and to determine whether a 200-digit integer is prime requires just 10 minutes. Even a 1000-digit integer may be checked for primality in a reasonable amount of time, one week. For more information about this test see [90] and [105].

---

## 8.5 Exercises

1. Show that 101 is prime using Theorem 8.18 with $x = 2$.

2. Show that 257 is prime using Corollary 8.18.1 with $x = 3$.

3. Show that if an integer $x$ exists such that

$$x^{2^{2^n}} \equiv 1 \pmod{F_n}$$

and

$$x^{2^{(2^n-1)}} \not\equiv 1 \pmod{F_n},$$

then the Fermat number $F_n = 2^{2^n} + 1$ is prime.

* 4. Let $n$ be a positive integer. Show that if the prime-power factorization of $n - 1$ is $n - 1 = p_1^{a_1} p_2^{a_2} \cdots p_t^{a_t}$ and for $j = 1, 2,..., t$, there exists an integer $x_j$ such that

$$x_j^{(n-1)/p_j} \not\equiv 1 \pmod{n}$$

and

$$x_j^{n-1} \equiv 1 \pmod{n},$$

then $n$ is prime.

* 5.  Let $n$ be a positive integer such that

$$n - 1 = m \prod_{j=1}^{r} q_j^{a_j}$$

where $m$ is a positive integer, $a_1, a_2,..., a_r$ are positive integers, and $q_1, q_2,..., q_r$ are relatively prime integers greater than one. Furthermore, let $b_1, b_2,..., b_r$ be positive integers such that there exist integers $x_1, x_2,..., x_r$ with

$$x_j^{n-1} \equiv 1 \pmod{n}$$

and

$$(x_j^{(n-1)/q_j} - 1, n) = 1$$

for $j = 1, 2,..., r$, where every prime factor of $q_j$ is greater than or equal to $b_j$ for $j = 1, 2,..., r$, and

$$n < (1 + \prod_{j=1}^{r} b_j^{a_j})^2.$$

Show that $n$ is prime.

## 8.5 Computer Projects

Write programs to show that a positive integer $n$ is prime using

1.  Theorem 8.18.
2.  Corollary 8.18.1.
3.  Exercise 4.
4.  Exercise 5.

## 8.6 Universal Exponents

Let $n$ be a positive integer with prime-power factorization

$$n = p_1^{t_1} p_2^{t_2} \cdots p_m^{t_m}.$$

If $a$ is an integer relatively prime to $n$, then Euler's theorem tells us that

$$a^{\phi(p^t)} \equiv 1 \pmod{p^t}$$

whenever $p^t$ is one of the prime powers occurring in the factorization of $n$ . As in the proof of Theorem 8.13, let

$$U = [\phi(p_1^{t_1}), \phi(p_2^{t_2}), ..., \phi(p_m^{t_m})].$$

the least common multiple of the integers $\phi(p_i^{t_i})$, $i = 1,2, ..., m$. Since

$$\phi(p_i^{t_i}) \mid U$$

for $i = 1,2,..., m$, using Theorem 8.1 we see that

$$a^U \equiv 1 (\mathrm{mod}\ p_i^{t_i})$$

for $i = 1,2, ..., m$. Hence, by Corollary 3.8.1 it follows that

$$a^U \equiv 1\ (\mathrm{mod}\ n).$$

This leads to the following definition.

**Definition.** A *universal exponent* of the positive integer $n$ is a positive integer $U$ such that

$$a^U \equiv 1\ (\mathrm{mod}\ n),$$

for all integers $a$ relatively prime to $n$.

**Example 8.19.** Since the prime power factorization of 600 is $2^3 \cdot 3 \cdot 5^2$, it follows that $U = [\phi(2^3), \phi(3), \phi(5^2)] = [2, 2, 20] = 20$ is a universal exponent of 600.                                                                                    □

From Euler's theorem, we know that $\phi(n)$ is a universal exponent. As we have already demonstrated, the integer $U = [\phi(p_1^{t_1}), \phi(p_2^{t_2}), ..., \phi(p_m^{t_m})]$ is also a universal exponent of $n = p_1^{t_1} p_2^{t_2} \cdots p_m^{t_m}$. We are interested in finding the *smallest* positive universal exponent of $n$.

**Definition.** The least universal exponent of the positive integer $n$ is called the *minimal universal exponent of $n$*, and is denoted by $\lambda(n)$.

We now find a formula for the minimal universal exponent $\lambda(n)$, based on the prime-power factorization of $n$.

First, note that if $n$ has a primitive root, then $\lambda(n) = \phi(n)$. Since powers of odd primes possess primitive roots, we know that

$$\lambda(p^t) = \phi(p^t),$$

whenever $p$ is an odd prime and $t$ is a positive integer. Similarly, we have $\lambda(2) = \phi(2) = 1$ and $\lambda(4) = \phi(4) = 2$, since both 2 and 4 have primitive roots. On the other hand, if $t \geqslant 3$, then we know by Theorem 8.11 that

$$a^{2^{t-2}} = 1\ (\mathrm{mod}\ 2^t)$$

and $\mathrm{ord}_t a = 2^{t-2}$, so that we can conclude that $\lambda(2^t) = 2^{t-2}$ if $t \geqslant 3$.

We have found $\lambda(n)$ when $n$ is a power of a prime. Next, we turn our attention to arbitrary positive integers $n$.

**Theorem 8.21.** Let $n$ be a positive integer with prime-power factorization

$$n = 2^{t_0} p_1^{t_1} p_2^{t_2} \cdots p_m^{t_m}.$$

Then $\lambda(n)$, the minimal universal exponent of $n$, is given by

$$\lambda(n) = [\lambda(2^{t_0}), \phi(p_1^{t_1}),..., \phi(p_m^{t_m})],$$

Moreover, there exists an integer $a$ such that $\text{ord}_n a = \lambda(n)$, the largest possible order of an integer modulo $n$.

*Proof.* Let $a$ be an integer with $(a, n) = 1$. For convenience, let

$$M = [\lambda(2^{t_0}), \phi(p_1^{t_1}), \phi(p_2^{t_2}),..., \phi(p_m^{t_m})].$$

Since $M$ is divisible by all of the integers $\lambda(2^{t_0})$, $\phi(p_1^{t_1}) = \lambda(p_1^{t_1})$, $\phi(p_2^{t_2}) = \lambda(p_2^{t_2}),..., \phi(p_m^{t_m}) = \lambda(p_m^{t_m})$, and since $a^{\lambda(p^t)} \equiv 1 \pmod{p^t}$ for all prime-powers in the factorization of $n$, we see that

$$a^M \equiv 1 \pmod{p^t},$$

whenever $p^t$ is a prime-power occurring in the factorization of $n$.

Consequently, by Corollary 3.8.1 we can conclude that

$$a^M \equiv 1 \pmod{n}.$$

The last congruence establishes the fact that $M$ is a universal exponent. We must now show that $M$ is the *least* universal exponent. To do this, we find an integer $a$ such that no positive power smaller than the $M$th power of $a$ is congruent to 1 modulo $n$. With this in mind, let $r_i$ be a primitive root of $p_i^{t_i}$.

We consider the system of simultaneous congruences

$$x \equiv 3 \pmod{2^{t_0}}$$
$$x \equiv r_1 \pmod{p_1^{t_1}}$$
$$x \equiv r_2 \pmod{p_2^{t_2}}$$

$$.$$
$$.$$
$$.$$

$$x \equiv r_m \pmod{p_m^{t_m}}.$$

By the Chinese remainder theorem, there is a simultaneous solution $a$ of this system which is unique modulo $n = 2^{t_0} p_1^{t_1} p_2^{t_2} \cdots p_m^{t_m}$; we will show that $\text{ord}_n a = M$. To prove this claim, assume that $N$ is a positive integer such that

$$a^N \equiv 1 \pmod{n}.$$

Then, if $p^t$ is a prime-power divisor of $n$, we have

$$a^N \equiv 1 \pmod{p^t},$$

so that

$$\text{ord}_{p^t} a \mid N.$$

But, since $a$ satisfies each of the $m + 1$ congruences of the system, we have

$$\text{ord}_{p^t} a = \lambda(p^t),$$

for each prime power in the factorization. Hence by Theorem 8.1 we have

$$\lambda(p^t) \mid N$$

for all prime powers $p^t$ in the factorization of $n$. Therefore, by Corollary 3.8.1, we know that $M = [\lambda(2^{t_0}), \lambda(p_1^{t_1}), \lambda(p_2^{t_2}),..., \lambda(p_m^{t_m})] \mid N$.

Since $a^M \equiv 1 \pmod{n}$ and $M \mid N$ whenever $a^N \equiv 1 \pmod{n}$, we can conclude that

$$\text{ord}_n a = M.$$

This shows that $M = \lambda(n)$ and simultaneously produces a positive integer $a$ with $\text{ord}_n a = \lambda(n)$.                                                              ∎

**Example 8.20.** Since the prime-power factorization of 180 is $2^2 \cdot 3^2 \cdot 5$, from Theorem 8.21 it follows that

$$\lambda (180) = [\phi(2^2), \phi(3^2), \phi(5)] = [2, 6, 4] = 12.$$

To find an integer $a$ with $\text{ord}_{180} a = 12$, first we find primitive roots modulo $3^2$ and 5. For instance, we take 2 and 3 as primitive roots modulo $3^2$ and 5, respectively. Then, using the Chinese remainder theorem, we find a solution of the system of congruences

$$a \equiv 3 \pmod 4$$
$$a \equiv 2 \pmod 9$$
$$a \equiv 3 \pmod 5,$$

obtaining $a \equiv 83 \pmod{180}$. From the proof of Theorem 8.21 we see that $\text{ord}_{180} 83 = 12$.                                                              □

**Example 8.21.** Let $n = 2^6 3^2 5 \cdot 7 \cdot 13 \cdot 17 \cdot 19 \cdot 37 \cdot 73$. Then, we have

$$\begin{aligned}
\lambda(n) &= [\lambda(2^6), \phi(3^2), \phi(5), \phi(17), \phi(19), \phi(37), \phi(73)] \\
&= [2^4, 2 \cdot 3, 2^2, 2^4, 2 \cdot 3^2, 2^2 3^2, 2^3 3^2] \\
&= 2^4 \cdot 3^2 \\
&= 144.
\end{aligned}$$

Hence, whenever $a$ is a positive integer relatively prime to $2^6 \cdot 3^2 \cdot 5 \cdot 17 \cdot 17 \cdot 19 \cdot 37 \cdot 73$ we know that $a^{144} \equiv 1 \pmod{2^6 \cdot 3^2 \cdot 5 \cdot 17 \cdot 19 \cdot 37 \cdot 37 \cdot 73}$.                                                              □

We now return to the Carmichael numbers that we discussed in Section 5.2. Recall that a Carmichael number is a composite integer that satisfies $b^{n-1} \equiv 1 \pmod n$ for all positive integers $b$ with $(b, n) = 1$. We proved that if $n = q_1 q_2 \cdots q_k$, where $q_1, q_2,..., q_k$ are distinct primes satisfying $(q_j - 1) \mid (n-1)$ for $j = 1, 2,...,k$, then $n$ is a Carmichael number. Here, we prove the converse of this result.

**Theorem 8.22.** If $n > 2$ is a Carmichael number, then $n = q_1 q_2 \cdots q_k$, where the $q_j$'s are distinct primes such that $(q_j - 1) \mid (n-1)$ for $j = 1, 2,..., k$.

*Proof.* If $n$ is a Carmichael number, then

$$b^{n-1} \equiv 1 \pmod{n}$$

for all positive integers $b$ with $(b, n) = 1$. Theorem 8.21 tells us that there is an integer $a$ with $\operatorname{ord}_n a = \lambda(n)$, where $\lambda(n)$ is the minimal universal exponent, and since $a^{n-1} \equiv 1 \pmod{n}$, Theorem 8.1 tells us that

$$\lambda(n) \mid (n-1).$$

Now $n$ must be odd, for if $n$ was even, then $n-1$ would be odd, but $\lambda(n)$ is even (since $n > 2$), contradicting the fact that $\lambda(n) \mid (n-1)$.

We now show that $n$ must be the product of distinct primes. Suppose $n$ has a prime-power factor $p^t$ with $t \geqslant 2$. Then

$$\lambda(p^t) = \phi(p^t) = p^{t-1}(p-1) \mid \lambda(n) = n-1.$$

This implies that $p \mid (n-1)$, which is impossible since $p \mid n$. Consequently, $n$ must be the product of distinct odd primes, say

$$n = q_1 q_2 \cdots q_k.$$

We conclude the proof by noting that

$$\lambda(q_i) = \phi(q_i) = (q_j-1) \mid \lambda(n) = n-1. \qquad \blacksquare$$

We can easily prove more about the prime factorizations of Carmichael numbers.

**Theorem 8.23.** A Carmichael number must have at least three different odd prime factors.

*Proof.* Let $n$ be a Carmichael number. Then $n$ cannot have just one prime factor, since it is composite, and is the product of distinct primes. So assume that $n = pq$, where $p$ and $q$ are odd primes with $p > q$. Then

$$n-1 = pq-1 = (p-1)q + (q-1) \equiv q-1 \not\equiv 0 \pmod{p-1},$$

which shows that $(p-1) \nmid (n-1)$. Hence, $n$ cannot be a Carmichael number if it has just two different prime factors. $\qquad \blacksquare$

## 8.6 Exercises

1. Find $\lambda(n)$, the minimal universal exponent of $n$, for the following values of $n$

   a) 100
   b) 144
   c) 222
   d) 884

   e) $2^4 \cdot 3^3 \cdot 5^2 \cdot 7$
   f) $2^5 \cdot 3^2 \cdot 5^2 \cdot 7^3 \cdot 11^2 \cdot 13 \cdot 17 \cdot 19$
   g) 10!
   h) 20!.

2. Find all positive integers $n$ such that $\lambda(n)$ is equal to

   a) 1
   b) 2
   c) 3

   d) 4
   e) 5
   f) 6.

3. Find the largest integer $n$ with $\lambda(n) = 12$.

4. Find an integer with the largest possible order modulo

   a) 12
   b) 15
   c) 20

   d) 36
   e) 40
   f) 63.

5. Show that if $m$ is a positive integer, then $\lambda(m)$ divides $\phi(m)$.

6. Show that if $m$ and $n$ are relatively prime positive integers, then $\lambda(mn) = [\lambda(m), \lambda(n)]$.

7. Let $n$ be the largest positive integer satisfying the equation $\lambda(n) = a$, where $a$ is a fixed positive integer. Show that if $m$ is another solution of $\lambda(m) = a$, then $m$ divides $n$.

8. Show that if $n$ is a positive integer, then there are exactly $\phi(\lambda(n))$ incongruent integers with maximal order modulo $n$.

9. Show that if $a$ and $m$ are relatively prime positive integers, then the solutions of the congruence $ax \equiv b \pmod{m}$ are the integers $x$ such that $x \equiv a^{\lambda(m)-1}b \pmod{m}$.

10. Show that if $c$ is a positive integer greater than one, then the integers $1^c, 2^c, ..., (m-1)^c$ form a complete system of residues modulo $m$ if and only if $m$ is square-free and $(c, \lambda(m)) = 1$.

\* 11. a) Show that if $c$ and $m$ are positive integers then the congruence $x^c \equiv x \pmod{m}$ has exactly

$$\prod_{j=1}^{r} (1 + (c-1, \phi(p_j^{a_j})))$$

incongruent solutions, where $m$ has prime-power factorization $m = p_1^{a_1} p_2^{a_2} \cdots p_r^{a_r}$.

   b) Show that $x^c \equiv x \pmod{m}$ has exactly $3^r$ solutions if and only if $(c-1, \phi(m)) = 2$.

12. Use Exercise 11 to show that there are always at least 9 plaintext messages that are not changed when enciphered using an RSA cipher.

* 13. Show that there are no Carmichael numbers of the form $3pq$ where $p$ and $q$ are primes.

* 14. Find all Carmichael numbers of the form $5pq$ where $p$ and $q$ are primes.

* 15. Show that there are only a finite number of Carmichael numbers of the form $n = pqr$, where $p$ is a fixed prime, and $q$ and $r$ are also primes.

16. Show that the deciphering exponent $d$ for an RSA cipher with enciphering key $(e,n)$ can be taken to be an inverse of $e$ modulo $\lambda(n)$.

## 8.6 Computer Projects

Write programs to do the following:

1. Find the minimal universal exponent of a positive integer.

2. Find an integer with order modulo $n$ equal to the minimal universal exponent of $n$.

3. Given a positive integer $M$, find all positive integers $n$ with minimal universal exponent equal to $M$.

4. Solve linear congruences using the method of Exercise 9.

---

## 8.7 Pseudo-Random Numbers

Numbers chosen randomly are often useful in computer simulation of complicated phenomena. To perform simulations, some method for generating random numbers is needed. There are various mechanical means for generating random numbers, but these are inefficient for computer use. Instead, a systematic method using computer arithmetic is preferable. One such method, called the *middle-square method*, introduced by John Von Neumann*, works as follows. To generate four-digit random numbers, we start with an arbitrary four-digit number, say 6139. We square this number to obtain 37687321, and we take the middle four digits 6873 as the second

---

* **JOHN VON NEUMANN** (1903-1957) was born in Budapest, Hungary. In 1930, after holding several positions at universities in Germany, he came to the United States. In 1933 Von Neumann became, along with Albert Einstein, one of the first members of the famous Institute for Advanced Study in Princeton, New Jersey. Von Neumann was one of the most versatile mathematical talents of the twentieth century. He invented the mathematical discipline known as game theory. Using game theory, he made many important discoveries in mathematical ecomonics. Von Neumann made fundamental contributions to the development of the first computers, and participated in the early development of atomic weapons.

random number. We iterate this procedure to obtain a sequence of random numbers, always squaring and removing the middle four-digits to obtain a new random number from the preceding one. (The square of a four-digit number has eight or fewer digits. Those with fewer than eight digits are considered eight-digit numbers by adding initial digits of 0.)

Sequences produced by the middle-square method are, in reality, not randomly chosen. When the initial four-digit number is known, the entire sequence is determined. However, the sequence of numbers produced appears to be random, and the numbers produced are useful for computer simulations. The integers in sequences that have been chosen in some methodical manner, but appear to be random, are called *pseudo-random numbers*.

It turns out that the middle-square method has some unfortunate weaknesses. The most undesirable feature of this method is that, for many choices of the initial integer, the method produces the same small set of numbers over and over. For instance, starting with the four-digit integer 4100 and using the middle-square method, we obtain the sequence 8100, 6100, 2100, 4100, 8100, 6100, 2100,... which only gives four different numbers before repeating.

The most commonly used method for generating pseudo-random numbers is called the *linear congruential method* which works as follows. A set of integers $m$, $a$, $c$, and $x_0$ is chosen so that $m > 0$, $2 \leqslant a \leqslant m$, $0 \leqslant c \leqslant m$, and $0 \leqslant x_0 \leqslant m$. The sequence of pseudo-random numbers is defined recursively by

$$x_{n+1} \equiv ax_n + c \pmod{m}, \quad 0 \leqslant x_{n+1} < m,$$

for $n = 0, 1, 2, 3,...$ . We call $m$ the *modulus*, $a$ the *multiplier*, $c$ the *increment*, and $x_0$ the *seed* of the pseudo-random number generator. The following examples illustrate the linear congruential method.

**Example 8.22.** With $m = 12$, $a = 3$, $c = 4$, and $x_0 = 5$, we obtain $x_1 \equiv 3 \cdot 5 + 4 \equiv 7 \pmod{12}$, so that $x_1 = 7$. Similarly, we find that $x_2 = 1$, since $x_2 \equiv 3 \cdot 7 + 4 \equiv 1 \pmod{12}$, $x_3 = 7$, since $x_3 \equiv 3 \cdot 1 + 4 \equiv 7 \pmod{12}$, and so on. Hence, the generator produces just three different integers before repeating. The sequence of pseudo-random numbers obtained is 5, 7, 1, 7, 1, 7, 1,... .

With $m = 9$, $a = 7$, $c = 4$, and $x_0 = 3$, we obtain the sequence 3, 7, 8, 6, 1, 2, 0, 4, 5, 3,... . This sequence contains 9 different numbers before repeating. □

The following theorem tells us how to find the terms of a sequence of pseudo-random numbers generated by the linear congruential method directly from the multiplier, the increment, and the seed.

**Theorem 8.25.** The terms of the sequence generated by the linear congruential method previously described are given by

$$x_k \equiv a^k x_0 + c(a^k - 1) / (a - 1) \pmod{m}, \ 0 \leqslant x_k < m.$$

*Proof.* We prove this result using mathematical induction. For $k = 1$, the formula is obviously true, since $x_1 \equiv ax_0 + c \pmod{m}$, $0 \leqslant x_1 < m$. Assume that the formula is valid for the $k$th term, so that

$$x_k \equiv a^k x_0 + c(a^k - 1)/(a - 1) \pmod{m}, \quad 0 \leqslant x_k < m.$$

Since

$$x_{k+1} \equiv ax_k + c \pmod{m}, \quad 0 \leqslant x_{k+1} < m,$$

we have

$$
\begin{aligned}
x_{k+1} &\equiv a(a^k x_0 + c(a^k - 1)/(a - 1)) + c \\
&\equiv a^{k+1} x_0 + c(a(a^k - 1)/(a - 1) + 1) \\
&\equiv a^{k+1} x_0 + c(a^{k+1} - 1)/(a - 1) \pmod{m},
\end{aligned}
$$

which is the correct formula for the $(k+1)$th term. This demonstrates that the formula is correct for all positive integers $k$. ∎

The *period length* of a linear-congruential pseudo-random number generator is the maximum length of the sequence obtained without repetition. We note that the longest possible period length for a linear congruential generator is the modulus $m$. The following theorem tells us when this maximum length is obtained.

**Theorem 8.26.** The linear congruential generator produces a sequence of period length $m$ if and only if $(c, m) = 1$, $a \equiv 1 \pmod{p}$ for all primes $p$ dividing $m$, and $a \equiv 1 \pmod{4}$ if $4 \mid m$.

Because the proof of Theorem 8.26 is complicated and quite lengthy we omit it. For the proof, the reader is referred to Knuth [80].

The case of the linear congruential generator with $c = 0$ is of special interest because of its simplicity. In this case, the method is called the *pure multiplicative congruential method*. We specify the modulus $m$, multiplier $a$, and seed $x_0$. The sequence of pseudo-random numbers is defined recursively by

$$x_{n+1} \equiv ax_n \pmod{m}, \ 0 < x_{n+1} < m.$$

In general, we can express the pseudo-random numbers generated in terms of the multiplier and seed:

$$x_n \equiv a^n x_0 \pmod{m}, \ 0 < x_{n+1} < m.$$

If $l$ is the period length of the sequence obtained using this pure multiplicative generator, then $l$ is the smallest positive integer such that

$$x_0 \equiv a^l x_0 \pmod{m}.$$

If $(x_0, m) = 1$, using Corollary 3.4.1 we have

$$a^l \equiv 1 \pmod{m}.$$

From this congruence, we know that the largest possible period length is $\lambda(m)$, where $\lambda(m)$ is the minimal universal exponent modulo $m$.

For many applications, the pure multiplicative generator is used with the modulus $m$ equal to the Mersenne prime $M_{31} = 2^{31} - 1$. When the modulus $m$ is a prime, the maximum period length is $m-1$, and this is obtained when $a$ is a primitive root of $m$. To find a primitive root of $M_{31}$ that can be used with good results, we first demonstrate that 7 is a primitive root of $M_{31}$.

**Theorem 8.27.**  The integer 7 is a primitive root of $M_{31} = 2^{31}-1$.

*Proof.* To show that 7 is a primitive root of $M_{31} = 2^{31}-1$, it is sufficient to show that

$$7^{(M_{31}-1)/q} \not\equiv 1 \pmod{M_{31}}$$

for all prime divisors $q$ of $M_{31}-1$. With this information, we can conclude that $\mathrm{ord}_{M_{31}} 7 = M_{31}-1$. To find the factorization of $M_{31}-1$, we note that

$$
\begin{aligned}
M_{31}-1 &= 2^{31} - 2 = 2(2^{30}-1) = 2(2^{15}-1)(2^{15}+1) \\
&= 2(2^5-1)(2^{10}+2^5+1)(2^5+1)(2^{10}-2^5+1) \\
&= 2 \cdot 3^2 \cdot 7 \cdot 11 \cdot 31 \cdot 151 \cdot 331.
\end{aligned}
$$

If we show that

$$7^{(M_{31}-1)/q} \not\equiv 1 \pmod{M_{31}}$$

for $q = 2, 3, 7, 11, 31, 151,$ and 331, then we know that 7 is a primitive root of $M_{31} = 2147483647$. Since

$$
\begin{aligned}
7^{(M_{31}-1)/2} &\equiv 2147483646 \not\equiv 1 \pmod{M_{31}} \\
7^{(M_{31}-1)/3} &\equiv 1513477735 \not\equiv 1 \pmod{M_{31}} \\
7^{(M_{31}-1)/7} &\equiv 120536285 \not\equiv 1 \pmod{M_{31}} \\
7^{(M_{31}-1)/11} &\equiv 1969212174 \not\equiv 1 \pmod{M_{31}} \\
7^{(M_{31}-1)/31} &\equiv 512 \not\equiv 1 \pmod{M_{31}} \\
7^{(M_{31}-1)/151} &\equiv 535044134 \not\equiv 1 \pmod{M_{31}} \\
7^{(M_{31}-1)/331} &\equiv 1761885083 \not\equiv 1 \pmod{M_{31}},
\end{aligned}
$$

we see that 7 is a primitive root of $M_{31}$.                                      ∎

In practice we do not want to use the primitive root 7 as the generator, since the first few integers generated are small. Instead, we find a larger primitive root using Corollary 8.4.1. We take a power of 7 where the exponent is relatively prime to $M_{31}-1$. For instance, since $(5, M_{31}-1) = 1$, Corollary 8.4.1 tells us that $7^5 = 16807$ is also a primitive root. Since

$(13, M_{31} - 1) = 1$, another possibility is to use $7^{13} \equiv 252246292 \pmod{M_{31}}$ as the multiplier.

We havely touched briefly on the important subject of pseudo-random numbers. For a thorough discussion of the generation and statistical properties of pseudo-random numbers see Knuth [80].

## 8.7 Exercises

1. Find the sequence of two-digit pseudo-random numbers generated using the middle-square method, taking 69 as the seed.

2. Find the first ten terms of the sequence of pseudo-random numbers generated by the linear congruential method with $x_0 = 6$ and $x_{n+1} \equiv 5x_n + 2 \pmod{19}$. What is the period length of this generator?

3. Find the period length of the sequence of pseudo-random numbers generated by the linear congruential method with $x_0 = 2$ and $x_{n+1} \equiv 4x_n + 7 \pmod{25}$.

4. Show that if either $a = 0$ or $a = 1$ is used for the multiplier in the generation of pseudo-random numbers by the linear congruential method, the resulting sequence would not be a good choice for a sequence of pseudo-random numbers.

5. Using Theorem 8.26, find those integers $a$ which give period length $m$, where $(c, m) = 1$, for the linear congruential generator $x_{n+1} \equiv ax_n + c \pmod{m}$, where

   a) $m = 1000$     c) $m = 10^6 - 1$
   b) $m = 30030$     d) $m = 2^{25} - 1$.

\* 6. Show that every linear congruential pseudo-random number generator can be simply expressed in terms of a linear congruential generator with increment $c = 1$ and seed 0, by showing that the terms generated by the linear congruential generator $x_{n+1} \equiv ax_n + c \pmod{m}$, with seed $x_0$, can be expressed as $x_n \equiv b \, y_n + x_0 \pmod{m}$, where $b \equiv (a-1) x_0 + c \pmod{m}$, $y_0 = 0$, and $y_{n+1} \equiv ay_n + 1 \pmod{m}$.

7. Find the period length of the pure multiplicative pseudo-random number generator $x_n \equiv cx_{n-1} \pmod{2^{31} - 1}$ when the multiplier $c$ is equal to

   a) 2     c) 4     e) 13.
   b) 3     d) 5

8. Show that the maximal possible period length for a pure multiplicative generator of the form $x_{n+1} \equiv ax_n \pmod{2^e}$, $e \geqslant 3$, is $2^{e-2}$. Show that this is obtained when $a \equiv \pm 3 \pmod{8}$.

9. Another way to generate pseudo-random numbers is to use the *Fibonacci generator.* Let $m$ be a positive integer. Two initial integers $x_0$ and $x_1$ both less than $m$ are specified and the rest of the sequence is generated recursively by the congruence $x_{n+1} \equiv x_n + x_{n-1} \pmod{m}$, $0 \leqslant x_{n+1} < m$.

   Find the first eight pseudo-random numbers generated by the Fibonacci generator with modulus $m = 31$ and initial values $x_0 = 1$ and $x_1 = 24$.

10. Find a good choice for the multiplier $a$ in the pure multiplicative pseudo-random number generator $x_{n+1} \equiv ax_n \pmod{101}$. (*Hint*: Find a primitive root of 101 that is not too small.)

11. Find a good choice for the multiplier $a$ in the pure multiplicative pseudo-random number generator $x_n \equiv ax_{n-1} \pmod{2^{25}-1}$. (*Hint*: Find a primitive root of $2^{25}-1$ and then take an appropriate power of this root.)

12. Find the multiplier $a$ and increment $c$ of the linear congruential pseudo-random number generator $x_{n+1} \equiv ax_n + c \pmod{1003}$, $0 \leqslant x_{n+1} < 1003$, if $x_0 = 1$, $x_2 = 402$, and $x_3 = 361$.

13. Find the multiplier $a$ of the pure multiplicative pseudo-random number generator $x_{n+1} \equiv ax_n \pmod{1000}$, $0 \leqslant x_{n+1} < 1000$, if 313 and 145 are consecutive terms generated.

## 8.7 Computer Projects

Write programs to generate pseudo-random numbers using the following generators:

1. The middle-sequence generator.

2. The linear congruential generator.

3. The pure multiplicative generator.

4. The Fibonacci generator (see Exercise 9).

## 8.8 An Application to the Splicing of Telephone Cables

An interesting application of the preceding material involves the splicing of telephone cables. We base our discussion on the exposition of Ore [30], who relates the contents of an original article by Lawther [99], reporting on work done for the Southwestern Bell Telephone Company.

To develop the application, we first make the following definition.

**Definition.** Let $m$ be a positive integer and let $a$ be an integer relatively prime to $m$. The $\pm 1$-*exponent of a modulo m* is the smallest positive integer $x$ such that

$$a^x \equiv \pm 1 \pmod{m}.$$

We are interested in determining the largest possible $\pm 1$-exponent of an integer modulo $m$; we denote this by $\lambda_0(m)$. The following two theorems relate the value of the maximal $\pm 1$-exponent $\lambda_0(m)$ to $\lambda(m)$, the minimal universal exponent modulo $m$.

First, we consider positive integers that possess primitive roots.

**Theorem 8.28.** If $m$ is a positive integer, $m > 2$, with a primitive root, then the maximal $\pm 1$-exponent $\lambda_0(m)$ equals $\phi(m)/2 = \lambda(m)/2$.

*Proof.* We first note that if $m$ has a primitive root, then $\lambda(m) = \phi(m)$. From Exercise 5 of Section 6.1, we know that $\phi(m)$ is even, so that $\phi(m)/2$ is an integer, if $m > 2$. Euler's Theorem tells us that

$$a^{\phi(m)} = (a^{\phi(m)/2})^2 \equiv 1 \pmod{m},$$

for all integers $a$ with $(a,m) = 1$. By Exercise 7 of Section 8.3, we know that when $m$ has a primitive root, the only solutions of $x^2 \equiv 1 \pmod{m}$ are $x \equiv \pm 1 \pmod{m}$. Hence,

$$a^{\phi(m)/2} \equiv \pm 1 \pmod{m}.$$

This implies that

$$\lambda_0(m) \leqslant \phi(m)/2.$$

Now let $r$ be a primitive root of modulo $m$ with $\pm 1$-exponent $e$. Then

$$r^e \equiv \pm 1 \pmod{m},$$

so that

$$r^{2e} \equiv 1 \pmod{m}.$$

Since $\text{ord}_m r = \phi(m)$, Theorem 8.1 tells us that $\phi(m) \mid 2e$, or equivalently, that $(\phi(m)/2) \mid e$. Hence, the maximum $\pm 1$-exponent $\lambda_0(m)$ is at least $\phi(m)/2$. However, we know that $\lambda(m) \leqslant \phi(m)/2$. Consequently, $\lambda_0(m) = \phi(m)/2 = \lambda(m)/2$. ∎

We now will find the maximal $\pm 1$-exponent of integers without primitive roots.

**Theorem 8.29.** If $m$ is a positive integer without a primitive root, then the maximal $\pm 1$-exponent $\lambda_0(m)$ equals $\lambda(m)$, the minimal universal exponent of $m$.

*Proof.* We first show that if $a$ is an integer of order $\lambda(m)$ modulo $m$ with $\pm 1$-exponent $e$ such that

$$a^{\lambda(m)/2} \not\equiv -1 \pmod{m},$$

then $e = \lambda(m)$. Consequently, once we have found such an integer $a$, we will have shown that $\lambda_0(m) = \lambda(m)$.

Assume that $a$ is an integer of order $\lambda(m)$ modulo $m$ with $\pm 1$ -exponent $e$ such that

$$a^{\lambda(m)/2} \not\equiv -1 \pmod{m}.$$

Since $a^e \equiv \pm 1 \pmod{m}$, it follows that $a^{2e} \equiv 1 \pmod{m}$. By Theorem 8.1 we know that $\lambda(m) \mid 2e$. Since $\lambda(m) \mid 2e$ and $e \leqslant \lambda(m)$, either $e = \lambda(m)/2$ or $e = \lambda(m)$. To see that $e \neq \lambda(m)/2$, note that $a^e \equiv \pm 1 \pmod{m}$, but $a^{\lambda(m)/2} \not\equiv 1 \pmod{m}$, since $\mathrm{ord}_m a = \lambda(m)$, and $a^{\lambda(m)/2} \not\equiv -1 \pmod{m}$, by hypothesis. Therefore, we can conclude that if $\mathrm{ord}_m a = \lambda(m)$, $a$ has $\pm 1$ - exponent $e$, and $a^e \equiv -1 \pmod{m}$, then $e = \lambda(m)$.

We now find an integer $a$ with the desired properties. Let the prime-power factorization of $m$ be $m = 2^{t_0} p_1^{t_1} p_2^{t_2} \cdots p_s^{t_s}$. We consider several cases.

We first consider those $m$ with at least two different odd prime factors. Among the prime-powers $p_i^{t_i}$ dividing $m$, let $p_j^{t_j}$ be one with the smallest power of 2 dividing $\phi(p_j^{t_j})$. Let $r_i$ be a primitive root of $p_i^{t_i}$ for $i = 1, 2, ..., s$. Let $a$ be an integer satisfying the simultaneous congruences

$$\begin{aligned} a &\equiv 3 \pmod{2^{t_0}}, \\ a &\equiv r_i \pmod{p_i^{t_i}} \text{ for all } i \text{ with } i \neq j, \\ a &\equiv r_j^2 \pmod{p_j^{t_j}}. \end{aligned}$$

Such an integer $a$ is guaranteed to exist by the Chinese remainder theorem. Note that

$$\mathrm{ord}_m a = [\lambda(2^{t_0}), \phi(p_i^{t_2}), ..., \phi(p_j^{t_j})/2, ..., \phi(p_s^{t_s})],$$

and, by our choice of $p_j^{t_j}$, we know that this least common multiple equals $\lambda(m)$. Since $a \equiv r_j^2 \pmod{p_j^{t_j}}$, it follows that $a^{\phi(p_j^{t_j})/2} \equiv r_j^{\phi(p_j^{t_j})} \equiv 1 \pmod{p_j^{t_j}}$. Because $\phi(p_j^{t_j})/2 \mid \lambda(m)/2$, we know that

$$a^{\lambda(m)/2} \equiv 1 \pmod{p_j^{t_j}},$$

so that

$$a^{\lambda(m)/2} \not\equiv -1 \pmod{m}.$$

Consequently, the $\pm 1$-exponent of $a$ is $\lambda(m)$.

The next case we consider deals with integers of the form $m = 2^{t_0} p^{t_1}$, where $p$ is an odd prime, $t_1 \geqslant 1$ and $t_0 \geqslant 2$, since $m$ has no primitive roots. When $t_0 = 2$ or 3, we have

$$\lambda(m) = [2, \phi(p_1^{t_1})] = \phi(p_1^{t_1}).$$

Let $a$ be a solution of the simultaneous congruences

$$a \equiv 1 \pmod 4$$
$$a \equiv r \pmod{p_1^{t_1}},$$

where $r$ is a primitive root of $p_1^{t_1}$. We see that $\text{ord}_m a = \lambda(m)$. Because

$$a^{\lambda(m)/2} \equiv 1 \pmod 4,$$

we know that

$$a^{\lambda(m)/2} \not\equiv -1 \pmod m.$$

Consequently, the $\pm 1$-exponent of $a$ is $\lambda(m)$.

When $t_0 \geqslant 4$, let $a$ be a solution of the simultaneous congruences

$$a \equiv 3 \pmod{2^{t_0}}$$
$$a \equiv r \pmod{p_1^{t_1}};$$

the Chinese remainder theorem tells us that such an integer exists. We see that $\text{ord}_m a = \lambda(m)$. Since $4 \mid \lambda(2^{t_0})$, we know that $4 \mid \lambda(m)$. Hence,

$$a^{\lambda(m)/2} \equiv 3^{\lambda(m)/2} \equiv (3^2)^{\lambda(m)/4} \equiv 1 \pmod 8.$$

Thus,

$$a^{\lambda(m)/2} \not\equiv -1 \pmod m,$$

so that the $\pm 1$-exponent of $a$ is $\lambda(m)$.

Finally, when $m = 2^{t_0}$ with $t_0 \geqslant 3$, from Theorem 8.12 we know that $\text{ord}_m 5 = \lambda(m)$, but

$$5^{\lambda(m)/2} \equiv (5^2)^{(\lambda(m)/4)} \equiv 1 \pmod 8.$$

Therefore, we see that

$$5^{\lambda(m)/2} \not\equiv -1 \pmod m;$$

we conclude that the $\pm 1$-exponent of 5 is $\lambda(m)$.

This finishes the argument since we have dealt with all cases where $m$ does not have a primitive root. ∎

We now develop a system for splicing telephone cables. Telephone cables are made up of concentric layers of insulated copper wire, as illustrated in Figure 8.1, and are produced in sections of specified length.

Telephone lines are constructed by splicing together sections of cable. When two wires are adjacent in the same layer in multiple sections of the cable, there are often problems with interference and crosstalk. Consequently, two wires adjacent in the same layer in one section should not be adjacent in the same layer in any nearby sections. For practical purpose, the splicing system should be simple. We use the following rules to describe the system. Wires in concentric layers are spliced to wires in the corresponding layers of the next

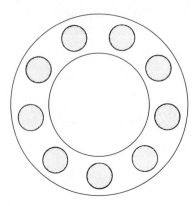

**Figure 8.1.** A cross-section of one layer of a telephone cable.

section, following identical splicing direction at each connection. In a layer with $m$ wires, we connect the wire in position $j$ in one section, where $1 \leqslant j \leqslant m$ to the wire in position $S(j)$ in the next section, where $S(j)$ is the least positive residue of $1 + (j-1)s$ modulo $m$. Here, $s$ is called the *spread* of the splicing system. We see that when a wire in one section is spliced to a wire in the next section, the adjacent wire in the first section is spliced to the wire in the next section in the position obtained by counting forward $s$ modulo $m$ from the position of the last wire spliced in this section. To have a one-to-one correspondence between wires of adjacent sections, we require that the spread $s$ be relatively prime to the number of wires $m$. This shows that if wires in positions $j$ and $k$ are sent to the same wire in the next section, then $S(j) = S(k)$ and

$$1 + (j-1)s \equiv 1 + (k-1)s \pmod{m},$$

so that $js \equiv ks \pmod{m}$. Since $(m, s) = 1$, from Corollary 3.4.1 we see that $j \equiv k \pmod{m}$, which is impossible.

**Example 8.23.** Let us connect 9 wires with a spread of 2. We have the correspondence

$$
\begin{array}{lll}
1 \rightarrow 1 & 2 \rightarrow 3 & 3 \rightarrow 5 \\
4 \rightarrow 7 & 5 \rightarrow 9 & 6 \rightarrow 2 \\
7 \rightarrow 4 & 8 \rightarrow 6 & 9 \rightarrow 8.
\end{array}
$$

$\square$

This is illustrated in Figure 8.2.

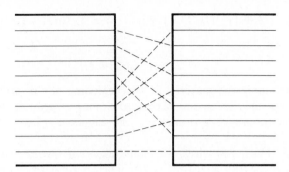

**Figure 8.2.** Splicing of 9 wires with spread of 2.

The following result tells us the correspondence of wires in the first section of cable to the wires in the $n$th section.

**Theorem 8.30.** Let $S_n(j)$ denote the position of the wire in the $n$th section spliced to the $j$th wire of the first section. Then

$$S_n(j) \equiv 1 + (j-1)s^{n-1} \pmod{m}.$$

*Proof.* For $n = 2$, by the rules for the splicing system, we have

$$S_2(j) \equiv 1 + (j-1)s \pmod{m},$$

so the proposition is true for $n = 2$. Now assume that

(8.4) $$S_n(j) \equiv 1 + (j-1)s^{n-1} \pmod{m}.$$

Then, in the next section, we have the wire in position $S_n(j)$ spliced to the wire in position

$$\begin{aligned} S_{n+1}(j) &\equiv 1 + (S_n(j)-1)s \\ &\equiv 1 + ((j-1)s^{n-1})s \\ &\equiv 1 + (j-1)s^n \pmod{m}. \end{aligned}$$

This shows that the proposition is true.                                   ∎

In a splicing system, we want to have wires adjacent in one section separated as long as possible in the following sections. After $n$ splices, Theorem 8.30 tells us that the adjacent wires in the $j$th and $(j + 1)$th positions are connected to wires in positions $S_n(j) \equiv 1 + (j-1)s^n \pmod{m}$ and $S_n(j+1) = 1 + js^n \pmod{m}$, respectively. These wires are adjacent in the $n$th section if, and only if,

$$S_n(j) - S_n(j+1) \equiv \pm 1 \pmod{m},$$

or equivalently,

$$(1 + (j-1)s^n) - (1+js^n) \equiv \pm 1 \pmod{m},$$

which holds if and only if

$$s^n \equiv \pm 1 \pmod{m}.$$

We can now apply the material at the beginning of this section. To keep adjacent wires in the first section separated as long as possible, we should pick for the spread $s$ an integer with maximal $\pm 1$-exponent $\lambda_0(m)$.

**Example 8.24.** With 100 wires, we should choose a spread $s$ so that the $\pm 1$-exponent of $s$ is $\lambda_0(100) = \lambda(100) = 20$. The appropriate computations show that $s = 3$ is such a spread.                                                             □

---

## 8.8 Exercises

    **1.** Find the maximal $\pm 1$-exponent of

        a) 17              d) 36
        b) 22              e) 99
        c) 24              f)  100.

    **2.** Find an integer with maximal $\pm 1$-exponent modulo

        a) 13              d) 25
        b) 14              e) 36
        c) 15              f)  60.

    **3.** Devise a splicing scheme for telephone cables containing

        a)   50 wires     b)   76 wires     c)   125 wires.

  **\* 4.** Show that using any splicing system of telephone cables with $m$ wires arranged in a concentric layer, adjacent wires in one section can be kept separated in at most $[(m-1)/2]$ successive sections of cable. Show that when $m$ is prime this upper limit is achieved using the system developed in this section.

## 8.8 Computer Projects

Write programs to do the following:

    **1.** Find maximal $\pm 1$-exponents.

    **2.** Develop a scheme for splicing telephone cables as described in this section.

# 9

# Quadratic Residues

## 9.1 Quadratic Residues

Let $p$ be an odd prime and $a$ an integer relatively prime to $p$. In this chapter we devote our attention to the question: Is $a$ a perfect square modulo $p$? We begin with a definition.

**Definition.** If $m$ is a positive integer, we say that the integer $a$ is a *quadratic residue of $m$* if $(a,m) = 1$ and the congruence $x^2 \equiv a \pmod{m}$ has a solution. If the congruence $x^2 \equiv a \pmod{m}$ has no solution, we say that $a$ is a *quadratic nonresidue of $m$*.

**Example 9.1.** To determine which integers are quadratic residues of 11, we compute the squares of the integers $1, 2, 3,...,10$. We find that $1^2 \equiv 10^2 \equiv 1 \pmod{11}$, $2^2 \equiv 9^2 \equiv 4 \pmod{11}$, $3^2 \equiv 8^2 \equiv 9 \pmod{11}$, $4^2 \equiv 7^2 \equiv 5 \pmod{11}$, and $5^2 \equiv 6^2 \equiv 3 \pmod{11}$. Hence, the quadratic residues of 11 are 1, 3, 4, 5, and 9; the integers 2, 6, 7, 8, and 10 are quadratic nonresidues of 11. □

Note that the quadratic residues of the positive integer $m$ are just the $k$th power residues of $m$ with $k=2$, as defined in Section 8.4. We will show that if $p$ is an odd prime, then there are exactly as many quadratic residues as quadratic nonresidues of $p$ among the integers $1, 2,...,p - 1$. To demonstrate this fact, we use the following lemma.

**Lemma 9.1.** Let $p$ be an odd prime and $a$ an integer not divisible by $p$. Then, the congruence

$$x^2 \equiv a \pmod{p}$$

has either no solutions or exactly two incongruent solutions modulo $p$.

*Proof.* If $x^2 \equiv a \pmod p$ has a solution, say $x = x_0$, then we can easily demonstrate that $x = -x_0$ is a second incongruent solution. Since $(-x_0)^2 = x_0^2 \equiv a \pmod p$, we see that $-x_0$ is a solution. We note that $x_0 \not\equiv -x_0 \pmod p$, for if $x_0 \equiv -x_0 \pmod p$, then we have $2x_0 \equiv 0 \pmod p$. This is impossible since $p$ is odd and $p \nmid x_0$ (since $x_0^2 \equiv a \pmod p$ and $p \nmid a$).

To show that there are no more than two incongruent solutions, assume that $x = x_0$ and $x = x_1$ are both solutions of $x^2 \equiv a \pmod p$. Then we have $x_0^2 \equiv x_1^2 \equiv a \pmod p$, so that $x_0^2 - x_1^2 = (x_0+x_1)(x_0-x_1) \equiv 0 \pmod p$. Hence, $p \mid (x_0+x_1)$ or $p \mid (x_0-x_1)$, so that $x_1 \equiv -x_0 \pmod p$ or $x_1 \equiv x_0 \pmod p$. Therefore, if there is a solution of $x^2 \equiv a \pmod p$, there are exactly two incongruent solutions. ∎

This leads us to the following theorem.

**Theorem 9.1.** If $p$ is an odd prime, then there are exactly $(p-1)/2$ quadratic residues of $p$ and $(p-1)/2$ quadratic nonresidues of $p$ among the integers $1, 2,...,p-1$.

*Proof.* To find all the quadratic residues of $p$ among the integers $1,2,...,p-1$ we compute the least positive residues modulo $p$ of the squares of the integers $1, 2,...,p-1$. Since there are $p-1$ squares to consider and since each congruence $x^2 \equiv a \pmod p$ has either zero or two solutions, there must be exactly $(p-1)/2$ quadratic residues of $p$ among the integers $1, 2,...,p-1$. The remaining $p-1 - (p-1)/2 = (p-1)/2$ positive integers less than $p-1$ are quadratic nonresidues of $p$. ∎

The special notation associated with quadratic residues is described in the following definition.

**Definition.** Let $p$ be an odd prime and $a$ an integer not divisible by $p$. The *Legendre symbol* $\left(\dfrac{a}{p}\right)$ is defined by

$$\left(\frac{a}{p}\right) = \begin{cases} 1 & \text{if } a \text{ is a quadratic residue of } p \\ -1 & \text{if } a \text{ is a quadratic nonresidue of } p. \end{cases}$$

This symbol is named after the French mathematician Adrien-Marie Legendre* who introduced the use of this notation.

**Example 9.2.** The previous example shows that the Legendre symbols $\left(\dfrac{a}{11}\right)$, $a = 1, 2,...,10$, have the following values:

$$\left(\frac{1}{11}\right) = \left(\frac{3}{11}\right) = \left(\frac{4}{11}\right) = \left(\frac{5}{11}\right) = \left(\frac{9}{11}\right) = 1,$$

$$\left(\frac{2}{11}\right) = \left(\frac{6}{11}\right) = \left(\frac{7}{11}\right) = \left(\frac{8}{11}\right) = \left(\frac{10}{11}\right) = -1. \qquad \square$$

We now present a criterion for deciding whether an integer is a quadratic residue of a prime. This criterion is useful in demonstrating properties of the Legendre symbol.

**Theorem 9.2.. Euler's Criterion.** Let $p$ be an odd prime and let $a$ be a positive integer not divisible by $p$. Then

$$\left(\frac{a}{p}\right) \equiv a^{(p-1)/2} \pmod{p}.$$

*Proof.* First, assume that $\left(\dfrac{a}{p}\right) = 1$. Then, the congruence $x^2 \equiv a \pmod{p}$ has a solution, say $x = x_0$. Using Fermat's little theorem, we see that

$$a^{(p-1)/2} = (x_0^2)^{(p-1)/2} = x_0^{p-1} \equiv 1 \pmod{p}.$$

Hence, if $\left(\dfrac{a}{p}\right) = 1$, we know that $\left(\dfrac{a}{p}\right) \equiv a^{(p-1)/2} \pmod{p}$.

---

* **ADRIEN-MARIE LEGENDRE** (1752-1833) was born into a well-to-do family. He was professor at the Ecole Militaire in Paris from 1775 to 1780. In 1795 he was appointed professor at the Ecole Normale. His memoir *Recherches d'Analyse Indeterminée* published in 1785 contains a discussion of the law of quadratic reciprocity, a statement of Dirichlet's theorem on primes in arithmetic progressions, and a discussion of the representation of positive integers as the sum of three squares. He established the $n=5$ case of Fermat's last theorem. Legendre wrote a textbook on geometry, *Eléments de géométrie* that was used for over 100 years, and served as a model for other textbooks. Legendre made fundamental discoveries in mathematical astronomy and geodesy. He gave the first treatment of the law of least squares.

Now consider the case where $\left(\dfrac{a}{p}\right) = -1$. Then, the congruence $x^2 \equiv a \pmod{p}$ has no solutions. By Theorem 3.10 for each integer $i$ such that $1 \leqslant i \leqslant p-1$, there is a unique integer $j$ with $1 \leqslant j \leqslant p-1$, such that $ij \equiv a \pmod{p}$. Furthermore, since the congruence $x^2 \equiv a \pmod{p}$ has no solutions, we know that $i \neq j$. Thus, we can group the integers $1, 2, ..., p-1$ into $(p-1)/2$ pairs each with product $a$. Multiplying these pairs together, we find that

$$(p-1)! \equiv a^{(p-1)/2} \pmod{p}.$$

Since Wilson's theorem tells us that $(p-1)! \equiv -1 \pmod{p}$, we see that

$$-1 \equiv a^{(p-1)/2} \pmod{p}.$$

In this case, we also have $\left(\dfrac{a}{p}\right) \equiv a^{(p-1)/2} \pmod{p}$.                    ∎

**Example 9.3.** Let $p = 23$ and $a = 5$. Since $5^{11} \equiv -1 \pmod{23}$, Euler's criterion tells us that $\left(\dfrac{5}{23}\right) = -1$. Hence, 5 is a quadratic nonresidue of 23.                    □

We now prove some properties of the Legendre symbol.

**Theorem 9.3.** Let $p$ be an odd prime and $a$ and $b$ integers not divisible by $p$. Then

    (i)   if $a \equiv b \pmod{p}$, then $\left(\dfrac{a}{p}\right) = \left(\dfrac{b}{p}\right)$.

    (ii)   $\left(\dfrac{a}{p}\right)\left(\dfrac{b}{p}\right) = \left(\dfrac{ab}{p}\right)$.

    (iii)   $\left(\dfrac{a^2}{p}\right) = 1$.

*Proof of (i).* If $a \equiv b \pmod{p}$, then $x^2 \equiv a \pmod{p}$ has a solution if and only if $x^2 \equiv b \pmod{p}$ has a solution. Hence, $\left(\dfrac{a}{p}\right) = \left(\dfrac{b}{p}\right)$.

*Proof of (ii).* By Euler's criterion we know that

$$\left(\dfrac{a}{p}\right) \equiv a^{(p-1)/2} \pmod{p}, \quad \left(\dfrac{b}{p}\right) \equiv b^{(p-1)/2} \pmod{p},$$

and

$$\left(\frac{ab}{p}\right) \equiv (ab)^{(p-1)/2} \pmod{p}.$$

Hence,

$$\left(\frac{a}{b}\right)\left(\frac{b}{p}\right) \equiv a^{(p-1)/2}b^{(p-1)/2} = (ab)^{(p-1)/2} \equiv \left(\frac{ab}{p}\right) \pmod{p}.$$

Since the only possible values of a Legendre symbol are $\pm 1$, we conclude that

$$\left(\frac{a}{p}\right)\left(\frac{b}{p}\right) = \left(\frac{ab}{p}\right).$$

*Proof of (iii).*  Since $\left(\frac{a}{p}\right) = \pm 1$ , from part (ii) it follows that

$$\left(\frac{a^2}{p}\right) = \left(\frac{a}{p}\right)\left(\frac{a}{p}\right) = 1. \qquad\blacksquare$$

Part (ii) of Theorem 9.3 has the following interesting consequence.  The product of two quadratic residues, or of two quadratic nonresidues, of a prime is a quadratic residue of that prime, whereas the product of a quadratic residue and a quadratic nonresidue is a quadratic nonresidue.

Using Euler's criterion, we can classify those primes having $-1$ as a quadratic residue.

**Theorem  9.4.**  If $p$ is an odd prime, then

$$\left(\frac{-1}{p}\right) = \begin{cases} 1 & \text{if } p \equiv 1 \pmod{4} \\ -1 & \text{if } p \equiv -1 \pmod{4}. \end{cases}$$

*Proof.*  By Euler's criterion we know that

$$\left(\frac{-1}{p}\right) \equiv (-1)^{(p-1)/2} \pmod{p}.$$

If $p \equiv 1 \pmod 4$, then $p = 4k + 1$ for some integer $k$.  Thus,

$$(-1)^{(p-1)/2} = (-1)^{2k} = 1,$$

so that $\left(\frac{-1}{p}\right) = 1$.  If $p \equiv 3 \pmod 4$, then $p = 4k + 3$ for some integer $k$.  Thus,

$$(-1)^{(p-1)/2} = (-1)^{2k+1} = -1,$$

so that $\left(\frac{-1}{p}\right) = -1.$ $\qquad\blacksquare$

The following elegant result of Gauss provides another criterion to determine whether an integer $a$ relatively prime to the prime $p$ is a quadratic residue of $p$.

**Lemma 9.2. Gauss' Lemma.** Let $p$ be an odd prime and $a$ an integer with $(a,p) = 1$. If $s$ is the number of least positive residues of the integers $a, 2a, 3a,...,((p-1)/2)a$ that are greater than $p/2$, then $\left(\dfrac{a}{p}\right) = (-1)^s$.

*Proof.* Let $u_1, u_2,...,u_s$ represent the least positive residues of the integers $a, 2a, 3a,...,((p-1)/2)a$ that are greater than $p/2$, and let $v_1, v_2,...,v_t$ be the least positive residues of these integers that are less than $p/2$. Since $(ja,p) = 1$ for all $j$ with $1 \leqslant j \leqslant (p-1)/2$, all of these least positive residues are in the set $1, 2,...,p - 1$.

We will show that $p-u_1, p-u_2, ..., p-u_s, v_1, v_2,...,v_t$ comprise the set of integers $1, 2,...,(p-1)/2$, in some order. To demonstrate this, it suffices to show that no two of these integers are congruent modulo $p$, since there are exactly $(p-1)/2$ numbers in the set, and all are positive integers not exceeding $(p-1)/2$.

It is clear that no two of the $u_i$'s are congruent modulo $p$ and that no two of the $v_j$'s are congruent modulo $p$; if a congruence of either of these two sorts held, we would have $ma \equiv na \pmod{p}$ where $m$ and $n$ are both positive integers not exceeding $(p-1)/2$. Since $p \nmid a$, this implies that $m \equiv n \pmod{p}$ which is impossible.

In addition, one of the integers $p - u_i$ cannot be congruent to a $v_j$, for if such a congruence held, we would have $ma \equiv p - na \pmod{p}$, so that $ma \equiv -na \pmod{p}$. Since $p \nmid a$, this implies that $m \equiv -n \pmod{p}$ . This is impossible because both $m$ and $n$ are in the set $1, 2,...,(p-1)/2$.

Now that we know that $p - u_1, p - u_2,...,p - u_s, v_1, v_2, ..., v_t$ are the integers $1, 2,...,(p-1)/2$, in some order, we conclude that

$$(p-u_1)(p-u_2) \cdots (p-u_s)v_1v_2 \cdots v_t \equiv \left(\frac{p-1}{2}\right)! \pmod{p},$$

which implies that

(9.1) $\qquad (-1)^s u_1 u_2 \cdots u_s v_1 v_2 \cdots v_t \equiv \left(\dfrac{p-1}{2}\right)! \pmod{p}.$

But, since $u_1, u_2,...,u_s, v_1, v_2,...,v_t$ are the least positive residues of $a, 2a,...,((p-1)/2)a$, we also know that

(9.2)
$$u_1 u_2 \cdots u_s v_1 v_2 \cdots v_t \equiv a \cdot 2a \cdots \left(\frac{p-1}{2}\right) a$$

$$= a^{\frac{p-1}{2}} \left[\frac{p-1}{2}\right]! \pmod{p}.$$

Hence, from (9.1) and (9.2), we see that

$$(-1)^s a^{\frac{p-1}{2}} \left[\frac{p-1}{2}\right]! \equiv \left[\frac{p-1}{2}\right]! \pmod{p}.$$

Because $(p, ((p-1)/2)!) = 1$, this congruence implies that

$$(-1)^s a^{\frac{p-1}{2}} \equiv 1 \pmod{p}.$$

By multiplying both sides by $(-1)^s$, we obtain

$$a^{\frac{p-1}{2}} \equiv (-1)^s \pmod{p}.$$

Since Euler's criterion tells us that $a^{\frac{p-1}{2}} \equiv \left[\frac{a}{p}\right] \pmod{p}$, it follows that

$$\left[\frac{a}{p}\right] \equiv (-1)^s \pmod{p},$$

establishing Gauss' lemma.  ■

**Example 9.4.** Let $a = 5$ and $p = 11$. To find $\left[\frac{5}{11}\right]$ by Gauss' lemma, we compute the least positive residues of $1 \cdot 5$, $2 \cdot 5$, $3 \cdot 5$, $4 \cdot 5$, and $5 \cdot 5$. These are 5, 10, 4, 9, and 3, respectively. Since exactly two of these are greater than $11/2$, Gauss' lemma tells us that $\left[\frac{5}{11}\right] = (-1)^2 = 1$.  □

Using Gauss' lemma, we can characterize all primes that have 2 as a quadratic residue.

**Theorem 9.5.** If $p$ is an odd prime, then

$$\left[\frac{2}{p}\right] = (-1)^{(p^2-1)/8}.$$

Hence, 2 is a quadratic residue of all primes $p \equiv \pm 1 \pmod 8$ and a quadratic nonresidue of all primes $p \equiv \pm 3 \pmod 8$.

*Proof.* By Gauss' lemma we know that if $s$ is the number of least positive residues of the integers

$$1\cdot2,\ 2\cdot2,\ 3\cdot2,\ ...,\ \left[\frac{p-1}{2}\right]\cdot2$$

that are greater than $p/2$, then $\left[\dfrac{2}{p}\right] = (-1)^s$. Since all these integers are less than $p$, we only need to count those greater than $p/2$ to find how many have least positive residue greater than $p/2$.

The integer $2j$, where $1 \leqslant j \leqslant (p-1)/2$, is less than $p/2$ when $j \leqslant p/4$. Hence, there are $[p/4]$ integers in the set less than $p/2$. Consequently, there are $s = \dfrac{p-1}{2} - [p/4]$ greater than $p/2$. Therefore, by Gauss' lemma we see that

$$\left[\frac{2}{p}\right] = (-1)^{\frac{p-1}{2}-[p/4]}.$$

To prove the theorem, we must show that

$$\frac{p-1}{2} - [\frac{p}{4}] \equiv (p^2-1)/8 \pmod 2.$$

To establish this, we need to consider the congruence class of $p$ modulo 8, since, as we will see, both sides of the above congruence depend only on the congruence class of $p$ modulo 8.

We first consider $(p^2-1)/8$. If $p \equiv \pm1 \pmod 8$, then $p = 8k \pm 1$ where $k$ is an integer, so that

$$(p^2-1)/8 = ((8k\pm1)^2-1)/8 = (64k^2\pm16k)/8 = 8k^2 \pm 2k \equiv 0 \pmod 2.$$

If $p \equiv \pm 3 \pmod 8$, then $p = 8k \pm 3$ where $k$ is an integer, so that

$$(p^2-1)/8 = ((8k \pm 3)^2-1)/8 = (64k^2 \pm 48k + 8)/8 = 8k^2 + 6k + 1$$
$$\equiv 1 \pmod 2.$$

Now consider $\dfrac{p-1}{2} - [p/4]$. If $p \equiv 1 \pmod 8$, then $p = 8k + 1$ for some integer $k$ and

$$\frac{p-1}{2} - [p/4] = 4k - [2k + 1/4] = 2k \equiv 0 \pmod 2;$$

if $p \equiv 3 \pmod 8$, then $p = 8k + 3$ for some integer $k$, and

$$\frac{p-1}{2} - [p/4] = 4k + 1 - [2k + 3/4] = 2k + 1 \equiv 1 \pmod 2;$$

if $p \equiv 5 \pmod 8$, then $p = 8k + 5$ for some integer $k$, and

$$\frac{p-1}{2} - [p/4] = 4k + 2 - [2k + 5/4] = 2k + 1 \equiv 1 \pmod 2;$$

if $p \equiv 7 \pmod 8$, then $p = 8k + 7$ for some integer $k$, and

$$\frac{p-1}{2} - [p/4] = 4k + 3 - [2k + 7/4] = 2k + 2 \equiv 0 \pmod 2.$$

Comparing the congruence classes modulo 2 of $\frac{p-1}{2} - [p/4]$ and $(p^2-1)/8$ for the four possible congruence classes of the odd prime $p$ modulo 8, we see that we always have $\frac{p-1}{2} - [p/4] \equiv (p^2-1)/8 \pmod 2$.

Hence, $(\frac{2}{p}) = (-1)^{(p^2-1)/8}$.

From the computations of the congruence class of $(p^2-1)/8 \pmod 2$, we see that $\left(\dfrac{2}{p}\right) = 1$ if $p \equiv \pm 1 \pmod 8$, while $\left(\dfrac{2}{p}\right) = -1$ if $p \equiv \pm 3 \pmod 8$. ∎

**Example 9.5.** By Theorem 9.5 we see that

$$\left(\frac{2}{7}\right) = \left(\frac{2}{17}\right) = \left(\frac{2}{23}\right) = \left(\frac{2}{31}\right) = 1,$$

while

$$\left(\frac{2}{3}\right) = \left(\frac{2}{5}\right) = \left(\frac{2}{11}\right) = \left(\frac{2}{13}\right) = \left(\frac{2}{19}\right) = \left(\frac{2}{29}\right) = -1.$$ □

We now present an example to show how to evaluate Legendre symbols.

**Example 9.6.** To evaluate $\left(\dfrac{317}{11}\right)$, we use part (i) of Theorem 9.3 to obtain

$$\left(\frac{317}{11}\right) = \left(\frac{9}{11}\right) = \left(\frac{3}{11}\right)^2 = 1, \text{ since } 317 \equiv 9 \pmod{11}.$$

To evaluate $\left(\dfrac{89}{13}\right)$, since $89 \equiv -2 \pmod{13}$, we have

$$\left(\frac{89}{13}\right) = \left(\frac{-2}{13}\right) = \left(\frac{-1}{13}\right)\left(\frac{2}{13}\right).$$ Because $13 \equiv 1 \pmod 4$, Theorem 9.4 tells us that $\left(\dfrac{-1}{13}\right) = 1$. Since $13 \equiv -3 \pmod 8$, we see from Theorem 9.5 that

$\left[\dfrac{2}{13}\right] = -1.$ Consequently, $\left[\dfrac{89}{13}\right] = -1.$

$\square$

In the next section, we state and prove a theorem of fundamental importance for the evaluation of Legendre symbols. This theorem is called *the law of quadratic reciprocity*.

The difference in the length of time needed to find primes and to factor is the basis of the RSA cipher discussed in Chapter 7. This difference is also the basis of a method to "flip coins" electronically that was invented by Blum [117]. Results about quadratic residues are used to develop this method.

Suppose that $n = pq$, where $p$ and $q$ are distinct odd primes and suppose that the congruence $x^2 \equiv a \pmod{n}$, $0 < a < n$, has a solution $x = x_0$. We show that there are exactly four incongruent solutions modulo $n$. To see this, let $x_0 \equiv x_1 \pmod{p}$, $0 < x_1 < p$, and let $x_0 \equiv x_2 \pmod{q}$, $0 < x_2 < q$. Then the congruence $x^2 \equiv a \pmod{p}$ has exactly two incongruent solutions, namely $x \equiv x_1 \pmod{p}$ and $x \equiv p - x_1 \pmod{p}$. Similarly the congruence $x^2 = a \pmod{q}$ has exactly two incongruent solutions, namely $x \equiv x_2 \pmod{q}$ and $x \equiv q - x_2 \pmod{q}$.

From the Chinese remainder theorem, there are exactly four incongruent solutions of the congruence $x^2 \equiv a \pmod{n}$; these four incongruent solutions are the unique solutions modulo $pq$ of the four sets of simultaneous congruences

(i)  $x \equiv x_1 \pmod{p}$          (iii)  $x \equiv p - x_1 \pmod{p}$
     $x \equiv x_2 \pmod{q}$                 $x \equiv x_2 \pmod{q}$

(ii)  $x \equiv x_1 \pmod{p}$          (iv)  $x \equiv p - x_1 \pmod{p}$
      $x \equiv q - x_2 \pmod{q}$            $x \equiv q - x_2 \pmod{q}$.

We denote solutions of (i) and (ii) by $x$ and $y$, respectively. Solutions of (iii) and (iv) are easily seen to be $n-y$ and $n-x$, respectively.

We also note that when $p \equiv q \equiv 3 \pmod 4$, the solutions of $x^2 \equiv a \pmod{p}$ and of $x^2 \equiv a \pmod{q}$ are $x \equiv \pm a^{(p+1)/4} \pmod{p}$ and $x \equiv \pm a^{(q+1)/4} \pmod{q}$, respectively. By Euler's criterion, we know that $a^{(p-1)/2} \equiv \left[\dfrac{a}{p}\right] = 1 \pmod{p}$ and $a^{(q-1)/2} \equiv \left[\dfrac{a}{q}\right] = 1 \pmod{q}$ (recall that we are assuming that $x^2 \equiv a \pmod{pq}$ has a solution, so that $a$ is a quadratic residue of both $p$ and $q$) . Hence,

$$(a^{(p+1)/4})^2 = a^{(p+1)/2} = a^{(p-1)/2} \cdot a \equiv a \pmod{p}$$

and

$$(a^{(q+1)/4})^2 = a^{(q+1)/2} = a^{(q-1)/2} \cdot a \equiv a \pmod{q}.$$

Using the Chinese remainder theorem, together with the explicit solutions just constructed, we can easily find the four incongruent solutions of $x^2 \equiv a \pmod{n}$. The following example illustrates this procedure.

**Example 9.7.** Suppose we know *a priori* that the congruence

$$x^2 \equiv 860 \pmod{11021}$$

has a solution. Since $11021 = 103 \cdot 107$, to find the four incongruent solutions we solve the congruences

$$x^2 \equiv 860 \equiv 36 \pmod{103}$$

and

$$x^2 \equiv 860 \equiv 4 \pmod{107}.$$

The solutions of these congruences are

$$x \equiv \pm 36^{(103+1)/4} \equiv \pm 36^{26} \equiv \pm 6 \pmod{103}$$

and

$$x \equiv \pm 4^{(107+1)/4} \equiv \pm 4^{27} \equiv \pm 2 \pmod{107},$$

respectively. Using the Chinese remainder theorem, we obtain $x \equiv \pm 212$, $\pm 109 \pmod{11021}$ as the solutions of the four systems of congruences described by the four possible choices of signs in the system of congruences $x \equiv \pm 6 \pmod{103}$, $x \equiv \pm 2 \pmod{107}$. $\square$

We can now describe a method for electronically flipping coins. Suppose that Bob and Alice are communicating electronically. Alice picks two distinct large primes $p$ and $q$, with $p \equiv q \equiv 3 \pmod{4}$. Alice sends Bob the integer $n = pq$. Bob picks, at random, a positive integer $x$ less than $n$ and sends to Alice the integer $a$ with $x^2 \equiv a \pmod{n}$, $0 < a < n$. Alice finds the four solutions of $x^2 \equiv a \pmod{n}$, namely $x$, $y$, $n-x$, and $n-y$. Alice picks one of these four solutions and sends it to Bob. Note that since $x + y \equiv 2x_1 \not\equiv 0 \pmod{p}$ and $x + y \equiv 0 \pmod{q}$, we have $(x+y,n) = q$, and similarly $(x+(n-y), n) = p$. Thus, if Bob receives either $y$ or $n-y$, he can rapidly factor $n$ by using the Euclidean algorithm to find one of the two prime factors of $n$. On the other hand, if Bob receives either $x$ or $n-x$, he has no way to factor $n$ in a reasonable length of time.

Consequently, Bob wins the coin flip if he can factor $n$, whereas Alice wins if Bob cannot factor $n$. From previous comments, we know that there is an equal chance for Bob to receive a solution of $x^2 \equiv a \pmod{n}$ that helps him rapidly factor $n$, or a solution of $x^2 \equiv a \pmod{n}$ that does not help him factor $n$. Hence, the coin flip is fair.

## 9.1 Exercises

1. Find all the quadratic residues of each of the following integers.

   a) 3     c) 13
   b) 5     d) 19

2. Find the value of the Legendre symbols $\left(\dfrac{j}{7}\right)$, for $j = 1,2,3,4,5$, and 6.

3. Evaluate the Legendre symbol $\left(\dfrac{7}{11}\right)$

   a) using Euler's criterion.
   b) using Gauss' lemma.

4. Let $a$ and $b$ be integers not divisible by the prime $p$. Show that there is either one or three quadratic residues among the integers $a$, $b$, and $ab$.

5. Show that if $p$ is an odd prime, then

$$\left(\frac{-2}{p}\right) = \begin{cases} 1 & \text{if } p \equiv 1 \text{ or } 3 \pmod 8 \\ -1 & \text{if } p \equiv -1 \text{ or } -3 \pmod 8. \end{cases}$$

6. Show that if the prime-power factorization of $n$ is

$$n = p_1^{2t_1+1} p_2^{2t_2+1} \cdots p_k^{2t_k+1} p_{k+1}^{2t_{k+1}} \cdots p_n^{2t_n}$$

   and $q$ is a prime not dividing $n$, then

$$\left(\frac{n}{q}\right) = \left(\frac{p_1}{q}\right)\left(\frac{p_2}{q}\right) \cdots \left(\frac{p_k}{q}\right).$$

7. Show that if $p$ is prime and $p \equiv 3 \pmod 4$, then $[(p-1)/2]! \equiv (-1)^t \pmod p$, where $t$ is the number of positive integers less than $p/2$ that are quadratic residues of $p$.

8. Show that if $b$ is a positive integer not divisible by the prime $p$, then

$$\left(\frac{b}{p}\right) + \left(\frac{2b}{p}\right) + \left(\frac{3b}{p}\right) + \cdots + \left(\frac{(p-1)b}{p}\right) = 0.$$

9. Let $p$ be prime and $a$ a quadratic residue of $p$. Show that if $p \equiv 1 \pmod 4$, then $-a$ is also a quadratic residue of $p$, while if $p \equiv 3 \pmod 4$, then $-a$ is a quadratic nonresidue of $p$.

10. Consider the quadratic congruence $ax^2 + bx + c \equiv 0 \pmod p$, where $p$ is prime and $a,b$, and $c$ are integers with $p \nmid a$.

    a) Let $p = 2$. Determine which quadratic congruences (mod 2) have solutions.

    b) Let $p$ be an odd prime and let $d = b^2 - 4ac$. Show that the congruence $ax^2 + bx + c \equiv 0 \pmod p$ is equivalent to the congruence $y^2 \equiv d \pmod p$, where $y = 2ax + b$. Conclude that if

$d \equiv 0 \pmod{p}$, then there is exactly one solution $x$ modulo $p$, if $d$ is a quadratic residue of $p$, then there are two incongruent solutions, while if $d$ is a quadratic nonresidue of $p$, then there are no solutions.

11. Find all solutions of the quadratic congruences

    a)   $x^2 + x + 1 \equiv 0 \pmod{7}$
    b)   $x^2 + 5x + 1 \equiv 0 \pmod{7}$
    c)   $x^2 + 3x + 1 \equiv 0 \pmod{7}$.

12. Show that if $p$ is prime and $p \geqslant 7$, then there are always two consecutive quadratic residues of $p$. (*Hint*: First show that at least one of $2, 5$, and $10$ is a quadratic residue of $p$.)

* 13. Show that if $p$ is prime and $p \geqslant 7$, then there are always two quadratic residues of $p$ that differ by 2.

* 14. Show that if $p$ is prime and $p \geqslant 7$, then there are always two quadratic residues of $p$ that differ by 3.

15. Show that if $a$ is a quadratic residue of the prime $p$, then the solutions of $x^2 \equiv a \pmod{p}$ are

    a)  $x \equiv \pm a^{n+1} \pmod{p}$, if $p = 4n + 3$.

    b)  $x \equiv \pm a^{n+1}$ or $\pm 2^{2n+1} a^{n+1} \pmod{p}$, if $p = 8n + 5$.

* 16. Show that if $p$ is a prime and $p = 8n + 1$, and $r$ is a primitive root modulo $p$, then the solutions of $x^2 \equiv \pm 2 \pmod{p}$ are given by

$$x \equiv \pm (r^{7n} \pm r^n) \pmod{p},$$

where the $\pm$ sign in the first congruence corresponds to the $\pm$ sign inside the parentheses in the second congruence.

17. Find all solutions of the congruence $x^2 \equiv 1 \pmod{15}$.

18. Let $p$ be an odd prime, $e$ a positive integer, and $a$ an integer relatively prime to $p$. Show that the congruence $x^2 \equiv a \pmod{p^e}$, has either no solutions or exactly two incongruent solutions modulo $p^e$.

* 19. Let $p$ be an odd prime, $e$ a positive integer, and $a$ an integer relatively prime to $p$. Show that there is a solution to the congruence $x^2 \equiv a \pmod{p^{e+1}}$ if and only if there is a solution to the congruence $x^2 \equiv a \pmod{p^e}$. Use Exercise 18 to conclude that the congruence $x^2 \equiv a \pmod{p^e}$ has no solutions if $a$ is a quadratic nonresidue of $p$, and exactly two incongruent solutions modulo $p$ if $a$ is a quadratic residue of $p$.

20. Let $n$ be an odd integer. Find the number of incongruent solutions modulo $n$ of the congruence $x^2 \equiv a \pmod{n}$, where $n$ has prime-power factorization $n = p_1^{t_1} p_2^{t_2} \cdots p_m^{t_m}$, in terms of the Legendre symbols $\left(\dfrac{a}{p_1}\right), \ldots, \left(\dfrac{a}{p_m}\right)$. (*Hint*: Use Exercise 19.)

**21.** Find the number of incongruent solutions of each of the following congruences.

a)   $x^2 \equiv 31 \pmod{75}$
b)   $x^2 \equiv 16 \pmod{105}$
c)   $x^2 \equiv 46 \pmod{231}$
d)   $x^2 \equiv 1156 \pmod{3^2 5^3 7^5 11^6}$

**\* 22.** Show that the congruence $x^2 \equiv a \pmod{2^e}$, where $e$ is an integer, $e \geqslant 3$, has either no solutions or exactly four incongruent solutions. (*Hint*: Use the fact that $(\pm x)^2 \equiv (2^{e-1} \pm x)^2 \pmod{2^e}$.)

**23.** Show that there are infinitely many primes of the four $4k + 1$. (*Hint*: Assume that $p_1, p_2, \ldots, p_n$ are the only such primes. Form $N = 4(p_1 p_2 \cdots p_n)^2 + 1$, and show, using Theorem 9.4, that $N$ has a prime factor of the form $4k + 1$ that is not one of $p_1, p_2, \ldots, p_n$.)

**\* 24.** Show that there are infinitely many primes of each of the following forms.

a)   $8k + 3$      b)   $8k + 5$      c)   $8k + 7$

(*Hint*: For each part, assume that there are only finitely many primes $p_1, p_2, \ldots, p_n$ of the particular form. For part (a) look at $(4p_1 p_2 \cdots p_n)^2 - 2$, for part (b), look at $(p_1 p_2 \cdots p_n)^2 + 2$, and for part (c), look at $(p_1 p_2 \cdots p_n)^2 + 4$. In each part, show that there is a prime factor of this integer of the required form not among the primes $p_1, p_2, \ldots, p_n$. Use Theorems 9.4 and 9.5.)

**25.** Show that if $p$ is an odd prime, then the congruence $x^2 \equiv a \pmod{p^n}$ has a solution for all positive integers $n$ if and only if $a$ is a quadratic residue of $p$.

**26.** Show that if $p$ is an odd prime with primitive root $r$, and $a$ is a positive integer not divisible by $p$, then $a$ is a quadratic residue of $p$ if and only if $\text{ind}_r a$ is even.

**27.** Show that every primitive root of an odd prime $p$ is a quadratic nonresidue of $p$.

**28.** Let $p$ be an odd prime. Show that there are $(p-1)/2 - \phi(p-1)$ quadratic nonresidues of $p$ that are not primitive roots of $p$.

**\* 29.** Let $p$ and $q = 2p + 1$ both be odd primes. Show that the $p-1$ primitive roots of $q$ are the quadratic residues of $q$, other than the nonresidue $2p$ of $q$.

**\* 30.** Show that if $p$ and $q = 4p + 1$ are both primes and if $a$ is a quadratic nonresidue of $q$ with $\text{ord}_q a \neq 4$, then $a$ is a primitive root of $q$.

**\* 31.** Show that a prime $p$ is a Fermat prime if and only if every quadratic nonresidue of $p$ is also a primitive root of $p$.

**\* 32.** Show that a prime divisor $p$ of the Fermat number $F_n = 2^{2^n} + 1$ must be of the form $2^{n+2}k + 1$. (*Hint*: Show that $\text{ord}_p 2 = 2^{n+1}$. Then show that $2^{(p-1)/2} \equiv 1 \pmod{p}$ using Theorem 9.5. Conclude that $2^{n+1} | (p-1)/2$.)

* **33.** a) Show that if $p$ is a prime of the form $4k + 3$ and $q = 2p + 1$ is prime, then $q$ divides the Mersenne number $M_p = 2^p - 1$. (*Hint*: Consider the Legendre symbol $\left(\dfrac{2}{q}\right)$.)

  b) From part (a), show that $23 \mid M_{11}$, $47 \mid M_{23}$, and $503 \mid M_{251}$.

* **34.** Show that if $n$ is a positive integer and $2n + 1$ is prime, and if $n \equiv 0$ or $3 \pmod 4$, then $2n + 1$ divides the Mersenne number $M_n = 2^n - 1$, while if $n \equiv 1$ or $2 \pmod 4$, then $2n + 1$ divides $M_n + 2 = 2^n + 1$. (*Hint*: Consider the Legendre symbol $\left(\dfrac{2}{2n+1}\right)$ and use Theorem 9.5.)

* **35.** Show that if $p$ is an odd prime, then
$$\sum_{j=1}^{p-2} \left(\frac{j(j+1)}{p}\right) = -1.$$
  (*Hint*: First show that $\left(\dfrac{j(j+1)}{p}\right) = \left(\dfrac{\bar{j}+1}{p}\right)$ where $\bar{j}$ is an inverse of $j$ modulo $p$).

* **36.** Let $p$ be an odd prime. Among pairs of consecutive positive integers less than $p$, let **(RR)**, **(RN)**, **(NR)**, and **(NN)** denote the number of pairs of two quadratic residues, of a quadratic residue followed by a quadratic nonresidue, of a quadratic nonresidue followed by a quadratic residue, and of two quadratic nonresidues, respectively.

  a) Show that
$$\begin{aligned}
\textbf{(RR)} + \textbf{(RN)} &= \tfrac{1}{2}(p - 2 - (-1)^{(p-1)/2}) \\
\textbf{(NR)} + \textbf{(NN)} &= \tfrac{1}{2}(p - 2 + (-1)^{(p-1)/2}) \\
\textbf{(RR)} + \textbf{(NR)} &= \tfrac{1}{2}(p - 1) - 1 \\
\textbf{(RN)} + \textbf{(NN)} &= \tfrac{1}{2}(p - 1).
\end{aligned}$$

  b) Using Exercise 35 show that
$$\sum_{j=1}^{p-2} \left(\frac{j(j+1)}{p}\right) = \textbf{(RR)} + \textbf{(NN)} - \textbf{(RN)} - \textbf{(NR)} = -1.$$

  c) From parts (a) and (b), find **(RR)**, **(RN)**, **(NR)**, and **(NN)**.

**37.** Use Theorem 8.16 to prove Theorem 9.1.

* **38.** Let $p$ and $q$ be odd primes. Show that 2 is a primitive root of $q$, if $q = 4p + 1$.

* **39.** Let $p$ and $q$ be odd primes. Show that 2 is a primitive root of $q$ if $p$ is of the form $4k + 1$ and $q = 2p + 1$.

* **40.** Let $p$ and $q$ be odd primes. Show that $-2$ is a primitive root of $q$ if $p$ is of the form $4k - 1$ and $q = 2p + 1$.

* **41.** Let $p$ and $q$ be odd primes. Show that $-4$ is a primitive root of $q$, if $q = 2p + 1$.

  **42.** Find the solutions of $x^2 \equiv 482 \pmod{2773}$ (note that $2773 = 47 \cdot 59$).

* **43.** In this exercise we develop a method for deciphering messages enciphered using a Rabin cipher. Recall that the relationship between a ciphertext block $C$ and the corresponding plaintext block $P$ in a Rabin cipher is $C \equiv P(P+b) \pmod{n}$, where $n = pq$, $p$ and $q$ are distinct odd primes, and $b$ is a positive integer less than $n$.

   a) Show that $C + a \equiv (P+b)^2 \pmod{n}$, where $a \equiv (\bar{2}b)^2 \pmod{n}$, and $\bar{2}$ is an inverse of 2 modulo $n$.

   b) Using the algorithm in the text for solving congruences of the type $x^2 \equiv a \pmod{n}$, together with part (a), show how to find a plaintext block $P$ from the corresponding ciphertext block $C$. Explain why there are four possible plaintext messages. (This ambiguity is a disadvantage of Rabin ciphers.)

   c) Decipher the ciphertext message 1819 0459 0803 that was enciphered using the Rabin cipher with $b = 3$ and $n = 47 \cdot 59 = 2773$.

  **44.** Let $p$ be an odd prime and let $C$ be the ciphertext obtained by modular exponentiation, with exponent $e$ and modulus $p$, from the plaintext $P$, i.e., $C \equiv P^e \pmod{p}$, $0 < C < n$, where $(e, p-1) = 1$. Show that $C$ is a quadratic residue of $p$ if and only if $P$ is a quadratic residue of $p$.

* **45.** a) Show that the second player in a game of electronic poker (see Section 7.3) can obtain an advantage by noting which cards have numerical equivalents that are quadratic residues modulo $p$. (*Hint*: Use Exercise 44.)

   b) Show that the advantage of the second player noted in part (a) can be eliminated if the numerical equivalents of cards that are quadratic nonresidues are all multiplied by a fixed quadratic nonresidue.

* **46.** Show that if the probing sequence for resolving collisions in a hashing scheme is $h_j(K) \equiv h(K) + aj + bj^2 \pmod{m}$, where $h(K)$ is a hashing function, $m$ is a positive integer, and $a$ and $b$ are integers with $(b, m) = 1$, then only half the possible file locations are probed. This is called the *quadratic search*.

## 9.1 Computer Projects

Write programs to do the following:

   **1.** Evaluate Legendre symbols using Euler's criterion.

   **2.** Evaluate Legendre symbols using Gauss' lemma.

   * 3. Flip coins electronically using the procedure described in this section.

   ** 4. Decipher messages that were enciphered using a Rabin cipher (see Exercise 43).

---

## 9.2 The Law of Quadratic Reciprocity

An elegant theorem of Gauss relates the two Legendre symbols $\left(\dfrac{p}{q}\right)$ and $\left(\dfrac{q}{p}\right)$, where $p$ and $q$ are both odd primes. This theorem, called the *law of quadratic reciprocity*, tells us whether the congruence $x^2 \equiv q \pmod{p}$ has solutions, once we know whether there are solutions of the congruence $x^2 \equiv p \pmod{q}$, where the roles of $p$ and $q$ are switched.

We now state this famous theorem.

**Theorem 9.6. The Law of Quadratic Reciprocity.** Let $p$ and $q$ be odd primes. Then

$$\left(\frac{p}{q}\right)\left(\frac{q}{p}\right) = (-1)^{\frac{p-1}{2}\cdot\frac{q-1}{2}}.$$

Before we prove this result, we will discuss its consequences and its use. We first note that the quantity $(p-1)/2$ is even when $p \equiv 1 \pmod 4$ and odd when $p \equiv 3 \pmod 4$. Consequently, we see that $\dfrac{p-1}{2}\cdot\dfrac{q-1}{2}$ is even if $p \equiv 1 \pmod 4$ or $q \equiv 1 \pmod 4$, while $\dfrac{p-1}{2}\cdot\dfrac{q-1}{2}$ is odd if $p \equiv q \equiv 3 \pmod 4$. Hence, we have

$$\left(\frac{p}{q}\right)\left(\frac{q}{p}\right) = \begin{cases} 1 & \text{if } p \equiv 1 \pmod 4 \text{ or } q \equiv 1 \pmod 4 \text{ (or both)} \\ -1 & \text{if } p \equiv q \equiv 3 \pmod 4. \end{cases}$$

Since the only possible values of $\left(\dfrac{p}{q}\right)$ and $\left(\dfrac{q}{p}\right)$ are $\pm 1$, we see that

$$\left(\frac{p}{q}\right) = \begin{cases} \left(\dfrac{q}{p}\right) & \text{if } p \equiv 1 \pmod 4 \text{ or } q \equiv 1 \pmod 4 \text{ (or both)} \\ -\left(\dfrac{q}{p}\right) & \text{if } p \equiv q \equiv 3 \pmod 4. \end{cases}$$

This means that if $p$ and $q$ are odd primes, then $\left(\dfrac{p}{q}\right) = \left(\dfrac{q}{p}\right)$ unless both $p$ and $q$ are congruent to 3 modulo 4, and in that case, $\left(\dfrac{p}{q}\right) = -\left(\dfrac{q}{p}\right)$.

**Example 9.8.** Let $p = 13$ and $q = 17$. Since $p \equiv q \equiv 1 \pmod 4$, the law of quadratic reciprocity tells us that $\left(\dfrac{13}{17}\right) = \left(\dfrac{17}{13}\right)$. By part (i) of Theorem 9.3 we know that $\left(\dfrac{17}{13}\right) = \left(\dfrac{4}{13}\right)$, and from part (iii) of Theorem 9.3, it follows that $\left(\dfrac{4}{13}\right) = \left(\dfrac{2^2}{13}\right) = 1$. Combining these equalities, we conclude that $\left(\dfrac{13}{17}\right) = 1$.

□

**Example 9.9.** Let $p = 7$ and $q = 19$. Since $p \equiv q \equiv 3 \pmod 4$, by the law of quadratic reciprocity, we know that $\left(\dfrac{7}{19}\right) = -\left(\dfrac{19}{7}\right)$. From part (i) of Theorem 9.3, we see that $\left(\dfrac{19}{7}\right) = \left(\dfrac{5}{7}\right)$. Again, using the law of quadratic reciprocity, since $5 \equiv 1 \pmod 4$ and $7 \equiv 3 \pmod 4$, we have $\left(\dfrac{5}{7}\right) = \left(\dfrac{7}{5}\right)$. By part (i) of Theorem 9.3 and Theorem 9.5, we know that $\left(\dfrac{7}{5}\right) = \left(\dfrac{2}{5}\right) = -1$. Hence $\left(\dfrac{7}{19}\right) = 1$.

□

We can use the law of quadratic reciprocity and Theorems 9.3 and 9.5 to evaluate Legendre symbols. Unfortunately, prime factorizations must be computed to evaluate Legendre symbols in this way.

**Example 9.10.** We will calculate $\left(\dfrac{713}{1009}\right)$ (note that 1009 is prime). We factor $713 = 23 \cdot 31$, so that by part (ii) of Theorem 9.3, we have

$$\left(\frac{713}{1009}\right) = \left(\frac{23 \cdot 31}{1009}\right) = \left(\frac{23}{1009}\right)\left(\frac{31}{1009}\right).$$

To evaluate the two Legendre symbols on the right side of this equality, we use the law of quadratic reciprocity. Since $1009 \equiv 1 \pmod 4$, we see that

$$\left(\frac{23}{1009}\right) = \left(\frac{1009}{23}\right), \quad \left(\frac{31}{1009}\right) = \left(\frac{1009}{31}\right).$$

Using Theorem 9.3, part (i), we have

$$\left(\frac{1009}{23}\right) = \left(\frac{20}{23}\right), \quad \left(\frac{1009}{31}\right) = \left(\frac{17}{31}\right).$$

By parts (ii) and (iii) of Theorem 9.3, it follows that

$$\left(\frac{20}{23}\right) = \left(\frac{2^2 5}{23}\right) = \left(\frac{2^2}{23}\right)\left(\frac{5}{23}\right) = \left(\frac{5}{23}\right).$$

The law of quadratic reciprocity, part (i) of Theorem 9.3, and Theorem 9.5 tell us that

$$\left(\frac{5}{23}\right) = \left(\frac{23}{5}\right) = \left(\frac{3}{5}\right) = \left(\frac{5}{3}\right) = \left(\frac{2}{3}\right) = -1.$$

Thus, $\left(\dfrac{23}{1009}\right) = -1.$

Likewise, using the law of quadratic reciprocity, Theorem 9.3, and Theorem 9.5, we find that

$$\left(\frac{17}{31}\right) = \left(\frac{31}{17}\right) = \left(\frac{14}{17}\right) = \left(\frac{2}{17}\right)\left(\frac{7}{17}\right) = \left(\frac{7}{17}\right) = \left(\frac{17}{7}\right) = \left(\frac{3}{7}\right)$$

$$= -\left(\frac{7}{3}\right) = -\left(\frac{4}{3}\right) = -\left(\frac{2^2}{3}\right) = -1.$$

Consequently, $\left(\dfrac{31}{1009}\right) = -1.$

Therefore, $\left(\dfrac{713}{1009}\right) = (-1)(-1) = 1.$ $\qquad\square$

We now present one of the many possible approaches for proving the law of quadratic reciprocity. Gauss, who first proved this result, found eight different proofs, and an article published a few years ago offered what was facetiously called the 152nd proof of the law of quadratic reciprocity. Before presenting the proof, we give a somewhat technical lemma, which we use in the proof of this important law.

**Lemma 9.3.** If $p$ is an odd prime and $a$ is an odd integer not divisible by $p$, then

$$\left(\frac{a}{p}\right) = (-1)^{T(a,p)},$$

where

$$T(a,p) = \sum_{j=1}^{(p-1)/2} [ja/p].$$

*Proof.* Consider the least positive residues of the integers $a, 2a,...,((p-1)/2)a$; let $u_1, u_2,..., u_s$ be those greater than $p/2$ and let $v_1, v_2,..., v_t$ be those less than $p/2$. The division algorithm tells us that

$$ja = p[ja/p] + \text{remainder},$$

where the remainder is one of the $u_j$'s or $v_j$'s. By adding the $(p-1)/2$ equations of this sort, we obtain

$$(9.3) \qquad \sum_{j=1}^{(p-1)/2} ja = \sum_{j=1}^{(p-1)/2} p[ja/p] + \sum_{j=1}^{s} u_j + \sum_{j=1}^{t} v_j.$$

As we showed in the proof of Gauss' lemma, the integers $p - u_1, ..., p - u_s$, $v_1, ..., v_t$ are precisely the integers $1, 2, ..., (p-1)/2$, in some order. Hence, summing all these integers, we obtain

$$(9.4) \qquad \sum_{j=1}^{(p-1)/2} j = \sum_{j=1}^{s} (p-u_j) + \sum_{j=1}^{t} v_j = ps - \sum_{j=1}^{s} u_j + \sum_{j=1}^{t} v_j.$$

Subtracting (9.4) from (9.3), we find that

$$\sum_{j=1}^{(p-1)/2} ja - \sum_{j=1}^{(p-1)/2} j = \sum_{j=1}^{(p-1)/2} p[ja/p] - ps + 2 \sum_{j=1}^{s} u_j$$

or equivalently, since $T(a,p) = \sum_{j=1}^{(p-1)/2} [ja/p]$,

$$(a-1) \sum_{j=1}^{(p-1)/2} j = pT(a,p) - ps + 2 \sum_{j=1}^{s} u_j.$$

Reducing this last equation modulo 2, since $a$ and $p$ are odd, yields

$$0 \equiv T(a,p) - s \pmod{2}.$$

Hence,

$$T(a,p) \equiv s \pmod{2}.$$

To finish the proof, we note that from Gauss' lemma

$$\left(\frac{a}{p}\right) = (-1)^s.$$

Consequently, since $(-1)^s = (-1)^{T(a,p)}$, it follows that

$$\left(\frac{a}{p}\right) = (-1)^{T(a,p)}. \qquad \blacksquare$$

Although Lemma 9.3 is used primarily as a tool in the proof of the law of quadratic reciprocity, it can also be used to evaluate Legendre symbols.

**Example 9.11.** To find $\left(\dfrac{7}{11}\right)$ using Lemma 9.3 we evaluate the sum

$$\sum_{j=1}^{5} [7_j/11] = [7/11] + [14/11] + [21/11] + [28/11] + [35/11]$$
$$= 0 + 1 + 1 + 2 + 3 = 7.$$

Hence, $\left[\dfrac{7}{11}\right] = (-1)^7 = -1$.

Likewise, to find $\left[\dfrac{11}{7}\right]$, we note that

$$\sum_{j=1}^{3} [11j/7] = [11/7] + [22/7] + [33/7] = 1 + 3 + 4 = 8,$$

so that $\left[\dfrac{11}{7}\right] = (-1)^8 = 1.$

$\square$

Before we present a proof of the law of quadratic reciprocity, we use an example to illustrate the method of proof.

Let $p = 7$ and $q = 11$. We consider pairs of integers $(x,y)$ with $1 \leqslant x \leqslant \dfrac{7-1}{2} = 3$ and $1 \leqslant y \leqslant \dfrac{11-1}{2} = 5$. There are 15 such pairs. We note that none of these pairs satisfy $11x = 7y$, since the equality $11x = 7y$ implies that $11 \mid 7y$, so that either $11 \mid 7$, which is absurd, or $11 \mid y$, which is impossible because $1 \leqslant y \leqslant 5$.

We divide these 15 pairs into two groups, depending on the relative sizes of $11x$ and $7y$.

The pairs of integers $(x,y)$ with $1 \leqslant x \leqslant 3$, $1 \leqslant y \leqslant 5$, and $11x > 7y$ are precisely those pairs satisfying $1 \leqslant x \leqslant 3$ and $1 \leqslant y \leqslant 11x/7$. For a fixed integer $x$ with $1 \leqslant x \leqslant 3$, there are $[11x/7]$ allowable values of $y$. Hence, the total number of pairs satisfying $1 \leqslant x \leqslant 3$, $1 \leqslant y \leqslant 5$, and $11x > 7y$ is

$$\sum_{j=1}^{3} [11/7] = [11/7] + [22/7] + [33/7] = 1 + 3 + 4 = 8;$$

these eight pairs are $(1,1)$, $(2,1)$, $(2,2)$, $(2,3)$, $(3,1)$, $(3,2)$, $(3,3)$ and $(3,4)$.

The pairs of integers $(x,y)$ with $1 \leqslant x \leqslant 3$, $1 \leqslant y \leqslant 5$, and $11x < 7y$ are precisely those pairs satisfying $1 \leqslant y \leqslant 5$ and $1 \leqslant x \leqslant 7y/11$. For a fixed integer $y$ with $1 \leqslant y \leqslant 5$, there are $[7y/11]$ allowable values of $x$. Hence, the total number of pairs satisfying $1 \leqslant x \leqslant 3$, $1 \leqslant y \leqslant 5$, and $11x < 7y$ is

$$\sum_{j=1}^{5} [7j/11] = [7/11] + [14/11] + [21/11] + [28/11] + [35/11]$$
$$= 0 + 1 + 1 + 2 + 3 = 7.$$

These seven pairs are $(1,2)$, $(1,3)$, $(1,4)$, $(1,5)$, $(2,4)$, $(2,5)$, and $(3,5)$.

Consequently, we see that

$$\frac{11-1}{2}\cdot\frac{7-1}{2} = 5\cdot3 = 15 = \sum_{j=1}^{3}[11j/7] + \sum_{j=1}^{5}[7j/11] = 8 + 7.$$

Hence,

$$(-1)^{\frac{11-1}{2}\cdot\frac{7-1}{2}} = (-1)^{\sum_{j=1}^{3}[11j/7] + \sum_{j=1}^{5}[7j/11]}$$

$$= (-1)^{\sum_{j=1}^{3}[11j/7]}\ (-1)^{\sum_{j=1}^{5}[7j/11]}.$$

Since Lemma 9.3 tells us that $\left(\dfrac{11}{7}\right) = (-1)^{\sum_{j=1}^{3}[11j/7]}$ and

$\left(\dfrac{7}{11}\right) = (-1)^{\sum_{j=1}^{5}[7j/11]}$, we see that $\left(\dfrac{7}{11}\right)\left(\dfrac{11}{7}\right) = (-1)^{\frac{7-1}{2}\cdot\frac{11-1}{2}}$.

This establishes the special case of the law of quadratic reciprocity when $p = 7$ and $q = 11$. □

We now prove the law of quadratic reciprocity, using the idea illustrated in the example.

*Proof.* We consider pairs of integers $(x,y)$ with $1 \leqslant x \leqslant (p-1)/2$ and $1 \leqslant y \leqslant (q-1)/2$. There are $\dfrac{p-1}{2}\cdot\dfrac{q-1}{2}$ such pairs. We divide these pairs into two groups, depending on the relative sizes of $qx$ and $py$.

First, we note that $qx \neq py$ for all of these pairs. For if $qx = py$, then $q \mid py$, which implies that $q \mid p$ or $q \mid y$. However, since $q$ and $p$ are distinct primes, we know that $q \nmid p$, and since $1 \leqslant y \leqslant (q-1)/2$, we know that $q \nmid y$.

To enumerate the pairs of integers $(x,y)$ with $1 \leqslant x \leqslant (p-1)/2$, $1 \leqslant y \leqslant (q-1)/2$, and $qx > py$, we note that these pairs are precisely those where $1 \leqslant x \leqslant (p-1)/2$ and $1 \leqslant y \leqslant qx/p$. For each fixed value of the integer $x$, with $1 \leqslant x \leqslant (p-1)/2$, there are $[qx/p]$ integers satisfying $1 \leqslant y \leqslant qx/p$. Consequently, the total number of pairs of integers $(x,y)$ with $1 \leqslant x \leqslant (p-1)/2$, $1 \leqslant y \leqslant (q-1)/2$, and $qx > py$ is $\sum_{j=1}^{(p-1)/2}[qj/p]$.

We now consider the pairs of integers $(x,y)$ with $1 \leqslant x \leqslant (p-1)/2$, $1 \leqslant y \leqslant (q-1)/2$, and $qx < py$. These pairs are precisely the pairs of integers $(x,y)$ with $1 \leqslant y \leqslant (q-1)/2$ and $1 \leqslant x \leqslant py/q$. Hence, for each fixed value of the integer $y$, where $1 \leqslant y \leqslant (q-1)/2$, there are exactly $[py/q]$ integers $x$ satisfying $1 \leqslant x \leqslant py/q$. This shows that the total number of pairs of integers $(x,y)$ with $1 \leqslant x \leqslant (p-1)/2$, $1 \leqslant y \leqslant (q-1)/2$, and $qx < py$ is $\sum_{j=1}^{(q-1)/2}[pj/q]$.

Adding the numbers of pairs in these classes, and recalling that the total number of such pairs is $\frac{p-1}{2} \cdot \frac{q-1}{2}$, we see that

$$\sum_{j=1}^{(p-1)/2} [qj/p] + \sum_{j=1}^{(q-1)/2} [pj/q] = \frac{p-1}{2} \cdot \frac{q-1}{2},$$

or using the notation of Lemma 9.3,

$$T(q,p) + T(p,q) = \frac{p-1}{2} \cdot \frac{q-1}{2}.$$

Hence,

$$(-1)^{T(q,p)+T(p,q)} = (-1)^{T(q,p)} (-1)^{T(p,q)} = (-1)^{\frac{p-1}{2} \cdot \frac{q-1}{2}}.$$

Lemma 9.2 tells us that $(-1)^{T(q,p)} = \left(\frac{q}{p}\right)$ and $(-1)^{T(p,q)} = \left(\frac{p}{q}\right)$. Hence

$$\left(\frac{p}{q}\right)\left(\frac{q}{p}\right) = (-1)^{\frac{p-1}{2} \cdot \frac{q-1}{2}}.$$

This concludes the proof of the law of quadratic reciprocity. ∎

The law of quadratic reciprocity has many applications. One use is to prove the validity of the following primality test for Fermat numbers.

**Theorem 9.7. Pepin's Test.** The Fermat number $F_m = 2^{2^m} + 1$ is prime if and only if

$$3^{(F_m-1)/2} \equiv -1 \pmod{F_m}.$$

*Proof.* We will first show that $F_m$ is prime if the congruence in the statement of the theorem holds. Assume that

$$3^{(F_m-1)/2} \equiv -1 \pmod{F_m}.$$

Then, by squaring both sides, we obtain

$$3^{F_m-1} \equiv 1 \pmod{F_m}.$$

From this congruence, we see that if $p$ is a prime dividing $F_m$, then

$$3^{F_m-1} \equiv 1 \pmod{p},$$

and hence,

$$\text{ord}_p 3 \mid (F_m-1) = 2^{2^m}.$$

Consequently, $\text{ord}_p 3$ must be a power of 2. However,

$$\text{ord}_p 3 \nmid 2^{2^m-1} = (F_m-1)/2,$$

since $3^{(F_m-1)/2} \equiv -1 \pmod{F_m}$. Hence, the only possibility is that $\text{ord}_p 3 = 2^{2^m} = F_m - 1$. Since $\text{ord}_p 3 = F_m-1 \leqslant p - 1$ and $p \mid F_m$, we see that $p = F_m$, and consequently, $F_m$ must be prime.

Conversely, if $F_m = 2^{2^m} + 1$ is prime for $m \geqslant 1$, then the law of quadratic reciprocity tells us that

(9.5)
$$\left(\frac{3}{F_m}\right) = \left(\frac{F_m}{3}\right) = \left(\frac{2}{3}\right) = -1,$$

since $F_m \equiv 1 \pmod 4$ and $F_m \equiv 2 \pmod 3$.

Now, using Euler's criterion, we know that

(9.6)
$$\left(\frac{3}{F_m}\right) \equiv 3^{(F_m-1)/2} \pmod{F_m}.$$

By the two equations involving $\left(\dfrac{3}{F_m}\right)$, (9.5) and (9.6), we conclude that

$$3^{(F_m-1)/2} \equiv -1 \pmod{F_m}.$$

This finishes the proof.                                                                    ∎

**Example 9.12.** Let $m = 2$. Then $F_2 = 2^{2^2} + 1 = 17$ and
$$3^{(F_2-1)/2} = 3^8 \equiv -1 \pmod{17}.$$

By Pepin's test, we see that $F_2 = 17$ is prime.

Let $m = 5$. Then $F_5 = 2^{2^5} + 1 = 2^{32} + 1 = 4294967297$. We note that
$$3^{(F_5-1)/2} = 3^{2^{31}} = 3^{2147483648} \equiv 10324303 \neq -1 \pmod{4294967297}.$$

Hence, by Pepin's test, we see that $F_5$ is composite.                                     □

---

## 9.2 Exercises

1. Evaluate each of the following Legendre symbols.

    a) $\left(\dfrac{3}{53}\right)$          d) $\left(\dfrac{31}{641}\right)$

    b) $\left(\dfrac{7}{79}\right)$          e) $\left(\dfrac{111}{991}\right)$

    c) $\left(\dfrac{15}{101}\right)$          f) $\left(\dfrac{105}{1009}\right)$

2. Using the law of quadratic reciprocity, show that if $p$ is an odd prime, then

$$\left(\frac{3}{p}\right) = \begin{cases} 1 & \text{if } p \equiv \pm 1 \pmod{12} \\ -1 & \text{if } p \equiv \pm 5 \pmod{12}. \end{cases}$$

3. Show that if $p$ is an odd prime, then

$$\left(\frac{-3}{p}\right) = \begin{cases} 1 & \text{if } p \equiv 1 \pmod 6 \\ -1 & \text{if } p \equiv -1 \pmod 6. \end{cases}$$

4. Find a congruence describing all primes for which 5 is a quadratic residue.

5. Find a congruence describing all primes for which 7 is a quadratic residue.

6. Show that there are infinitely many primes of the form $5k + 4$. (*Hint*: Let $n$ be a positive integer and form $Q = 5(n!)^2 - 1$. Show that $Q$ has a prime divisor of the form $5k + 4$ greater than $n$. To do this, use the law of quadratic reciprocity to show that if a prime $p$ divides $Q$, then $\left(\frac{p}{5}\right) = 1$.)

7. Use Pepin's test to show that the following Fermat numbers are primes.

   a)  $F_1 = 5$     b)  $F_3 = 257$     c)  $F_4 = 65537$

* 8. Use Pepin's test to conclude that 3 is a primitive root of every Fermat prime.

* 9. In this exercise we give another proof of the law of quadratic reciprocity. Let $p$ and $q$ be distinct odd primes. Let **R** be the interior of the rectangle with vertices $\mathbf{O} = (0,0)$, $\mathbf{A} = (p/2,0)$, $\mathbf{B} = (q/2,0)$, and $\mathbf{C} = (p/2,q/2)$ as shown.

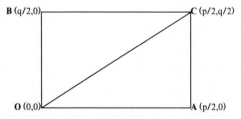

B (q/2,0)                               C (p/2,q/2)

O (0,0)                                 A (p/2,0)

   a)  Show that the number of lattice points (points with integer coordinates) in **R** is $\dfrac{p-1}{2} \cdot \dfrac{q-1}{2}$.

   b)  Show that there are no lattice points on the diagonal connecting **O** and **C**.

   c)  Show that the number of lattice points in the triangle with vertices **O**, **A, C** is $\displaystyle\sum_{j=1}^{(p-1)/2} [jq/p]$.

d)  Show that the number of lattice points in the triangle with vertices **O**,
    **B**, and **C** is $\displaystyle\sum_{j=1}^{(q-1)/2} [jp/q]$.

e)  Conclude from parts (a), (b), (c), and (d) that

$$\sum_{j=1}^{(p-1)/2} [jq/p] + \sum_{j=1}^{(q-1)/2} [jp/q] = \frac{p-1}{2} \cdot \frac{q-1}{2}.$$

Derive the law of quadratic reciprocity using this equation and
Lemma 9.2.

## 9.2  Computer Projects

Write programs to do the following:

1.  Evaluate Legendre symbols, using the law of quadratic reciprocity.

2.  Determine whether Fermat numbers are prime using Pepin's test.

## 9.3  The Jacobi symbol

In this section we define the Jacobi symbol, named after the German
mathematician Carl Jacobi* who introduced this symbol. The Jacobi symbol
is a generalization of the Legendre symbol studied in the previous two
sections. Jacobi symbols are useful in the evaluation of Legendre symbols and
in the definition of a type of pseudoprime.

**Definition.**  Let $n$ be an odd positive integer with prime factorization
$n = p_1^{t_1} p_2^{t_2} \cdots p_m^{t_m}$ and let $a$ be an integer relatively prime to $n$. Then, the
*Jacobi symbol* $\left(\dfrac{a}{n}\right)$ is defined by

$$\left(\frac{a}{n}\right) = \left(\frac{a}{p_1^{t_1} p_2^{t_2} \cdots p_m^{t_m}}\right) = \left(\frac{a}{p_1}\right)^{t_1} \left(\frac{a}{p_2}\right)^{t_2} \cdots \left(\frac{a}{p_m}\right)^{t_m},$$

where the symbols on the right-hand side of the equality are Legendre
symbols.

---

\*  **CARL GUSTAV JACOB JACOBI (1804-1851)** was born into a well-to-do German banking
   family. Jacobi received an excellent early education at home. He studied at the Univeristy of
   Berlin, mastered mathematics through the texts of Euler, and obtained his doctorate in 1825.
   In 1826 he became a lecturer at the University of Königsberg; he was appointed a professor
   there in 1831. Besides his work in number theory, Jacobi made important contributions to
   analysis, geometry, and mechanics. He was also interested in the history of mathematics and
   was a catalyst in the publication of the collected works of Euler, a job still not completed
   although it started more than 125 years ago!

**Example 9.13.** From the definition of the Jacobi symbol, we see that

$$\left(\frac{2}{45}\right) = \left(\frac{2}{3^2 \cdot 5}\right) = \left(\frac{2}{3}\right)^2 \left(\frac{2}{5}\right) = (-1)^2(-1) = -1,$$

and

$$\left(\frac{109}{385}\right) = \left(\frac{109}{5 \cdot 7 \cdot 11}\right) = \left(\frac{109}{5}\right)\left(\frac{109}{7}\right)\left(\frac{109}{11}\right) = \left(\frac{4}{5}\right)\left(\frac{4}{7}\right)\left(\frac{10}{11}\right)$$

$$= \left(\frac{2}{5}\right)^2 \left(\frac{2}{7}\right)^2 \left(\frac{-1}{11}\right) = (-1)^2 \ 1^2(-1) = -1.$$ □

When $n$ is prime, the Jacobi symbol is the same as the Legendre symbol. However, when $n$ is composite, the value of the Jacobi symbol $\left(\frac{a}{n}\right)$ does *not* tell us whether the congruence $x^2 \equiv a \pmod{n}$ has solutions. We do know that if the congruence $x^2 \equiv a \pmod{n}$ has solutions, then $\left(\frac{a}{n}\right) = 1$. To see this, note that if $p$ is a prime divisor of $n$ and if $x^2 \equiv a \pmod{n}$ has solutions, then the congruence $x^2 \equiv a \pmod{p}$ also has solutions. Thus, $\left(\frac{a}{p}\right) = 1$. Consequently, $\left(\frac{a}{n}\right) = \prod_{j=1}^{m} \left(\frac{a}{p_j}\right)^{t_j} = 1$. To see that it is possible that $\left(\frac{a}{n}\right) = 1$ when there are no solutions to $x^2 \equiv a \pmod{n}$, let $a = 2$ and $n = 15$. Note that $\left(\frac{2}{15}\right) = \left(\frac{2}{3}\right)\left(\frac{2}{5}\right) = (-1)(-1) = 1$. However, there are no solutions to $x^2 \equiv 2 \pmod{15}$, since the congruences $x^2 \equiv 2 \pmod 3$ and $x^2 \equiv 2 \pmod 5$ have no solutions.

We now show that the Jacobi symbol enjoys some properties similar to those of the Legendre symbol.

**Theorem 9.8.** Let $n$ be an odd positive integer and let $a$ and $b$ be integers relatively prime to $n$. Then

(i)    if $a \equiv b \pmod{n}$, then $\left(\frac{a}{n}\right) = \left(\frac{b}{n}\right)$,

(ii)    $\left(\frac{ab}{n}\right) = \left(\frac{a}{n}\right)\left(\frac{b}{n}\right)$,

(iii)    $\left(\frac{-1}{n}\right) = (-1)^{(n-1)/2}$,

(iv) $\qquad \left[\dfrac{2}{n}\right] = (-1)^{(n^2-1)/8}$ .

*Proof.* In the proof of all four parts of this theorem we use the prime factorization $n = p_1^{t_1} p_2^{t_2} \cdots p_m^{t_m}$.

*Proof of (i).* We know that if $p$ is a prime dividing $n$, then $a \equiv b \pmod{p}$. Hence, from Theorem 9.3 (i), we have $\left[\dfrac{a}{b}\right] = \left[\dfrac{b}{p}\right]$. Consequently, we see that

$$\left[\frac{a}{n}\right] = \left[\frac{a}{p_1}\right]^{t_1} \left[\frac{a}{p_2}\right]^{t_2} \cdots \left[\frac{a}{p_m}\right]^{t_m} = \left[\frac{b}{p_1}\right]^{t_1} \left[\frac{b}{p_2}\right]^{t_2} \cdots \left[\frac{b}{p_m}\right]^{t_m} = \left[\frac{b}{n}\right].$$

*Proof of (ii).* From Theorem 9.3 (ii), we know that $\left[\dfrac{ab}{p_i}\right] = \left[\dfrac{a}{p_i}\right]\left[\dfrac{b}{p_i}\right]$. Hence,

$$\begin{aligned}
\left[\frac{ab}{n}\right] &= \left[\frac{ab}{p_1}\right]^{t_1} \left[\frac{ab}{p_2}\right]^{t_2} \cdots \left[\frac{ab}{p_m}\right]^{t_m} \\
&= \left[\frac{a}{p_1}\right]^{t_1} \left[\frac{b}{p_1}\right]^{t_1} \left[\frac{a}{p_2}\right]^{t_2} \left[\frac{b}{p_2}\right]^{t_2} \cdots \left[\frac{a}{p_m}\right]^{t_m} \left[\frac{b}{p_m}\right]^{t_m} \\
&= \left[\frac{a}{n}\right]\left[\frac{b}{n}\right].
\end{aligned}$$

*Proof of (iii).* Theorem 9.4 tells us that if $p$ is prime, then $\left[\dfrac{-1}{p}\right] = (-1)^{(p-1)/2}$. Consequently,

$$\begin{aligned}
\left[\frac{-1}{n}\right] &= \left[\frac{-1}{p_1}\right]^{t_1} \left[\frac{-1}{p_2}\right]^{t_2} \cdots \left[\frac{-1}{p_m}\right]^{t_m} \\
&= (-1)^{t_1(p_1-1)/2 + t_2(p_2-1)/2 + \cdots + t_m(p_m-1)/2}.
\end{aligned}$$

From the prime factorization of $n$, we have

$$n = (1 + (p_1-1))^{t_1}(1 + (p_2-1))^{t_2} \cdots (1 + (p_m-1))^{t_m}.$$

Since $(p_i-1)$ is even, it follows that

$$(1 + (p_i-1))^{t_i} \equiv 1 + t_i(p_i-1) \pmod{4}$$

and

$$(1 + t_i(p_i-1))(1 + t_j(p_j-1)) \equiv 1 + t_i(p_j-1) + t_j(p_j-1) \pmod{4}.$$

Therefore,

$$n \equiv 1 + t_1(p_1-1) + t_2(p_2-1) + \cdots + t_m(p_m-1) \pmod 4.$$

This implies that

$$(n-1)/2 \equiv t_1(p_1-1)/2 + t_2(p_2-1)/2 + \cdots + t_m(p_m-1)/2 \pmod 2.$$

Combining this congruence for $(n-1)/2$ with the expression for $\left(\dfrac{-1}{n}\right)$ shows

that $\left(\dfrac{-1}{n}\right) = (-1)^{\frac{n-1}{2}}$ .

*Proof of* (iv) . If $p$ is prime, then $\left(\dfrac{2}{p}\right) = (-1)^{(p^2-1)/8}$. Hence,

$$\left(\frac{2}{n}\right) = \left(\frac{2}{\cdot p_1}\right)^{t_1} \left(\frac{2}{p_2}\right)^{t_2} \cdots \left(\frac{2}{p_m}\right)^{t_m} = (-1)^{t_1(p_1^2-1)/8 + t_2(p_2^2-1)/8 + \cdots + t_m(p_m^2-1)/8}.$$

As in the proof of (iii), we note that

$$n^2 = (1 + (p_1^2-1))^{t_1}(1 + (p_2^2-1))^{t_2} \cdots (1 + (p_m^2-1))^{t_m}.$$

Since $p_j^2-1 \equiv 0 \pmod 8$, we see that

$$(1 + (p_i^2-1))^{t_i} \equiv 1 + t_i(p_i^2-1) \pmod{64}$$

and

$$(1+t_i(p_i^2-1))(1 + t_j(p_j^2-1)) \equiv 1 + t_i(p_i^2-1) + t_j(p_j^2-1) \pmod{64}.$$

Hence,

$$n^2 \equiv 1 + t_1(p_1^2-1) + t_2(p_2^2-1) + \cdots + t_m(p_m^2-1) \pmod{64}.$$

This implies that

$$(n^2-1)/8 \equiv t_1(p_1^2-1)/8 + t_2(p_2^2-1)/8 + \cdots + t_m(p_m^2-1)/8 \pmod 8.$$

Combining this congruence for $(n^2 - 1)/8$ with the expression for $\left(\dfrac{2}{n}\right)$ tells

us that $\left(\dfrac{2}{n}\right) = (-1)^{(n^2-1)/8}$ . ∎

We now demonstrate that the reciprocity law holds for the Jacobi symbol as well as the Legendre symbol.

**Theorem 9.9**. **The Reciprocity Law for Jacobi Symbols.** Let $n$ and $m$ be relatively prime odd positive integers. Then

$$\left(\frac{n}{m}\right)\left(\frac{m}{n}\right) = (-1)^{\frac{m-1}{2}\frac{n-1}{2}}.$$

*Proof.* Let the prime factorizations of $m$ and $n$ be $m = p_1^{a_1}p_2^{a_2}\cdots p_s^{a_s}$ and $n = q_1^{b_1}q_2^{b_2}\cdots q_r^{b_r}$. We see that

$$\left(\frac{m}{n}\right) = \prod_{i=1}^{r}\left(\frac{m}{q_i}\right)^{b_i} = \prod_{i=1}^{r}\prod_{j=1}^{s}\left(\frac{p_j}{q_j}\right)^{b_i a_j}$$

and

$$\left(\frac{n}{m}\right) = \prod_{j=1}^{s}\left(\frac{n}{p_j}\right)^{a_j} = \prod_{j=1}^{s}\prod_{i=1}^{r}\left(\frac{q_i}{p_j}\right)^{a_j b_i}.$$

Thus,

$$\left(\frac{m}{n}\right)\left(\frac{n}{m}\right) = \prod_{i=1}^{r}\prod_{j=1}^{s}\left[\left(\frac{p_j}{q_i}\right)\left(\frac{q_i}{p_j}\right)\right]^{a_j b_i}.$$

By the law of quadratic reciprocity we know that

$$\left(\frac{p_j}{q_i}\right)\left(\frac{q_i}{p_j}\right) = (-1)^{\left[\frac{p_j-1}{2}\right]\left[\frac{q_i-1}{2}\right]}.$$

Hence,

$$(9.7)\quad \left(\frac{m}{n}\right)\left(\frac{n}{m}\right) = \prod_{i=1}^{r}\prod_{j=1}^{s}(-1)^{a_j\left[\frac{p_j-1}{2}\right]b_i\left[\frac{q_i-1}{2}\right]} = (-1)^{\sum_{i=1}^{r}\sum_{j=1}^{s}a_j\left[\frac{p_j-1}{2}\right]b_i\left[\frac{q_i-1}{2}\right]}.$$

We note that

$$\sum_{i=1}^{r}\sum_{j=1}^{s}a_j\left[\frac{p_j-1}{2}\right]b_i\left[\frac{q_i-1}{2}\right] = \sum_{j=1}^{s}a_j\left[\frac{p_j-1}{2}\right]\sum_{i=1}^{r}b_i\left[\frac{q_i-1}{2}\right].$$

As we demonstrated in the proof of Theorem 9.8 (iii),

$$\sum_{j=1}^{s}a_j\left[\frac{p_j-1}{2}\right] \equiv \frac{m-1}{2} \pmod 2$$

and

$$\sum_{i=1}^{r}b_i\left[\frac{q_i-1}{2}\right] \equiv \frac{n-1}{2} \pmod 2.$$

Thus,

$$(9.8) \qquad \sum_{i=1}^{r} \sum_{j=1}^{s} a_j \left( \frac{p_j-1}{2} \right) b_i \left( \frac{q_i-1}{2} \right) \equiv \frac{m-1}{2} \cdot \frac{n-1}{2} \pmod{2}.$$

Therefore, by equations (9.7) and (9.8) we can conclude that

$$\left( \frac{m}{n} \right) \left( \frac{n}{m} \right) = (-1)^{\frac{m-1}{2} \cdot \frac{n-1}{2}}.$$

∎

We now develop an efficient algorithm for evaluating Jacobi symbols. Let $a$ and $b$ be relatively prime positive integers with $a < b$. Let $R_0 = a$ and $R_1 = b$. Using the division algorithm and factoring out the highest power of two dividing the remainder, we obtain

$$R_0 = R_1 q_1 + 2^{s_1} R_2,$$

where $s_1$ is a nonnegative integer and $R_2$ is an odd positive integer less than $R_1$. When we successively use the division algorithm, and factor out the highest power of two dividing remainders, we obtain

$$R_1 = R_2 q_2 + 2^{s_2} R_3$$
$$R_2 = R_3 q_3 + 2^{s_3} R_4$$

.

.

.

$$R_{n-3} = R_{n-2} q_{n-2} + 2^{s_{n-2}} R_{n-1}$$
$$R_{n-2} = R_{n-1} q_{n-1} + 2^{s_{n-1}} \cdot 1,$$

where $s_j$ is a nonnegative integer and $R_j$ is an odd positive integer less than $R_{j-1}$ for $j = 2,3,...,n-1$. Note that the number of divisions required to reach the final equation does not exceed the number of divisions required to find the greatest common divisor of $a$ and $b$ using the Euclidean algorithm.

We illustrate this sequence of equations with the following example.

**Example 9.14.** Let $a = 401$ and $b = 111$. Then

$$401 = 111 \cdot 3 + 2^2 \cdot 17$$
$$111 = 17 \cdot 6 + 2^0 \cdot 9$$
$$17 = 9 \cdot 1 + 2^3 \cdot 1.$$

□

Using the sequence of equations we have described, together with the properties of the Jacobi symbol, we prove the following theorem, which gives an algorithm for evaluating Jacobi symbols.

**Theorem 9.10.** Let $a$ and $b$ be positive integers with $a > b$. Then

$$\left(\frac{a}{b}\right) = (-1)^{s_1\frac{R_1^2-1}{8}+\cdots+s_{n-1}\frac{R_{n-1}^2-1}{8}+\frac{R_1-1}{2}\cdot\frac{R_2-1}{2}+\cdots+\frac{R_{n-2}-1}{2}\cdot\frac{R_{n-1}-1}{2}},$$

where the integers $R_j$ and $s_j, j = 1,2,...,n-1$, are as previously described.

*Proof.* From the first equation and (i), (ii) and (iv) of Theorem 9.8, we have

$$\left(\frac{a}{b}\right) = \left(\frac{R_0}{R_1}\right) = \left(\frac{2^{s_1}R_2}{R_1}\right) = \left(\frac{2}{R_1}\right)^{s_1}\left(\frac{R_2}{R_1}\right) = (-1)^{s_1\frac{R_1^2-1}{8}}\left(\frac{R_2}{R_1}\right).$$

Using Theorem 9.9, the reciprocity law for Jacobi symbols, we have

$$\left(\frac{R_2}{R_1}\right) = (-1)^{\frac{R_1-1}{2}\cdot\frac{R_2-1}{2}}\left(\frac{R_1}{R_2}\right),$$

so that

$$\left(\frac{a}{b}\right) = (-1)^{\frac{R_1-1}{2}\cdot\frac{R_2-1}{2}+s_1\frac{R_1^2-1}{8}}\left(\frac{R_1}{R_2}\right).$$

Similarly, using the subsequent divisions, we find that

$$\left(\frac{R_{j-1}}{R_j}\right) = (-1)^{\frac{R_j-1}{2}\cdot\frac{R_{j+1}-1}{2}+s_j\cdot\frac{R_j^2-1}{8}}\left(\frac{R_j}{R_{j+1}}\right)$$

for $j = 2,3,...,n-1$. When we combine all the equalities, we obtain the desired expression for $\left(\frac{a}{b}\right)$. ∎

The following example illustrates the use of Theorem 9.10.

**Example 9.15.** To evaluate $\left(\frac{401}{111}\right)$, we use the sequence of divisions in the previous example and Theorem 9.10. This tells us that

$$\left(\frac{401}{111}\right) = (-1)^{2\cdot\frac{111^2-1}{8}+0\cdot\frac{17^2-1}{8}+3\cdot\frac{9^2-1}{8}+\frac{111-1}{2}\cdot\frac{17-1}{2}+\frac{17-1}{2}\cdot\frac{9-1}{2}} = 1.$$
□

The following corollary describes the computational complexity of the algorithm for evaluating Jacobi symbols given in Theorem 9.10.

**Corollary 9.10.1.** Let $a$ and $b$ be relatively prime positive integers with $a > b$. Then the Jacobi symbol $\left(\frac{a}{b}\right)$ can be evaluated using $O((\log_2 b)^3)$ bit operations.

*Proof.* To find $\left(\frac{a}{b}\right)$ using Theorem 9.10, we perform a sequence of $O(\log_2 b)$

divisions. To see this, note that the number of divisions does not exceed the number of divisions needed to find $(a,b)$ using the Euclidean algorithm. Thus, by Lamé's theorem we know that $O(\log_2 b)$ divisions are needed. Each division can be done using $O((\log_2 b)^2)$ bit operations. Each pair of integers $R_j$ and $s_j$ can be found using $O(\log_2 b)$ bit operations once the appropriate division has been carried out.

Consequently, $O((\log_2 b)^3)$ bit operations are required to find the integers $R_j, s_j$, $j = 1,2,...,n-1$ from $a$ and $b$. Finally, to evaluate the exponent of $-1$ in the expression for $\left(\dfrac{a}{b}\right)$ in Theorem 9.10, we use the last three bits in the binary expansions of $R_j, j = 1,2,...,n-1$ and the last bit in the binary expansions of $s_j, j = 1,2,...,n-1$. Therefore, we use $O(\log_2 b)$ additional bit operations to find $\left(\dfrac{a}{b}\right)$. Since $O((\log_2 b)^3) + O(\log_2 b) = O((\log_2 b)^3)$, the corollary holds. ∎

---

## 9.3 Exercises

1. Evaluate each of the following Jacobi symbols.

   a) $\left(\dfrac{5}{21}\right)$   d) $\left(\dfrac{1009}{2307}\right)$

   b) $\left(\dfrac{27}{101}\right)$   e) $\left(\dfrac{2663}{3299}\right)$

   c) $\left(\dfrac{111}{1001}\right)$   f) $\left(\dfrac{10001}{20003}\right)$

2. For which positive integers $n$ that are relatively prime to 15 does the Jacobi symbol $\left(\dfrac{15}{n}\right)$ equal 1?

3. For which positive integers $n$ that are relatively prime to 30 does the Jacobi symbol $\left(\dfrac{30}{n}\right)$ equal 1?

4. Let $a$ and $b$ be relatively prime integers such that $b$ is odd and positive and $a = (-1)^s 2^t q$ where $q$ is odd. Show that
$$\left(\frac{a}{b}\right) = (-1)^{\frac{b-1}{2} \cdot s + \frac{b^2-1}{8} \cdot t} \left(\frac{q}{b}\right).$$

5. Let $n$ be an odd square-free positive integer. Show that there is an integer $a$ such that $(a,n) = 1$ and $\left(\dfrac{a}{n}\right) = -1$.

**6.** Let $n$ be an odd square-free positive integer.

a) Show that $\sum \left(\dfrac{k}{n}\right) = 0$, where the sum is taken over all $k$ in a reduced set of residues modulo $n$. (*Hint*: Use Exercise 5.)

b) From part (a), show that the number of integers in a reduced set of residues modulo $n$ such that $\left(\dfrac{k}{n}\right) = 1$ is equal to the number with $\left(\dfrac{k}{n}\right) = -1$.

* **7.** Let $a$ and $b = r_0$ be relatively prime odd positive integers such that

$$a = r_0 q_1 + \epsilon_1 r_1$$
$$r_0 = r_1 q_2 + \epsilon_2 r_2$$
$$\cdot$$
$$\cdot$$
$$\cdot$$
$$r_{n-1} = r_{n-1} q_{n-1} + \epsilon_n r_n$$

where $q_i$ is a nonnegative even integer, $\epsilon_i = \pm 1$, $r_i$ is a positive integer with $r_i < r_{i-1}$, for $i = 1, 2, ..., n_j$ , and $r_n = 1$. These equations are obtained by successively using the modified division algorithm given in Exercise 16 of Section 1.5.

a) Show that the Jacobi symbol $\left(\dfrac{a}{b}\right)$ is given by

$$\left(\frac{a}{b}\right) = (-1)^{\left[\frac{r_0-1}{2}\cdot\frac{\epsilon_1 r_1-1}{2} + \frac{r_1-1}{2}\cdot\frac{\epsilon_2 r_2-1}{2} + \cdots + \frac{r_{n-1}-1}{2}\cdot\frac{\epsilon_n r_n-1}{2}\right]}.$$

b) Show that the Jacobi symbol $\left(\dfrac{a}{b}\right)$ is given by

$$\left(\frac{a}{b}\right) = (-1)^T,$$

where $T$ is the number of integers $i$, $1 \leqslant i \leqslant n$, with $r_{i-1} \equiv \epsilon_i r_i \equiv 3 \pmod 4$.

* **8.** Show that if $a$ and $b$ are odd integers and $(a,b) = 1$, then the following reciprocity law holds for the Jacobi symbol:

$$\left(\frac{a}{|b|}\right)\left(\frac{b}{|a|}\right) = \begin{cases} -(-1)^{\frac{a-1}{2}\cdot\frac{b-1}{2}} & \text{if } a < 0 \text{ and } b < 0 \\ (-1)^{\frac{a-1}{2}\cdot\frac{b-1}{2}} & \text{otherwise.} \end{cases}$$

In Exercises 9-15 we deal with the *Kronecker symbol* which is defined as follows. Let $a$ be a positive integer that is not a perfect square such that $a \equiv 0$ or $1 \pmod 4$. We define

$$\left(\frac{a}{2}\right) = \begin{cases} 1 \text{ if } a \equiv 1 \pmod 8 \\ -1 \text{ if } a \equiv 5 \pmod 8. \end{cases}$$

$$\left(\frac{a}{p}\right) = \text{the Legendre symbol } \left(\frac{a}{p}\right) \text{ if } p \text{ is an odd prime such that } p \nmid a.$$

$$\left(\frac{a}{n}\right) = \prod_{j=1}^{r} \left(\frac{a}{p_j}\right)^{t_j} \text{ if } (a,n) = 1 \text{ and } n = \prod_{j=1}^{r} p_j^{t_j} \text{ is the prime factorization of } n.$$

**9.** Evaluate each of the following Kronecker symbols,

a) $\left(\dfrac{5}{12}\right)$    b) $\left(\dfrac{13}{20}\right)$    c) $\left(\dfrac{101}{200}\right)$

For Exercises 10–15 let $a$ be a positive integer that is not a perfect square such that $a \equiv 0$ or $1 \pmod 4$.

**10.** Show that $\left(\dfrac{a}{2}\right) = \left(\dfrac{2}{|a|}\right)$ if $2 \nmid a$, where the symbol on the right is a Jacobi symbol.

**11.** Show that if $n_1$ and $n_2$ are positive integers and if $(a, n_1 n_2) = 1$, then
$$\left(\frac{a}{n_1 n_2}\right) = \left(\frac{a}{n_1}\right) \cdot \left(\frac{a}{n_2}\right).$$

**\* 12.** Show that if $n$ is a positive integer relatively prime to $a$ and if $a$ is odd, then $\left(\dfrac{a}{n}\right) = \left(\dfrac{n}{|a|}\right)$, while if $a$ is even, and $a = 2^s t$ where $t$ is odd, then
$$\left(\frac{a}{n}\right) = \left(\frac{2}{n}\right)^s (-1)^{\frac{t-1}{2} \cdot \frac{n-1}{2}} \left(\frac{n}{|t|}\right).$$

**\* 13.** Show that if $n_1$ and $n_2$ are positive integers relatively prime to $a$ and $n_1 \equiv n_2 \pmod{|a|}$, then $\left(\dfrac{a}{n_1}\right) = \left(\dfrac{a}{n_2}\right)$.

**\* 14.** Show that if $a \neq 0$, then there exists a positive integer $n$ with $\left(\dfrac{a}{n}\right) = -1$.

**\* 15.** Show that if $a \neq 0$, then $\left(\dfrac{a}{|a|-1}\right) = \begin{cases} 1 \text{ if } a > 0 \\ -1 \text{ if } a < 0. \end{cases}$

## 9.3 Computer Projects

Write programs to do the following:

**1.** Evaluate Jacobi symbols using the method of Theorem 9.7.

**2.** Evaluate Jacobi symbols using Exercises 4 and 7.

**3.** Evaluate Kronecker symbols (defined in the preamble to Exercise 9).

## 9.4 Euler Pseudoprimes

Let $p$ be an odd prime number and let $b$ be an integer not divisible by $p$. By Euler's criterion, we know that

$$b^{(p-1)/2} \equiv \left(\frac{b}{p}\right) \pmod{p}.$$

Hence, if we wish to test the positive integer $n$ for primality, we can take an integer $b$, with $(b, n) = 1$, and determine whether

$$b^{(n-1)/2} \equiv \left(\frac{b}{n}\right) \pmod{n},$$

where the symbol on the right-hand side of the congruence is the Jacobi symbol. If we find that this congruence fails, then $n$ is composite.

**Example 9.16.** Let $n = 341$ and $b = 2$. We calculate that $2^{170} \equiv 1 \pmod{341}$. Since $341 \equiv -3 \pmod{8}$, using Theorem 9.8 (iv), we see that $\left(\frac{2}{341}\right) = -1$. Consequently, $2^{170} \not\equiv \left(\frac{2}{341}\right) \pmod{341}$. This demonstrates that 341 is not prime. □

Thus, we can define a type of pseudoprime based on Euler's criterion.

**Definition.** An odd, composite, positive integer $n$ that satisfies the congruence

$$b^{(n-1)/2} \equiv \left(\frac{b}{n}\right) \pmod{n},$$

where $b$ is a positive integer is called an *Euler pseudoprime to the base b*.

An Euler pseudoprime to the base $b$ is a composite integer that masquerades as a prime by satisfying the congruence given in the definition.

**Example 9.17.** Let $n = 1105$ and $b = 2$. We calculate that $2^{552} \equiv 1 \pmod{1105}$. Since $1105 \equiv 1 \pmod{8}$, we see that $\left(\frac{2}{1105}\right) = 1$. Hence, $2^{552} \equiv \left(\frac{2}{1105}\right) \pmod{1105}$. Because 1105 is composite, it is an Euler pseudoprime to the base 2. □

The following theorem shows that every Euler pseudoprime to the base $b$ is a pseudoprime to this base.

**Theorem 9.11**. If $n$ is an Euler pseudoprime to the base $b$, then $n$ is a pseudoprime to the base $b$.

*Proof.* If $n$ is an Euler pseudoprime to the base $b$, then

$$b^{(n-1)/2} \equiv \left( \frac{b}{n} \right) \pmod{n}.$$

Hence, by squaring both sides of this congruence, we find that

$$(b^{(n-1)/2})^2 \equiv \left( \frac{b}{n} \right)^2 \pmod{n}.$$

Since $\left( \dfrac{b}{n} \right) = \pm 1$, we see that $b^{n-1} \equiv 1 \pmod{n}$. This means that $n$ is a pseudoprime to the base $b$. ∎

Not every pseudoprime is an Euler pseudoprime. For example, the integer 341 is not an Euler pseudoprime to the base 2, as we have shown, but is a pseudoprime to this base.

We know that every Euler pseudoprime is a pseudoprime. Next, we show that every strong pseudoprime is an Euler pseudoprime.

**Theorem 9.12**. If $n$ is a strong pseudoprime to the base $b$, then $n$ is an Euler pseudoprime to this base .

*Proof.* Let $n$ be a strong pseudoprime to the base $b$. Then if $n - 1 = 2^s t$, where $t$ is odd, either $b^t \equiv 1 \pmod{n}$ or $b^{2^r t} \equiv -1 \pmod{n}$ where $0 \leqslant r \leqslant s-1$. Let $n = \prod_{i=1}^{m} p_i^{a_i}$ be the prime-power factorization of $n$.

First, consider the case where $b^t \equiv 1 \pmod{n}$. Let $p$ be a prime divisor of $n$. Since $b^t \equiv 1 \pmod{p}$, we know that $\mathrm{ord}_p b | t$. Because $t$ is odd, we see that $\mathrm{ord}_p b$ is also odd. Hence, $\mathrm{ord}_p b \mid (p-1)/2$, since $\mathrm{ord}_p b$ is an odd divisor of the even integer $\phi(p) = p - 1$. Therefore,

$$b^{(p-1)/2} \equiv 1 \pmod{p}.$$

Consequently, by Euler's criterion, we have $\left( \dfrac{b}{p} \right) = 1$.

To compute the Jacobi symbol $\left( \dfrac{b}{n} \right)$, we note that $\left( \dfrac{b}{p} \right) = 1$ for all primes $p$ dividing $n$. Hence,

$$\left(\frac{b}{n}\right) = \left(\frac{b}{\prod\limits_{i=1}^{m}p_i^{a_i}}\right) = \prod_{i=1}^{m}\left(\frac{b}{p_i}\right)^{a_i} = 1.$$

Since $b^t \equiv 1 \pmod{n}$, we know that $b^{(n-1)/2} = (b^t)^{2^{s-1}} \equiv 1 \pmod{n}$. Therefore, we have

$$b^{(n-1)/2} \equiv \left(\frac{b}{n}\right) \equiv 1 \pmod{n}.$$

We conclude that $n$ is an Euler pseudoprime to the base $b$.

Next, consider the case where

$$b^{2^r t} \equiv -1 \pmod{n}$$

for some $r$ with $0 \leqslant r \leqslant s - 1$. If $p$ is a prime divisor of $n$, then

$$b^{2^r t} \equiv -1 \pmod{p}.$$

Squaring both sides of this congruence, we obtain

$$b^{2^{r+1}t} \equiv 1 \pmod{p}.$$

This implies that $\mathrm{ord}_p b \mid 2^{r+1}t$, but that $\mathrm{ord}_p b \nmid 2^r t$. Hence,

$$\mathrm{ord}_p b = 2^{r+1}c,$$

where $c$ is an odd integer. Since $\mathrm{ord}_p b \mid (p-1)$ and $2^{r+1} \mid \mathrm{ord}_p b$, it follows that $2^{r+1} \mid (p-1)$.

Therefore, we have $p = 2^{r+1}d + 1$, where $d$ is an integer. Since

$$b^{(\mathrm{ord}_p b)/2} \equiv -1 \pmod{p},$$

we have

$$\left(\frac{b}{p}\right) \equiv b^{(p-1)/2} = b^{(\mathrm{ord}_p b/2)\,((p-1)/\mathrm{ord}_p b)}$$

$$\equiv (-1)^{(p-1)/\mathrm{ord}_p b} = (-1)^{(p-1)/2^{r+1}c} \pmod{p}.$$

Because $c$ is odd, we know that $(-1)^c = -1$. Hence,

(9.9) $$\left(\frac{b}{p}\right) = (-1)^{(p-1)/2^{r+1}} = (-1)^d,$$

recalling that $d = (p-1)/2^{r+1}$. Since each prime $p_i$ dividing $n$ is of the form $p_i = 2^{r+1}d_i + 1$, it follows that

$$n = \prod_{i=1}^{m} p_i^{a_i}.$$

$$= \prod_{i=1}^{m} (2^{r+1} d_i + 1)^{a_i}$$

$$\equiv \prod_{i=1}^{m} (1 + 2^{r+1} a_i d_i)$$

$$\equiv 1 + 2^{r+1} \sum_{i=1}^{m} a_i d_i \pmod{2^{2r+2}}.$$

Therefore,

$$t\, 2^{s-1} = (n-1)/2 \equiv 2^r \sum_{i=1}^{m} a_i d_i \pmod{2^{r+1}}.$$

This congruence implies that

$$t\, 2^{s-1-r} \equiv \sum_{i=1}^{m} a_i d_i \pmod{2}$$

and

$$(9.10) \quad b^{(n-1)/2} = (b^{2^r t})^{2^{s-1-r}} \equiv (-1)^{2^{s-1-r}} = (-1)^{\sum_{i=1}^{m} a_i d_i} \pmod{n}.$$

On the other hand, from (9.9), we have

$$\left(\frac{b}{n}\right) = \prod_{i=1}^{m} \left(\frac{b}{p_i}\right)^{a_i} = \prod_{i=1}^{m} ((-1)^{d_i})^{a_i} = \prod_{i=1}^{m} (-1)^{a_j d_j} = (-1)^{\sum_{i=1}^{m} a_i d_i}.$$

Therefore, combining the previous equation with (9.10), we see that

$$b^{(n-1)/2} \equiv \left(\frac{b}{n}\right) \pmod{n}.$$

Consequently, $n$ is an Euler pseudoprime to the base $b$. ∎

Although every strong pseudoprime to the base $b$ is an Euler pseudoprime to this base, note that not every Euler pseudoprime to the base $b$ is a strong pseudoprime to the base $b$, as the following example shows.

**Example 9.18.** We have previously shown that the integer 1105 is an Euler pseudoprime to the base 2. However, 1105 is not a strong pseudoprime to the base 2 since

$$2^{(1105-1)/2} = 2^{552} \equiv 1 \pmod{1105},$$

while

$$2^{(1105-1)/2^2} = 2^{276} \equiv 781 \not\equiv \pm 1 \pmod{1105}. \qquad \square$$

Although an Euler pseudoprime to the base $b$ is not always a strong pseudoprime to this base, when certain extra conditions are met, an Euler pseudoprime to the base $b$ is, in fact, a strong pseudoprime to this base. The following two theorems give results of this kind.

**Theorem 9.13.** If $n \equiv 3 \pmod 4$ and $n$ is an Euler pseudoprime to the base $b$, then $n$ is a strong pseudoprime to the base $b$.

*Proof.* From the congruence $n \equiv 3 \pmod 4$, we know that $n-1 = 2 \cdot t$ where $t = (n-1)/2$ is odd. Since $n$ is an Euler pseudoprime to the base $b$, it follows that

$$b^t = b^{(n-1)/2} \equiv \left(\frac{b}{n}\right) \pmod n.$$

Since $\left(\dfrac{b}{n}\right) = \pm 1$, we know that either $b^t \equiv 1 \pmod n$ or $b^t \equiv -1 \pmod n$. Hence, one of the congruences in the definition of a strong pseudoprime to the base $b$ must hold. Consequently, $n$ is a strong pseudoprime to the base $b$.  ∎

**Theorem 9.14.** If $n$ is an Euler pseudoprime to the base $b$ and $\left(\dfrac{b}{n}\right) = -1$, then $n$ is a strong pseudoprime to the base $b$.

*Proof.* We write $n-1 = 2^s t$, where $t$ is odd and $s$ is a positive integer. Since $n$ is an Euler pseudoprime to the base $b$, we have

$$b^{2^{s-1}t} = b^{(n-1)/2} \equiv \left(\frac{b}{n}\right) \pmod n.$$

But since $\left(\dfrac{b}{n}\right) = -1$, we see that

$$b^{t2^{s-1}} \equiv -1 \pmod n.$$

This is one of the congruences in the definition of a strong pseudoprime to the base $b$. Since $n$ is composite, it is a strong pseudoprime to the base $b$.  ∎

Using the concept of Euler pseudoprimality, we will develop a probabilistic primality test. This test was first suggested by Solovay and Strassen [110].

Before presenting the test, we give some helpful lemmata.

**Lemma 9.4.** If $n$ is an odd positive integer that is not a perfect square, then there is at least one integer $b$ with $1 < b < n$, $(b,n) = 1$, and $\left(\dfrac{b}{n}\right) = -1$, where $\left(\dfrac{b}{n}\right)$ is the Jacobi symbol.

*Proof.* If $n$ is prime, the existence of such an integer $b$ is guaranteed by Theorem 9.1. If $n$ is composite, since $n$ is not a perfect square, we can write $n = rs$ where $(r,s) = 1$ and $r = p^e$, with $p$ an odd prime and $e$ an odd positive integer.

Now let $t$ be a quadratic nonresidue of the prime $p$; such a $t$ exists by Theorem 9.1. We use the Chinese remainder theorem to find an integer $b$ with $1 < b < n$, $(b,n) = 1$, and such that $b$ satisfies the two congruences

$$b \equiv t \pmod{r}$$
$$b \equiv 1 \pmod{s}.$$

Then,

$$\left(\frac{b}{r}\right) = \left(\frac{b}{p^e}\right) = \left(\frac{b}{p}\right)^e = (-1)^e = -1,$$

and $\left(\dfrac{b}{s}\right) = 1$. Since $\left(\dfrac{b}{n}\right) = \left(\dfrac{b}{r}\right)\left(\dfrac{b}{s}\right)$, it follows that $\left(\dfrac{b}{n}\right) = -1$. ∎

**Lemma 9.5.** Let $n$ be an odd composite integer. Then there is at least one integer $b$ with $1 < b < n$, $(b,n) = 1$, and

$$b^{(n-1)/2} \not\equiv \left(\frac{b}{n}\right) \pmod{n}.$$

*Proof.* Assume that for all positive integers not exceeding $n$ and relatively prime to $n$, that

(9.11) $$b^{(n-1)/2} \equiv \left(\frac{b}{n}\right) \pmod{n}.$$

Squaring both sides of this congruence tells us that

$$b^{n-1} \equiv \left(\frac{b}{n}\right)^2 \equiv (\pm 1)^2 = 1 \pmod{n},$$

if $(b,n) = 1$. Hence, $n$ must be a Carmichael number. Therefore, by Theorem 8.22 we know that $n = q_1 q_2 \cdots q_r$, where $q_1, q_2, ..., q_r$ are distinct odd primes.

We will now show that

$$b^{(n-1)/2} \equiv 1 \pmod{n}$$

for all integers $b$ with $1 \leqslant b \leqslant n$ and $(b,n) = 1$. Suppose that $b$ is an integer such that

$$b^{(n-1)/2} \equiv -1 \pmod{n}.$$

We use the Chinese remainder theorem to find an integer $a$ with $1 < a < n$, $(a,n) = 1$, and

$$a \equiv b \pmod{q_1}$$
$$a \equiv 1 \pmod{q_2 q_3 \cdots q_r}.$$

Then we observe that

(9.12) $$a^{(n-1)/2} \equiv b^{(n-1)/2} \equiv -1 \pmod{q_1},$$

while

(9.13) $$a^{(n-1)/2} \equiv 1 \pmod{q_2 q_3 \cdots q_r}.$$

From congruences (9.12) and (9.13), we see that

$$a^{(n-1)/2} \not\equiv \pm 1 \pmod{n},$$

contradicting congruence (9.11). Hence, we must have

$$b^{(n-1)/2} \equiv 1 \pmod{n},$$

for all $b$ with $1 \leqslant b \leqslant n$ and $(b,n) = 1$. Consequently, from the definition of an Euler pseudoprime, we know that

$$b^{(n-1)/2} \equiv \left[ \frac{b}{n} \right] = 1 \pmod{n}$$

for all $b$ with $1 \leqslant b \leqslant n$ and $(b,n) = 1$. However, Lemma 9.4 tells us that this is impossible. Hence, the original assumption is false. There must be at least one integer $b$ with $1 < b < n$, $(b,n) = 1$, and

$$b^{(n-1)/2} \not\equiv \left[ \frac{b}{n} \right] \pmod{n}.$$   ∎

We can now state and prove the theorem that is the basis of the probabilistic primality test.

**Theorem 9.15.** Let $n$ be an odd composite integer. Then, the number of positive integers less then $n$, relatively prime to $n$, that are bases to which $n$ is an Euler pseudoprime, is less than $\phi(n)/2$.

*Proof.* By Lemma 9.5 we know that there is an integer $b$ with $1 < b < n$, $(b,n) = 1$, and

(9.14) $$b^{(n-1)/2} \not\equiv \left[ \frac{b}{n} \right] \pmod{n}.$$

Now, let $a_1, a_2, ..., a_m$ denote the positive integers less than $n$ satisfying $1 \leqslant a_j \leqslant n$, $(a_j, n) = 1$, and

(9.15) $$a_j^{(n-1)/2} \equiv \left[ \frac{a_j}{n} \right] \pmod{n},$$

for $j = 1, 2, ..., m$.

Let $r_1, r_2, \ldots, r_m$ be the least positive residues of the integers $ba_1, ba_2, \ldots, ba_m$ modulo $n$. We note that the integers $r_j$ are distinct and $(r_j, n) = 1$ for $j = 1, 2, \ldots, m$. Furthermore,

$$(9.16) \qquad\qquad r_j^{(n-1)/2} \not\equiv \left[ \frac{r_j}{n} \right] \pmod{n}.$$

For, if it were true that

$$r_j^{(n-1)/2} \equiv \left[ \frac{r_j}{n} \right] \pmod{n},$$

then we would have

$$(ba_j)^{(n-1)/2} \equiv \left[ \frac{ba_j}{n} \right] \pmod{n}.$$

This would imply that,

$$b^{(n-1)/2} a_j^{(n-1)/2} \equiv \left[ \frac{b}{n} \right] \left[ \frac{a_j}{n} \right] \pmod{n},$$

and since (9.15) holds, we would have

$$b^{(n-1)/2} \equiv \left[ \frac{b}{n} \right],$$

contradicting (9.14).

Since $a_j$, $j = 1, 2, \ldots, m$, satisfies the congruence (9.15) while $r_j$, $j = 1, 2, \ldots, m$, does not, as (9.16) shows, we know these two sets of integers share no common elements. Hence, looking at the two sets together, we have a total of $2m$ distinct positive integers less than $n$ and relatively prime to $n$. Since there are $\phi(n)$ integers less than $n$ that are relatively prime to $n$, we can conclude that $2m < \phi(n)$, so that $m < \phi(n)/2$. This proves the theorem. ∎

By Theorem 9.15, we see that if $n$ is an odd composite integer, when an integer $b$ is selected at random from the integers $1, 2, \ldots, n-1$, the probability that $n$ is an Euler pseudoprime to the base $b$ is less than $1/2$. This leads to the following probabilistic primality test.

**Theorem 9.16. The Solovay-Strassen Probabilistic Primality Test.** Let $n$ be a positive integer. Select, at random, $k$ integers $b_1, b_2, \ldots, b_k$ from the integers $1, 2, \ldots, n-1$. For each of these integers $b_j, j = 1, 2, \ldots, k$, determine whether

$$b_j^{(n-1)/2} \equiv \left[ \frac{b_j}{n} \right] \pmod{n}.$$

If any of these congruences fails, then $n$ is composite. If $n$ is prime then all these congruences hold. If $n$ is composite, the probability that all $k$

congruences hold is less than $1/2^k$. Therefore, if $n$ passes this test $n$ is "almost certainly prime."

Since every strong pseudoprime to the base $b$ is an Euler pseudoprime to this base, more composite integers pass the Solovay-Strassen probabilistic primality test than the Rabin probabilistic primality test, although both require $O(k(\log_2 n)^3)$ bit operations.

## 9.4 Exercises

1. Show that the integer 561 is an Euler pseudoprime to the base 2.

2. Show that the integer 15841 is an Euler pseudoprime to the base 2, a strong pseudoprime to the base 2 and a Carmichael number.

3. Show that if $n$ is an Euler pseudoprime to the bases $a$ and $b$, then $n$ is an Euler pseudoprime to the base $ab$.

4. Show that if $n$ is an Euler pseudoprime to the base $b$, then $n$ is also an Euler pseudoprime to the base $n-b$.

5. Show that if $n \equiv 5 \pmod 8$ and $n$ is an Euler pseudoprime to the base 2, then $n$ is a strong pseudoprime to the base 2.

6. Show that if $n \equiv 5 \pmod{12}$ and $n$ is an Euler pseudoprime to the base 3, then $n$ is a strong pseudoprime to the base 3.

7. Find a congruence condition that guarantees that an Euler pseudoprime to the base 5 satisfying this congruence condition is a strong pseudoprime to the base 5.

** 8. Let the composite positive integer $n$ have prime-power factorization $n = p_1^{a_1} p_2^{a_2} \cdots p_m^{a_m}$, where $p_j = 1 + 2^{k_j} q_j$ for $j = 1,2,...,m$, where $k_1 \leqslant k_2 \leqslant \cdots \leqslant k_m$, and where $n = 1 + 2^k q$. Show that $n$ is an Euler pseudoprime to exactly

$$\delta_n \prod_{j=1}^{m} ((n-1)/2, p_j-1)$$

different bases $b$ with $1 \leqslant b < n$, where

$$\delta_n = \begin{cases} 2 & \text{if } k_1 = k \\ 1/2 & \text{if } k_j < k \text{ and } a_j \text{ is odd for some } j \\ 1 & \text{otherwise.} \end{cases}$$

## 9.4 Computer Projects

Write programs to do the following:

1. Determine if an integer passes the test for Euler pseudoprimes to the base $b$.

2. Perform the Solovay-Strassen probabilistic primality test.

# 10

# Decimal Fractions and Continued Fractions

## 10.1 Decimal Fractions

In this chapter we discuss the representation of rational and irrational numbers as decimal fractions and continued fractions. We first consider base $b$ expansions of real numbers, where $b$ is a positive integer, $b > 1$. Let $\alpha$ be a positive real number, and let $a = [\alpha]$ be the integer part of $\alpha$, so that $\gamma = \alpha - [\alpha]$ is the fractional part of $\alpha$ and $\alpha = a + \gamma$ with $0 \leqslant \gamma < 1$. By Theorem 1.10 the integer $a$ has a unique base $b$ expansion. We now show that the fractional part $\gamma$ also has a unique base $b$ expansion.

**Theorem 10.1.** Let $\gamma$ be a real number with $0 \leqslant \gamma < 1$, and let $b$ be a positive integer, $b > 1$. Then $\gamma$ can be uniquely written as

$$\gamma = \sum_{j=1}^{\infty} c_j / b^j,$$

where the coefficients $c_j$ are integers with $0 \leqslant c_j \leqslant b-1$ for $j = 1,2,...$, with the restriction that for every positive integer $N$ there is an integer $n$ with $n \geqslant N$ and $c_n \neq b-1$.

In the proof of Theorem 10.1 we deal with infinite series. We will use the following formula for the sum of the terms of an infinite geometric series.

**Theorem 10.2.** Let $a$ and $r$ be real numbers with $|r| < 1$. Then

$$\sum_{j=0}^{\infty} ar^j = a/(1-r).$$

Most books on calculus or mathematical analysis contain a proof of Theorem 10.2 (see [89] for instance).

We can now prove Theorem 10.1.

*Proof.* We first let

$$c_1 = [b\gamma],$$

so that $0 \leqslant c_1 \leqslant b-1$, since $0 \leqslant b\gamma < b$. In addition, let

$$\gamma_1 = b\gamma - c_1 = b\gamma - [b\gamma],$$

so that $0 \leqslant \gamma_1 < 1$ and

$$\gamma = \frac{c_1}{b} + \frac{\gamma_1}{b}.$$

We recursively define $c_k$ and $\gamma_k$ for $k = 2, 3, \ldots$, by

$$c_k = [b\gamma_{k-1}]$$

and

$$\gamma_k = b\gamma_{k-1} - c_k$$

so that $0 \leqslant c_k \leqslant b-1$, since $0 \leqslant b\gamma_{k-1} < b$, and $0 \leqslant \gamma_k < 1$. Then, it follows that

$$\gamma = \frac{c_1}{b} + \frac{c_2}{b^2} + \cdots + \frac{c_n}{b^n} + \frac{\gamma_n}{b^n}.$$

Since $0 \leqslant \gamma_n < 1$, we see that $0 \leqslant \gamma_n/b^n < 1/b^n$. Consequently,

$$\lim_{n \to \infty} \gamma_n/b^n = 0.$$

Therefore, we can conclude that

$$\gamma = \lim_{n \to \infty} \left[ \frac{c_1}{b} + \frac{c_2}{b^2} + \cdots + \frac{c_n}{b^n} \right]$$

$$= \sum_{j=1}^{\infty} c_j/b^j.$$

To show that this expansion is unique, assume that

$$\gamma = \sum_{j=1}^{\infty} c_j/b^j = \sum_{j=1}^{\infty} d_j/b^j,$$

where $0 \leqslant c_j \leqslant b-1$ and $0 \leqslant d_j \leqslant b-1$, and, for every positive integer $N$, there are integers $n$ and $m$ with $c_n \neq b-1$ and $d_m \neq b-1$. Assume that $k$ is the smallest index for which $c_k \neq d_k$, and assume that $c_k > d_k$ (the case $c_k < d_k$ is handled by switching the roles of the two expansions). Then

$$0 = \sum_{j=1}^{\infty} (c_j - d_j)/b^j = (c_k - d_k)/b^k + \sum_{j=k+1}^{\infty} (c_j - d_j)/b^j,$$

so that

(10.1) $$(c_k - d_k)/b^k = \sum_{j=k+1}^{\infty} (d_j - c_j)/b^j.$$

Since $c_k > d_k$, we have

(10.2) $$(c_k - d_k)/b^k \geqslant 1/b^k,$$

while

(10.3) $$\sum_{j=k+1}^{\infty} (d_j - c_j)/b^j \leqslant \sum_{j=k+1}^{\infty} (b-1)/b^j$$
$$= (b-1)\, \frac{1/b^{k+1}}{1 - 1/b}$$
$$= 1/b^k,$$

where we have used Theorem 10.2 to evaluate the sum on the right-hand side of the inequality. Note that equality holds in (10.3) if and only if $d_j - c_j = b-1$ for all $j$ with $j \geqslant k + 1$, and this occurs if and only if $d_j = b-1$ and $c_j = 0$ for $j \geqslant k+1$. However, such an instance is excluded by the hypotheses of the theorem. Hence, the inequality in (10.3) is strict, and therefore (10.2) and (10.3) contradict (10.1). This shows that the base $b$ expansion of $\alpha$ is unique. ∎

The unique expansion of a real number in the form $\sum_{j=1}^{\infty} c_j/b^j$ is called the *base b expansion* of this number and is denoted by $(.c_1 c_2 c_3 ...)_b$.

To find the base $b$ expansion $(.c_1 c_2 c_3 ...)_b$ of a real number $\gamma$, we can use the recursive formula for the digits given in the proof of Theorem 10.1, namely

$$c_k = [b\gamma_{k-1}], \quad \gamma_k = b\gamma_{k-1} - [b\gamma_{k-1}],$$

where $\gamma_0 = \gamma$, for $k = 1, 2, 3, ...$ .

**Example 10.1.** Let $(.c_1 c_2 c_3 ...)_b$ be the base 8 expansion of $1/6$. Then

$$c_1 = [8 \cdot \frac{1}{6}] = 1, \qquad \gamma_1 = 8 \cdot \frac{1}{6} - 1 = \frac{1}{3},$$

$$c_2 = [8 \cdot \frac{1}{3}] = 2, \qquad \gamma_2 = 8 \cdot \frac{1}{3} - 2 = \frac{2}{3},$$

$$c_3 = [8 \cdot \frac{2}{3}] = 5, \qquad \gamma_3 = 8 \cdot \frac{2}{3} - 5 = \frac{1}{3},$$

$$c_4 = [8 \cdot \frac{1}{3}] = 2, \qquad \gamma_4 = 8 \cdot \frac{1}{3} - 2 = \frac{2}{3},$$

$$c_5 = [8 \cdot \frac{2}{3}] = 5, \qquad \gamma_5 = 8 \cdot \frac{2}{3} - 5 = \frac{1}{3},$$

and so on. We see that the expansion repeats and hence,

$$1/6 = (.1252525...)_8.$$

$\square$

We will now discuss base $b$ expansions of rational numbers. We will show that a number is rational if and only if its base $b$ expansion is periodic or terminates.

**Definition.** A base $b$ expansion $(.c_1c_2c_3...)_b$ is said to *terminate* if there is a positive integer $n$ such that $c_n = c_{n+1} = c_{n+2} = \cdots = 0$.

**Example 10.2.** The decimal expansion of $1/8$, $(.125000...)_{10} = (.125)_{10}$, terminates. Also, the base 6 expansion of $4/9$, $(.24000...)_6 = (.24)_6$, terminates. $\square$

To describe those real numbers with terminating base $b$ expansion, we prove the following theorem.

**Theorem 10.3.** The real number $\alpha$, $0 \leqslant \alpha < 1$, has a terminating base $b$ expansion if and only if $\alpha$ is rational and $\alpha = r/s$, where $0 \leqslant r < s$ and every prime factor of $s$ also divides $b$.

*Proof.* First, suppose that $\alpha$ has a terminating base $b$ expansion,

$$\alpha = (.c_1c_2...c_n)_b.$$

Then

$$\alpha = \frac{c_1}{b} + \frac{c_2}{b^2} + \cdots + \frac{c_n}{b^n}$$

$$= \frac{c_1b^{n-1} + c_2b^{n-2} + \cdots + c_n}{b^n},$$

so that $\alpha$ is rational, and can be written with a denominator divisible only by primes dividing $b$.

Conversely, suppose that $0 \leqslant \alpha < 1$, and

$$\alpha = r/s,$$

where each prime dividing $s$ also divides $b$. Hence, there is a power of $b$, say $b^N$, that is divisible by $s$ (for instance, take $N$ to be the largest exponent in the prime-power factorization of $s$). Then

$$b^N \alpha = b^N r/s = ar,$$

where $sa = b^N$, and $a$ is a positive integer since $s \mid b^n$. Now let $(a_m a_{m-1} ... a_1 a_0)_b$ be the base $b$ expansion of $ar$. Then

$$\alpha = ar/b^N = \frac{a_m b^m + a_{m-1} b^{m-1} + \cdots + a_1 b + a_0}{b^N}$$

$$= a_m b^{m-N} + a_{m-1} b^{m-1-N} + \cdots + a_1 b^{1-N} + a_0 b^{-N}$$

$$= (.00...a_m a_{m-1}...a_1 a_0)_b.$$

Hence, $\alpha$ has a terminating base $b$ expansion. ∎

Note that every terminating base $b$ expansion can be written as a nonterminating base $b$ expansion with a tail-end consisting entirely of the digit $b-1$, since $(.c_1 c_2 \cdots c_m)_b = (.c_1 c_2 \cdots c_m - 1 \, b - 1 \, b - 1...)_b$. For instance, $(.12)_{10} = (.11999...)_{10}$. This is why we require in Theorem 10.1 that for every integer $N$ there is an integer $n$, such that $n > N$ and $c_n \neq b-1$; without this restriction base $b$ expansions would not be unique.

A base $b$ expansion that does not terminate may be *periodic*, for instance

$$1/3 = (.333...)_{10},$$
$$1/6 = (.1666...)_{10},$$

and

$$1/7 = (.142857142857142857...)_{10}.$$

**Definition.** A base $b$ expansion $(.c_1 c_2 c_3 ...)_b$ is called *periodic* if there are positive integers $N$ and $k$ such that $c_{n+k} = c_n$ for $n \geqslant N$.

We denote by $(.c_1 c_2 ... c_{N-1} \overline{c_N ... c_{N+k-1}})_b$ the periodic base $b$ expansion $(.c_1 c_2 ... c_{N-1} c_N ... c_{N+k-1} c_N ... c_{N+k-1} c_N ...)_b$. For instance, we have

$$1/3 = (.\overline{3})_{10},$$
$$1/6 = (.1\overline{6})_{10},$$

and

$$1/7 = (.\overline{142857})_{10}.$$

Note that the periodic parts of the decimal expansions of 1/3 and 1/7 begin immediately, while in the decimal expansion of 1/6 the digit 1 proceeds the periodic part of the expansion. We call the part of a periodic base $b$ expansion preceding the periodic part the *pre-period*, and the periodic part the *period*, where we take the period to have minimal possible length.

**Example 10.3.** The base 3 expansion of 2/45 is $(.00\overline{1012})_3$. The pre-period is $(00)_3$ and the period is $(1012)_3$.                                    □

The next theorem tells us that the rational numbers are those real numbers with periodic or terminating base $b$ expansions. Moreover, the theorem gives the lengths of the pre-period and periods of base $b$ expansions of rational numbers.

**Theorem 10.4.** Let $b$ be a positive integer. Then a periodic base $b$ expansion represents a rational number. Conversely, the base $b$ expansion of a rational number either terminates or is periodic. Further, if $0 < \alpha < 1$, $\alpha = r/s$, where $r$ and $s$ are relatively prime positive integers, and $s = TU$ where every prime factor of $T$ divides $b$ and $(U,b) = 1$, then the period length of the base $b$ expansion of $\alpha$ is $\text{ord}_U b$, and the pre-period length is $N$, where $N$ is the smallest positive integer such that $T|b^N$.

*Proof.* First, suppose that the base $b$ expansion of $\alpha$ is periodic, so that

$$\alpha = (.c_1 c_2 ... c_N \overline{c_{N+1} ... c_{N+k}})_b$$

$$= \frac{c_1}{b} + \frac{c_2}{b^2} + \cdots + \frac{c_N}{b^N} + \left( \sum_{j=0}^{\infty} \frac{1}{b^{jk}} \right) \left( \frac{c_{N+1}}{b^{N+1}} + \cdots + \frac{c_{N+k'}}{b^{N+k}} \right)$$

$$= \frac{c_1}{b} + \frac{c_2}{b^2} + \cdots + \frac{c_N}{b^N} + \left( \frac{b^k}{b^k-1} \right) \left( \frac{c_{N+1}}{b^{N+1}} + \cdots + \frac{c_{N+k}}{b^{N+k}} \right),$$

where we have used Theorem 10.2 to see that

$$\sum_{j=0}^{\infty} \frac{1}{b^{jk}} = \frac{1}{1 - \dfrac{1}{b^k}} = \frac{b^k}{b^k-1}.$$

Since $\alpha$ is the sum of rational numbers, it is rational.

Conversely, suppose that $0 < \alpha < 1$, $\alpha = r/s$, where $r$ and $s$ are relatively prime positive integers, $s = TU$, where every prime factor of $T$ divides $b$, $(U,b) = 1$, and $N$ is the smallest integer such that $T|b^N$.

Since $T|b^N$, we have $aT = b^N$, where $a$ is a positive integer. Hence

(10.4)                          $$b^N \alpha = b^N \frac{r}{TU} = \frac{ar}{U}.$$

Furthermore, we can write

(10.5)
$$\frac{ar}{U} = A + \frac{C}{U},$$

where $A$ and $C$ are integers with

$$0 \leqslant A < b^N, \quad 0 < C < U,$$

and $(C,U) = 1$. (The inequality for $A$ follows since $0 < b^N \alpha = \frac{ar}{U} < b^N$, which results from the inequality $0 < \alpha < 1$ when both sides are multiplied by $b^N$) . The fact that $(C,U) = 1$ follows easily from the condition $(r,s) = 1$. By Theorem 1.10 $A$ has a base $b$ expansion $A = (a_n a_{n-1}...a_1 a_0)_b$.

If $U = 1$, then the base $b$ expansion of $\alpha$ terminates as shown above. Otherwise, let $v = \mathrm{ord}_U b$. Then,

(10.6)
$$b^v \frac{C}{U} = \frac{(tU+1)C}{U} = tC + \frac{C}{U},$$

where $t$ is an integer, since $b^v \equiv 1 \pmod{U}$. However, we also have

(10.7)
$$b^v \frac{C}{U} = b^v \left[ \frac{c_1}{b} + \frac{c_2}{b^2} + \cdots + \frac{c_v}{b^v} + \frac{\gamma_v}{b^v} \right],$$

where $(.c_1 c_2 c_3...)_b$ is the base $b$ expansion of $\frac{C}{U}$, so that

$$c_k = [b\gamma_{k-1}], \quad \gamma_k = b\gamma_{k-1} - [b\gamma_{k-1}]$$

where $\gamma_0 = \frac{C}{U}$, for $k = 1,2,3,...$ . From (10.7) we see that

(10.8)
$$b^v \frac{C}{U} = \left[ c_1 b^{v-1} + c_2 b^{v-2} + \cdots + c_v \right] + \gamma_v.$$

Equating the fractional parts of (10.6) and (10.8), noting that $0 \leqslant \gamma_v < 1$, we find that

$$\gamma_v = \frac{C}{U}.$$

Consequently, we see that

$$\gamma_v = \gamma_0 = \frac{C}{U},$$

so that from the recursive definition of $c_1, c_2,...$ we can conclude that $c_{k+v} = c_k$ for $k = 1,2,3,...$ . Hence $\frac{C}{U}$ has a periodic base $b$ expansion

$$\frac{C}{U} = (.\overline{c_1 c_2 ... c_v})_b.$$

Combining (10.4) and (10.5), and inserting the base $b$ expansions of $A$ and

$\dfrac{C}{U}$, we have

(10.9) $$b^N \alpha = (a_n a_{n-1}...a_1 a_0 . \overline{c_1 c_2...c_v})_b.$$

Dividing both sides of (10.9) by $b^N$, we obtain

$$\alpha = (.00...a_n a_{n-1}...a_1 a_0 \overline{c_1 c_2...c_v})_b,$$

(where we have shifted the decimal point in the base $b$ expansion of $b^N \alpha$ $N$ spaces to the left to obtain the base $b$ expansion of $\alpha$). In this base $b$ expansion of $\alpha$, the pre-period $(.00...a_n a_{n-1}...a_1 a_0)_b$ is of length $N$, beginning with $N - (n+1)$ zeros, and the period length is $v$.

We have shown that there is a base $b$ expansion of $\alpha$ with a pre-period of length $N$ and a period of length $v$. To finish the proof, we must show that we cannot regroup the base $b$ expansion of $\alpha$, so that either the pre-period has length less than $N$, or the period has length less than $v$. To do this, suppose that

$$\alpha = (.c_1 c_2...c_M \overline{c_{M+1}...c_{M+k}})_b$$

$$= \frac{c_1}{b} + \frac{c_2}{b^2} + \cdots + \frac{c_M}{b^M} + \left[\frac{b^k}{b^k-1}\right]\left[\frac{c_{M+1}}{b^{M+1}} + \cdots + \frac{c_{M+k}}{b^{M+k}}\right]$$

$$= \frac{(c_1 b^{M-1} + c_2 b^{M-2} + \cdots + c_M)(b^k-1) + (c_{M+1} b^{k-1} + \cdots + c_{M+k})}{b^M(b^k-1)}.$$

Since $\alpha = r/s$, with $(r,s) = 1$, we see that $s|b^M(b^k-1)$. Consequently, $T|b^M$ and $U|(b^k-1)$. Hence, $M \geqslant N$, and $v|k$ (by Theorem 8.1, since $b^k \equiv 1 \pmod{U}$ and $v = \mathrm{ord}_U b$). Therefore, the pre-period length cannot be less than $N$ and the period length cannot be less than $v$. ∎

We can use Theorem 10.4 to determine the lengths of the pre-period and period of decimal expansions. Let $\alpha = r/s$, $0 < \alpha < 1$, and $s = 2^{s_1} 5^{s_2} t$, where $(t, 10) = 1$. Then by Theorem 10.4 the pre-period has length max $(s_1, s_2)$ and the period has length $\mathrm{ord}_t 10$.

**Example 10.4.** Let $\alpha = 5/28$. Since $28 = 2^2 \cdot 7$, Theorem 10.4 tells us that the pre-period has length 2 and the period has length $\mathrm{ord}_7 10 = 6$. Since $5/28 = (.17\overline{857142})$, we see that these lengths are correct. □

Note that the pre-period and period lengths of a rational number $r/s$, in lowest terms, depends only on the denominator $s$, and not on the numerator $r$.

We observe that by Theorem 10.4 a base $b$ expansion that is not terminating and is not periodic represents an irrational number.

**Example 10.5.** The number with decimal expansion

$$\alpha = .10100100010000... \, ,$$

consisting of a one followed by a zero, a one followed by two zeros, a one followed by three zeroes, and so on, is irrational because this decimal expansion does not terminate, and is not periodic.                    □

The number $\alpha$ in the above example is concocted so that its decimal expansion is clearly not periodic. To show that naturally occurring numbers such as $e$ and $\pi$ are irrational, we cannot use Theorem 10.4 because we do not have explicit formulae for the decimal digits of these numbers. No matter how many decimal digits of their expansions we compute, we still cannot conclude that they are irrational from this evidence, because the period could be longer than the number of digits we have computed.

## 10.1 Exercises

1. Find the decimal expansions of each of the following numbers.

   a) 2/5            d) 8/15
   b) 5/12           e) 1/111
   c) 12/13          f) 1/1001

2. Find the base 8 expansions of each of the following numbers.

   a) 1/3            d) 1/6
   b) 1/4            e) 1/12
   c) 1/5            f) 1/22

3. Find the fraction, in lowest terms, represented by each of the following expansions.

   a) .12      b) .1$\overline{2}$      c) $\overline{.12}$

4. Find the fraction, in lowest terms, represented by each of the following expansions.

   a) $(.123)_7$                c) $(.\overline{17})_{11}$
   b) $(.01\overline{3})_6$     d) $(.\overline{ABC})_{16}$

5. For which positive integers $b$ does the base $b$ expansion of 11/210 terminate?

6. Find the pre-period and period lengths of the decimal expansions of each of the following rational numbers.

   a) 7/12           d) 10/23
   b) 11/30          e) 13/56
   c) 1/75           f) 1/61

7.  Find the pre-period and period lengths of the base 12 expansions of each
    of the following rational numbers.

    a) 1/4                              d) 5/24
    b) 1/8                              e) 17/132
    c) 7/10                             f) 7/360

8.  Let $b$ be a positive integer. Show that the period length of the base $b$
    expansion of $1/m$ is $m - 1$ if and only if $m$ is prime and $b$ is a primitive
    root of $m$.

9.  For which primes $p$ does the decimal expansion of $1/p$ have period length
    equal to each of the following integers?

    a) 1                               d) 4
    b) 2                               e) 5
    c) 3                               f) 6

10. Find the base $b$ expansions of each of the following numbers.

    a) $1/(b-1)$                       b) $1/(b+1)$

11. Let $b$ be an integer with $b > 2$. Show that the base $b$ expansion of
    $1/(b-1)^2$ is $(.0123...b-3 \ b-1)_b$.

12. Show that the real number with base $b$ expansion

    $$(.0123...b-1 \ 101112...)_b,$$

    constructed by successively listing the base $b$ expansions of the integers, is
    irrational.

13. Show that

    $$\frac{1}{b} + \frac{1}{b^4} + \frac{1}{b^9} + \frac{1}{b^{16}} + \frac{1}{b^{25}} + \cdots$$

    is irrational, whenever $b$ is a positive integer larger than one.

14. Let $b_1, b_2, b_3, ...$ be an infinite sequence of positive integers greater than
    one. Show that every real number can be represented as

    $$c_0 + \frac{c_1}{b_1} + \frac{c_2}{b_1 b_2} + \frac{c_3}{b_1 b_2 b_3} + \cdots,$$

    where $c_0, c_1, c_2, c_3, ...$ are integers such that $0 \leqslant c_k < b_k$ for
    $k = 1, 2, 3, ....$

15. Show that every real number has an expansion

    $$c_0 + \frac{c_1}{1!} + \frac{c_2}{2!} + \frac{c_3}{3!} + \cdots$$

    where $c_0, c_1, c_2, c_3, ...$ are integers and $0 \leqslant c_k < k$ for $k = 1, 2, 3, ....$

16. Show that every rational number has a terminating expansion of the type described in Exercise 15.

* 17. Suppose that $p$ is a prime and the base $b$ expansion of $1/p$ is $(.c_1 c_2 ... c_{p-1})_b$, so that the period length of the base $b$ expansion of $1/p$ is $p - 1$. Show that if $m$ is a positive integer with $1 \leqslant m < p$, then

$$m/p = (.\overline{c_{k+1} ... c_{p-1} c_1 c_2 ... c_{k-1} c_k})_b,$$

where $k = \text{ind}_b m$ modulo $p$.

* 18. Show that if $p$ is prime and $1/p = (.\overline{c_1 c_2 ... c_k})_b$ has an even period length, $k = 2t$, then $c_j + c_{j+t} = b-1$ for $j = 1, 2, ..., t$.

## 10.1 Computer Projects

Write computer programs to do the following:

1. Find the base $b$ expansion of a rational number, where $b$ is a positive integer.

2. Find the numerator and denominator of a rational number in lowest terms from its base $b$ expansion.

3. Find the pre-period and period lengths of the base $b$ expansion of a rational number, where $b$ is a positive integer.

## 10.2 Finite Continued Fractions

Using the Euclidean algorithm we can express rational numbers as *continued fractions*. For instance, the Euclidean algorithm produces the following sequence of equations:

$$62 = 2 \cdot 23 + 16$$
$$23 = 1 \cdot 16 + 7$$
$$16 = 2 \cdot 7 + 2$$
$$7 = 3 \cdot 2 + 1.$$

When we divide both sides of each equation by the divisor of that equation, we obtain

$$\frac{62}{23} = 2 + \frac{16}{23} = 2 + \frac{1}{23/16}$$

$$\frac{23}{16} = 1 + \frac{7}{16} = 1 + \frac{1}{16/7}$$

$$\frac{16}{7} = 2 + \frac{2}{7} = 2 + \frac{1}{7/2}$$

$$\frac{7}{2} = 3 + \frac{1}{2}.$$

By combining these equations, we find that

$$\frac{62}{23} = 2 + \frac{1}{23/16}$$

$$= 2 + \cfrac{1}{1 + \cfrac{1}{16/7}}$$

$$= 2 + \cfrac{1}{1 + \cfrac{1}{2 + 7/2}}$$

$$= 2 + \cfrac{1}{1 + \cfrac{1}{2 + \cfrac{1}{3 + \cfrac{1}{2}}}}.$$

The final expression in the above string of equations is a continued fraction expansion of 62/23.

We now define continued functions.

**Definition.** A *finite continued fraction* is an expression of the form

$$a_0 + \cfrac{1}{a_1 + \cfrac{1}{a_2 + \cfrac{\phantom{x}}{\phantom{x} + \cfrac{1}{a_{n-1} + \cfrac{1}{a_n}}}}},$$

where $a_0, a_1, a_2, ..., a_n$ are real numbers with $a_1, a_2, a_3, ..., a_n$ positive. The real numbers $a_1, a_2, ..., a_n$ are called the *partial quotients* of the continued fraction. The continued fraction is called *simple* if the real numbers

$a_0, a_1, ..., a_n$ are all integers.

Because it is cumbersome to fully write out continued fractions, we use the notation $[a_0; a_1, a_2, ..., a_n]$ to represent the continued fraction in the above definition.

We will now show that every finite simple continued fraction represents a rational number. Later we will demonstrate that every rational number can be expressed as a finite simple continued fraction.

**Theorem 10.5.** Every finite simple continued fraction represents a rational number.

*Proof.* We will prove the theorem using mathematical induction. For $n = 1$ we have

$$[a_o; a_1] = a_0 + \frac{1}{a_1} = \frac{a_0 a_1 + 1}{a_0}$$

which is rational. Now assume that for the positive integer $k$ the simple continued fraction $[a_0; a_1, a_2, ..., a_k]$ is rational whenever $a_0, a_1, ..., a_k$ are integers with $a_1, ..., a_k$ positive. Let $a_0, a_1, ..., a_{k+1}$ be integers with $a_1, ..., a_{k+1}$ positive. Note that

$$[a_0; a_1, ..., a_{k+1}] = a_0 + \frac{1}{[a_1; a_2, ..., a_k, a_{k+1}]}.$$

By the induction hypothesis, $[a_1; a_2, ..., a_k, a_{k+1}]$ is rational; hence, there are integers $r$ and $s$, with $s \neq 0$, such that this continued fraction equals $r/s$. Then

$$[a_o; a_1, ..., a_k, a_{k+1}] = a_0 + \frac{1}{r/s} = \frac{a_0 r + s}{r},$$

which is again a rational number.                                               ∎

We now show, using the Euclidean algorithm, that every rational number can be written as a finite simple continued fraction.

**Theorem 10.6.** Every rational number can be expressed by a finite simple continued fraction.

*Proof.* Let $x = a/b$ where $a$ and $b$ are integers with $b > 0$. Let $r_0 = a$ and $r_1 = b$. Then the Euclidean algorithm produces the following sequence of equations:

$$
\begin{aligned}
r_0 &= r_1 q_1 + r_2 & 0 &< r_2 < r_1, \\
r_1 &= r_2 q_2 + r_3 & 0 &< r_3 < r_2, \\
r_2 &= r_3 q_3 + r_4 & 0 &< r_4 < r_3,
\end{aligned}
$$

$$
\begin{aligned}
&\quad\cdot \\
&\quad\cdot \\
&\quad\cdot
\end{aligned}
$$

$$
\begin{aligned}
r_{n-3} &= r_{n-2} q_{n-2} + r_{n-1} & 0 &< r_{n-1} < r_{n-2}, \\
r_{n-2} &= r_{n-1} q_{n-1} + r_n & 0 &< r_n < r_{n-1} \\
r_{n-1} &= r_n q_n.
\end{aligned}
$$

In these equations $q_2, q_3, \ldots, q_n$ are positive integers. Writing these equations in fractional form we have

$$
\frac{a}{b} = \frac{r_0}{r_1} = q_1 + \frac{r_2}{r_1} = q_1 + \frac{1}{r_1/r_2}
$$

$$
\frac{r_1}{r_2} = q_2 + \frac{r_3}{r_2} = q_2 + \frac{1}{r_2/r_3}
$$

$$
\frac{r_2}{r_3} = q_3 + \frac{r_4}{r_3} = q_3 + \frac{1}{r_3/r_4}
$$

$$
\begin{aligned}
&\quad\cdot \\
&\quad\cdot \\
&\quad\cdot
\end{aligned}
$$

$$
\frac{r_{n-3}}{r_{n-2}} = \frac{r_{n-1}}{r_{n-2}} = q_{n-2} + \frac{1}{r_{n-2}/r_{n-1}}
$$

$$
\frac{r_{n-2}}{r_{n-1}} = q_{n-1} + \frac{r_n}{r_{n-1}} = q_{n-1} + \frac{1}{r_{n-1}/r_n}
$$

$$
\frac{r_{n-1}}{r_n} = q_n.
$$

Substituting the value of $r_1/r_2$ from the second equation into the first equation, we obtain

(10.10)
$$
\frac{a}{b} = q_1 + \cfrac{1}{q_2 + \cfrac{1}{r_2/r_3}}.
$$

Similarly, substituting the value of $r_2/r_3$ from the third equation into (10.10) we obtain

$$
\frac{c}{b} = q_1 + \cfrac{1}{q_2 + \cfrac{1}{q_3 + \cfrac{1}{r_3/r_4}}}.
$$

Continuing in this manner, we find that

$$\frac{a}{b} = q_1 + \cfrac{1}{q_2 + \cfrac{1}{q_3 + \cfrac{\ddots}{\quad + q_{n-1} + \cfrac{1}{q_n}}}}.$$

Hence $\dfrac{a}{b} = [q_1; q_2, ..., q_n]$. This shows that every rational number can be written as a finite simple continued fraction. ■

We note that continued fractions for rational numbers are not unique. From the identity

$$a_n = (a_n - 1) + \frac{1}{1},$$

we see that

$$[a_0; a_1, a_2, ..., a_{n-1}, a_n] = [a_0; a_1, a_2, ..., a_{n-1}, a_n - 1, 1]$$

whenever $a_n > 1$.

**Example 10.6.** We have

$$\frac{7}{11} = [0; 1, 1, 1, 3] = [0; 1, 1, 1, 2, 1].$$ □

In fact, it can be shown that every rational number can be written as a finite simple continued fraction in exactly two ways, one with an odd number of terms, the other with an even number (see Exercise 8 at the end of this section).

Next, we will discuss the numbers obtained from a finite continued fraction by cutting off the expression at various stages.

**Definition.** The continued fractions $[a_0; a_1, a_2, ..., a_k]$, where $k$ is a nonnegative integer less than $n$, is called the *kth convergent* of the continued fraction $[a_0; a_1, a_2, ..., a_n]$. The $k$th convergent is denoted by $C_k$.

In our subsequent work, we will need some properties of the convergents of a continued fraction. We now develop these properties, starting with a formula for the convergents.

**Theorem 10.7.** Let $a_0, a_1, a_2, ..., a_n$ be real numbers, with $a_1, a_2, ..., a_n$ positive. Let the sequences $p_0, p_1, ..., p_n$ and $q_0, q_1, ..., q_n$ be defined recursively by

$$p_0 = a_0 \qquad\qquad q_0 = 1$$
$$p_1 = a_0 a_1 + 1 \qquad\qquad q_1 = a_1$$

and

$$p_k = a_k p_{k-1} + p_{k-2} \qquad\qquad q_k = a_k q_{k-1} + q_{k-2}$$

for $k = 2, 3, ..., n$. Then the $k$th convergent $C_k = [a_0; a_1, ..., a_k]$ is given by

$$C_k = p_k / q_k.$$

*Proof.* We will prove this theorem using mathematical induction. For $k = 0$ we have

$$C_0 = [a_0] = a_0/1 = p_0/q_0.$$

For $k = 1$, we see that

$$C_1 = [a_0; a_1] = a_0 + \frac{1}{a_1} = \frac{a_0 a_1 + 1}{a_1} = \frac{p_1}{q_1}.$$

Hence, the theorem is valid for $k = 0$ and $k = 1$.

Now assume that the theorem is true for the positive integer $k$ where $2 \leqslant k < n$. This means that

(10.11) $$\qquad C_k = [a_0; a_1, ..., a_k] = \frac{p_k}{q_k} = \frac{a_k p_{k-1} + p_{k-2}}{a_k q_{k-1} + q_{k-2}}.$$

Because of the way in which the $p_j$'s and $q_j$'s are defined, we see that the real numbers $p_{k-1}, p_{k-2}, q_{k-1}$, and $q_{k-2}$ depend only on the partial quotients $a_0, a_1, ..., a_{k-1}$. Consequently, we can replace the real number $a_k$ by $a_k + 1/a_{k+1}$ in (10.11), to obtain

$$C_{k+1} = [a_0; a_1, ..., a_k, a_{k+1}] = [a_0; a_1, ..., a_{k-1}, a_k + \frac{1}{a_{k+1}}]$$

$$= \frac{\left( a_k + \dfrac{1}{a_{k+1}} \right) p_{k-1} + p_{k-2}}{\left( a_k + \dfrac{1}{a_{k+1}} \right) q_{k-1} + q_{k-2}}$$

$$= \frac{a_{k+1}(a_k p_{k-1} + p_{k-2}) + p_{k-1}}{a_{k+1}(a_k q_{k-1} + q_{k-1}) + q_{k-1}}$$

$$= \frac{a_{k+1} p_k + p_{k-1}}{a_{k+1} q_k + q_{k-1}}$$

$$= \frac{p_{k+1}}{q_{k+1}}.$$

This finishes the proof by induction. ∎

We illustrate how to use Theorem 10.7 with the following example.

**Example 10.7.** We have $173/55 = [3;6,1,7]$. We compute the sequences $p_j$ and $q_j$ for $j = 0,1,2,3$, by

$$
\begin{aligned}
p_0 &= 3 & q_0 &= 1 \\
p_1 &= 3 \cdot 6 + 1 = 19 & q_1 &= 6 \\
p_2 &= 1 \cdot 19 + 3 = 22 & q_2 &= 1 \cdot 6 + 1 = 7 \\
p_3 &= 7 \cdot 22 + 19 = 173 & q_3 &= 7 \cdot 7 + 6 = 55.
\end{aligned}
$$

Hence, the convergents of the above continued fraction are

$$
\begin{aligned}
C_0 &= p_0/q_0 = 3/1 = 3 \\
C_1 &= p_1/q_1 = 19/6 \\
C_2 &= p_2/q_2 = 22/7 \\
C_3 &= p_3/q_3 = 173/55.
\end{aligned}
$$
□

We now state and prove another important property of the convergents of a continued fraction.

**Theorem 10.8.** Let $k$ be a positive integer, $k \geqslant 1$. Let the $k$th convergent of the continued fraction $[a_0;a_1,...,a_n]$ be $C_k = p_k/q_k$, where $p_k$ and $q_k$ are as defined in Theorem 10.7. Then

$$
p_k q_{k-1} - p_{k-1} q_k = (-1)^{k-1}.
$$

*Proof.* We use mathematical induction to prove the theorem. For $k = 1$ we have

$$
p_1 q_0 - p_0 q_1 = (a_0 a_1 + 1) \cdot 1 - a_0 a_1 = 1.
$$

Assume the theorem is true for an integer $k$ where $1 \leqslant k < n$, so that

$$
p_k q_{k-1} - p_{k-1} q_k = (-1)^{k-1}.
$$

Then, we have

$$
\begin{aligned}
p_{k+1} q_k - p_k q_{k+1} &= (a_{k+1} p_k + p_{k-1}) q_k - p_k (a_{k+1} q_k + q_{k-1}) \\
&= p_{k-1} q_k - p_k q_{k-1} = -(-1)^{k-1} = (-1)^k,
\end{aligned}
$$

so that the theorem is true for $k + 1$. This finishes the proof by induction. ∎

We illustrate this theorem with the example we used to illustrate Theorem 10.7.

**Example 10.8.** For the continued fraction $[3;6,1,7]$ we have

$$p_0 q_1 - p_1 q_0 = 3 \cdot 6 - 19 \cdot 1 = -1$$
$$p_1 q_2 - p_2 q_1 = 19 \cdot 7 - 22 \cdot 6 = 1$$
$$p_2 q_3 - p_3 q_2 = 22 \cdot 55 - 173 \cdot 7 = -1. \qquad \square$$

As a consequence of Theorem 10.8, we see that the convergents $p_k/q_k$ for $k = 1, 2, \ldots$ are in lowest terms. Corollary 10.8.1 demonstrates this.

**Corollary 10.8.1.** Let $C_k = p_k/q_k$ be the $k$th convergent of the simple continued fraction $[a_0; a_1, \ldots, a_n]$, where the integers $p_k$ and $q_k$ are as defined in Theorem 10.9. Then the integers $p_k$ and $q_k$ are relatively prime.

*Proof.* Let $d = (p_k, q_k)$. By Theorem 10.8 we know that

$$p_k q_{k-1} - q_k p_{k-1} = (-1)^{k-1}.$$

Hence

$$d \mid (-1)^{k-1}.$$

Therefore $d = 1$.      ∎

We also have the following useful corollary of Theorem 10.8.

**Corollary 10.8.2.** Let $C_k = p_k/q_k$ be the $k$th convergent of the simple continued fraction $[a_0; a_1, a_2, \ldots, a_k]$. Then

$$C_k - C_{k-1} = \frac{(-1)^{k-1}}{q_k q_{k-1}}$$

for all integers $k$ with $1 \leqslant k \leqslant n$. Also,

$$C_k - C_{k-2} = \frac{a_k (-1)^k}{q_k q_{k-2}}$$

for all integers $k$ with $2 \leqslant k \leqslant n$.

*Proof.* By Theorem 10.8 we know that $p_k q_{k-1} - q_k p_{k-1} = (-1)^{k-1}$.

We obtain the first identity,

$$C_k - C_{k-1} = \frac{p_k}{q_k} - \frac{p_{k-1}}{q_{k-1}} = \frac{(-1)^{k-1}}{q_k q_{k-1}},$$

by dividing both sides by $q_k q_{k-1}$.

To obtain the second identity, note that

$$C_k - C_{k-2} = \frac{p_k}{q_k} - \frac{p_{k-2}}{q_{k-2}} = \frac{p_k q_{k-2} - p_{k-2} q_k}{q_k q_{k-2}}.$$

Since $p_k = a_k p_{k-1} + p_{k-2}$ and $q_k = a_k q_{k-1} + q_{k-2}$, we see that the numerator of the fraction on the right is

$$p_k q_{k-2} - p_{k-2} q_k = (a_k p_{k-1} + p_{k-2}) q_{k-2} - p_{k-2}(a_k q_{k-1} + q_{k-2})$$
$$= a_k(p_{k-1}q_{k-2} - p_{k-2}q_{k-1})$$
$$= a_k(-1)^{k-2},$$

using Theorem 10.8 to see that $p_{k-1}q_{k-2} - p_{k-2}q_{k-1} = (-1)^{k-2}$.

Therefore, we find that

$$C_k - C_{k-2} = \frac{a_k(-1)^k}{q_k q_{k-2}}.$$

This is the second identity of the corollary. ∎

Using Corollary 10.8.2 we can prove the following theorem which is useful when developing infinite continued fractions.

**Theorem 10.9.** Let $C_k$ be the $k$th convergent of the finite simple continued fraction $[a_0;a_1,a_2,..., a_n]$. Then

$$C_1 > C_3 > C_5 > \cdots ,$$
$$C_0 < C_2 < C_4 < \cdots ,$$

and every odd-numbered convergent $C_{2j+1}, j = 0,1,2,...$ is greater than every even-numbered convergent $C_{2j}, j = 0,1,2,...$ .

*Proof.* Since Corollary 10.8.2 tells us that, for $k = 2,3,...,n$,

$$C_k - C_{k-2} = \frac{a_k(-1)^k}{q_k q_{k-2}},$$

we know that

$$C_k < C_{k-2}$$

when $k$ is odd, and

$$C_k > C_{k-2}$$

when $k$ is even. Hence

$$C_1 > C_3 > C_5 > \cdots ,$$

and

$$C_0 < C_2 < C_4 < \cdots .$$

To show that every odd-numbered convergent is greater than every even-numbered convergent, note that from Corollary 10.8.2 we have

$$C_{2m} - C_{2m-1} = \frac{(-1)^{2m-1}}{q_{2m}q_{2m-1}} < 0,$$

so that $C_{2m-1} > C_{2m}$. To compare $C_{2k}$ and $C_{2j-1}$, we see that

$$C_{2j-1} > C_{2j+2k-1} > C_{2j+2k} > C_{2k}.$$

so that every odd-numbered convergent is greater than every even-numbered convergent.                                                                 ∎

**Example 10.9.** Consider the finite simple continued fraction $[2;3,1,1,2,4]$. Then the convergents are

$$
\begin{aligned}
C_0 &= & 2/1 &= 2 \\
C_1 &= & 7/3 &= 2.3333... \\
C_2 &= & 9/4 &= 2.25 \\
C_3 &= & 16/7 &= 2.2857... \\
C_4 &= & 41/18 &= 2.2777... \\
C_5 &= & 180/79 &= 2.2784...\ .
\end{aligned}
$$

We see that

$$C_0 = 2 < C_2 = 2.25 < C_4 = 2.2777\cdots$$
$$< C_5 = 2.2784... < C_3 = 2.2857... < C_1 = 2.3333...\ .$$                □

## 10.2 Exercises

1.  Find the rational number, expressed in lowest terms, represented by each of the following simple continued fractions.

    a)  $[2;7]$              e)  $[1;1]$
    b)  $[1;2,3]$            f)  $[1;1,1]$
    c)  $[0;5,6]$            g)  $[1;1,1,1]$
    d)  $[3;7,15,1]$         h)  $[1;1,1,1,1]$

2.  Find the simple continued fraction expansion not terminating with the partial quotient one, of each of the following rational numbers.

    a)  $6/5$               d)  $5/999$
    b)  $22/7$              e)  $-43/1001$
    c)  $19/29$             f)  $873/4867$

3.  Find the convergents of each of the continued fractions found in Exercise 2.

4.  Let $f_k$ denote the $k$th Fibonaccci number. Find the simple continued fraction, terminating with the partial quotient of one, of $f_{k+1}/f_k$, where $k$ is a positive integer.

5. Show that if the simple continued fraction expression of the rational number $\alpha$, $\alpha > 1$, is $[a_0;a_1,...,a_k]$, then the simple continued fraction expression of $1/\alpha$ is $[0;a_0,a_1,...,a_k]$.

6. Show that if $a_0 \neq 0$, then

$$p_k/p_{k-1} = [a_k;a_{k-1}, \ldots ,a_1,a_0]$$

and

$$q_k/q_{k-1} = [a_k;a_{k-1},...,a_2,a_1],$$

where $C_{k-1} = p_{k-1}/q_{k-1}$ and $C_k = p_k/q_k, k \geqslant 1$, are successive convergents of the continued fraction $[a_0;a_1,...,a_n]$. (*Hint*: Use the relation $p_k = a_k p_{k-1} + p_{k-2}$ to show that $p_k/p_{k-1} = a_k + 1/(p_{k-1}/p_{k-2})$.

☛ 7. Show that $q_k \geqslant f_k$ for $k=1,2,...$ where $C_k = p_k/q_k$ is the $k$th convergent of the simple continued fraction $[a_0;a_1,...,a_n]$ and $f_k$ denotes the $k$th Fibonacci number.

8. Show that every rational number has exactly two finite simple continued fraction expansions.

* 9. Let $[a_0;a_1,a_2,...,a_n]$ be the simple continued fraction expansion of $r/s$ where $(r,s) = 1$ and $r \geqslant 1$. Show that this continued fraction is symmetric, *i.e.* $a_0 = a_n, a_1 = a_{n-1}, a_2 = a_{n-2},...$, if and only if $s \mid (r^2+1)$ if $n$ is odd and $s \mid (r^2-1)$ if $n$ is even. (*Hint*: Use Exercise 6 and Theorem 10.8).

* 10. Explain how finite continued fractions for rational numbers, with both plus and minus signs allowed, can be generated from the division algorithm given in Exercise 6 of Section 1.5.

11. Let $a_0,a_1,a_2,...,a_k$ be real numbers with $a_1,a_2,...$ positive and let $x$ be a positive real number. Show that $[a_0;a_1,...,a_k] < [a_0;a_1,...,a_k+x]$ if $k$ is odd and $[a_0;a_1,...,a_k] > [a_0;a_1,...,a_k+x]$ if $k$ is even.

## 10.2 Computer Projects

Write programs to do the following:

1. Find the simple continued fraction expansion of a rational number

2. Find the convergents of a finite simple continued fraction.

---

## 10.3 Infinite Continued Fractions

Suppose that we have an infinite sequence of positive integers $a_0,a_1,a_2,...$. How can we define the infinite continued fraction $[a_0,a_1,a_2,...]$? To make sense of infinite continued fractions, we need a result from mathematical analysis. We state the result below, and refer the reader to a mathematical analysis book, such as Rudin [89], for a proof.

**Theorem 10.10.** Let $x_0, x_1, x_2, \ldots$ be a sequence of real numbers such that $x_0 < x_1 < x_2 < \ldots$ and $x_k < U$ for $k = 0, 1, 2, \ldots$ for some real number $U$, or $x_0 > x_1 > x_2 > \ldots$ and $x_k > L$ for $k = 0, 1, 2, \ldots$ for some real number $L$. Then the terms of the sequence $x_0, x_1, x_2, \ldots$ tend to a limit $x$, *i.e.* there exists a real number $x$ such that

$$\lim_{k \to \infty} x_k = x.$$

Theorem 10.10 tells us that the terms of an infinite sequence tend to a limit in two special situations, when the terms of the sequence are increasing and all less than an upper bound, and when the terms of the sequence are decreasing and all are greater than a lower bound.

We can now define infinite continued fractions as limits of finite continued fractions, as the following theorem shows.

**Theorem 10.11.** Let $a_0, a_1, a_2, \ldots$ be an infinite sequence of integers with $a_1, a_2, \ldots$ positive, and let $C_k = [a_0; a_1, a_2, \ldots, a_k]$. Then the convergents $C_k$ tend to a limit $\alpha$, *i.e*

$$\lim_{k \to \infty} C_k = \alpha.$$

Before proving Theorem 10.11 we note that the limit $\alpha$ described in the statement of the theorem is called the value of the *infinite simple continued fraction* $[a_0; a_1, a_2, \ldots]$.

To prove Theorem 10.11 we will show that the infinite sequence of even-numbered convergents is increasing and has an upper bound and that the infinite sequence of odd-numbered convergents is decreasing and has a lower bound. We then show that the limits of these two sequences, guaranteed to exist by Theorem 10.10, are in fact equal.

We now will prove Theorem 10.11.

*Proof.* Let $m$ be an even positive integer. By Theorem 10.9 we see that

$$C_1 > C_3 > C_5 > \cdots > C_{m-1},$$
$$C_0 < C_2 < C_4 < \cdots < C_m,$$

and $C_{2j} < C_{2k+1}$ whenever $2j \leqslant m$ and $2k + 1 < m$. By considering all possible values of $m$, we see that

$$C_1 > C_3 > C_5 > \cdots > C_{2n-1} > C_{2n+1} > \cdots,$$
$$C_0 < C_2 < C_4 < \cdots < C_{2n-2} < C_{2n} < \cdots,$$

and $C_{2j} > C_{2k+1}$ for all positive integers $j$ and $k$. We see that the hypotheses of Theorem 10.10 are satisfied for each of the two sequences $C_1, C_3, C_2, \ldots$ and $C_0, C_2, C_4, \ldots$. Hence, the sequence $C_1, C_3, C_5, \ldots$ tends to a limit $\alpha_1$ and the sequence $C_0, C_2, C_4, \dots$ tends to a limit $\alpha_2$, *i.e.*

$$\lim_{n \to \infty} C_{2n+1} = \alpha_1$$

and

$$\lim_{n \to \infty} C_{2n} = \alpha_2.$$

Our goal is to show that these two limits $\alpha_1$ and $\alpha_2$ are equal. Using Corollary 10.8.2 we have

$$C_{2n+1} - C_{2n} = \frac{p_{2n+1}}{q_{2n+1}} - \frac{p_{2n}}{q_{2n}} = \frac{(-1)^{(2n+1)-1}}{q_{2n+1}q_{2n}} = \frac{1}{q_{2n+1}q_{2n}}.$$

Since $q_k \geqslant k$ for all positive integers $k$ (see Exercise 7 of Section 10.2), we know that

$$\frac{1}{q_{2n+1}q_{2n}} < \frac{1}{(2n+1)(2n)},$$

and hence

$$C_{2n+1} - C_{2n} = \frac{1}{q_{2n+1}q_{2n}}$$

tends to zero, *i.e.*

$$\lim_{n \to \infty} (C_{2n+1} - C_{2n}) = 0.$$

Hence, the sequences $C_1, C_3, C_5, \dots$ and $C_0, C_2, C_4, \dots$ have the same limit, since

$$\lim_{n \to \infty} (C_{2n+1} - C_{2n}) = \lim_{n \to \infty} C_{2n+1} - \lim_{n \to \infty} C_{2n} = 0.$$

Therefore $\alpha_1 = \alpha_2$, and we conclude that all the convergents tend to the limit $\alpha = \alpha_1 = \alpha_2$. This finishes the proof of the theorem. ∎

Previously, we showed that rational numbers have finite simple continued fractions. Next, we will show that the value of any infinite simple continued fraction is irrational.

**Theorem 10.12.** Let $a_0, a_1, a_2, \dots$ be integers with $a_1, a_2, \dots$ positive. Then $[a_0; a_1, a_2, \dots]$ is irrational.

*Proof.* Let $\alpha = [a_0; a_1, a_2, \dots]$ and let

$$C_k = p_k/q_k = [a_0; a_1, \dots, a_k]$$

denote the $k$th convergent of $\alpha$. When $n$ is a positive integer, Theorem 10.11 shows that $C_{2n} < \alpha < C_{2n+1}$, so that

$$0 < \alpha - C_{2n} < C_{2n+1} - C_{2n}.$$

However, by Corollary 10.8.2 we know that

$$C_{2n+1} - C_{2n} = \frac{1}{q_{2n+1} q_{2n}},$$

this means that

$$0 < \alpha - C_{2n} = \alpha - \frac{p_{2n}}{q_{2n}} < \frac{1}{q_{2n+1} q_{2n}}.$$

and therefore we have

$$0 < \alpha q_{2n} - p_{2n} < 1/q_{2n+1}.$$

Assume that $\alpha$ is rational, so that $\alpha = a/b$ where $a$ and $b$ are integers with $b \neq 0$. Then

$$0 < \frac{a q_{2n}}{b} - p_{2n} < \frac{1}{q_{2n+1}},$$

and by multiplying this inequality by $b$ we see that

$$0 < a q_{2n} - b p_{2n} < \frac{b}{q_{2n+1}}.$$

Note that $a q_{2n} - b p_{2n}$ is an integer for all positive integers $n$. However, since $q_{2n+1} > 2n+1$, for each integer $n$ there is an integer $n_0$ such that $q_{2n_0+1} > b$, so that $b/q_{2n_0+1} < 1$. This is a contradiction, since the integer $a q_{2n_0} - b p_{2n_0}$ cannot be between 0 and 1. We conclude that $\alpha$ is irrational. ∎

We have demonstrated that every infinite simple continued fraction represents an irrational number. We will now show that every irrational number can be uniquely expressed by an infinite simple continued fraction, by first constructing such a continued fraction, and then by showing that it is unique.

**Theorem 10.13.** Let $\alpha = \alpha_0$ be an irrational number and define the sequence $a_0, a_1, a_2,...$ recursively by

$$a_k = [\alpha_k], \quad \alpha_{k+1} = 1/(\alpha_k - a_k)$$

for $k = 0, 1, 2,....$ Then $\alpha$ is the value of the infinite, simple continued fraction $[a_0; a_1, a_2,...]$.

*Proof.* From the recursive definition given above, we see that $a_k$ is an integer for every $k$. Further, we can easily show using mathematical induction that $\alpha_k$ is irrational for every $k$. We first note that $\alpha_0 = \alpha$ is irrational. Next, if we assume that $\alpha_k$ is irrational, then we can easily see that $\alpha_{k+1}$ is also irrational, since the relation

$$\alpha_{k+1} = 1/(\alpha_k - a_k)$$

implies that

(10.12)
$$\alpha_k = a_k + \frac{1}{\alpha_{k+1}},$$

and if $\alpha_{k+1}$ were rational, then $\alpha_k$ would also be rational. Now, since $\alpha_k$ is irrational and $a_k$ is an integer, we know that $\alpha_k \neq a_k$, and

$$a_k < \alpha_k < a_k + 1,$$

so that

$$0 < \alpha_k - a_k < 1.$$

Hence,

$$\alpha_{k+1} = 1/(\alpha_k - a_k) > 1,$$

and consequently,

$$a_{k+1} = [\alpha_{k+1}] \geqslant 1$$

for $k = 0, 1, 2, \ldots$ . This means that all the integers $a_1$, $a_2$,... are positive.

Note that by repeatedly using (10.12) we see that

$$\alpha = \alpha_0 = a_0 + \frac{1}{\alpha_1} = [a_0;\alpha_1]$$

$$= a_0 + \cfrac{1}{a_1 + \cfrac{1}{\alpha_2}} = [a_0;a_1,\alpha_2]$$

$$\vdots$$

$$= a_0 + \cfrac{1}{a_1 + \cfrac{1}{a_2 + \cfrac{\ddots}{\quad + a_k + \cfrac{1}{\alpha_{k+1}}}}} = [a_0;a_1,a_2,\ldots,a_k,\alpha_{k+1}].$$

What we must now show is that the value of $[a_0;a_1,a_2,\ldots,a_k,\alpha_{k+1}]$ tends to $\alpha$ as $k$ tends to infinity, *i.e.*, as $k$ grows without bound. By Theorem 10.7 we see that

$$\alpha = [a_0;a_1,\ldots,a_k,\alpha_{k+1}] = \frac{\alpha_{k+1}p_k + p_{k+1}}{\alpha_{k+1}q_k + q_{k-1}},$$

where $C_j = p_j/q_j$ is the $j$th convergent of $[a_0;a_1 a_2,\ldots]$. Hence

$$\alpha - C_k = \frac{\alpha_{k+1}p_k + p_{k-1}}{\alpha_{k+1}q_k + q_{k-1}} - \frac{p_k}{q_k}$$

$$= \frac{-(p_k q_{k-1} - p_{k-1}q_k)}{(\alpha_{k+1}q_k + q_{k-1})q_k}$$

$$= \frac{(-1)^{k-1}}{(\alpha_{k+1}q_k + q_{k-1})q_k},$$

where we have used Theorem 10.8 to simplify the numerator on the right-hand side of the second equality. Since

$$\alpha_{k+1}q_k + q_{k-1} > a_{k+1}q_k + q_{k-1} = q_{k+1},$$

we see that

$$|\alpha - C_k| < \frac{1}{q_k q_{k+1}}.$$

Since $q_k > k$ (from Exercise 7 of Section 10.2), we note that $1/(q_k q_{k+1})$ tends to zero as $k$ tends to infinity. Hence, $C_k$ tends to $\alpha$ as $k$ tends to infinity, or phrased differently, the value of the infinite simple continued fraction $[a_0;a_1,a_2,...]$ is $\alpha$.    ∎

To show that the infinite simple continued fraction that represents an irrational number is unique, we prove the following theorem.

**Theorem 10.14.** If the two infinite simple continued fractions $[a_0;a_1,a_2,...]$ and $[b_0;b_1,b_2,...]$ represents the same irrational number, then $a_k = b_k$ for $k = 0,1,2,...$ .

*Proof.* Suppose that $\alpha = [a_0;a_1,a_2,...]$. Then, since $C_0 = a_0$ and $C_1 = a_0 + 1/a_1$, Theorem 10.9 tells us that

$$a_0 < \alpha < a_0 + 1/a_1,$$

so that $a_0 = [\alpha]$. Further, we note that

$$[a_0;a_1,a_2,...] = a_0 + \frac{1}{[a_1;a_2,a_3,...]},$$

since

$$\alpha = [a_0;a_1,a_2,...] = \lim_{k \to \infty} [a_0;a_1,a_2,...,a_k]$$

$$= \lim_{k \to \infty} \left( a_0 + \cfrac{1}{[a_1;a_2,a_3,...,a_k]} \right)$$

$$= a_0 + \cfrac{1}{\lim_{k \to \infty} [a_1;a_2,...,a_k]}$$

$$= a_0 + \cfrac{1}{[a_1;a_2,a_3,...]}.$$

Suppose that

$$[a_0;a_1,a_2,...] = [b_0;b_1,b_2,...].$$

Our remarks show that

$$a_0 = b_0 = [\alpha]$$

and that

$$a_0 + \cfrac{1}{[a_1;a_2,...]} = b_0 + \cfrac{1}{[b_1;b_2,...]},$$

so that

$$[a_1;a_2,...] = [b_1;b_2,...].$$

Now assume that $a_k = b_k$, and that $[a_{k+1};a_{k+2},...] = [b_{k+1};b_{k+2},...]$. Using the same argument, we see that $a_{k+1} = b_{k+1}$, and

$$a_{k+1} + \cfrac{1}{[a_{k+2};a_{k+3},...]} = b_{k+1} + \cfrac{1}{[b_{k+1};b_{k+3},..]},$$

which implies that

$$[a_{k+2};a_{k+3},...] = [b_{k+2};b_{k+3},...].$$

Hence, by mathematical induction we see that $a_k = b_k$ for $k = 0,1,2,...$ .  ∎

To find the simple continued fraction expansion of a real number, we use the algorithm given in Theorem 10.13. We illustrate this procedure with the following example.

**Example 10.10.** Let $\alpha = \sqrt{6}$. We find that

$$a_0 = [\sqrt{6}] = 2, \quad \alpha_1 = \frac{1}{\sqrt{6}-2} = \frac{\sqrt{6}+2}{2},$$

$$a_1 = [\frac{\sqrt{6}+2}{2}] = 2, \quad \alpha_2 = \frac{1}{(\frac{\sqrt{6}+2}{2})-2} = \sqrt{6}+2,$$

$$a_2 = [\sqrt{6}+2] = 4, \quad \alpha_3 = \frac{1}{(\sqrt{6}+2)-4} = \frac{\sqrt{6}+2}{2} = \alpha_1.$$

Since $\alpha_3 = \alpha_1$, we see that $a_3 = a_1$, $a_4 = a_2,...,$ and so on. Hence

$$\sqrt{6} = [2;2,4,2,4,2,4,...].$$

The simple continued fraction of $\sqrt{6}$ is periodic. We will discuss periodic simple continued fractions in the next section.                                     □

The convergents of the infinite simple continued fraction of an irrational number are good approximations to $\alpha$. In fact, if $p_k/q_k$ is the $j$th convergent of this continued fraction, then from the proof of Theorem 10.13 we know that

$$|\alpha - p_k/q_k| < 1/q_k q_{k+1},$$

so that

$$|\alpha - p_k/q_k| < 1/q_k^2,$$

since $q_k < q_{k+1}$.

The next theorem and corollary show that the convergents of the simple continued fraction of $\alpha$ are the best rational approximations to $\alpha$, in the sense that $p_k/q_k$ is closer to $\alpha$ than any other rational number with a denominator less than $q_k$.

**Theorem 10.15.** Let $\alpha$ be an irrational number and let $p_j/q_j$, $j = 1,2,...,$ be the convergents of the infinite simple continued fraction of $\alpha$. If $r$ and $s$ are integers with $s > 0$ and if $k$ is a positive integer such that

$$|s\alpha-r| < |q_k\alpha-p_k|,$$

then $s \geqslant q_{k+1}$.

*Proof.* Assume that $|s\alpha-r| < |q_k\alpha-p_k|$, but that $1 \leqslant s < q_{k+1}$. We consider the simultaneous equations

$$p_k x + p_{k+1} y = r$$
$$q_k x + q_{k+1} y = s.$$

By multiplying the first equation by $q_k$ and the second by $p_k$, and then subtracting the second from the first, we find that

$$(p_{k+1}q_k - p_k q_{k+1})y = rq_k - sp_k.$$

By Theorem 10.8 we know that $p_{k+1}q_k - p_k q_{k+1} = (-1)^k$, so that

$$y = (-1)^k (rq_k - sp_k).$$

Similarly, multiplying the first equation by $q_{k+1}$ and the second by $p_{k+1}$ and then subtracting the first from the second, we find that

$$x = (-1)^k (sp_{k+1} - rq_{k+1}).$$

We note that $x \neq 0$ and $y \neq 0$. If $x = 0$ then $sp_{k+1} = rq_{k+1}$. Since $(p_{k+1}, q_{k+1}) = 1$, Lemma 2.3 tells us that $q_{k+1} | s$, which implies that $q_{k+1} \leqslant s$, contrary to our assumption. If $y = 0$, then $r = p_k x$ and $s = q_k x$, so that

$$|s\alpha - r| = |x| |q_k\alpha - p_k| \geqslant |q_k\alpha - p_k|,$$

since $|x| \geqslant 1$, contrary to our assumption.

We will now show that $x$ and $y$ have opposite signs. First, suppose that $y < 0$. Since $q_k x = s - q_{k+1} y$, we know that $x > 0$, because $q_k x > 0$ and $q_k > 0$. When $y > 0$, since $q_{k+1} y \geqslant q_{k+1} > s$, we see that $q_k x = s - q_{k+1} y < 0$, so that $x < 0$.

By Theorem 10.9 we know that either $p_k/q_k < \alpha < p_{k+1}/q_{k+1}$ or that $p_{k+1}/q_{k+1} < \alpha < p_k/q_k$. In either case, we easily see that $q_k\alpha - p_k$ and $q_{k+1}\alpha - p_{k+1}$ have opposite signs.

From the simultaneous equations we started with, we see that

$$\begin{aligned}|s\alpha - r| &= |(q_k x + q_{k+1} y)\alpha - (p_k x + p_{k+1} y)| \\ &= |x(q_k\alpha - p_k) + y(q_{k+1}\alpha - p_{k+1})|.\end{aligned}$$

Combining the conclusions of the previous two paragraphs, we see that $x(q_k\alpha - p_k)$ and $y(q_{k+1}\alpha - p_{k+1})$ have the same sign, so that

$$\begin{aligned}|s\alpha - r| &= |x| |q_k\alpha - p_k| + |y| |q_{k+1}\alpha - p_{k+1}| \\ &\geqslant |x| |q_k\alpha - p_k| \\ &\geqslant |q_k\alpha - p_k|,\end{aligned}$$

since $|x| \geqslant 1$. This contradicts our assumption.

We have shown that our assumption is false, and consequently, the proof is complete. ∎

**Corollary 10.15.1.** Let $\alpha$ be an irrational number and let $p_j/q_j$, $j = 1,2,\ldots$ be the convergents of the infinite simple continued fraction of $\alpha$. If $r/s$ is a rational number, where $r$ and $s$ are integers with $s > 0$ and if $k$ is a positive integer such that

$$|\alpha - r/s| < |\alpha - p_k/q_k|,$$

then $s > q_k$.

*Proof.* Suppose that $s \leqslant q_k$ and that

$$|\alpha - r/s| < |\alpha - p_k/q_k|.$$

By multiplying these two inequalities, we find that

$$s|\alpha - r/s| < q_k|\alpha - p_k/q_k|$$

so that

$$|s\alpha - r| < |q_k\alpha - p_k|,$$

violating the conclusion of Theorem 10.15. ∎

**Example 10.11.** The simple continued fraction of $\pi$ is $\pi = [3;7,15,1,292,1,1,1,2,1,3,...]$. Note that there is no discernible pattern in the sequence of partial quotients. The convergents of this continued fraction are the best rational approximations to $\pi$. The first five are 3, 22/7, 333/106, 355/113, and 103993/33102. We conclude from Corollary 10.15.1 that 22/7 is the best rational approximation of $\pi$ with denominator less than 106, that 335/113 is the best rational approximation of $\pi$ with denominator less than 33102, and so on. □

Finally, we conclude this section with a result that shows that any sufficiently close rational approximation to an irrational number must be a convergent of the infinite simple continued fraction expansion of this number.

**Theorem 10.16.** If $\alpha$ is an irrational number and if $r/s$ is a rational number in lowest terms, where $r$ and $s$ are integers with $s > 0$, such that

$$|\alpha - r/s| < 1/2s^2,$$

then $r/s$ is a convergent of the simple continued fraction expansion of $\alpha$.

*Proof.* Assume that $r/s$ is not a convergent of the simple continued fraction expansion of $\alpha$. Then, there are successive convergents $p_k/q_k$ and $p_{k+1}/q_{k+1}$ such that $q_k \leqslant s < q_{k+1}$. By Theorem 10.15 we see that

$$|q_k\alpha - p_k| \leqslant |s\alpha - r| = s|\alpha - r/s| < 1/2s.$$

Dividing by $q_k$ we obtain

$$|\alpha - p_k/q_k| < 1/2sq_k.$$

Since we know that $|sp_k - rq_k| \geqslant 1$ (we know that $sp_k - rq_k$ is a nonzero integer since $r/s \neq p_k/q_k$), it follows that

$$\frac{1}{sq_k} \leq \frac{|sp_k - rq_k|}{sq_k}$$

$$= \left| \frac{p_k}{q_k} - \frac{r}{s} \right|$$

$$\leq \left| \alpha - \frac{p_k}{q_k} \right| + \left| \alpha - \frac{r}{s} \right|$$

$$< \frac{1}{2sq_k} + \frac{1}{2s^2}$$

(where we have used the triangle inequality to obtain the second inequality above). Hence, we see that

$$1/2sq_k < 1/2s^2.$$

Consequently,

$$2sq_k > 2s^2,$$

which implies that $q_k > s$, contradicting the assumption. ∎

---

## 10.3 Exercises

1. Find the simple continued fractions of each of the following real numbers.

   a) $\sqrt{2}$     c) $\sqrt{5}$
   b) $\sqrt{3}$     d) $(1+\sqrt{5})/2$

2. Find the first five partial quotients of the simple continued fractions of each of the following real numbers.

   a) $\sqrt[3]{2}$     c) $(e-1)/(e+1)$
   b) $2\pi$     d) $(e^2-1)/(e^2+1)$

3. Find the best rational approximation to $\pi$ with a denominator less than 10000.

4. The infinite simple continued fraction expansion of the number $e$ is

   $$e = [2;1,2,1,1,4,1,1,6,1,1,8,...].$$

   a) Find the first eight convergents of the continued fraction of $e$.

   b) Find the best rational approximation to $e$ having a denominator less than 100.

* 5. Let $\alpha$ be an irrational number with simple continued fraction expansion $\alpha = [a_0;a_1,a_2,...]$. Show that the simple continued fraction of $-\alpha$ is $[-a_0-1;1,a,-1,a_2,a_3,...]$ if $a_1 > 1$ and $[-a_0-1;a_2+1,a_3, \cdots ]$ if

$a_1 = 1$.

**\* 6.** Show that if $p_k/q_k$ and $p_{k+1}/q_{k+1}$ are consecutive convergents of the simple continued fraction of an irrational number $\alpha$, then

$$|\alpha - p_k/q_k| < 1/2q_k^2$$

or

$$|\alpha - p_{k+1}/q_{k+1}| < 1/2q_{k+1}^2.$$

(*Hint* : First show that $|\alpha - p_{k+1}/q_{k+1}| + |\alpha - p_k/q_k| = |p_{k+1}/q_{k+1} - p_k/q_k| = 1/q_k q_{k+1}$.)

**☛ 7.** Let $\alpha$ be an irrational number, $\alpha > 1$. Show that the $k$ th convergent of the simple continued fraction of $1/\alpha$ is the reciprocal of the $(k-1)$th convergent of the simple continued fraction of $\alpha$.

**\* 8.** Let $\alpha$ be an irrational number, and let $p_j/q_j$ denote the $j$ th convergent of the simple continued fraction expansion of $\alpha$. Show that at least one of any three consecutive convergents satisfies the inequality

$$|\alpha - p_j/q_j| < 1/(\sqrt{5}q_j^2).$$

Conclude that there are infinitely many rational numbers $p/q$, where $p$ and $q$ are integers with $q \neq 0$, such that

$$|\alpha - p/q| < 1/(\sqrt{5}q^2).$$

**\* 9.** Show that if $\alpha = (1+\sqrt{5})/2$ , and $c > \sqrt{5}$, then there are only a finite number of rational numbers $p/q$ , where p and q are integers, $q \neq 0$, such that

$$|\alpha - p/q| < 1/(cq^2).$$

(*Hint*: Consider the convergents of the simple continued fraction expansion of $\sqrt{5}$.)

If $\alpha$ and $\beta$ are two real numbers, we say that $\beta$ is equivalent to $\alpha$ if there are integers $a,b,c,$ and $d$, such that $ad - bc = \pm 1$ and $\beta = \dfrac{a\alpha+b}{c\alpha+d}$.

**10.** Show that a real number $\alpha$ is equivalent to itself.

**11.** Show that if $\alpha$ and $\beta$ are real numbers with $\beta$ equivalent to $\alpha$ , then $\alpha$ is equivalent to $\beta$ . Hence, we can say that two numbers $\alpha$ and $\beta$ are equivalent.

**12.** Show that if $\alpha,\beta,$ and $\lambda$ are real numbers such that $\alpha$ and $\beta$ are equivalent and $\beta$ and $\lambda$ are equivalent, then $\alpha$ and $\lambda$ are equivalent.

**13.** Show that any two rational numbers are equivalent.

**\* 14.** Show that two irrational numbers $\alpha$ and $\beta$ are equivalent if and only if the tails of their simple continued fractions agree, i.e., $\alpha = [a_0;a_1,a_2,...,a_j,c_1,c_2,c_3,...]$, $\beta = [b_0;b_1,b_2,...,b_k,c_1,c_2,c_3,...]$, where $a_i, i=0,1,2,...,j$, $b_i, i=0,1,2,...,k$ and $c_i, i = 1,2,3,...$ are integers, all positive except perhaps $a_0$ and $b_0$ .

Let $\alpha$ be an irrational number, and let the simple continued fraction expansion of $\alpha$ be $\alpha = [a_0;a_1,a_2,...]$. Let $p_k/q_k$ denote, as usual, the $k$ th convergent of this continued fraction. We define the *pseudoconvergnts* of this continued fraction to be

$$p_{k,t}/q_{k,t} = (tp_{k-1} + p_{k-2})/(tq_{k-1} + q_{k-2}),$$

where $k$ is a positive integer, $k \geqslant 2$ , and $t$ is an integer with $0 < t < a_k$.

15. Show that each pseudoconvergent is in lowest terms

* 16. Show that the sequence of rational numbers $p_{k,2}/q_{k,2},..., p_{k,a_k-1}/q_{k,a_k-1}$, $p_k/q_k$ is increasing if $k$ is even, and decreasing if $k$ is odd.

* 17. Show that if $r$ and $s$ are integers with $s > 0$ such that

$$|\alpha - r/s| \leqslant |\alpha - p_{k,t}/q_{k,t}|$$

where $k$ is a positive integer and $0 < t < a_k$, then $s > q_{k,t}$ or $r/s = p_{k-1}/q_{k-1}$.

18. Find the pseudoconvergents of the simple continued fraction of $\pi$ for $k = 2$.

## 10.3 Computer Projects

Write programs to do the following:

1. Find the simple continued fraction of a real number.

2. Find the best rational approximations to an irrational number.

---

## 10.4 Periodic Continued Fractions

We call the infinite simple continued fraction $[a_0;a_1,a_2,...]$ *periodic* if there are positive integers $N$ and $k$ such that $a_n = a_{n+k}$ for all positive integers $n$ with $n \geqslant N$. We use the notation

$$[a_0;a_1,a_2,...,a_{N-1},\overline{a_N,a_{N+1},...,a_{N+k-1}}]$$

to express the periodic infinite simple continued fraction

$$[a_0;a_1,a_2,...,a_{N-1},a_N,a_{N+1},...,a_{N+k-1},a_N,a_{N+1},...].$$

For instance, $[1;2,\overline{3,4}]$ denotes the infinite simple continued fraction $[1;2,3,4,3,4,3,4,...]$.

In Section 10.1 we showed that the base $b$ expansion of a number is periodic if and only if the number is rational. To characterize those irrational numbers with periodic infinite simple continued fractions, we need the following definition.

**Definition.** The real number $\alpha$ is said to be a *quadratic irrational* if $\alpha$ is irrational and if $\alpha$ is a root of a quadratic polynomial with integer coefficients, *i.e.*

$$A\alpha^2 + B\alpha + C = 0,$$

where $A, B,$ and $C$ are integers and $A \neq 0$.

**Example 10.12.** Let $\alpha = 2 + \sqrt{3}$. Then $\alpha$ is irrational, for if $\alpha$ were rational, then by Theorem 10.1, $\alpha - 2 = \sqrt{3}$ would be rational, contradicting Theorem 10.2. Next, note that

$$\alpha^2 - 4\alpha + 1 = (7 + 4\sqrt{3}) - 4(2 + \sqrt{3}) + 1 = 0.$$

Hence $\alpha$ is a quadratic irrational.                                             $\square$

We will show that the infinite simple continued fraction of an irrational number is periodic if and only if this number is a quadratic irrational. Before we do this, we first develop some useful results about quadratic irrationals.

**Lemma 10.1.** The real number $\alpha$ is a quadratic irrational if and only if there are integers $a, b,$ and $c$ with $b > 0$ and $c \neq 0$, such that $b$ is not a perfect square and

$$\alpha = (a + \sqrt{b})/c.$$

*Proof.* If $\alpha$ is a quadratic irrational, then $\alpha$ is irrational, and there are integers $A, B,$ and $C$ such that $A\alpha^2 + B\alpha + C = 0$. From the quadratic formula, we know that

$$\alpha = \frac{-B \pm \sqrt{B^2 - 4AC}}{2A}.$$

Since $\alpha$ is a real number, we have $B^2 - 4AC > 0$, and since $\alpha$ is irrational, $B^2 - 4AC$ is not a perfect square and $A \neq 0$. By either taking $a = -B$, $b = B^2 - 4AC$, $c = 2A$ or $a = B$, $b = B^2 - 4AC$, $c = -2A$, we have our desired representation of $\alpha$.

Conversely, if

$$\alpha = (a + \sqrt{b})/c,$$

where $a, b,$ and $c$ are integers with $b > 0$, $c \neq 0$, and $b$ not a perfect square, then by Theorems 10.1 and 10.2, we can easily see that $\alpha$ is irrational. Furthermore, we note that

$$c^2\alpha^2 - 2ac\alpha + (a^2 - b) = 0,$$

so that $\alpha$ is a quadratic irrational.                                             ■

The following lemma will be used when we show that periodic simple continued fractions represent quadratic irrationals.

**Lemma 10.2.** If $\alpha$ is a quadratic irrational and if $r,s,t,$ and $u$ are integers, then $(r\alpha+s)/(t\alpha+u)$ is either rational or a quadratic irrational.

*Proof.* From Lemma 10.1, there are integers $a,b,$ and $c$ with $b > 0$, $c \neq 0$, and $b$ not a perfect square such that

$$\alpha = (a+\sqrt{b})/c.$$

Thus

$$\frac{r\alpha+s}{t\alpha+u} = \left[\frac{r(a+\sqrt{b})}{c} + s\right] \Big/ \left[\frac{t(a+\sqrt{b})}{c} + u\right]$$

$$= \frac{(ar+cs)+r\sqrt{b}}{(at+cu)+t\sqrt{b}}$$

$$= \frac{[(ar+cs)+r\sqrt{b}][(at+cu)-t\sqrt{b}]}{[(at+cu)+t\sqrt{b}][(at+cu)-t\sqrt{b}]}$$

$$= \frac{[(ar+cs)(at+cu)-rtb]+[r(at+cu)-t(ar+cs)]\sqrt{b}}{(at+cu)^2-t^2b}.$$

Hence, by Lemma 10.1 $(r\alpha+s)/(t\alpha+u)$ is a quadratic irrational, unless the coefficient of $\sqrt{b}$ is zero, which would imply that this number is rational. ∎

In our subsequent discussions of simple continued fractions of quadratic irrationals we will use the notion of the conjugate of a quadratic irrational.

**Definition.** Let $\alpha = (a+\sqrt{b})/c$ be a quadratic irrational. Then the *conjugate* of $\alpha$, denoted by $\alpha'$, is defined by $\alpha' = (a-\sqrt{b})/c$.

**Lemma 10.3.** If the quadratic irrational $\alpha$ is a root of the polynomial $Ax^2 + Bx + C = 0$, then the other root of this polynomial is $\alpha'$, the conjugate of $\alpha$.

*Proof.* From the quadratic formula, we see that the two roots of $Ax^2 + Bx + C = 0$ are

$$\frac{-B \pm \sqrt{B^2-4AC}}{2A}.$$

If $\alpha$ is one of these roots, then $\alpha'$ is the other root, because the sign of $\sqrt{B^2-4AC}$ is reversed to obtain $\alpha'$ from $\alpha$. ∎

The following lemma tells us how to find the conjugates of arithmetic expressions involving quadratic irrationals.

**Lemma 10.4.** If $\alpha_1 = (a_1+b_1\sqrt{d})/c_1$ and $\alpha_2 = (a_2+b_2\sqrt{d})/c_2$ are rational or quadratic irrationals, then

(i)     $(\alpha_1+\alpha_2)' = \alpha_1' + \alpha_2'$

(ii)    $(\alpha_1 - \alpha_2)' = \alpha_1' - \alpha_2'$

(iii)    $(\alpha_1 \alpha_2)' = \alpha_1' \alpha_2'$

(iv)    $(\alpha_1 / \alpha_2)' = \alpha_1' / \alpha_2'.$

The proof of (iv) will be given here; the proofs of the other parts are easier. These appear at the end of this section as problems for the reader.

*Proof of (iv).* Note that

$$\alpha_1 / \alpha_2 = \frac{(a_1 + b_1 \sqrt{d}) / c_1}{(a_2 + b_2 \sqrt{d}) / c_2}$$

$$= \frac{c_2(a_1 + b_1 \sqrt{d})(a_2 - b_2 \sqrt{d})}{c_1(a_2 + b_2 \sqrt{d})(a_2 - b_2 \sqrt{d})}$$

$$= \frac{(c_2 a_1 a_2 - c_2 b_1 b_2 d) + (c_2 a_2 b_1 - c_2 a_1 b_2) \sqrt{d}}{c_1(a_2^2 - b_2^2 d)}.$$

While

$$\alpha_1' / \alpha_2' = \frac{(a_1 - b_1 \sqrt{d}) / c_2}{(a_2 - b_2 \sqrt{d}) / c_2}$$

$$= \frac{c_2(a_1 - b_1 \sqrt{d})(a_2 + b_2 \sqrt{d})}{c_1(a_2 - b_2 \sqrt{d})(a_2 + b_2 \sqrt{d})}$$

$$= \frac{(c_2 a_1 a_2 - c_2 b_1 b_2 d) - (c_2 a_2 b_1 - c_2 a_1 b_2) \sqrt{d}}{c_1(a_2^2 - b_2^2 d)}.$$

Hence $(\alpha_1 / \alpha_2)' = \alpha_1' / \alpha_2'.$ □

The fundamental result about periodic simple continued fractions is Lagrange's Theorem. (Note that this theorem is different than Lagrange's theorem on polynomial congrunces discussed in Chapter 8. In this chapter we do not refer to that result.)

**Theorem 10.17. Lagrange's Theorem.** The infinite simple continued fraction of an irrational number is periodic if and only if this number is a quadratic irrational.

We first prove that a periodic continued fraction represents a quadratic irrational. The converse, that the simple continued fraction of a quadratic irrational is periodic, will be proved after a special algorithm for obtaining the continued fraction of a quadratic irrational is developed.

*Proof.* Let the simple continued fraction of $\alpha$ be periodic, so that

$$\alpha = [a_0; a_1, a_2, ..., a_{N-1}, \overline{a_N, a_{N+1}, ..., a_{N+k}}]$$

Now let

$$\beta = \overline{[a_N;a_{N+1},...,a_{N+k}]}.$$

Then

$$\beta = [a_N;a_{N+1},...,a_{N+k},\beta],$$

and by Theorem 10.5 it follows that

(10.13)
$$\beta = \frac{\beta p_k + p_{k-1}}{\beta q_k + q_{k-1}},$$

where $p_k/q_k$ and $p_{k-1}/q_{k-1}$ are convergents of $[a_N;a_{N+1},...,a_{N+k}]$. Since the simple continued fraction of $\beta$ is infinite, $\beta$ is irrational, and by (10.13) we have

$$q_k\beta^2 + (q_{k-1}-p_k)\beta - p_{k-1} = 0,$$

so that $\beta$ is a quadratic irrational. Now note that

$$\alpha = [a_0;a_1,a_2,...,a_{N-1},\beta],$$

so that from Theorem 10.9 we have

$$\alpha = \frac{\beta p_{N-1} + p_{N-2}}{\beta q_{N-1} + q_{N-2}},$$

where $p_{N-1}/q_{N-1}$ and $p_{N-2}/q_{N-2}$ are convergents of $[a_0;a_1,a_2,...,a_{N-1}]$. Since $\beta$ is a quadratic irrational, Lemma 10.2 tells us that $\alpha$ is also a quadratic irrational (we know that $\alpha$ is irrational because it has an infinite simple continued fraction expansion). ∎

To develop an algorithm for finding the simple continued fraction of a quadratic irrational, we need the following lemma.

**Lemma 10.5.** If $\alpha$ is a quadratic irrational, then $\alpha$ can be written as

$$\alpha = (P+\sqrt{d})/Q,$$

where $P,Q$, and $d$ are integers, $Q \neq 0$, $d > 0$, $d$ is not a perfect square, and $Q \mid (d-P^2)$.

*Proof.* Since $\alpha$ is a quadratic irrational, Lemma 10.1 tells us that

$$\alpha = (a+\sqrt{b})/c,$$

where $a,b$, and $c$ are integers, $b > 0$, and $c \neq 0$. We multiply both the numerator and denominator of this expression for $\alpha$ by $|c|$ to obtain

$$\alpha = \frac{a|c|+\sqrt{bc^2}}{c|c|},$$

(where we have used the fact that $|c| = \sqrt{c^2}$). Now let $P = a|c|$, $Q = c|c|$, and $d = bc^2$. Then $P,Q$, and $d$ are integers, $Q \neq 0$ since $c \neq 0$, $d > 0$ (since

$b > 0$), $d$ is not a perfect square since $b$ is not a perfect square, and finally $Q|(d-P^2)$ since $d-P^2 = bc^2-a^2c^2 = c^2(b-a^2) = \pm Q(b-a^2)$.                    ∎

We now present an algorithm for finding the sample continued fractions of quadratic irrationals.

**Theorem 10.18.** Let $\alpha$ be a quadratic irrational, so that by Lemma 10.5 there are integers $P_0, Q_0$, and $d$ such that

$$\alpha = (P_0+\sqrt{d})/Q_0,$$

where $Q_0 \neq 0, d > 0$, $d$ is not a perfect square, and $Q_0 | (d-P_0^2)$. Recursively define

$$\alpha_k = (P_k + \sqrt{d})/Q_k,$$
$$a_k = [\alpha_k],$$
$$P_{k+1} = a_k Q_k - P_k,$$
$$Q_{k+1} = (d-P_{k+1}^2)/Q_k,$$

for $k = 0,1,2,....$ Then $\alpha = [a_0;a_1,a_2,...]$.

*Proof.* Using mathematical induction, we will show that $P_k$ and $Q_k$ are integers with $Q_k \neq 0$ and $Q_k | (d-P_k^2)$, for $k = 0,1,2,....$ First, note that this assertion is true for $k = 0$ from the hypotheses of the theorem. Now assume that $P_k$ and $Q_k$ are integers with $Q_k \neq 0$ and $Q_k | (d-P_k^2)$. Then

$$P_{k+1} = a_k Q_k - P_k$$

is also an integer. Further,

$$\begin{aligned} Q_{k+1} &= (d-P_{k+1}^2)/Q_k \\ &= [d-(a_k Q_k-P_k)^2]/Q_k \\ &= (d-P_k^2)/Q_k + (2a_k P_k-a_k^2 Q_k). \end{aligned}$$

Since $Q_k | (d-P_k^2)$, by the induction hypothesis, we see that $Q_{k+1}$ is an integer, and since $d$ is not a perfect square, we see that $d \neq P_k^2$, so that $Q_{k+1} = (d-P_{k+1}^2)/Q_k \neq 0$. Since

$$Q_k = (d-P_{k+1}^2)/Q_{k+1},$$

we can conclude that $Q_{k+1} | (d-P_{k+1}^2)$. This finishes the inductive argument.

To demonstrate that the integers $a_0,a_1,a_2,...$ are the partial quotients of the simple continued fraction of $\alpha$, we use Theorem 10.13. If we can show that

$$\alpha_{k+1} = 1/(\alpha_k-a_k),$$

for $k = 0,1,2,...$, then we know that $\alpha = [a_0;a_1,a_2,...]$. Note that

$$\alpha_k - a_k = \frac{P_k + \sqrt{d}}{Q_k} - a_k$$

$$= [\sqrt{d} - (a_k Q_k - P_k)]/Q_k$$

$$= (\sqrt{d} - P_{k+1})/Q_k$$

$$= (\sqrt{d} - P_{k+1})(\sqrt{d} + P_{k+1})/Q_k(\sqrt{d} + P_{k+1})$$

$$= (d - P_{k+1}^2)/Q_k(\sqrt{d} + P_{k+1})$$

$$= Q_k Q_{k+1}/Q_k(\sqrt{d} + P_{k+1})$$

$$= Q_{k+1}/(\sqrt{d} + P_{k+1})$$

$$= 1/\alpha_{k+1},$$

where we have used the defining relation for $Q_{k+1}$ to replace $d - P_{k+1}^2$ with $Q_k Q_{k+1}$. Hence, we can conclude that $\alpha = [a_0; a_1, a_2, \ldots]$. ∎

We illustrate the use of the algorithm given in Theorem 10.18 with the following example.

**Example 10.13.** Let $\alpha = (3+\sqrt{7})/2$. Using Lemma 10.5, we write

$$\alpha = (6+\sqrt{28})/4$$

where we set $P_0 = 6$, $Q_0 = 4$, and $d = 28$. Hence $a_0 = [\alpha] = 2$, and

| | | | | | |
|---|---|---|---|---|---|
| $P_1$ | = | $2 \cdot 4 - 6 = 2,$ | $\alpha_1$ | = | $(2+\sqrt{28})/6,$ |
| $Q_1$ | = | $(28-2^2)/4 = 6,$ | $a_1$ | = | $[(2+\sqrt{28})/6] = 1,$ |
| | | | | | |
| $P_2$ | = | $1 \cdot 6 - 2 = 4,$ | $\alpha_2$ | = | $(4+\sqrt{28})/2,$ |
| $Q_2$ | = | $(28-4^2)/6 = 2,$ | $a_2$ | = | $[(4+\sqrt{28})/2] = 4,$ |
| | | | | | |
| $P_3$ | = | $4 \cdot 2 - 4 = 4,$ | $\alpha_3$ | = | $(4+\sqrt{28})/6,$ |
| $Q_3$ | = | $(28-4^2)/2 = 6$ | $a_3$ | = | $[(4+\sqrt{28})/6] = 1,$ |
| | | | | | |
| $P_4$ | = | $1 \cdot 6 - 4 = 2,$ | $\alpha_4$ | = | $(2+\sqrt{28})/4,$ |
| $Q_4$ | = | $(28-2^2)/6 = 4,$ | $a_4$ | = | $[(2+\sqrt{28})/4] = 1,$ |
| | | | | | |
| $P_5$ | = | $1 \cdot 4 - 2 = 2,$ | $\alpha_5$ | = | $(2+\sqrt{28})/6,$ |
| $Q_5$ | = | $(28-2^2)/4 = 6,$ | $a_5$ | = | $[(2+\sqrt{28})/6] = 1,$ |

and so, with repetition, since $P_1 = P_5$ and $Q_1 = Q_5$. Hence, we see that

$$(3+\sqrt{7})/2 = [2;\underline{1,4,1,1,1,4,1,1},...]$$
$$= [2;\overline{1,4,1,1}]. \qquad \square$$

We now finish the proof of Lagrange's Theorem by showing that the simple continued fraction expansion of a quadratic irrational is periodic.

*Proof (continued).* Let $\alpha$ be a quadratic irrational, so that by Lemma 10.5 we can write $\alpha$ as

$$\alpha = (P_0 + \sqrt{d})/Q_0.$$

Furthermore, by Theorem 10.18 we have $\alpha = [a_0;a_1,a_2,...]$ where

$$\alpha_k = (P_k + \sqrt{d})/Q_k,$$
$$a_k = [\alpha_k],$$
$$P_{k+1} = a_kQ_k - P_k,$$
$$Q_{k+1} = (d-P_{k+1}^2)/Q_k,$$

for $k = 0,1,2,...$.

Since $\alpha = [a_0;a_1,a_2,...,\alpha_k]$, Theorem 10.9 tells us that

$$\alpha = (p_{k-1}\alpha_k + p_{k-2})/(q_{k-1}\alpha_k + q_{k-2}).$$

Taking conjugates of both sides of this equation, and using Lemma 10.4, we see that

(10.14) $$\alpha' = (p_{k-1}\alpha_k' + p_{k-2})/(q_{k-1}\alpha_k' + q_{k-2}).$$

When we solve (10.14) for $\alpha_k'$, we find that

$$\alpha_k' = \frac{-q_{k-2}}{q_{k-1}}\left(\frac{\alpha' - \dfrac{p_{k-2}}{q_{k-2}}}{\alpha' - \dfrac{p_{k-1}}{q_{k-1}}}\right).$$

Note that the convergents $p_{k-2}/q_{k-2}$ and $p_{k-1}/q_{k-1}$ tend to $\alpha$ as $k$ tends to infinity, so that

$$\left(\alpha' - \frac{p_{k-2}}{q_{k-2}}\right)\bigg/\left(\alpha' - \frac{p_{k-1}}{q_{k-1}}\right)$$

tends to 1. Hence, there is an integer $N$ such that $\alpha_k' < 0$ for $k \geq N$. Since $\alpha_k > 0$ for $k \geq 1$, we have

$$\alpha_k - \alpha_k' = \frac{P_k + \sqrt{d}}{Q_k} - \frac{P_k - \sqrt{d}}{Q_k} = \frac{2\sqrt{d}}{Q_k} > 0,$$

so that $Q_k > 0$ for $k \geq N$.

Since $Q_kQ_{k+1} = d - P_{k+1}^2$, we see that for $k \geq N$,

$$Q_k \leqslant Q_k Q_{k+1} = d - P_{k+1}^2 \leqslant d.$$

Also for $k \geqslant N$, we have

$$P_{k+1}^2 \leqslant d = P_{k+1}^2 - Q_k Q_{k+1},$$

so that

$$-\sqrt{d} < P_{k+1} < \sqrt{d}.$$

From the inequalities $0 \leqslant Q_k \leqslant d$ and $-\sqrt{d} < P_{k+1} < \sqrt{d}$, that hold for $k \geqslant N$, we see that there are only a finite number of possible values for the pair of integers $P_k, Q_k$ for $k > N$. Since there are infinitely many integers $k$ with $k \geqslant N$, there are two integers $i$ and $j$ such that $P_i = P_j$ and $Q_i = Q_j$ with $i < j$. Hence, from the defining relation for $\alpha_k$, we see that $\alpha_i = \alpha_j$. Consequently, we can see that $a_i = a_j, a_{i+1} = a_{j+1}, a_{i+2} = a_{j+2}, \dots$. Hence

$$\alpha = [a_0; a_1, a_2, \dots, a_{i-1}, a_i, a_{i+1}, \dots, a_{j-1}, a_i, a_{i+1}, \dots, a_{j-1}, \dots]$$

$$= [a_0; a_1, a_2, \dots, a_{i-1}, \overline{a_j, a_{i+1}, \dots, a_{j-1}}].$$

This shows that $\alpha$ has a periodic simple continued fraction. ∎

Next, we investigate those periodic simple continued fractions that are *purely periodic, i.e.* those without a pre-period.

**Definition.** The continued fraction $[a_0; a_1, a_2, \dots]$ is called *purely periodic* if there is an integer $n$ such that $a_k = a_{n+k}$, for $k = 0, 1, 2, \dots$, so that

$$[a_0; a_1, a_2, \dots] = [\overline{a_0; a_1, a_2, a_3, \dots, a_{n-1}}].$$

**Example 10.14.** The continued fraction $[\overline{2;3}] = (1+\sqrt{3})/2$ is purely periodic while $[2; \overline{2,4}] = \sqrt{6}$ is not. □

The next definition and theorem describe those quadratic irrationals with purely periodic simple continued fractions.

**Definition.** A quadratic irrational $\alpha$ if called *reduced* if $\alpha > 1$ and $-1 < \alpha' < 0$, where $\alpha'$ is the conjugate of $\alpha$.

**Theorem 10.19.** The simple continued fraction of the quadratic irrational $\alpha$ is purely periodic if and only if $\alpha$ is reduced. Further, if $\alpha$ is reduced and $\alpha = [\overline{a_0; a_1, a_2, \dots, a_n}]$ then the continued fraction of $-1/\alpha'$ is $[\overline{a_n; a_{n-1}, \dots, a_0}]$.

*Proof.* First, assume that $\alpha$ is a reduced quadratic irrational. Recall from Theorem 10.15 that the partial fractions of the simple continued fraction of $\alpha$ are given by

$$a_k = [\alpha_k], \quad \alpha_{k+1} = 1/(\alpha_k - a_k),$$

for $k = 0, 1, 2, \dots$, where $\alpha_0 = \alpha$. We see that

$$1/\alpha_{k+1} = \alpha_k - a_k,$$

and taking conjugates, using Lemma 10.4, we see that

(10.15) $$1/\alpha'_{k+1} = \alpha'_k - a_k.$$

We can prove, by mathematical induction, that $-1 < \alpha'_k < 0$ for $k = 0,1,2,\ldots$. First, note that since $\alpha_0 = \alpha$ is reduced, $-1 < \alpha'_0 < 0$. Now assume that $-1 < \alpha'_k < 0$. Then, since $a_k \geq 1$ for $k = 0,1,2,\ldots$ (note that $a_0 \geq 1$ since $\alpha > 1$), we see from (10.15) that

$$1/\alpha'_{k+1} < -1,$$

so that $-1 < \alpha'_{k+1} < 0$. Hence, $-1 < \alpha'_k < 0$ for $k = 0,1,2,\ldots$ .

Next, note that from (10.15) we have

$$\alpha'_k = a_k + 1/\alpha'_{k+1},$$

and since $-1 < \alpha'_k < 0$, it follows that

$$-1 < a_k + 1/\alpha'_{k+1} < 0.$$

Consequently,

$$-1 - 1/\alpha'_{k+1} < a_k < -1/\alpha'_{k+1},$$

so that

$$a_k = [-1/\alpha'_{k+1}].$$

Since $\alpha$ is a quadratic irrational, the proof of Lagrange's Theorem shows that there are nonnegative integers $i$ and $j$, $i < j$, such that $\alpha_i = \alpha_j$, and hence with $-1/\alpha'_i = -1/\alpha'_j$. Since $a_{i-1} = [-1/\alpha'_i]$ and $a_{j-1} = [-1/\alpha'_j]$, we see that $a_{i-1} = a_{j-1}$. Furthermore, since $\alpha_{i-1} = a_{i-1} + 1/\alpha_i$ and , $\alpha_{j-1} = a_{j-1} + 1/\alpha_j$ we also see that $\alpha_{i-1} = \alpha_{j-1}$. Continuing this argument, we see that $\alpha_{i-2} = \alpha_{j-2}, \alpha_{i-3} = \alpha_{j-3},\ldots$, and finally, that $\alpha_0 = \alpha_{j-i}$. Since

$$\alpha_0 = \alpha = [a_0;a_1,\ldots,a_{j-i-1},\alpha_{j-i}]$$
$$= [a_o;a_1,\ldots,a_{j-i-1},\alpha_0]$$
$$= [\overline{a_0;a_1,\ldots,a_{j-i-1}}],$$

we see that the simple continued fraction of $\alpha$ is purely periodic.

To prove the converse, assume that $\alpha$ is a quadratic irrational with a purely periodic continued fraction $\alpha = [\overline{a_0;a_1,a_2,\ldots,a_k}]$. Since $\alpha = [a_0;a_1,a_2,\ldots,a_k,\alpha]$, Theorem 10.9 tells that

(10.16) $$\alpha = \frac{\alpha p_k + p_{k-1}}{\alpha q_k + q_{k-1}},$$

where $p_{k-1}/q_{k-1}$ and $p_k/q_k$ are the $(k-1)$th and $k$th convergents of the continued fraction expansion of $\alpha$. From (10.16), we see that

(10.17) $$q_k \alpha^2 + (q_{k-1} - p_k)\alpha - p_{k-1} = 0.$$

Now, let $\beta$ be the quadratic irrational such that $\beta = [\overline{a_k; a_{k-1}, \dots, a_1, a_0}]$, *i.e.* with the period of the simple continued fraction for $\alpha$ reversed. Then $\beta = [a_k; a_{k-1}, \dots, a_1, a_0, \beta]$, so that by Theorem 10.9, it follows that

(10.18) $$\beta = \frac{\beta p_k' + p_{k-1}'}{\beta q_k' + q_{k-1}'},$$

where $p_{k-1}'/q_{k-1}'$ and $p_k'/q_k'$ are the $(k-1)$th and $k$th convergents of the continued fraction expansion of $\beta$. Note, however, from problem 6 of Section 10.2, that

$$p_k/p_{k-1} = [a_n; a_{n-1}, \dots, a_1, a_0] = p_k'/q_k'$$

and

$$q_k/q_{k-1} = [a_n; a_{n-1}, \dots, a_2, a_1] = p_{k-1}'/q_{k-1}'.$$

Since $p_{k-1}'/q_{k-1}'$ and $p_k'/q_k'$ are convergents, we know that they are in lowest terms. Also, $p_k/p_{k-1}$ and $q_k/q_{k-1}$ are in lowest terms, since Theorem 10.10 tells us that $p_k q_{k-1} - p_{k-1} q_k = (-1)^{k-1}$. Hence,

$$p_k' = p_k, \quad q_k' = p_{k-1}$$

and

$$p_{k-1}' = q_k, \quad q_{k-1}' = q_{k-1}.$$

Inserting these values into (10.18), we see that

$$\beta = \frac{\beta p_k + q_k}{\beta p_{k-1} + q_{k-1}}.$$

Therefore, we know that

$$p_{k-1}\beta^2 + (q_{k-1} - p_k)\beta - q_k = 0.$$

This implies that

(10.19) $$q_k(-1/\beta)^2 + (q_{k-1} - p_k)(-1/\beta) - p_{k-1} = 0.$$

By (10.17) and (10.19) we see that the two roots of the quadratic equation

$$q_k x^2 + (q_{k-1} - p_k)x - p_{k-1} = 0$$

are $\alpha$ and $-1/\beta$, so that by the quadratic equation, we have $\alpha' = -1/\beta$. Since $\beta = [a_n; a_{n-1}, \dots, a_1, a_0]$, we see that $\beta > 1$, so that $-1 < \alpha' = -1/\beta < 0$. Hence, $\alpha$ is a reduced quadratic irrational.

Furthermore, note that since $\beta = -1/\alpha'$, it follows that

$$-1/\alpha' = [\overline{a_n; a_{n-1}, \dots, a_1, a_0}]. \qquad \blacksquare$$

We now find the form of the periodic simple continued fraction of $\sqrt{D}$, where $D$ is a positive integer that is not a perfect square. Although $\sqrt{D}$ is not reduced, since its conjugate $-\sqrt{D}$ is not between $-1$ and $0$, the quadratic irrational $[\sqrt{D}] + \sqrt{D}$ is reduced, since its conjugate, $[\sqrt{D}] - \sqrt{D}$, does lie between $-1$ and $0$. Therefore, from Theorem 10.21, we know that the continued fraction of $[\sqrt{D}] + \sqrt{D}$ is purely periodic. Since the initial partial quotient of the simple continued fraction of $[\sqrt{D}] + \sqrt{D}$ is $[[\sqrt{D}] + \sqrt{D}] = 2[\sqrt{D}] = 2a_0$, where $a_0 = [\sqrt{D}]$, we can write

$$[\sqrt{D}] + \sqrt{D} = [\overline{2a_0; a_1, a_2, ..., a_n}]$$
$$= [2a_0; a_1, a_2, ..., a_n, 2a_0, a_1, ..., a_n].$$

Subtracting $a_0 = \sqrt{D}$ from both sides of this equality, we find that

$$\sqrt{D} = [a_0; a_1, a_2, ..., 2a_0, a_1, a_2, ... 2a_0, ...]$$
$$= [a_0; \overline{a_1, a_2, ..., a_n, 2a_0}].$$

To obtain even more information about the partial quotients of the continued fraction of $\sqrt{D}$, we note that from Theorem 10.21, the simple continued fraction expansion of $-1/([\sqrt{D}] - \sqrt{D})$ can be obtained from that for $[\sqrt{D}] + \sqrt{D}$, by reversing the period, so that

$$1/(\sqrt{D} - [\sqrt{D}]) = [\overline{a_n; a_{n-1}, ..., a_1, 2a_0}].$$

But also note that

$$\sqrt{D} - [\sqrt{D}] = [0; \overline{a_1, a_2, ..., a_n 2a_0}],$$

so that by taking reciprocals, we find that

$$1/(\sqrt{D} - [\sqrt{D}]) = [\overline{a_1; a_2, ..., a_n, 2a_0}].$$

Therefore, when we equate these two expressions for the simple continued fraction of $1/(\sqrt{D} - [\sqrt{D}])$, we obtain

$$a_1 = a_n, a_2 = a_{n-1}, ..., a_n = a_1,$$

so that the periodic part of the continued fraction for $\sqrt{D}$ is symmetric from the first to the penultimate term.

In conclusion, we see that the simple continued fraction of $\sqrt{D}$ has the form

$$\sqrt{D} = [a_0; \overline{a_1, a_2, ..., a_2, a_1, 2a_0}].$$

We illustrate this with some examples.

**Example 10.15.** Note that

$$\sqrt{23} = [4; \overline{1, 3, 1, 8}]$$
$$\sqrt{31} = [5; \overline{1, 1, 3, 5, 3, 1, 1, 10}]$$
$$\sqrt{46} = [6; \overline{1, 2, 1, 1, 2, 6, 2, 1, 1, 2, 1, 12}]$$
$$\sqrt{76} = [8; \overline{1, 2, 1, 1, 5, 4, 5, 1, 1, 2, 1, 16}]$$

and

$$\sqrt{97} = [9;\overline{1,5,1,1,1,1,1,1,5,1,18}],$$

where each continued fraction has a pre-period of length 1 and a period ending with twice the first partial quotient which is symmetric from the first to the next to the last term. □

The simple continued fraction expansions of $\sqrt{d}$ for positive integers $d$ such that $d$ is not a perfect square and $d < 100$ can be found in Table 5 of the Appendix.

## 10.4 Exercises

1. Find the simple continued fractions of each of the following numbers.

   a) $\sqrt{7}$    d) $\sqrt{47}$
   b) $\sqrt{11}$   e) $\sqrt{59}$
   c) $\sqrt{23}$   f) $\sqrt{94}$

2. Find the simple continued fractions of each of the following numbers.

   a) $(1+\sqrt{3})/2$

   b) $(14+\sqrt{37})/3$

   c) $(13-\sqrt{2})7$

3. Find the quadratic irrational with each of the following simple continued fraction expansions.

   a) $[2;1,\overline{5}]$

   b) $[2;\overline{1,5}]$

   c) $[\overline{2;1,5}]$

4. a) Let $d$ be a positive integer. Show that the simple continued fraction of $\sqrt{d^2+1}$ is $[d;\overline{2d}]$.

   b) Use part (a) to find the simple continued fractions of $\sqrt{101}, \sqrt{290}$, and $\sqrt{2210}$.

5. Let d be a integer, $d \geqslant 2$.

   a) Show that the simple continued fraction of $\sqrt{d^2-1}$ is $[d-1;\overline{1,2d-2}]$.

   b) Show that the simple continued fraction of $\sqrt{d^2-d}$ is $[d-1;\overline{2,2d-2}]$.

   c) Use parts (a) and (b) to find the simple continued fractions of $\sqrt{99}$, $\sqrt{110}, \sqrt{272}$, and $\sqrt{600}$.

6. a) Show that if d is an integer, $d \geqslant 3$, then the simple continued fraction of $\sqrt{d^2-2}$ is $[d-1;\overline{1,d-2,1,2d-2}]$.

b) Show that if d is a positive integer, then the simple continued fraction of $\sqrt{d^2+2}$ is $[d;\overline{d,2d}]$.

c) Find the simple continued fraction expansions of $\sqrt{47}$, $\sqrt{51}$, and $\sqrt{187}$.

7. Let d be an odd positive integer.

a) Show that the simple continued fraction of $\sqrt{d^2+4}$ is $[d;\overline{(d-1)/2,1,1,(d-1)/2,2d}]$, if $d > 1$.

b) Show that the simple continued fraction of $\sqrt{d^2-4}$ is $[d-1;\overline{1,(d-3)/2,2,(d-3)/2,1,2d-2}]$, if $d > 3$.

8. Show that the simple continued fraction of $\sqrt{d}$, where d is a positive integer, has period length one if and only if $d = a^2+1$ where $a$ is a nonnegative integer.

9. Show that the simple continued fraction of $\sqrt{d}$, where d is a positive integer, has period length two if and only if $d = a^2 + b$ where $a$ and $b$ are integers, $b > 1$, and $b|\,2a$.

10. Prove that if $\alpha_1 = (a_1+b_1\sqrt{d})/c_1$ and $\alpha_2 = (a_2+b_2\sqrt{d})/c_2$ are quadratic irrationals, then

a) $(\alpha_1+\alpha_2)' = \alpha_1' + \alpha_2'$

b) $(\alpha_1-\alpha_2)' = \alpha_1' - \alpha_2'$

c) $(\alpha_1\alpha_2)' = \alpha_1'\cdot\alpha_2'$.

11. Which of the following quadratic irrationals have purely periodic continued fractions?

a)  $1 + \sqrt{5}$        d)  $(11 - \sqrt{10})/9$
b)  $2 + \sqrt{8}$        e)  $(3 + \sqrt{23})/2$
c)  $4 + \sqrt{17}$       f)  $(17 + \sqrt{188})/3$

12. Suppose that $\alpha = (a+\sqrt{b})/c$, where $a,b$, and $c$ are integers, $b > 0$, and $b$ is not a perfect square. Show that is a reduced quadratic irrational if and only if $0 < a < \sqrt{b}$ and $\sqrt{b}-a < c < \sqrt{b} + a < 2\sqrt{b}$.

13. Show that if $\alpha$ is a reduced quadratic irrational, then $-1/\alpha'$ is also a reduced quadratic irrational.

* 14. Let $k$ be a positive integer. Show that there are infinitely many positive integers $D$, such that the simple continued fraction expansion of $\sqrt{D}$ has a period of length $k$. (*Hint*: Let $a_1 = 2$, $a_2 = 5$, and for $k \geqslant 3$ let $a_k = 2a_{k-1} + a_{k-2}$. Show that if $D = (ta_k + 1)^2 + 2a_{k-1} + 1$, where $t$ is a nonnegative integer, then $\sqrt{D}$ has a period of length $k + 1$.)

* 15. Let $k$ be a positive integer. Let $D_k = (3^k+1)^2 + 3$. Show that the simple continued fraction of $\sqrt{D_k}$ has a period of length $6k$.

## 10.4 Computer Projects

Write computer programs to do the following:

* **1.** Find the quadratic irrational that is the value of a periodic simple continued fraction.

* **2.** Find the periodic simple continued fraction expansion of a quadratic irrational.

---

## 10.5 Factoring Using Continued Fractions

We can factor the positive integer $n$ if we can find positive integers $x$ and $y$ such that $x^2 - y^2 = n$ and $x - y \neq 1$. This is basis of the Fermat factorization method discussed in Section 2.4. However, it is possible to factor $n$ if we can find positive integers $x$ and $y$ that satisfy the weaker condition

(10.20) $\qquad x^2 \equiv y^2 \pmod{n}, \; 0 < x < y < n, \text{ and } x + y = n.$

To see this, note that if (10.20) holds, then $n$ divides $x^2 - y^2 = (x + y)(x - y)$, and $n$ divides neither $x - y$ nor $x + y$. It follows that $(n, x - y)$ and $(n, x + y)$ are divisors of $n$ that do not equal 1 or $n$. We can find these divisors rapidly using the Euclidean algorithm.

**Example 10.16.** Note that $29^2 - 17^2 = 841 - 289 = 552 \equiv 0 \pmod{69}$. Since $29^2 - 17^2 = (19 - 17)(19 + 17) \equiv 0 \pmod{69}$, both $(19-7,69) = (12,69)$ and $(19+7,69) = (26,69)$ are divisors of 69 not equal to either 1 or 69; using the Euclidean algorithm we find that these factors are $(12,69) = 3$ and $(26,69) = 13$. ☐

The continued fraction expansion of $\sqrt{n}$ can be used to find solutions of the congruence $x^2 \equiv y^2 \pmod{n}$. The following theorem is the basis for this.

**Theorem 10.20.** Let $n$ be a positive integer that is not a perfect square. Define $\alpha_k = (P_k + \sqrt{n})/Q_k$, $a_k = [\alpha_k]$, $P_{k+1} = a_k Q_k - P_k$, and $Q_{k+1} = (n - P_{k+1}^2)/Q_k$, for $k = 0,1,2,\ldots$ where $\alpha_0 = \sqrt{n}$. Furthermore, let $p_k/q_k$ denote the $k$th convergent of the simple continued fraction expansion of $\sqrt{n}$. Then

$$p_k^2 - nq_k^2 = (-1)^{k-1} Q_{k+1}.$$

The proof of Theorem 10.20 depends on the following useful lemma.

**Lemma 10.5.** Let $r + s\sqrt{n} = t + u\sqrt{n}$ where $r$, $s$, $t$, and $u$ are rational numbers and $n$ is a positive integer that is not a perfect square. Then $r = t$ and $s = u$.

*Proof.* Since $r + s\sqrt{n} = t + u\sqrt{n}$, we see that if $s \neq u$ then

$$\sqrt{n} = \frac{r-t}{u-s}.$$

Since $(r-t)/(u-s)$ is rational and $\sqrt{n}$ is irrational it follows that $s = u$, and consequently $r = t$. ∎

We can now prove Theorem 10.20.

*Proof.* Since $\sqrt{n} = \alpha_0 = [a_0; a_1, a_2, \ldots, a_k, \alpha_{k+1}]$, Theorem 10.7 tells us that

$$\sqrt{n} = \frac{\alpha_{k+1} p_k + p_{k-1}}{\alpha_{k+1} q_k + q_{k-1}}.$$

Since $\alpha_{k+1} = (P_{k+1} + \sqrt{n})/Q_{k+1}$ we have

$$\sqrt{n} = \frac{(P_{k+1} + \sqrt{n})p_k + Q_{k+1} p_{k-1}}{(P_{k+1} + \sqrt{n})q_k + Q_{k+1} q_{k-1}}.$$

Therefore, we see that

$$nq_k + (P_{k+1}q_k + Q_{k+1}q_{k-1})\sqrt{n} = (P_{k+1}p_k + Q_{k+1}p_{k-1}) + p_k\sqrt{n}.$$

By Lemma 10.5 we see that $nq_k = P_{k+1}p_k + Q_{k+1}p_{k-1}$ and $P_{k+1}q_k + Q_{k+1}q_{k-1} = p_k$. When we multiply the first of these two equations by $q_k$ and the second by $p_k$, subtract the first from the second, and then simplify, we obtain

$$p_k^2 - nq_k^2 = (p_k q_{k-1} - p_{k-1}q_k)Q_{k+1} = (-1)^{k-1}Q_{k+1},$$

where we have used Theorem 10.8 to complete the proof. ∎

We now outline the technique known as the *continued fraction algorithm* for factoring an integer $n$. Suppose that the terms $p_k$, $q_k$, $P_k$, $Q_k$, $a_k$, and $\alpha_k$ have their usual meanings in the computation of the continued fraction expansion of $\sqrt{n}$. By Theorem 10.20 it follows that for every nonnegative integer $k$

$$p_k^2 \equiv (-1)^{k-1}Q_{k+1} \pmod{n}$$

where $p_k$ and $Q_{k+1}$ are as defined in the statement of the theorem. Now suppose that $k$ is odd and that $Q_{k+1}$ is a square, that is, $Q_{k+1} = s^2$ where $s$ is a positive integer. Then $p_k^2 \equiv s^2 \pmod{n}$ and we may be able to use this congruence of two squares modulo $n$ to find factors of $n$. Summarizing, to factor $n$ we carry out the algorithm described in Theorem 10.8 to find the continued fraction expansion of $\sqrt{n}$. We look for squares among the terms with even index in the sequence $\{Q_k\}$. Each such occurrence may lead to a nonproper factor of $n$ (or may just lead to the factorization $n = 1 \cdot n$). We illustrate this technique with several examples.

**Example 10.17.** We can factor 1037 using the continued fraction algorithm. Take $\alpha = \sqrt{1037} = (0 + \sqrt{1037})/1$ with $P_0 = 0$ and $Q_1 = 1$ and generate the terms $P_k$, $Q_k$, $\alpha_k$, and $a_k$. We look for squares among the terms with even index in the sequence $\{Q_k\}$. We find that $Q_1 = 13$ and $Q_2 = 49$. Since $49 = 7^2$ is a square, and the index of $Q_2$ is even, we examine the congruence $p_1^2 \equiv (-1)^2 Q_2 \pmod{1037}$. Computing the terms of the sequence $\{p_k\}$, we find that $p_1 \equiv 129$. This gives the congruence $129^2 \equiv 49 \pmod{1037}$. Hence $129^2 - 7^2 = (129 - 7)(129 + 7) \equiv 0 \pmod{1037}$. This produces the factors $(129 - 7, 1037) = (122, 1037) = 61$ and $(129 + 7, 1037) = (136, 1037) = 17$ of 1037. □

**Example 10.18.** We can use the continued fraction algorithm to find factors of 1000009. (We follow computations of Riesel [54].) We have $Q_1 = 9$, $Q_2 = 445$, $Q_3 = 873$, and $Q_4 = 81$. Since $81 = 9^2$ is square, we examine the congruence $p_3^2 \equiv (-1)^4 Q_4 \pmod{1000009}$. However, $p_3 = 2000009 \equiv -9 \pmod{1000009}$, so that $p_3 + 9$ is divisible by 1000009. It follows that we do not get any propers factors of 1000009 from this.

We continue until we reach another square in the sequence $\{Q_k\}$ with $k$ even. This happens when $k = 18$ with $Q_{18} = 16$. Calcuating $p_{17}$ gives $p_{17} = 494881$. From the congruence $p_{17}^2 \equiv (-1)^{18} Q_{18} \pmod{1000009}$ we have $494881^2 \equiv 2^2 \pmod{1000009}$. It follows that $(494881 - 4, 1000009) = (494887, 1000009) = 293$ and $(494881 + 4, 1000009) = (494885, 1000009) = 3413$ are factors of 1000009. □

More powerful techniques based on continued fraction expansions are known. These are described in Dixon [92], Guy [95], Riesel [54] and Wagstaff and Smith [111]. We describe one such generalization in the exercises.

## 10.5 Exercises

1. Find factors of 119 using the congruence $19^2 \equiv 2^2 \pmod{119}$.

2. Factor 1537 using the continued fraction algorithm.

3. Factor the integer 13290059 using the continued fraction algorithm. (*Hint*: Use a computer program to generate the integers $Q_k$ for the continued fraction for $\sqrt{13290059}$. You will need more than 50 terms.)

4. Let $n$ be a positive integer and let $p_1$, $p_2$, ..., and $p_m$ be primes. Suppose that there exist integers $x_1$, $x_2$,...$x_r$ such that

$$x_1^2 \equiv (-1)^{e_{01}} p_1^{e_{11}} \cdots p_m^{e_{m1}} \pmod{n},$$
$$x_2^2 \equiv (-1)^{e_{02}} p_1^{e_{12}} \cdots p_m^{e_{m2}} \pmod{n},$$

.

.

.

$$x_r^2 \equiv (-1)^{e_{0r}} p_1^{e_{1r}} \cdots p_m^{e_{mr}} \pmod{n},$$

where

$$e_{01} + e_{02} + \cdots + e_{0r} = 2e_0$$
$$e_{11} + e_{12} + \cdots + e_{1r} = 2e_1$$

.

.

.

$$e_{m1} + e_{m2} + \cdots + e_{mr} = 2e_r.$$

Show that $x^2 \equiv y^2 \pmod{n}$ where $x = x_1 x_2 \cdots x_r$ and $y = (-1)^{e_0} p_1^{e_1} \cdots p_r^{e_r}$. Explain how to factor $n$ using this information. Here the primes $p_1, \ldots, p_r$, together with -1, are called the *factor base*.

5. Show that 143 can be factored by setting $x_1 = 17$ and $x_2 = 19$, taking the factor base to be $\{3, 5\}$.

6. Let $n$ be a positive integer and let $p_1, p_2, \ldots, p_r$ be primes. Suppose that $Q_{k_i} = \prod_{j=1}^{m} p_i^{k_{ij}}$ for $i = 1, \ldots, t$ where the integers $Q_j$ have their usual meaning with respect to the continued fraction of $\sqrt{n}$. Explain how $n$ can be factored if $\sum_{i=1}^{t} k_i$ is even and $\sum_{i=1}^{t} k_{ij}$ is even for $j = 1, 2, \ldots, r$.

7. Show that 12007001 can be factoring using the continued fraction expansion of $\sqrt{12007001}$ with factor base -1,2,31,71, and 97. (*Hint*: Use the factorizations $Q_1 = 2^3 \cdot 97$, $Q_{12} = 2^4 \cdot 71$, $Q_{28} = 2^{11}$, $Q_{34} = 31 \cdot 97$, and $Q_{41} = 31 \cdot 71$, and show that $p_0 p_{11} p_{27} p_{33} p_{40} = 9815310$.)

8. Factor 197209 using the continued fraction expansion of $\sqrt{197209}$ and factor base 2,3, and 5.

## 10.5  Computer Projects

Write programs to do the following things.

* 1. Factor positive integers using the continued fraction algorithm.

** 2. Factor positive integers using factor bases and continued fraction expansions (see the Exercise 6).

# 11

# Some Nonlinear Diophantine Equations

## 11.1 Pythagorean Triples

The Pythagorean theorem tells us that the sum of the squares of the lengths of the legs of a right triangle equals the square of the length of the hypotheneuse. Conversely, any triangle for which the sum of the squares of the lengths of the two shortest sides equals the square of the third side is a right triangle. Consequently, to find all right triangles with integral side lengths, we need to find all triples of positive integers $x, y, z$ satisfying the diophantine equation

$$(11.1) \qquad\qquad x^2 + y^2 = z^2.$$

Triples of positive integers satisfying this equation are called *Pythagorean triples* after the ancient Greek mathematician Pythagoras*.

---

* **PYTHAGORAS (c. 572 - c. 500 B.C.E.)** was born on the Greek island of Samos. After extensive travels and studies, Pythagoras founded his famous school at the Greek port of Crotona in what is now southern Italy. Besides being an academy devoted to the study of mathematics, philosophy, and science, the school was the site of a brotherhood sharing secret rites. The Pythagoreans, as the members of this brotherhood were called, published nothing and ascribed all their discoveries to Pythagoras himself. However, it is believed that Pythagoras himself discoveried what is now called the Pythagorean Theorem, namely that $a^2 + b^2 = c^2$ where $a$, $b$, and $c$ are the lengths of the two legs and of the hypotenuse of a right triangle, respectively. The Pythagoreans believed that the key to understanding the world lay with natural numbers and form. Their central tenant was "Everything is Number." Because of their fascination with the natural numbers, the Pythagoreans made many discoveries in number theory. In particular they studied perfect numbers and amicable numbers for the mystical properties they felt these numbers possessed.

**Example 11.1.** The triples 3,4,5; 6,8,10; and 5,12,13 are Pythagorean triples because $3^2 + 4^2 = 5^2$, $6^2 + 8^2 = 10^2$, and $5^2 + 12^2 = 13^2$.                    □

Unlike most nonlinear diophantine equations, it is possible to explicitly describe all the integral solutions of (11.1). Before developing the result describing all Pythagorean triples, we need a definition.

**Definition.** A Pythagorean triple $x,y,z$ is called *primitive* if $(x,y,z) = 1$.

**Example 11.2.** The Pythagorean triples 3,4,5 and 5,12,13 are primitive, whereas the Pythagorean triple 6,8,10 is not.                    □

Let $x,y,z$ be a Pythagorean triple with $(x,y,z) = d$. Then there are integers $x_1,y_1,z_1$ with $x = dx_1, y = dy_1, z = dz_1$ and $(x_1,y_1,z_1) = 1$. Furthermore, because

$$x^2 + y^2 = z^2,$$

we have

$$(x/d)^2 + (y/d)^2 = (z/d)^2,$$

so that

$$x_1^2 + y_1^2 = z_1^2.$$

Hence, $x_1,y_1,z_1$ is a primitive Pythagorean triple, and the original triple $x,y,z$ is simply an integral multiple of this primitive Pytgagorean triple.

Also note that any integral multiple of a primitive (or for that matter any) Pythagorean triple is again a Pythagorean triple. If $x_1,y_1,z_1$ is a primitive Pythagorean triple, then we have

$$x_1^2 + y_1^2 = z_1^2,$$

and hence,

$$(dx_1)^2 + (dy_1)^2 = (dz_1)^2,$$

so that $dx_1,dy_1,dz_1$ is a Pythagorean triple.

Consequently, all Pythagorean triples can be found by forming integral multiples of primitive Pythagorean triples. To find all primitive Pythagorean triples, we need some lemmata. The first lemma tells us that any two integers of a primitive Pythagorean triple are relatively prime.

**Lemma 11.1.** If $x,y,z$ is a primitive Pythagorean triple, then $(x,y) = (x,z) = (y,z) = 1$.

*Proof.* Suppose $x,y,z$ is a primitive Pythagorean triple and $(x,y) > 1$. Then, there is a prime $p$ such that $p \mid (x,y)$, so that $p \mid x$ and $p \mid y$. Since $p \mid x$ and $p \mid y$, we know that $p \mid (x^2 + y^2) = z^2$. Because $p \mid z^2$, we can conclude that $p \mid z$. This is a contradiction since $(x,y,z) = 1$. Therefore, $(x,y) = 1$. In a similar manner we can easily show that $(x,z) = (y,z) = 1$. ∎

Next, we establish a lemma about the parity of the integers of a primitive Pythagorean triple.

**Lemma 11.2.** If $x,y,z$ is a primitive Pythagorean triple, then $x$ is even and $y$ is odd or $x$ is odd and $y$ is even.

*Proof.* Let $x,y,z$ be a primitive Pythagorean triple. By Lemma 11.1 we know that $(x,y) = 1$, so that $x$ and $y$ cannot both be even. Also $x$ and $y$ cannot both be odd. If $x$ and $y$ were both odd, then we would have

$$x^2 \equiv y^2 \equiv 1 \pmod 4,$$

so that

$$z^2 = x^2 + y^2 \equiv 2 \pmod 4.$$

This is impossible. Therefore, $x$ is even and $y$ is odd, or *vice versa*. ∎

The final lemma that we need is a consequence of the fundamental theorem of arithmetic. It tells us that two relatively prime integers that multiply together to give a square must both be squares.

**Lemma 11.3.** If $r,s$, and $t$ are positive integers such that $(r,s) = 1$ and $rs = t^2$, then there are integers $m$ and $n$ such that $r = m^2$ and $s = n^2$.

*Proof.* If $r = 1$ or $s = 1$, then the lemma is obviously true, so we may suppose that $r > 1$ and $s > 1$. Let the prime-power factorizations of $r,s$, and $t$ be

$$r = p_1^{a_1} p_2^{a_2} \cdots p_u^{a_u},$$
$$s = p_{u+1}^{a_{u+1}} p_{u+2}^{a_{u+2}} \cdots p_v^{a_v},$$

and

$$t = q_1^{b_1} q_2^{b_2} \cdots q_k^{b_k}.$$

Since $(r,s) = 1$, the primes occurring in the factorizations of $r$ and $s$ are distinct. Since $rs = t^2$, we have

$$p_1^{a_1} p_2^{a_2} \cdots p_u^{a_u} p_{u+1}^{a_{u+1}} p_{u+2}^{a_{u+2}} \cdots p_v^{a_v} = q_1^{2b_1} q_2^{2b_2} \cdots q_k^{2b_k}.$$

From the fundamental theorem of arithmetic, the prime-powers occurring on the two sides of the above equation are the same. Hence, each $p_i$ must be equal to $q_j$ for some $j$ with matching exponents, so that $a_i = 2b_j$. Consequently, every exponent $a_i$ is even, and therefore $a_i/2$ is an integer. We see that $r = m^2$ and $s = n^2$, where $m$ and $n$ are the integers

$$m = p_1^{a_1/2} p_2^{a_2/2} \cdots p_u^{a_u/2}$$

and

$$n = p_{u+1}^{a_u+1/2} p_{u+2}^{a_u+2/2} \quad \cdots \quad p_v^{a_v/2}.$$                    ∎

We can now prove the desired result that describes all primitive Pythagorean triples.

**Theorem 11.1.** The positive integers $x, y, z$ form a primitive Pythagorean triple, with $y$ even, if and only if there are relatively prime positive integers $m$ and $n$, $m > n$, with $m$ odd and $n$ even or $m$ even and $n$ odd, such that

$$x = m^2 - n^2$$
$$y = 2mn$$
$$z = m^2 + n^2.$$

*Proof.* Let $x, y, z$ be a primitive Pythagorean triple. Lemma 11.2 tells us that $x$ is odd and $y$ is even, or *vice versa*. Since we have assumed that $y$ is even, $x$ and $z$ are both odd. Hence, $z + x$ and $z - x$ are both even, so that there are positive integers $r$ and $s$ with $r = (z+x)/2$ and $s = (z-x)/2$.

Since $x^2 + y^2 = z^2$, we have $y^2 = z^2 - x^2 = (z+x)(z-x)$. Hence,

$$\left(\frac{y}{2}\right)^2 = \left(\frac{z+x}{2}\right)\left(\frac{z-x}{2}\right) = rs.$$

We note that $(r,s) = 1$. To see this, let $(r,s) = d$. Since $d \mid r$ and $d \mid s$, $d \mid (r+s) = z$ and $d \mid (r-s) = x$. This means that $d \mid (x,z) = 1$, so that $d = 1$.

Using Lemma 11.3, we see that there are integers $m$ and $n$ such that $r = m^2$ and $s = n^2$. Writing $x, y,$ and $z$ in terms of $m$ and $n$, we have

$$x = r - s = m^2 - n^2,$$
$$y = \sqrt{4rs} = \sqrt{4m^2n^2} = 2mn,$$

and

$$z = r + s = m^2 + n^2.$$

We also see that $(m,n) = 1$, since any common divisor of $m$ and $n$ must also divide $x = m^2 - n^2$, $y = 2mn$, and $z = m^2 + n^2$, and we know that $(x,y,z) = 1$. We also note that $m$ and $n$ cannot both be odd, for if they were, then $x, y,$ and $z$ would all be even, contradicting the condition $(x,y,z) = 1$. Since $(m,n) = 1$ and $m$ and $n$ cannot both be odd, we see $m$ is even and $n$ is odd, or *vice versa*. This shows that every primitive Pythagorean triple has the appropriate form.

To see that every triple

$$x = m^2 - n^2$$
$$y = 2mn$$
$$z \equiv m^2 + n^2,$$

where $m$ and $n$ are positive integers, $m > n$, $(m,n) = 1$, and $m \not\equiv n \pmod 2$, forms a primitive Pythagorean triple, first note that

$$x^2 + y^2 = (m^2-n^2)^2 + (2mn)^2$$
$$= (m^4-2m^2n^2+n^4) + 4m^2n^2$$
$$= m^4 + 2m^2n^2 + n^4$$
$$= (m^2+n^2)^2$$
$$= z^2.$$

To see that these values of $x, y$, and $z$ are mutually relatively prime, assume that $(x,y,z) = d > 1$. Then, there is a prime $p$ such that $p \mid (x,y,z)$. We note that $p \neq 2$, since $x$ is odd (because $x = m^2-n^2$ where $m^2$ and $n^2$ have opposite parity). Also, note that because $p \mid x$ and $p \mid z$, $p \mid (z+x) = 2m^2$ and $p \mid (z-x) = 2n^2$. Hence $p \mid m$ and $p \mid n$, contradicting the fact that $(m,n) = 1$. Therefore, $(x,y,z) = 1$, and $x, y, z$ is a primitive Pythagorean triple. This concludes the proof.  ∎

The following example illustrates the use of Theorem 11.1 to produce Pythagorean triples.

**Example 11.3.** Let $m = 5$ and $n = 2$, so that $(m,n) = 1$, $m \not\equiv n \pmod 2$, and $m > n$. Hence, Theorem 11.1 tells us that

$$x = m^2-n^2 = 5^2-2^2 = 21$$
$$y = 2mn = 2 \cdot 5 \cdot 2 = 20$$
$$z = m^2+n^2 = 5^2+2^2 = 29$$

is a primitive Pythagorean triple.  □

We list the primitive Pythagorean triples generated using Theorem 11.1 with $m \leqslant 6$ in Table 11.1.

| $m$ | $n$ | $x = m^2-n^2$ | $y = 2mn$ | $z = m^2+n^2$ |
|---|---|---|---|---|
| 2 | 1 | 3 | 4 | 5 |
| 3 | 2 | 5 | 12 | 13 |
| 4 | 1 | 15 | 8 | 17 |
| 4 | 3 | 7 | 24 | 25 |
| 5 | 2 | 21 | 20 | 29 |
| 5 | 4 | 9 | 40 | 41 |
| 6 | 1 | 35 | 12 | 37 |
| 6 | 5 | 11 | 60 | 61 |

**Table 11.1.** Some Primitive Pythagorean Triples.

## 11.1 Exercises

1. Find all

    a) primitive Pythagorean triples $x,y,z$ with $z \leqslant 40$.

    b) Pythagorean triples $x,y,z$ with $z \leqslant 40$.

2. Show that if $x,y,z$ is a primitive Pythagorean triple, then either $x$ or $y$ is divisible by 3.

3. Show that if $x,y,z$ is a Pythagorean triple, then exactly one of $x,y$, and $z$ is divisible by 5.

4. Show that if $x,y,z$ is a Pythagorean triple, then at least one of $x,y$, and $z$ is divisible by 4.

5. Show that every positive integer greater than three is part of at least one Pythagorean triple.

6. Let $x_1 = 3$, $y_1 = 4$, $z_1 = 5$, and let $x_n, y_n, z_n$, for $n = 2,3,4, ...,$ be defined recursively by

$$x_{n+1} = 3x_n + 2z_n + 1$$
$$y_{n+1} = 3x_n + 2z_n + 2$$
$$z_{n+1} = 4x_n + 3z_n + 2.$$

    Show that $x_n, y_n, z_n$ is a Pythagorean triple.

7. Show that if $x,y,z$ is a Pythagorean triple with $y = x + 1$, then $x,y,z$ is one of the Pythagorean triples given in problem 6.

8. Find all solutions in positive integers of the diophantine equation $x^2 + 2y^2 = z^2$.

9. Find all solutions in positive integers of the diophantine equation $x^2 + 3y^2 = z^2$

\* 10. Find all solutions in positive integers of the diophantine equation $w^2 + x^2 + y^2 = z^2$.

11. Find all Pythagorean triples containing the integer 12.

12. Find formulae for the integers of all Pythagorean triples $x,y,z$ with $z = y+1$.

13. Find formulae for the integers of all Pythagorean triples $x,y,z$ with $z = y + 2$.

\* 14. Show that the number of Pythagorean triples $x,y,z$ (with $x^2 + y^2 = z^2$) with a fixed integer $x$ is $(\tau(x^2)-1)/2$ if $x$ is odd, and $(\tau(x^2/4)-1)/2$ if $x$ is even.

15. Find all solutions in positive integers of the diophantine equation $x^2 + py^2 = z^2$, where $p$ is a prime.

## 11.1 Computer Projects

Write programs to do the following:

      **1.**   Find all Pythagorean triples $x,y,z$ with $x,y$, and $z$ less than a given bound.

      **2.**   Find all Pythagorean triples containing a given integer.

---

## 11.2  Fermat's Last Theorem

In the previous section, we showed that the diophantine equation $x^2 + y^2 = z^2$ has infinitely many solutions in nonzero integers $x$, $y$, $z$. What happens when we replace the exponent two in this equation with an integer greater than two? Next to the discussion of the equation $x^2 + y^2 = z^2$ in his copy of the works of Diophantus, Fermat wrote in the margin:

"However, it is impossible to write a cube as the sum of two cubes, a fourth power as the sum of two fourth powers and in general any power the sum of two similar powers. For this I have discovered a truly wonderful proof, but the margin is too small to contain it."

Since Fermat made this statement many people have searched for a proof of this assertion without success. Even though no correct proof has yet been discovered, the following conjecture is known as *Fermat's last theorem*.

**Fermat's Last Theorem.**    The diophantine equation

$$x^n + y^n = z^n$$

has no solutions in nonzero integers $x$, $y$, $z$ when $n$ is an integer with $n \geqslant 3$.

Currently, we know that Fermat's last theorem is true for all positive integers $n$ with $3 \leqslant n \leqslant 125000$. In this section, we will show that the special case of Fermat's last theorem with $n = 4$ is true. That is, we will show that the diophantine equation

$$x^4 + y^4 = z^4$$

has no solutions in nonzero integers $x$, $y$, $z$. Note that if we could also show that the diophantine equations

$$x^p + y^p = z^p$$

has no solutions in nonzero integers $x,y,z$ whenever $p$ is an odd prime, then we would know that Fermat's last theorem is true (see Exercise 2 at the end of this section).

The proof we will give of the special case of $n = 4$ uses the *method of infinite descent* devised by Fermat. This method is an offshoot of the well-ordering property, and shows that a diophantine equation has no solutions by showing that for every solution there is a "smaller" solution, contradicting the well-ordering property.

Using the method of infinite descent we will show that the diophantine equation $x^4 + y^4 = z^2$. has no solutions in nonzero integers $x$, $y$, and $z$. This is stronger than showing that Fermat's last theorem is true for $n = 4$, because any solution of $x^4 + y^4 = z^4 = (z^2)^2$ gives a solution of $x^4 + y^4 = z^2$.

**Theorem 11.2.** The diophantine equation

$$x^4 + y^4 = z^2$$

has no solutions in nonzero integers $x,y,z$.

*Proof.* Assume that the above equation has a solution in nonzero integers $x,y,z$. Since we may replace any number of the variables with their negatives without changing the validity of the equation, we may assume that $x,y,z$ are positive integers.

We may also suppose that $(x,y) = 1$. To see this, let $(x,y) = d$. Then $x = dx_1$ and $y = dy_1$, with $(x_1,y_1) = 1$, where $x_1$ and $y_1$ are positive integers. Since $x^4 + y^4 = z^2$, we have

$$(dx_1)^4 + (dy_1)^4 = z^2,$$

so that

$$d^4(x_1^4 + y_1^4) = z^2.$$

Hence $d^4 \mid z^2$, and, by problem 32 of Section 2.2, we know that $d^2 \mid z$. Therefore, $z = d^2z_1$, where $z_1$ is a positive integer. Thus,

$$d^4(x_1^4 + y_1^4) = (d^2z_1)^2 = d^4z_1^2,$$

so that

$$x_1^4 + y_1^4 = z_1^2.$$

This gives a solution of $x^4 + y^4 = z^2$ in positive integers $x = x_1, y = y_1, z = z_1$ with $(x_1,y_1) = 1$.

So suppose that $x = x_0, y = y_0$, $z = z_0$ is a solution of $x^4 + y^4 = z^2$, where $x_0$, $y_0$, and $z_0$ are positive integers with $(x_0,y_0) = 1$. We will show that there is another solution in positive integers $x = x_1$, $y = y_1$, $z = z_1$ with $(x_1,y_1) = 1$, such that $z_1 < z_0$.

Since $x_0^4 + y_0^4 = z_0^2$, we have

$$(x_0^2)^2 + (y_0^2)^2 = z_0^2,$$

so that $x_0^2, y_0^2, z_0$ is a Pythagorean triple. Furthermore, we have

$(x_0^2, y_0^2) = 1$, for if $p$ is a prime such that $p \mid x_0^2$ and $p \mid y_0^2$ , then $p \mid x_0$ and $p \mid y_0$, contradicting the fact that $(x_0, y_0) = 1$. Hence, $x_0^2, y_0^2, z_0$ is a primitive Pythagorean triple, and by Theorem 11.1, we know that there are positive integers $m$ and $n$ with $(m,n)$, $m \not\equiv n$ (mod 2) , and

$$
\begin{aligned}
x_0^2 &= m^2 - n^2 \\
y_0^2 &= 2mn \\
z_0 &= m^2 + n^2,
\end{aligned}
$$

where we have interchanged $x_0^2$ and $y_0^2$, if necessary, to make $y_0^2$ the even integer of this pair.

From the equation for $x_0^2$, we see that

$$
x_0^2 + n^2 = m^2.
$$

Since $(m,n) = 1$, it follows that $x_0, n, m$ is a primitive Pythagorean triple. Again using Theorem 11.1, we see that there are positive integers $r$ and $s$ with $(r,s) = 1$ , $r \not\equiv s$ (mod 2), and

$$
\begin{aligned}
x_0 &= r^2 - s^2 \\
n &= 2rs \\
m &= r^2 + s^2.
\end{aligned}
$$

Since $m$ is odd and $(m,n) = 1$, we know that $(m,2n) = 1$. We note that because $y_0^2 = (2n)m$, Lemma 11.3 tells us that there are positive integers $z_1$ and $w$ with $m = z_1^2$ and $2n = w^2$. Since $w$ is even, $w = 2v$ where $v$ is a positive integer, so that

$$
v^2 = n/2 = rs.
$$

Since $(r,s) = 1$, Lemma 11.3 tells us that there are positive integers $x_1$ and $y_1$ such that $r = x_1^2$ and $s = y_1^2$. Note that since $(r,s) = 1$, it easily follows that $(x_1, y_1) = 1$. Hence,

$$
x_1^4 + y_1^4 = z_1^2
$$

where $x_1, y_1, z_1$ are positive integers with $(x_1, y_1) = 1$. Moreover, we have $z_1 < z_0$, because

$$
z_1 \leqslant z_1^4 = m^2 < m^2 + n^2 = z_0.
$$

To complete the proof, assume that $x^4 + y^4 = z^2$ has at least one integral solution. By the well-ordering property, we know that among the solutions in positive integers, there is a solution with the smallest value $z_0$ of the variable $z$. However, we have shown that from this solution we can find another solution with a smaller value of the variable $z$, leading to a contradiction. This completes the proof by the method of infinite descent.  ∎

Readers interested in the history of Fermat's last theorem and how investigations relating to this conjecture led to the genesis of the theory of algebraic numbers are encouraged to consult the books of Edwards [14] and

Ribenboim [33]. A great deal of research relating to Fermat's last theorem is underway. Recently, the German mathematician Faltings established a result that shows that for a fixed positive integer $n$, $n \geqslant 3$, the diophantine equation $x^n + y^n = z^n$ has at most a finite number of solutions where $x, y$, and $z$ are integers and $(x, y) = 1$.

## 11.2 Exercises

1. Show that if $x, y, z$ is a Pythagorean triple and $n$ is an integer $n > 2$, then $x^n + y^n \neq z^n$.

2. Show that Fermat's last theorem is a consequence of Theorem 11.2, and the assertion that $x^p + y^p = z^p$ has no solutions in nonzero integers when $p$ is an odd prime.

3. Using Fermat's little theorem, show that if $p$ is prime and
   a) if $x^{p-1} + y^{p-1} = z^{p-1}$, then $p \mid xyz$.
   b) if $x^p + y^p = z^p$, then $p \mid (x+y-z)$.

4. Show that the diophantine equation $x^4 - y^4 = z^2$ has no solutions in nonzero integers using the method of infinite descent.

5. Using Exercise 4, show that the area of a right triangle with integer sides is never a perfect square.

* 6. Show that the diophantine equation $x^4 + 4y^4 = z^2$ has no solutions in nonzero integers.

* 7. Show that the diophantine equation $x^4 - 8y^4 = z^2$ has no solutions in nonzero integers.

* 8. Show that the diophantine equation $x^4 + 3y^4 = z^4$ has infinitely many solutions.

9. Show that in a Pythagorean triple there is at most one perfect square.

10. Show that the diophantine equation $x^2 + y^2 = z^3$ has infinitely many integer solutions by showing that for each positive integer $k$ the integers $x = 3k^2 - 1$, $y = k(k^2 - 3)$, $z = k^2 + 1$ form a solution.

## 11.2 Computer Projects

1. Write a computer program to search for solutions of diophantine equations such as $x^n + y^n = z^n$.

## 11.3 Sums of Squares

Mathematicians throughout history have been interested in problems about the representation of integers as sums of squares. Diophantus, Fermat, Euler, and Lagrange are some of the mathematicians who made important contributions to the solution of such problems. In this section we discuss two questions of this kind: Which integers are the sum of two squares? and What is the least integer $n$ such that every positive integer is the sum of $n$ squares?

We begin by considering the first question. Not every positive integer is the sum of two squares. In fact, $n$ is not the sum of two squares if it is of the form $4k+3$. To see this, note that since $a^2 \equiv 0$ or $1$ for every integer $a$, $x^2 + y^2 \equiv 0,1,$ or $2 \pmod 4$.

To conjecture which integers are the sum of two squares we first examine some small positive integers.

**Example 11.4**. Among the first 20 positive integers note that

| | |
|---|---|
| $1 = 0^2 + 1^2,$ | 11 is not the sum of two squares, |
| $2 = 1^2 + 1^2,$ | 12 is not the sum of two squares, |
| 3 is not the sum of two squares, | $13 = 3^2 + 2^2,$ |
| $4 = 2^2 + 0^2,$ | 14 is not the sum of two squares, |
| $5 = 1^2 + 2^2,$ | 15 is not the sum of two squares, |
| 6 is not the sum of two squares, | $16 = 4^2 + 0^2,$ |
| 7 is not the sum of two squares, | $17 = 4^2 + 1^2,$ |
| $8 = 2^2 + 2^2,$ | $18 = 3^2 + 3^2,$ |
| $9 = 3^2 + 0^2,$ | 19 is not the sum of two squares, |
| $10 = 3^2 + 1^2,$ | $20 = 2^2 + 4^2.$ |

$\square$

It is not immediately obvious from the evidence in Example 11.4 which integers are the sum of two squares. (Can you see anything in common among those positive integer not representable as the sum of two squares?)

We now begin a discussion which will show that the prime factorization on an integer determines whether this integer is the sum of two squares. There are two reasons for this. The first is that the product of two integers that are sums of two squares is again the sum of two squares. The second reason is that a prime is representable as the sum of two sqaures if and only if it is not of the form $4k + 3$. We will prove both of these results. Then we will state and prove the theorem that specifies which integers are the sum of two squares.

The proof that the product of sums of two squares is again the sum of two squares relies on an important algebraic identity which we will use several times in this section.

**Theorem 11.3**. If $m$ and $n$ are both sums of two squares then $mn$ is also the sum of two squares.

*Proof.* Let $m = a^2 + b^2$ and $n = c^2 + d^2$. Then

(11.2)      $mn = (a^2 + b^2)(c^2 + d^2) = (ac + bd)^2 + (ad - bc)^2.$

The reader can easily verify this identity by expanding all the terms.      ∎

**Example 11.5.** Since $5 = 2^2 + 1^2$ and $13 = 3^2 + 2^2$, it follows from (11.2) that

$$65 = 5 \cdot 13 = (2^2 + 1^2)(3^2 + 2^2) =$$
$$(2 \cdot 3 + 1 \cdot 2)^2 + (2 \cdot 2 - 1 \cdot 3)^2 = 8^2 + 1^2$$      □

One crucial result is that every prime of the form $4k+1$ is the sum of two squares. To prove this result we will need the following lemma.

**Lemma 11.4.** If $p$ is a prime of the form $4k+1$ where $k$ is an integer, then there exist integers $x$ and $y$ such that $x^2 + y^2 = kp$ for some positive integer $k$ with $k < p$.

*Proof.* By Theorem 9.4 we know that -1 is a quadratic residue of $p$. Hence there is an integer $a$, $a < p$ such that $a^2 \equiv -1 \pmod{p}$. It follows that $a^2 + 1 = kp$ for some positive integer $k$. Hence $x^2 + y^2 = kp$ where $x = a$ and $y = 1$. From the inequality $kp = x^2 + y^2 < (p-1)^2 + 1 < p^2$ we see that $k < p$.      ∎

We can now prove the following theorem which tells us that all primes not of the form $4k+3$ are the sum of two squares.

**Theorem 11.4.** If $p$ is a prime not of the form $4k+3$ then there are integers $x$ and $y$ such that $x^2 + y^2 = p$.

*Proof.* Note that 2 is the sum of two squares since $1^2 + 1^2 = 2$. Now suppose that $p$ is a prime of the form $4k+1$. Let $m$ be the smallest positive integer such that $x^2 + y^2 = mp$ has a solution in integers $x$ and $y$. By Lemma 11.4 there is such an integer less than $p$; by the well-ordering property a least such integer exists. We will show that $m=1$.

Assume that $m > 1$. Let $a$ and $b$ be defined by

$$a \equiv x \pmod{m}, \ b \equiv y \pmod{m}$$

and

$$-m/2 < a \leqslant m/2, \ -m/2 < b \leqslant m/2.$$

It follows that $a^2 + b^2 \equiv x^2 + y^2 = mp \equiv 0 \pmod{m}$. Hence there is an integer $k$ such that

$$a^2 + b^2 = km.$$

We have

$$(a^2 + b^2)(x^2 + y^2) = (km)(mp) = km^2p.$$

By equation (11.2) we have

$$(a^2 + b^2)(x^2 + y^2) = (ax + by)^2 + (ay - bx)^2$$

Furthermore, since $a \equiv x \pmod{m}$ and $b \equiv y \pmod{m}$ we have

$$ax + by \equiv x^2 + y^2 \equiv 0 \pmod{m}$$
$$ay - bx \equiv xy - yx \equiv 0 \pmod{m}.$$

Hence $(ax+by)/m$ and $(ay-bx)/m$ are integers, so that

$$\left( \frac{ax + by}{m} \right)^2 + \left( \frac{ax - by}{m} \right)^2 = km^2p/m^2 = kp$$

is the sum of two squares. If we show that $0 < k < m$ this will contradict the choice of $m$ as the minimum positive such that $x^2 + y^2 = mp$ has a solution in integers. We know that $a^2 + b^2 = km$, $-m/2 < a < m/2$, and $-m/2 < b < m/2$. Hence $a^2 < m^2/4$ and $b^2 < m^2/4$. We have

$$0 \leqslant km < 2(m^2/4) = m^2/2.$$

Consequently $0 \leqslant k < m/2$. It follows that $k < m$. All that remains to do is show that $k \neq 0$. If $k = 0$ we have $a^2 + b^2 = 0$. This implies that $a = b = 0$, so that $x \equiv y \equiv 0 \pmod{m}$, which shows that $m \mid x$ and $m \mid y$. Since $x^2 + y^2 = mp$ this implies that $m^2 \mid mp$ which implies that $m \mid p$. Since $m$ is less than $p$, this implies that $m = 1$ which is what we wanted to prove.                                                                                            ∎

We can now put all the pieces together and prove the fundamental result that classifies the positive integers which are representable as the sum of two squares.

**Theorem 11.5.** The positive integer $n$ is the sum of two squares if and only if each prime factor of $n$ of the form $4k+3$ occurs to an even power in the prime factorization of $n$.

*Proof.* Suppose that in the prime factorization of $n$ there are no primes of the form $4k+3$ that appear to an odd power. Write $n = t^2u$ where $u$ is the product of primes. No primes of the form $4k+3$ appear in $u$. By Theorem 11.4 each prime in $u$ can be written as the sum of two squares. Applying Theorem 11.3 one fewer time than then number of different primes in $u$ shows that $u$ is also the sum of two squares, say

$$u = x^2 + y^2.$$

It then follows that $n$ is also the sum of two squares, namely

$$n = (tx)^2 + (ty)^2.$$

Now suppose that there is a prime $p$, $p \equiv 3 \pmod 4$, that occurs in the prime factorization of $n$ to an odd power, say the $(2j+1)$th power. Furthermore, suppose that $n$ is the sum of two squares, that is,

$$n = x^2 + y^2.$$

Let $(x,y)=d$, $a = x/d$, $b = y/d$, and $m = n/d^2$. It follows that $(a,b) = 1$ and

$$a^2 + b^2 = m.$$

Suppose that $p^k$ is the largest power of $p$ dividing $d$. Then $m$ is divisible by $p^{2j-2k+1}$ and $2j-2k+1$ is at least 1 since it is nonnegative. Hence $p \mid m$. We know that $p$ does not divide $a$, for if $p \mid a$, then $p \mid b$ since $b^2 = m - a^2$ and $(a,b) = 1$.

Thus there is an integer $z$ such $az \equiv b \pmod p$. It follows that

$$a^2 + b^2 = a^2 + (az)^2 = a^2(1 + z^2).$$

Since $a^2 + b^2 = m$ and $p \mid m$, we see that

$$a^2(1 + z^2) \equiv 0 \pmod p.$$

Because $(a,p) = 1$, it follows that $1 + z^2 \equiv 0 \pmod p$. This implies that $z^2 \equiv -1 \pmod p$ which is impossible since -1 is not a quadratic residue of $p$ since $p \equiv 3 \pmod 4$. This contradiction shows that $n$ could not have been the sum of two squares.  ∎

Since there are positive integers not representable as the sum of two squares, we can ask whether every positive integer is the sum of three squares. The answer is no, since it is impossible to write 7 as the sum of three squares (as the reader should show). Since three squares do not suffice, we ask whether four squares do. The answer to this is yes, as we will show. Fermat wrote that he had a proof of this fact, although he never published it (and most historians of mathematics believe he actually had such a proof). Euler was unable to find a proof, although he made substantial progress towards a solution. It was in 1770 when Lagrange presented the first published solution.

The proof that every positive integer is the sum of four squares depends on the following lemma which shows that the product of two integers both representable as the sum of four squares can also be so represented. Just as for the analogous result for two squares there is an important algebraic identity used in the proof.

**Theorem 11.6.** If $m$ and $n$ are positive integers that are both the sum of four squares, then $mn$ is also the sum of four squares.

*Proof.* Let $m = a^2 + b^2 + c^2 + d^2$ and $n = e^2 + f^2 + g^2 + h^2$. That $mn$ is also the sum of four squares follows from the following algebraic identity:

(11.3)
$$mn = (a^2 + b^2 + c^2 + d^2)(e^2 + f^2 + g^2 + h^2)$$
$$= (ae + bf + cg + dh)^2 + (af - be + ch - dg)^2 +$$
$$(ag - bh - ce + df)^2 + (ah + bg - cf - de)^2.$$

The reader can easily verify this identity by multiplying all the terms.  ∎

We illustrate the use of Theorem 11.6 with an example.

**Example 11.6.** Since $7 = 2^2 + 1^2 + 1^2 + 1^2$ and $10 = 3^2 + 1^2 + 0^2 + 0^2$ from (11.3) it follows that

$$70 = 10 \cdot 7 = (2^2 + 1^2 + 1^2 + 1^2 + 1^2)(3^2 + 1^2 + 0^2 + 0^2)$$
$$= (2 \cdot 3 + 1 \cdot 1 + 1 \cdot 0 + 1 \cdot 0)^2 + (2 \cdot 1 - 1 \cdot 3 + 1 \cdot 0 - 1 \cdot 0)^2 +$$
$$(2 \cdot 0 - 1 \cdot 0 + 1 \cdot 3 - 1 \cdot 1)^2 + (2 \cdot 0 + 1 \cdot 0 + 1 \cdot 1 + 1 \cdot 3)^2$$
$$= 7^2 + 1^2 + 2^2 + 4^2. \qquad \square$$

We will now begin our work to show that every prime is the sum of four squares. We begin with a lemma.

**Lemma 11.5.** If $p$ is an odd prime then there exist an integer $k$, $k < p$, such that
$$kp = x^2 + y^2 + z^2 + w^2$$
has a solution in integers $x, y$, $z$, and $w$.

*Proof.* We will first show that there are integers $x$ and $y$ such that

$$x^2 + y^2 + 1 \equiv 0 \pmod{p}$$

with $0 \leqslant x < p/2$ and $0 \leqslant y < p/2$.

Let

$$S = \{0^2, 1^2, \ldots, (\frac{p-1}{2})^2\}$$

and

$$T = \{-1 - 0^2, -1 - 1^2, \ldots, -1 - (\frac{p-1}{2})^2\}.$$

No two elements of $S$ are congruent modulo $p$ (since $x^2 \equiv y^2 \pmod{p}$ implies that $x \equiv \pm y \pmod{p}$). Likewise, no two elements of $T$ are congruent modulo $p$. It is easy to see that the set $S \cup T$ contains $p+1$

distinct integers.    By the pigeonhole principle there two integers in this union
that are congruent modulo $p$. It follows that there are integers $x$ and $y$ with
$x^2 \equiv -1-y^2 \pmod{p}$ with $0 \leqslant x \leqslant (p-1)/2$ and $0 \leqslant y < (p-1)/2$. We
have

$$x^2 + y^2 + 1 \equiv 0 \pmod{p}$$

Hence, $x^2 + y^2 + 1^2 + 0^2 = kp$ for some integer $k$.   Since $x^2 + y^2 + 1 <$
$2(p-1)/2^2 + 1 < p^2$, it follows that $k < p$.                                    ■

We can now prove that every prime is the sum of four squares.

**Theorem 11.7.**  Let $p$ be a prime. Then the equation $x^2 + y^2 + z^2 + w^2 = p$
has a solution where $x,y,z,$ and $w$ are integers.

*Proof.*    The  result  is  true  when  $p = 2$  the  result  is  true,  since
$2 = 1^2 + 1^2 + 0^2 + 0^2$.  Now assume that $p$ is an odd prime. Let $m$ be the
smallest  integer  such  that  $x^2 + y^2 + z^2 + w^2 = mp$  has  a solution where
$x,y,z,$ and $w$ are integers.  (By Lemma 11.5 such integers exist and by the
well-ordering property there is a minimal such integer.)   The theorem will
follow if we can show that $m=1$. To do this we assume that $m>1$ and find a
smaller such such integer.

   If $m$ is even, then either all of $x$, $y$, $z$, and $w$ all odd, all are even, or two are
odd and two are even.  In all these cases we can rearrage these integers (if
necessary) so that $x \equiv y \pmod 2$ and $z \equiv w \pmod 2$. It then follows that
$(x-y)/2$, $(x+y)/2$, $(z-w)/2$, and $(z+w)/2$ are integers and

$$\left[\frac{x-y}{2}\right]^2 + \left[\frac{x+y}{2}\right]^2 + \left[\frac{z-w}{2}\right]^2 + \left[\frac{z+w}{2}\right]^2 = (m/2)p.$$

This contradicts the minimality of $m$.

   Now suppose that $m$ is odd and $m > 1$. Let $a$, $b$, $c$, and $d$ be integers such
that

   $a \equiv x \pmod m,\ b \equiv y \pmod m,\ c \equiv z \pmod m,\ d = w \pmod m,$

and

   $-m/2 < a < m/2,\ -m/2 < b < m/2,\ -m/2 < c < m/2,\ -m/2 < d < m/2.$

We have

$$a^2 + b^2 + c^2 + d^2 \equiv x^2 + y^2 + z^2 + w^2 \pmod m.$$

Hence

$$a^2 + b^2 + c^2 + d^2 = km.$$

for some integer $k$ and

$$0 \leqslant a^2 + b^2 + c^2 + d^2 < 4(m/2)^2 = m^2.$$

Consequently, $0 \leqslant k < m$. If $k = 0$ we have $a = b = c = d = 0$ so that $x \equiv y \equiv z \equiv w \equiv 0 \pmod{m}$. From this it follows that $m^2 \mid mp$ which is impossible since $1 < m < p$. It follow that $k > 0$.

We have

$$(x^2 + y^2 + z^2 + w^2)(a^2 + b^2 + c^2 + d^2) = mp \cdot km = m^2 kp.$$

But by the identity in the proof of Theorem 11.6 we have

$$(ax + by + cz + dw)^2 + (bx - ay + dz - cw)^2 +$$
$$(cx - dy - az + bw)^2 + (dx + cy - bz - aw)^2 = m^2 kp.$$

Each of the four terms being squared in divisible by $m$ since

$$ax + by + cz + dw \equiv x^2 + y^2 + z^2 + w^2 \equiv 0 \pmod{m}$$
$$bx - ay + dz - cw \equiv yx - xy + wz - zw \equiv 0 \pmod{m}$$
$$cx - dy - az + bw \equiv zx - wy - xz + yw \equiv 0 \pmod{m}$$
$$dx + cy - bz - aw \equiv wx + zy - yz - xw \equiv 0 \pmod{m}.$$

Let $X$, $Y$, $Z$, and $W$ be the integers obtained by dividing these quantities by $m$, that is

$$X = (ax + by + cz + dw)/m$$
$$Y = (bx - ay + dz - cw)/m$$
$$Z = (cx - dy - az + bw)/m$$
$$W = (dx + cy - bz - aw)/m.$$

It then follows that

$$X^2 + Y^2 + Z^2 + W^2 = m^2 kp/m^2 = kp.$$

But this contradicts the choice of $m$. Hence $m$ must be 1.  ∎

We now can state and prove the fundamental theorem about representations as sums of four squares.

**Theorem 11.8.** Every positive integer is the sum of the squares of four integers.

*Proof.* Suppose $n$ is a positive integer. Then by the fundamental theorem of arithmetic $n$ is the product of primes. By Theorem 11.7 each of these prime factors can be written as the sum of four squares. Applying Theorem 11.6 a sufficient number of times it follows that $n$ is also the sum of four squares. ∎

We have shown that every positive integer can be written as the sum of four squares. As we have mentioned this theorem was originally proved by Lagrange in 1770. Around the same time the English mathematician Waring generalized this problem. He stated, but did not prove, that every positive integer is the sum of 9 cubes of nonnegative integers, the sum of 19 fourth powers of nonnegative integers, and so on. We can phrase this conjecture in the following way.

**Waring's Problem.** If $k$ is a positive integer is there an integer $g(k)$ such that every positive intger can be written as the sum of $g(k)$ $k$th powers of nonnegative integers, and no smaller number of $k$th powers will suffice?

Lagrange's theorem shows that we can take $g(2) = 4$ (since there are integers that are not the sum of three squares). In the nineteenth century mathematicians showed that such an integer $g(k)$ exists for $3 \leqslant k \leqslant 8$ and $k = 10$. But it was not until 1906 that the famous German mathematician David Hilbert showed that for every positive integer $k$ there is a constant $g(k)$ such that every positive integer may be expressed as the sum of $g(k)$ $k$th powers of nonnegative integers. Hilbert's proof is extremely complicated and is not constructive, so that it gives no formula for $g(k)$. It is now known that $g(3) = 9$, $g(4) \geqslant 19$, $g(5) = 37$, and

$$g(k) = [(3/2)^k] + 2^k - 2$$

for $6 \leqslant k \leqslant 200000$. Proofs of these results rely on nonelementary results from analytical number theory. There still many unanswered questions about the values of $g(k)$. The interested reader can learn about recent results on Waring's problem by consulting the numerous articles on this problem described in [25].

---

## 11.3 Exercises

1.  Given that $13 = 3^2 + 2^2$, $29 = 5^2 + 2^2$, and $50 = 7^2 + 1^2$, write each of the following integers as the sum of two squares.

a) $377 = 13 \cdot 29$     c) $1450 = 29 \cdot 50$
b) $650 = 13 \cdot 50$     d) $18850 = 13 \cdot 29 \cdot 50$

2. Determine whether each of the following integers can be written as the sum of two squares.

   a) 19     d) 45     g) 99
   b) 25     e) 65     h) 999
   c) 29     f) 80     i) 1000

3. Represent each of the following integers as the sum of two squares.

   a) 34       d) 490
   b) 90       e) 21658
   c) 100      f) 324608

4. Show that a positive integer is the difference of two squares if and only if it is not of the form $4k + 2$ where $k$ is an integer.

5. Represent each of the following integers as the sum of three squares if possible.

   a) 3       d) 18
   b) 7       e) 23
   c) 11      f) 28

6. Show that the positive integer $n$ is not the sum of three squares of integers if $n$ is of the form $8k + 7$ where $k$ is an integer.

7. Show that the positive integer $n$ is not the sum of three squares of integers if $n$ is of the form $4^m (8k + 7)$ where $m$ and $k$ are integers.

8. Prove or disprove that the sum of two integers representable as the sum of three squares of integers is also so representable.

9. Given that $7 = 2^2 + 1^2 + 1^2 + 1^2$, $15 = 3^2 + 2^2 + 1^2 + 1^2$, and $34 = 4^2 + 4^2 + 1^2 + 1^2$ write each of the following integers as the sum of four squares.

   a) $105 = 7 \cdot 15$     c) $238 = 7 \cdot 34$
   b) $510 = 15 \cdot 34$    d) $3570 = 7 \cdot 15 \cdot 34$

10. Write each of the following positive integers as the sum of four squares.

    a) 6       d) 89
    b) 12      e) 99
    c) 21      f) 555

**11.** Show that every integer $n$, $n \geqslant 170$, is the sum of the squares of five positive integers. (*Hint*: Write $m = n - 169$ as the sum of the squares of four integers, and use the fact that $169 = 13^2 = 12^2 + 5^2 = 12^2 + 4^2 + 3^2 = 10^2 + 8^2 + 2^2 + 1^2$.)

**12.** Show that the only positive integers that are not expressible as the sum of five squares of positive integers are 1,2,3,4,6,7,9,10,12,15,18,33. (*Hint*: Use Exercise 11, show that each of these integers cannot be expressed as stated, and then show all remaining positive integers less than 170 can be expressed as stated.)

**\* 13.** Show that there are arbitrarily large integers that are not the sum of the squares of four positive integers.

We outline a second proof for Theorem 11.8 in Exercises 13-14.

**\* 14.** Show that if $p$ is prime and $a$ is an integer not divisible by $p$, then there exist integers $x$ and $y$ such that $ax \equiv y \pmod{p}$ with $0 < |x| < \sqrt{p}$ and $0 < |y| < \sqrt{p}$. This result is called *Thue's Lemma*. (*Hint*: Use the pigeonhole principle to show that there are two integers of the form $au - v$ with $0 \leqslant u \leqslant [\sqrt{p}]$ and $0 \leqslant v \leqslant [\sqrt{p}]$ that are congruent modulo $p$. Construct $x$ and $y$ from the two values of $u$ and the two values of $v$, respectively.)

**15.** Use Exercise 14 to prove Theorem 11.8. (*Hint*: Show that there is an integer $a$ with $a^2 \equiv -1 \pmod{p}$. Then apply Thue's Lemma with this value of $a$.)

**16.** Show that 23 is the sum of nine cubes of nonnegative integers but not the sum of eight cubes of nonnegative integers.

Exercises 17-21 give an elementary proof that $g(4) \leqslant 50$.

**17.** Show that

$$\sum_{1 \leqslant i < j \leqslant 4} \left[ (x_i + x_j)^4 + (x_i - x_j)^4 \right] = 6 \left( \sum_{k-1}^{4} x_k^2 \right)^2.$$

(*Hint*: Start with the identity $(x_i + x_j)^4 + (x_i - x_j)^4 = 2x_i^4 + 12x_i^2 x_j^2 + 2x_j^4$.)

**18.** Show from Exercise 17 that every integer of the form $6n^2$, where $n$ is a positive integer, is the sum of 12 fourth powers.

**19.** Use Exercise 18 and the fact that every positive integer is the sum of four squares to show that every positive integer of the form $6m$, where $m$ is a positive integer, can be written as the sum of 48 fourth powers.

**20.** Show that the integers $k = 0,1,2,81,16$, and 17 form a complete system of residues modulo 6 each of which is the sum of at most two fourth powers. Show from this that every integer $n$ with $n > 81$ can be written as $6m + k$ where $m$ is a positive integer and $k$ comes from this complete system of residues. Conclude from this that every integer $n$ with $n > 81$

is the sum of 50 fourth powers.

**21.** Show that every positive integer $n$ with $n \leqslant 81$ is the sum of at most 50 fourth powers. (*Hint:* For $51 \leqslant n \leqslant 81$ start by using three terms equal to $2^4$.) Conclude from this exercise and Exercise 20 that $g(4) \leqslant 50$.

## 11.3 Computer Projects

Write programs to do the following:

* **1.** Determine whether a positive integer $n$ can be represented as the sum of two squares and so represent it if possible.

* **2.** Represent a positive integer $n$ as the sum of four squares.

---

## 11.4 Pell's Equation

In this section, we study diophantine equations of the form

$$(11.4) \qquad x^2 - dy^2 = n,$$

where $d$ and $n$ are fixed integers. When $d < 0$ and $n < 0$, there are no solutions of (11.4). When $d < 0$ and $n > 0$, there can be at most a finite number of solutions, since the equation $x^2 - dy^2 = n$ implies that $|x| \leqslant \sqrt{n}$ and $|y| \leqslant \sqrt{n/|d|}$. Also, note that when $d$ is a perfect square, say $d = D^2$, then

$$x^2 - dy^2 = x^2 - D^2y = (x+Dy)(x-Dy) = n \ .$$

Hence, any solution of (11.4), when $d$ is a perfect square, corresponds to a simultaneous solution of the equations

$$x + Dy = a$$
$$x - Dy = b,$$

where $a$ and $b$ are integers such that $n = ab$. In this case, there are only a finite number of solutions, since there is at most one solution in integers of these two equations for each factorization $n = ab$ .

For the rest of this section, we are interested in the diophantine equation $x^2 - dy^2 = n$, where $d$ and $n$ are integers and $d$ is a positive integer which is not a perfect square. As the following theorem shows, the simple continued fraction of $\sqrt{d}$ is very useful for the study of this equation.

**Theorem 11.9.** Let $d$ and $n$ be integers such that $d > 0$, $d$ is not a perfect square, and $|n| < \sqrt{d}$. If $x^2 - dy^2 = n$, then $x/y$ is a convergent of the simple continued fraction of $\sqrt{d}$.

*Proof.* First consider the case where $n > 0$. Since $x^2 - dy^2 = n$, we see that

(11.4)                                 $$(x + y\sqrt{d})(x - y\sqrt{d}) = n.$$

From (11.4), we see that $x - y\sqrt{d} > 0$, so that $x > y\sqrt{d}$. Consequently,

$$\frac{x}{y} - \sqrt{d} > 0,$$

and since $0 < n < \sqrt{d}$, we see that

$$\frac{x}{y} - \sqrt{d} = \frac{(x - \sqrt{d}\,y)}{y}$$

$$= \frac{x^2 - dy^2}{y(x + y\sqrt{d})}$$

$$< \frac{n}{y(2y\sqrt{d})}$$

$$< \frac{\sqrt{d}}{2y^2\sqrt{d}}$$

$$= \frac{1}{2y^2}.$$

Since $0 < \dfrac{x}{y} - \sqrt{d} < \dfrac{1}{2y^2}$, Theorem 10.16 tells us that $x/y$ must be a convergent of the simple continued fraction of $\sqrt{d}$.

When $n < 0$, we divide both sides of $x^2 - dy^2 = n$ by $-d$, to obtain

$$y^2 - (1/d)x^2 = -n/d .$$

By a similar argument to that given when $n > 0$ , we see that $y/x$ is a convergent of the simple continued fraction expansion of $1/\sqrt{d}$. Therefore, from Exercise 7 of Section 10.3, we know that $x/y = 1/(y/x)$ must be a convergent of the simple continued fraction of $\sqrt{d} = 1/(1/\sqrt{d})$.    ■

We have shown that solutions of the diophantine equation $x^2 - dy^2 = n$, where $|n| < \sqrt{d}$, are given by the convergents of the simple continued fraction expansion of $\sqrt{d}$. We will restate Theorem 10.20 here since it will help us use these convergents to find solutions of this diophantine equation.

**Theorem 10.20.** Let $d$ be a positive integer that is not a perfect square. Define    $\alpha_k = (P_k + \sqrt{d})/Q_k$,    $a_k = [\alpha_k]$,    $P_{k+1} = a_k Q_k - P_k$,    and $Q_{k+1} = (d - P_{k+1}^2)/Q_k$,    for $k = 0,1,2,...$  where $\alpha_0 = \sqrt{d}$. Furthermore, let $p_k/q_k$ denote the $k$th convergent of the simple continued fraction expansion of $\sqrt{d}$. Then

$$p_k^2 - dq_k^2 = (-1)^{k-1}Q_{k+1}.$$

The special case of the diophantine equation $x^2 - dy^2 = n$ with $n = 1$ is called *Pell's equation*. We will use Theorems 11.9 and 10.20 to find all solutions of Pell's equation and the related equation $x^2 - dy^2 = -1$.

**Theorem 11.10.** Let $d$ be a positive integer that is not a perfect square. Let $p_k/q_k$ denote the $k$th convergent of the simple continued fraction of $\sqrt{d}$, $k = 1,2,3,...$ and let $n$ be the period length of this continued fraction. Then, when $n$ is even, the positive solutions of the diophantine equation $x^2 - dy^2 = 1$ are $x = p_{jn-1}$, $y = q_{jn-1}$, $j = 1,2,3,...$, and the diophantine equation $x^2 - dy^2 = -1$ has no solutions. When $n$ is odd, the positive solutions of $x^2 - dy^2 = 1$ are $x = p_{2jn-1}, y = q_{2jn-1}$, $j = 1,2,3,...$ and the solutions of $x^2 - dy^2 = -1$ are $x = p_{(2j-1)n-1}$, $y = q_{(2j-1)n-1}$, $j = 1,2,3,...$.

*Proof.* Theorem 11.9 tells us that if $x_0, y_0$ is a positive solution of $x^2 - dy^2 = \pm 1$, then $x_0 = p_k$, $y_0 = q_k$ where $p_k/q_k$ is a convergent of the simple continued fraction of $\sqrt{d}$. On the other hand, from Theorem 10.20 we know that

$$p_k^2 - dq_k^2 = (-1)^{k-1} Q_{k+1},$$

where $Q_{k+1}$ is as defined in the statement of Theorem 10.20.

Because the period of the continued expansion of $\sqrt{d}$ is $n$, we know that $Q_{jn} = Q_0 = 1$ for $j = 1,2,3,...$, (since $\sqrt{d} = \dfrac{P_0 + \sqrt{d}}{Q_0}$). Hence,

$$p_{jn-1}^2 - d\, q_{jn-1}^2 = (-1)^{jn} Q_{nj} = (-1)^{jn}.$$

This equation shows that when $n$ is even $p_{jn-1}, q_{jn-1}$ is a solution of $x^2 - dy^2 = 1$ for $j = 1,2,3,...$, and when $n$ is odd, $p_{2jn-1}, q_{2jn-1}$ is a solution of $x^2 - dy^2 = 1$ and $p_{2(j-1)n-1}, q_{2(j-1)n-1}$ is a solution of $x^2 - dy^2 = -1$ for $j = 1,2,3,...$ .

To show that the diophantine equations $x^2 - dy^2 = 1$ and $x^2 - dy^2 = -1$ have no solutions other than those already found, we will show that $Q_{k+1} = 1$ implies that $n|k$ and that $Q_j \neq -1$ for $j = 1,2,3...$ .

We first note that if $Q_{k+1} = 1$, then

$$\alpha_{k+1} = P_{k+1} + \sqrt{d}.$$

Since $\alpha_{k+1} = [a_{k+1}; a_{k+2},...]$, the continued fraction expansion of $\alpha_{k+1}$ is purely periodic. Hence, Theorem 10.19 tells us that $-1 < \overline{\alpha_{k+1}} = P_{k+1} - \sqrt{d} < 0$. This implies that $P_{k+1} = [\sqrt{d}]$, so that $\alpha_k = \alpha_0$, and $n|k$.

To see that $Q_j \neq -1$ for $j = 1,2,3,...$, note that $Q_j = -1$ implies that $\alpha_j = -P_j - \sqrt{d}$. Since $\alpha_j$ has a purely periodic simple continued fraction expansion, we know that

$$-1 < \alpha_j' = -P_j + \sqrt{d} < 0$$

and

$$\alpha_j = -P_j - \sqrt{d} > 1.$$

From the first of these inequalities, we see that $P_j > -\sqrt{d}$ and, from the second, we see that $P_j < -1-\sqrt{d}$. Since these two inequalities for $p_j$ are contradictory, we see that $Q_j \neq -1$.

Since we have found all solutions of $x^2-dy^2 = 1$ and $x^2-dy^2 = -1$, where $x$ and $y$ are positive integers, we have completed the proof. ∎

We illustrate the use of Theorem 11.5 with the following examples.

**Example 11.7.** Since the simple continued fraction of $\sqrt{13}$ is $[3;\overline{1,1,1,1,6}]$ the positive solutions of the diophantine equation $x^2 - 13y^2 = 1$ are $p_{10j-1}, q_{10j-1}$, $j = 1,2,3,...$ where $p_{10j-1}/q_{10j-1}$ is the $(10j-1)$th convergent of the simple continued fraction expansion of $\sqrt{13}$. The least positive solution is $p_9 = 649$, $q_9 = 180$. The positive solutions of the diophantine equation $x^2-13y^2 = -1$ are $p_{10j-6}, q_{10j-6}, j = 1,2,3,...$; the least positive solution is $p_4 = 18, q_4 = 5$. □

**Example 11.8.** Since the continued fraction of $\sqrt{14}$ is $[3;\overline{1,2,1,6}]$, the positive solutions of $x^2 - 14y^2 = 1$ are $p_{4j-1}, q_{4j-1}$, $j = 1,2,3,...$ where $p_{4j-1}/q_{4j-1}$ is the $j$th convergent of the simple continued fraction expansion of $\sqrt{14}$. The least positive solution is $p_3 = 15$, $q_3 = 4$. The diophantine equation $x^2 - 14y^2 = -1$ has no solutions, since the period length of the simple continued fraction expansion of $\sqrt{14}$ is even. □

We conclude this section with the following theorem that shows how to find all the positive solutions of Pell's equation $x^2 - dy^2 = 1$ from the least positive solution, without finding subsequent convergents of the continued fraction expansion of $\sqrt{d}$.

**Theorem 11.11.** Let $x_1, y_1$ be the least positive solution of the diophantine equation $x^2 - dy^2 = 1$, where $d$ is a positive integer that is not a perfect square. Then all positive solutions $x_k, y_k$ are given by

$$x_k + y_k\sqrt{d} = (x_1 + y_1 \sqrt{d})^k$$

for $k = 1,2,3,...$ . (Note that $x_k$ and $y_k$ are determined by the use of Lemma 11.4).

*Proof.* We need to show that $x_k, y_k$ is a solution for $k = 1,2,3,...$ and that every solution is of this form.

To show that $x_k, y_k$ is a solution, first note that by taking conjugates, it follows that $x_k - y_k\sqrt{d} = (x_1 - y_1\sqrt{d})^k$, because from Lemma 10.4, the conjugate of a power is the power of the conjugate. Now, note that

$$x_k^2 - dy_k^2 = (x_k + y_k\sqrt{d})(x_k - y_k\sqrt{d})$$
$$= (x_1 + y_1\sqrt{d})^k(x_1 - y_1\sqrt{d})^k$$
$$= (x_1^2 - dy_1^2)^k$$
$$= 1.$$

Hence $x_k, y_k$ is a solution for $k = 1, 2, 3, \ldots$ .

To show that every positive solution is equal to $x_k, y_k$ for some positive integer $k$, assume that $X, Y$ is a positive solution different from $x_k, y_k$ for $k = 1, 2, 3, \ldots$ . Then there is an integer $n$ such that

$$(x_1 + y_1\sqrt{d})^n < X + Y\sqrt{d} < (x_1 + y_1\sqrt{d})^{n+1}.$$

When we multiply this inequality by $(x_1 + y_1\sqrt{d})^{-n}$, we obtain

$$1 < (x_1 - y_1\sqrt{d})^n(X + Y\sqrt{d}) < x_1 + y_1\sqrt{d},$$

since $x_1^2 - dy_1^2 = 1$ implies that $x_1 - y_1\sqrt{d} = (x_1 + y_1\sqrt{d})^{-1}$.

Now let

$$s + t\sqrt{d} = (x_1 - y_1\sqrt{d})^n(X + Y\sqrt{d}),$$

and note that

$$s^2 - dt^2 = (s - t\sqrt{d})(s + t\sqrt{d})$$
$$= (x_1 + y_1\sqrt{d})^n(X - Y\sqrt{d})(x_1 - y_1\sqrt{d})^n(X + Y\sqrt{d})$$
$$= (x_1^2 - dy_1^2)^n(X^2 - dY^2)$$
$$= 1.$$

We see that $s, t$ is a solution of $x^2 - dy^2 = 1$, and furthermore, we know that $1 < s + t\sqrt{d} < x_1 + y_1\sqrt{d}$. Moreover, since we know that $s + t\sqrt{d} > 1$, we see that $0 < (s + t\sqrt{d})^{-1} < 1$. Hence

$$s = \frac{1}{2}[(s + t\sqrt{d}) + (s - t\sqrt{d})] > 0$$

and

$$t = \frac{1}{2\sqrt{d}}[(s + t\sqrt{d}) - (s - t\sqrt{d})] > 0.$$

This means that $s, t$ is a positive solution, so that $s \geq x_1$, and $t \geq y_1$, by the choice of $x_1, y_1$ as the smallest positive solution. But this contradicts the inequality $s + t\sqrt{d} < x_1 + y_1\sqrt{d}$. Therefore $X, Y$ must be $x_k, y_k$ for some choice of $k$. ∎

The following example illustrates the use of Theorem 11.11.

**Example 11.9.** From a previous example we know that the least positive solution of the diophantine equation $x^2 - 13y^2 = 1$ is $x_1 = 649$, $y_1 = 180$. Hence, all positive solutions are given by $x_k, y_k$ where

$$x_k + y_k\sqrt{13} = (649 + 180\sqrt{13})^k .$$

For instance, we have

$$x_2 + y_2\sqrt{13} = 842361 + 233640\sqrt{13}$$

Hence $x_2 = 842361$, $y_2 = 233640$ is the least positive solution of $x^2 - 13y^2 = 1$, other than $x_1 = 649$, $y_1 = 180$.                                                      □

---

## 11.4 Exercises

1. Find all the solutions where $x$ and $y$ are integers of each of the following equations.

   a)  $x^2 + 3y^2 = 4$

   b)  $x^2 + 5y^2 = 7$

   c)  $2x^2 + 7y^2 = 30$

2. Find all the solutions where $x$ and $y$ are integer of each of the following equations.

   a)  $x^2 - y^2 = 8$

   b)  $x^2 - 4y^2 = 40$

   c)  $4x^2 - 9y^2 = 100$

3. For which of the following values of $n$ does the diophantine equation $x^2 - 31y^2 = n$ have a solution?

   | | | | |
   |---|---|---|---|
   | a) | 1 | d) | −3 |
   | b) | −1 | e) | 4 |
   | c) | 2 | f) | −45 |

4. Find the least positive solution in integers of each of the diophantine equations.

   a)  $x^2 - 29y^2 = -1$       b)  $x^2 - 29y^2 = 1$

5. Find the three smallest positive solutions of the diophantine equation $x^2 - 37y^2 = 1$.

6. For each of the following values of $d$ determine whether the diophantine equation $x^2 - dy^2 = -1$ has solutions in integers.

   | | | | |
   |---|---|---|---|
   | a) | 2 | e) | 17 |
   | b) | 3 | f) | 31 |
   | c) | 6 | g) | 41 |
   | d) | 13 | h) | 50 |

7. The least positive solution of the diophantine equation $x^2 - 61y^2 = 1$ is $x_1 = 1766319049$, $y_1 = 226153980$. Find the least positive solution other than $x_1, y_1$.

* 8. Show that if $p_k/q_k$ is a convergent of the simple continued fraction expansion of $sqrt{d}$ then $|p_k^2 - dq_k^2| < 1 + 2\sqrt{d}$.

9. Show that if $d$ is a positive integer divisible by a prime of the form $4k + 3$, then the diophantine equation $x^2 - dy^2 = -1$ has no solutions.

10. Let $d$ and $n$ be positive integers.

   a) Show that if $r,s$ is a solution of the diophantine equation $x^2 - dy^2 = 1$ and $X,Y$ is a solution of the diophantine equation $x^2 - dy^2 = n$ then $Xr \pm dYs$, $Xs \pm Yr$ is also a solution of $x^2 - dy^2 = n$.

   b) Show that the diophantine equation $x^2 - dy^2 = n$ either has no solutions, or infinitely many solutions.

11. Find those right triangles having legs with lengths that are consecutive integers. (*Hint*: Use Theorem 11.1 to write the lengths of the legs as $x = s^2 - t^2$ and $y = 2st$, where $s$ and $t$ are positive integers such that $(s,t) = 1$, $s > t$ and $s$ and $t$ have opposite parity. Then $x - y = \pm 1$ implies that $(s - t)^2 - 2t^2 = \pm 1$.)

12. Show that the diophantine equation $x^4 - 2y^4 = 1$ has no solutions.

13. Show that the diophantine equation $x^4 - 2y^2 = -1$ has no solutions.

## 11.4 Computer Projects

Write programs to do the following:

1. Find those integers $n$ with $|n| < \sqrt{d}$ such that the diophantine equation $x^2 - dy^2 = n$ has no solutions.

2. Find the least positive solutions of the diophantine equations $x^2 - dy^2 = 1$ and $x^2 - dy^2 = -1$.

3. Find the solutions of Pell's equation from the least positive solution (see Theorem 11.11).

# Appendix

## Table 1. Factor Table.

The least prime factor of each odd positive integer less than 10000 and not divisible by five is given in the table. The initial digits of the integer are listed to the side and the last digit is at the top of the column. Primes are indicated with a dash.

| | 1 | 3 | 7 | 9 | | 1 | 3 | 7 | 9 | | 1 | 3 | 7 | 9 | | 1 | 3 | 7 | 9 |
|---|---|---|---|---|---|---|---|---|---|---|---|---|---|---|---|---|---|---|---|
| 0 | — | — | — | 3 | 40 | — | 13 | 11 | — | 80 | 3 | 11 | 3 | — | 120 | — | 3 | 17 | 3 |
| 1 | — | — | — | — | 41 | 3 | 7 | 3 | — | 81 | — | 3 | 19 | 3 | 121 | 7 | — | — | 23 |
| 2 | 3 | — | 3 | — | 42 | — | 3 | 7 | 3 | 82 | — | — | — | — | 122 | 3 | — | 3 | — |
| 3 | — | 3 | — | 3 | 43 | — | — | 19 | — | 83 | 3 | 7 | 3 | — | 123 | — | 3 | — | 3 |
| 4 | — | — | — | 7 | 44 | 3 | — | 3 | — | 84 | 29 | 3 | 7 | 3 | 124 | 17 | 11 | 29 | — |
| 5 | 3 | — | 3 | — | 45 | 11 | 3 | — | 3 | 85 | 23 | — | — | — | 125 | 3 | 7 | 3 | — |
| 6 | — | 3 | — | 3 | 46 | — | — | — | 7 | 86 | 3 | — | 3 | 11 | 126 | 13 | 3 | 7 | 3 |
| 7 | — | — | 7 | — | 47 | 3 | 11 | 3 | — | 87 | 13 | 3 | — | 3 | 127 | 31 | 19 | — | — |
| 8 | 3 | — | 3 | — | 48 | 13 | 3 | — | 3 | 88 | — | — | — | 7 | 128 | 3 | — | 3 | — |
| 9 | 7 | 3 | — | 3 | 49 | — | 17 | 7 | — | 89 | 3 | 19 | 3 | 29 | 129 | — | 3 | — | 3 |
| 10 | — | — | — | — | 50 | 3 | — | 3 | — | 90 | 17 | 3 | — | 3 | 130 | — | — | — | 7 |
| 11 | 3 | — | 3 | 7 | 51 | 7 | 3 | 11 | 3 | 91 | — | 11 | 7 | — | 131 | 3 | 13 | 3 | — |
| 12 | 11 | 3 | — | 3 | 52 | — | — | 17 | 23 | 92 | 3 | 13 | 3 | — | 132 | — | 3 | — | 3 |
| 13 | — | 7 | — | — | 53 | 3 | 13 | 3 | 7 | 93 | 7 | 3 | — | 3 | 133 | 11 | 31 | 7 | 13 |
| 14 | 3 | 11 | 3 | — | 54 | — | 3 | — | 3 | 94 | — | 23 | — | 13 | 134 | 3 | 17 | 3 | 19 |
| 15 | — | 3 | — | 3 | 55 | 19 | 7 | — | 13 | 95 | 3 | 8 | 3 | 7 | 135 | 7 | 3 | 23 | 3 |
| 16 | 7 | — | — | 13 | 56 | 3 | — | 3 | — | 96 | 31 | 3 | — | 3 | 136 | — | 29 | — | 37 |
| 17 | 3 | — | 3 | — | 57 | — | 3 | — | 3 | 97 | — | 7 | — | 11 | 137 | 3 | — | 3 | 7 |
| 18 | — | 3 | 11 | 3 | 58 | 7 | 11 | — | 19 | 98 | 3 | — | 3 | 23 | 138 | — | 3 | 19 | 3 |
| 19 | — | — | — | — | 59 | 3 | — | 3 | — | 99 | — | 3 | — | 3 | 139 | 13 | 7 | 11 | — |
| 20 | 3 | 7 | 3 | 11 | 60 | — | 3 | — | 3 | 100 | 7 | 17 | 19 | — | 140 | 3 | 23 | 3 | — |
| 21 | — | 3 | 7 | 3 | 61 | 13 | — | — | — | 101 | 3 | — | 3 | — | 141 | 17 | 3 | 13 | 3 |
| 22 | 13 | — | — | — | 62 | 3 | 7 | 3 | 17 | 102 | — | 3 | 13 | 3 | 142 | 7 | — | — | — |
| 23 | 3 | — | 3 | — | 63 | — | 3 | 7 | 3 | 103 | — | — | 17 | — | 143 | 3 | — | 3 | — |
| 24 | — | 3 | 13 | 3 | 64 | — | — | — | 11 | 104 | 3 | 7 | 3 | — | 144 | 11 | 3 | — | 3 |
| 25 | — | 11 | — | 7 | 65 | 3 | — | 3 | — | 105 | — | 3 | 7 | 3 | 145 | — | — | 31 | — |
| 26 | 3 | — | 3 | — | 66 | — | 3 | 23 | 3 | 106 | — | — | 11 | — | 146 | 3 | 7 | 3 | 13 |
| 27 | — | 3 | — | 3 | 67 | 11 | — | — | 7 | 107 | 3 | 29 | 3 | 13 | 147 | — | 3 | 7 | 3 |
| 28 | — | — | 7 | 17 | 68 | 3 | — | 3 | 13 | 108 | 23 | 3 | — | 3 | 148 | — | — | — | — |
| 29 | 3 | — | 3 | 13 | 69 | — | 3 | 17 | 3 | 109 | — | — | — | 7 | 149 | 3 | — | 3 | — |
| 30 | 7 | 3 | — | 3 | 70 | — | 19 | 7 | — | 110 | 3 | — | 3 | — | 150 | 19 | 3 | 11 | 3 |
| 31 | — | — | — | 11 | 71 | 3 | 23 | 3 | — | 111 | 11 | 3 | — | 3 | 151 | — | 17 | 37 | 7 |
| 32 | 3 | 17 | 3 | 7 | 72 | 7 | 3 | — | 3 | 112 | 19 | — | 7 | — | 152 | 3 | — | 3 | 11 |
| 33 | — | 3 | — | 3 | 73 | 17 | — | 11 | — | 113 | 3 | 11 | 3 | 17 | 153 | — | 3 | 29 | 3 |
| 34 | 11 | 7 | — | — | 74 | 3 | — | 3 | 7 | 114 | 7 | 3 | 31 | 3 | 154 | 23 | — | 7 | — |
| 35 | 3 | — | 3 | — | 75 | — | 3 | — | 3 | 115 | — | — | 13 | 19 | 155 | 3 | — | 3 | — |
| 36 | 19 | 3 | — | 3 | 76 | — | 7 | 13 | — | 116 | 3 | — | 3 | 7 | 156 | 7 | 3 | — | 3 |
| 37 | 7 | — | 13 | — | 77 | 3 | — | 3 | 19 | 117 | — | 3 | 11 | 3 | 157 | — | 11 | 19 | — |
| 38 | 3 | — | 3 | — | 78 | 11 | 3 | — | 3 | 118 | — | 7 | — | 29 | 158 | 3 | — | 3 | 7 |
| 39 | 17 | 3 | — | 3 | 79 | 7 | 13 | — | 17 | 119 | 3 | — | 3 | 11 | 159 | 37 | 3 | — | 3 |

**Table 1**. (Continued).

| | 1 | 3 | 7 | 9 | | 1 | 3 | 7 | 9 | | 1 | 3 | 7 | 9 | | 1 | 3 | 7 | 9 |
|---|---|---|---|---|---|---|---|---|---|---|---|---|---|---|---|---|---|---|---|
| 160 | — | 7 | — | — | 200 | 3 | — | 3 | 7 | 240 | 7 | 3 | 29 | 3 | 280 | — | — | 7 | 53 |
| 161 | 3 | — | 3 | — | 201 | — | 3 | — | 3 | 241 | — | 19 | — | 41 | 281 | 3 | 29 | 3 | — |
| 162 | — | 3 | — | 3 | 202 | 43 | 7 | — | — | 242 | 3 | — | 3 | 7 | 282 | 7 | 3 | 11 | 3 |
| 163 | 7 | 23 | — | 11 | 203 | 3 | 19 | 3 | — | 243 | 11 | 3 | — | 3 | 283 | 19 | — | — | 17 |
| 164 | 3 | 31 | 3 | 17 | 204 | 13 | 3 | 23 | 3 | 244 | — | 7 | — | 31 | 284 | 3 | — | 3 | 7 |
| 165 | 13 | 3 | — | 3 | 205 | 7 | — | 11 | 29 | 245 | 3 | 11 | 3 | — | 285 | — | 3 | — | 3 |
| 166 | 11 | — | — | — | 206 | 3 | — | 3 | — | 246 | 23 | 3 | — | 3 | 286 | — | 7 | 47 | 19 |
| 167 | 3 | 7 | 3 | 23 | 207 | 19 | 3 | 31 | 3 | 247 | 7 | — | — | 37 | 287 | 3 | 13 | 3 | — |
| 168 | 41 | 3 | 7 | 3 | 208 | — | — | — | — | 248 | 3 | 13 | 3 | 19 | 288 | 43 | 3 | — | 3 |
| 169 | 19 | — | — | — | 209 | 3 | 7 | 3 | — | 249 | 47 | 3 | 11 | 3 | 289 | 7 | 11 | — | 13 |
| 170 | 3 | 13 | 3 | — | 210 | 11 | 3 | 7 | 3 | 250 | 41 | — | 23 | 13 | 290 | 3 | — | 3 | — |
| 171 | 29 | 3 | 17 | 3 | 211 | — | — | 29 | 13 | 251 | 3 | 7 | 3 | 11 | 291 | 41 | 3 | — | 3 |
| 172 | — | — | 11 | 7 | 212 | 3 | 11 | 3 | — | 252 | — | 3 | 7 | 3 | 292 | 23 | 37 | — | 29 |
| 173 | 3 | — | 3 | 37 | 213 | — | 3 | — | 3 | 253 | — | 17 | 43 | — | 293 | 3 | 7 | 3 | — |
| 174 | — | 3 | — | 3 | 214 | — | — | 19 | 7 | 254 | 3 | — | 3 | — | 294 | 17 | 3 | 7 | 3 |
| 175 | 17 | — | 7 | — | 215 | 3 | — | 3 | 17 | 255 | — | 3 | — | 3 | 295 | 13 | — | — | 11 |
| 176 | 3 | 41 | 3 | 29 | 216 | — | 3 | 11 | 3 | 256 | 13 | 11 | 17 | 7 | 296 | 3 | — | 3 | — |
| 177 | 7 | 3 | — | 3 | 217 | 13 | 41 | 7 | — | 257 | 3 | 31 | 3 | — | 297 | — | 3 | 13 | 3 |
| 178 | 13 | — | — | — | 218 | 3 | 37 | 3 | 11 | 258 | 29 | 3 | 13 | 3 | 298 | 11 | 19 | 29 | 7 |
| 179 | 3 | 11 | 3 | 7 | 219 | 7 | 3 | 13 | 3 | 259 | — | — | 7 | 23 | 299 | 3 | 41 | 3 | — |
| 180 | — | 3 | 13 | 3 | 220 | 31 | — | — | 47 | 260 | 3 | 19 | 3 | — | 300 | — | 3 | 31 | 3 |
| 181 | — | 7 | 23 | 17 | 221 | 3 | — | 3 | 7 | 261 | 7 | 3 | — | 3 | 301 | — | 23 | 7 | — |
| 182 | 3 | — | 3 | 31 | 222 | — | 3 | 17 | 3 | 262 | — | 43 | 37 | 11 | 302 | 3 | — | 3 | 13 |
| 183 | — | 3 | 11 | 3 | 223 | 23 | 7 | — | — | 263 | 3 | — | 3 | 7 | 303 | 7 | 3 | — | 3 |
| 184 | 7 | 19 | — | 43 | 224 | 3 | — | 3 | 13 | 264 | 19 | 3 | — | 3 | 304 | — | 17 | 11 | — |
| 185 | 3 | 17 | 3 | 11 | 225 | — | 3 | 37 | 3 | 265 | 11 | 7 | — | — | 305 | 3 | 43 | 3 | 7 |
| 186 | — | 3 | — | 3 | 226 | 7 | 31 | — | — | 266 | 3 | — | 3 | 17 | 306 | — | 3 | — | 3 |
| 187 | — | — | — | — | 227 | 3 | — | 3 | 43 | 267 | — | 3 | — | 3 | 307 | 37 | 7 | 17 | — |
| 188 | 3 | 7 | 3 | — | 228 | — | 3 | — | 3 | 268 | 7 | — | — | — | 308 | 3 | — | 3 | — |
| 189 | 31 | 3 | 7 | 3 | 229 | 29 | — | — | 11 | 269 | 3 | — | 3 | — | 309 | 11 | 3 | 19 | 3 |
| 190 | — | 11 | — | 23 | 230 | 3 | 7 | 3 | — | 270 | 37 | 3 | — | 3 | 310 | 7 | 29 | 13 | — |
| 191 | 3 | — | 3 | 19 | 231 | — | 3 | 7 | 3 | 271 | — | — | 11 | — | 311 | 3 | 11 | 3 | — |
| 192 | 17 | 3 | 41 | 3 | 232 | 11 | 23 | 13 | 17 | 272 | 3 | 7 | 3 | — | 312 | — | 3 | 53 | 3 |
| 193 | — | — | 13 | 7 | 233 | 3 | — | 3 | — | 273 | — | 3 | 7 | 3 | 313 | 31 | 13 | — | 43 |
| 194 | 3 | 29 | 3 | — | 234 | — | 3 | — | 3 | 274 | — | 13 | 41 | — | 314 | 3 | 7 | 3 | 47 |
| 195 | — | 3 | 19 | 3 | 235 | — | 13 | — | 7 | 275 | 3 | — | 3 | 31 | 315 | 23 | 3 | 7 | 3 |
| 196 | 37 | 13 | 7 | 11 | 236 | 3 | 17 | 3 | 23 | 276 | 11 | 3 | — | 3 | 316 | 29 | — | — | — |
| 197 | 3 | — | 3 | — | 237 | — | 3 | — | 3 | 277 | 17 | 47 | — | 7 | 317 | 3 | 19 | 3 | 11 |
| 198 | 7 | 3 | — | 3 | 238 | — | — | 7 | — | 278 | 3 | 11 | 3 | — | 318 | — | 3 | — | 3 |
| 199 | 11 | — | — | — | 239 | 3 | — | 3 | — | 279 | — | 3 | — | 3 | 319 | — | 31 | 23 | 7 |

## Table 1. (Continued).

| | 1 | 3 | 7 | 9 | | 1 | 3 | 7 | 9 | | 1 | 3 | 7 | 9 | | 1 | 3 | 7 | 9 |
|---|---|---|---|---|---|---|---|---|---|---|---|---|---|---|---|---|---|---|---|
| 320 | 3 | — | 3 | — | 360 | 13 | 3 | — | 3 | 400 | — | — | — | 19 | 440 | 3 | 7 | 3 | — |
| 321 | 13 | 3 | — | 3 | 361 | 23 | — | — | 7 | 401 | 3 | — | 3 | — | 441 | 11 | 3 | 7 | 3 |
| 322 | — | 11 | 7 | — | 362 | 3 | — | 3 | 19 | 402 | — | 3 | — | 3 | 442 | — | — | 19 | 43 |
| 323 | 3 | 53 | 3 | 41 | 363 | — | 3 | — | 3 | 403 | 29 | 37 | 11 | 7 | 443 | 3 | 11 | 3 | 23 |
| 324 | 7 | 3 | 17 | 3 | 364 | 11 | — | 7 | 41 | 404 | 3 | 13 | 3 | — | 444 | — | 3 | — | 3 |
| 325 | — | — | — | — | 365 | 3 | 13 | 3 | — | 405 | — | 3 | — | 3 | 445 | — | 61 | — | 7 |
| 326 | 3 | 13 | 3 | 7 | 366 | 7 | 3 | 19 | 3 | 406 | 31 | 17 | 7 | 13 | 446 | 3 | — | 3 | 41 |
| 327 | — | 3 | 29 | 3 | 367 | — | — | — | 13 | 407 | 3 | — | 3 | — | 447 | 17 | 3 | 11 | 3 |
| 328 | 17 | 7 | 19 | 11 | 368 | 3 | 29 | 3 | 7 | 408 | 7 | 3 | 61 | 3 | 448 | — | — | 7 | 67 |
| 329 | 3 | 37 | 3 | — | 369 | — | 3 | — | 3 | 409 | — | — | 17 | — | 449 | 3 | — | 3 | 11 |
| 330 | — | 3 | — | 3 | 370 | — | 7 | 11 | — | 410 | 3 | 11 | 3 | 7 | 450 | 7 | 3 | — | 3 |
| 331 | 7 | — | 31 | — | 371 | 3 | 47 | 3 | — | 411 | — | 3 | 23 | 3 | 451 | 13 | — | — | — |
| 332 | 3 | — | 3 | — | 372 | 61 | 3 | — | 3 | 412 | 13 | 7 | — | — | 452 | 3 | — | 3 | 7 |
| 333 | — | 3 | 47 | 3 | 373 | 7 | — | 37 | — | 413 | 3 | — | 3 | — | 453 | 23 | 3 | 13 | 3 |
| 334 | 13 | — | — | 17 | 374 | 3 | 19 | 3 | 23 | 414 | 41 | 3 | 11 | 3 | 454 | 19 | 7 | — | — |
| 335 | 3 | 7 | 3 | — | 375 | 11 | 3 | 13 | 3 | 415 | 7 | — | — | — | 455 | 3 | 29 | 3 | 47 |
| 336 | — | 3 | 7 | 3 | 376 | — | 53 | — | — | 416 | 3 | 23 | 3 | 11 | 456 | — | 3 | — | 3 |
| 337 | — | — | 11 | 31 | 377 | 3 | 7 | 3 | — | 417 | 43 | 3 | — | 3 | 457 | 7 | 17 | 23 | 19 |
| 338 | 3 | 17 | 3 | — | 378 | 19 | 3 | 7 | 3 | 418 | 37 | 47 | 53 | 59 | 458 | 3 | — | 3 | 13 |
| 339 | — | 3 | 43 | 3 | 379 | 17 | — | — | 29 | 419 | 3 | 7 | 3 | 13 | 459 | — | 3 | — | 3 |
| 340 | 19 | 41 | — | 7 | 380 | 3 | — | 3 | 31 | 420 | — | 3 | 7 | 3 | 460 | 43 | — | 17 | 11 |
| 341 | 3 | — | 3 | 13 | 381 | 37 | 3 | 11 | 3 | 421 | — | 11 | — | — | 461 | 3 | 7 | 3 | 31 |
| 342 | 11 | 3 | 23 | 3 | 382 | — | — | 43 | 7 | 422 | 3 | 41 | 3 | — | 462 | — | 3 | 7 | 3 |
| 343 | 47 | — | 7 | 19 | 383 | 3 | — | 3 | 11 | 423 | — | 3 | 19 | 3 | 463 | 11 | 41 | — | — |
| 344 | 3 | 11 | 3 | — | 384 | 23 | 3 | — | 3 | 424 | — | — | 31 | 7 | 464 | 3 | — | 3 | — |
| 345 | 7 | 3 | — | 3 | 385 | — | — | 7 | 17 | 425 | 3 | — | 3 | — | 465 | — | 3 | — | 3 |
| 346 | — | — | — | — | 386 | 3 | — | 3 | 53 | 426 | — | 3 | 17 | 3 | 466 | 59 | — | 13 | 7 |
| 347 | 3 | 23 | 3 | 7 | 387 | 7 | 3 | — | 3 | 427 | — | — | 7 | 11 | 467 | 3 | — | 3 | — |
| 348 | 59 | 3 | 11 | 3 | 388 | — | 11 | 13 | — | 428 | 3 | — | 3 | — | 468 | 31 | 3 | 43 | 3 |
| 349 | — | 7 | 13 | — | 389 | 3 | 17 | 3 | 7 | 429 | 7 | 3 | — | 3 | 469 | — | 13 | 7 | 37 |
| 350 | 3 | 31 | 3 | 11 | 390 | 47 | 3 | — | 3 | 430 | 11 | 13 | 59 | 31 | 470 | 3 | — | 3 | 17 |
| 351 | — | 3 | — | 3 | 391 | — | 7 | — | — | 431 | 3 | 19 | 3 | 7 | 471 | 7 | 3 | 53 | 3 |
| 352 | 7 | 13 | — | — | 392 | 3 | — | 3 | — | 432 | 29 | 3 | — | 3 | 472 | — | — | 29 | — |
| 353 | 3 | — | 3 | — | 393 | — | 3 | 31 | 3 | 433 | 61 | 7 | — | — | 473 | 3 | — | 3 | 7 |
| 354 | — | 3 | — | 3 | 394 | 7 | — | — | 11 | 434 | 3 | 43 | 3 | — | 474 | 11 | 3 | 47 | 3 |
| 355 | 53 | 11 | — | — | 395 | 3 | 59 | 3 | 37 | 435 | 19 | 3 | — | 3 | 475 | — | 7 | 67 | — |
| 356 | 3 | 7 | 3 | 43 | 396 | 17 | 3 | — | 3 | 436 | 7 | — | 11 | 17 | 476 | 3 | 11 | 3 | 19 |
| 357 | — | 3 | 7 | 3 | 397 | 11 | 29 | 41 | 23 | 437 | 3 | — | 3 | 29 | 477 | 13 | 3 | 17 | 3 |
| 358 | — | — | 17 | 37 | 398 | 3 | 7 | 3 | — | 438 | 13 | 3 | 41 | 3 | 478 | 7 | — | — | — |
| 359 | 3 | — | 3 | 59 | 399 | 13 | 3 | 7 | 3 | 439 | — | 23 | — | 53 | 479 | 3 | — | 3 | — |

**Table 1**. (Continued).

| | 1 | 3 | 7 | 9 | | 1 | 3 | 7 | 9 | | 1 | 3 | 7 | 9 | | 1 | 3 | 7 | 9 |
|---|---|---|---|---|---|---|---|---|---|---|---|---|---|---|---|---|---|---|---|
| 480 | — | 3 | 11 | 3 | 520 | 7 | 11 | 41 | — | 560 | 3 | 13 | 3 | 71 | 600 | 17 | 3 | — | 3 |
| 481 | 17 | — | — | 61 | 521 | 3 | 13 | 3 | 17 | 561 | 31 | 3 | 41 | 3 | 601 | — | 7 | 11 | 13 |
| 482 | 3 | 7 | 3 | 11 | 522 | 23 | 3 | — | 3 | 562 | 7 | — | 17 | 13 | 602 | 3 | 19 | 3 | — |
| 483 | — | 3 | 7 | 3 | 523 | — | — | — | 13 | 563 | 3 | 43 | 3 | — | 603 | 37 | 3 | — | 3 |
| 484 | 47 | 29 | 37 | 13 | 524 | 3 | 7 | 3 | 29 | 564 | — | 3 | — | 3 | 604 | 7 | — | — | 23 |
| 485 | 3 | 23 | 3 | 43 | 525 | 59 | 3 | 7 | 3 | 565 | — | — | — | — | 605 | 3 | — | 3 | 73 |
| 486 | — | 3 | 31 | 3 | 526 | — | 19 | 23 | 11 | 566 | 3 | 7 | 3 | — | 606 | 11 | 3 | — | 3 |
| 487 | — | 11 | — | 7 | 527 | 3 | — | 3 | — | 567 | 53 | 3 | 7 | 3 | 607 | 13 | — | 59 | — |
| 488 | 3 | 19 | 3 | — | 528 | — | 3 | 17 | 3 | 568 | 13 | — | 11 | — | 608 | 3 | 7 | 3 | — |
| 489 | 67 | 3 | 59 | 3 | 529 | 11 | 67 | — | 7 | 569 | 3 | — | 3 | 41 | 609 | — | 3 | 7 | 3 |
| 490 | 13 | — | 7 | — | 530 | 3 | — | 3 | — | 570 | — | 3 | 13 | 3 | 610 | — | 17 | 31 | 41 |
| 491 | 3 | 17 | 3 | — | 531 | 47 | 3 | 13 | 3 | 571 | — | 29 | — | 7 | 611 | 3 | — | 3 | 29 |
| 492 | 7 | 3 | 13 | 3 | 532 | 17 | — | 7 | 73 | 572 | 3 | 59 | 3 | 17 | 612 | — | 3 | 11 | 3 |
| 493 | — | — | — | 11 | 533 | 3 | — | 3 | 19 | 573 | 11 | 3 | — | 3 | 613 | — | — | 17 | 7 |
| 494 | 3 | — | 3 | 7 | 534 | 7 | 3 | — | 3 | 574 | — | — | 7 | — | 614 | 3 | — | 3 | 11 |
| 495 | — | 3 | — | 3 | 535 | — | 53 | 11 | 23 | 575 | 3 | 11 | 3 | 13 | 615 | — | 3 | 47 | 3 |
| 496 | 11 | 7 | — | — | 536 | 3 | 31 | 3 | 7 | 576 | 7 | 3 | 73 | 3 | 616 | 61 | — | 7 | 31 |
| 497 | 3 | — | 3 | 13 | 537 | 41 | 3 | 19 | 3 | 577 | 29 | 23 | 53 | — | 617 | 3 | — | 3 | 37 |
| 498 | 17 | 3 | — | 3 | 538 | — | 7 | — | 17 | 578 | 3 | — | 3 | 7 | 618 | 7 | 3 | 23 | 3 |
| 499 | 7 | — | 19 | — | 539 | 3 | — | 3 | — | 579 | — | 3 | 11 | 3 | 619 | 41 | 11 | — | — |
| 500 | 3 | — | 3 | — | 540 | 11 | 3 | — | 3 | 580 | — | 7 | — | 37 | 620 | 3 | — | 3 | 7 |
| 501 | — | 3 | 29 | 3 | 541 | 7 | — | — | — | 581 | 3 | — | 3 | 11 | 621 | — | 3 | — | 3 |
| 502 | — | — | 11 | 47 | 542 | 3 | 11 | 3 | 61 | 582 | — | 3 | — | 3 | 622 | — | 7 | 13 | — |
| 503 | 3 | 7 | 3 | — | 543 | — | 3 | — | 3 | 583 | 7 | 19 | 13 | — | 623 | 3 | 23 | 3 | 17 |
| 504 | 71 | 3 | 7 | 3 | 544 | — | — | 13 | — | 584 | 3 | — | 3 | — | 624 | 79 | 3 | — | 3 |
| 505 | — | 31 | 13 | — | 545 | 3 | 7 | 3 | 53 | 585 | — | 3 | — | 3 | 625 | 7 | 13 | — | 11 |
| 506 | 3 | 61 | 3 | 37 | 546 | 43 | 3 | 7 | 3 | 586 | — | 11 | — | — | 626 | 3 | — | 3 | — |
| 507 | 11 | 3 | — | 3 | 547 | — | 13 | — | — | 587 | 3 | 7 | 3 | — | 627 | — | 3 | — | 3 |
| 508 | — | 13 | — | 7 | 548 | 3 | — | 3 | 11 | 588 | — | 3 | 7 | 3 | 628 | 11 | 61 | — | 19 |
| 509 | 3 | 11 | 3 | — | 549 | 17 | 3 | 23 | 3 | 589 | 43 | 71 | — | 17 | 629 | 3 | 7 | 3 | — |
| 510 | — | 3 | — | 3 | 550 | — | — | — | 7 | 590 | 3 | — | 3 | 19 | 630 | — | 3 | 7 | 3 |
| 511 | 19 | — | 7 | — | 551 | 3 | 37 | 3 | — | 591 | 23 | 3 | 61 | 3 | 631 | — | 59 | — | 71 |
| 512 | 3 | 47 | 3 | 23 | 552 | — | 3 | — | 3 | 592 | 31 | — | — | 7 | 632 | 3 | — | 3 | — |
| 513 | 7 | 3 | 11 | 3 | 553 | — | 11 | 7 | 29 | 593 | 3 | 17 | 3 | — | 633 | 13 | 3 | — | 3 |
| 514 | 53 | 37 | — | 19 | 554 | 3 | 23 | 3 | 31 | 594 | 13 | 3 | 19 | 3 | 634 | 17 | — | 11 | 7 |
| 515 | 3 | — | 3 | 7 | 555 | 7 | 3 | — | 3 | 595 | 11 | — | 7 | 59 | 635 | 3 | — | 3 | — |
| 516 | 13 | 3 | — | 3 | 556 | 67 | — | 19 | — | 596 | 3 | 67 | 3 | 47 | 636 | — | 3 | — | 3 |
| 517 | — | 7 | 31 | — | 557 | 3 | — | 3 | 7 | 597 | 7 | 3 | 43 | 3 | 637 | 23 | — | 7 | — |
| 518 | 3 | 71 | 3 | — | 558 | — | 3 | 37 | 3 | 598 | — | 31 | — | 53 | 638 | 3 | 13 | 3 | — |
| 519 | 29 | 3 | — | 3 | 559 | — | 7 | 29 | 11 | 599 | 3 | 13 | 3 | 7 | 639 | 7 | 3 | — | 3 |

**Table 1**. (Continued).

| | 1 | 3 | 7 | 9 | | 1 | 3 | 7 | 9 | | 1 | 3 | 7 | 9 | | 1 | 3 | 7 | 9 |
|---|---|---|---|---|---|---|---|---|---|---|---|---|---|---|---|---|---|---|---|
| 640 | 37 | 19 | 43 | 13 | 680 | 3 | — | 3 | 11 | 720 | 19 | 3 | — | 3 | 760 | 11 | — | — | 7 |
| 641 | 3 | 11 | 3 | 7 | 681 | 7 | 3 | 17 | 3 | 721 | — | — | 7 | — | 761 | 3 | 23 | 3 | 19 |
| 642 | — | 3 | — | 3 | 682 | 19 | — | — | — | 722 | 3 | 31 | 3 | — | 762 | — | 3 | 29 | 3 |
| 643 | 59 | 7 | 41 | 47 | 683 | 3 | — | 3 | 7 | 723 | 7 | 3 | — | 3 | 763 | 13 | 17 | 7 | — |
| 644 | 3 | 17 | 3 | — | 684 | — | 3 | 41 | 3 | 724 | 13 | — | — | 11 | 764 | 3 | — | 3 | — |
| 645 | — | 3 | 11 | 3 | 685 | 13 | 7 | — | 19 | 725 | 3 | — | 3 | 7 | 765 | 7 | 3 | 13 | 3 |
| 646 | 7 | 23 | 29 | — | 686 | 3 | — | 3 | — | 726 | 53 | 3 | 13 | 3 | 766 | 47 | 79 | 11 | — |
| 647 | 3 | — | 3 | 11 | 687 | — | 3 | 13 | 3 | 727 | 11 | 7 | 19 | 29 | 767 | 3 | — | 3 | 7 |
| 648 | — | 3 | 13 | 3 | 688 | 7 | — | 71 | 83 | 728 | 3 | — | 3 | 37 | 768 | — | 3 | — | 3 |
| 649 | — | 43 | 73 | 67 | 689 | 3 | 61 | 3 | — | 729 | 23 | 3 | — | 3 | 769 | — | 7 | 43 | — |
| 650 | 3 | 7 | 3 | 23 | 690 | 67 | 3 | — | 3 | 730 | 7 | 67 | — | — | 770 | 3 | — | 3 | 13 |
| 651 | 17 | 3 | 7 | 3 | 691 | — | 31 | — | 11 | 731 | 3 | 71 | 3 | 13 | 771 | 11 | 3 | — | 3 |
| 652 | — | 11 | 61 | — | 692 | 3 | 7 | 3 | 13 | 732 | — | 3 | 17 | 3 | 772 | 7 | — | — | 59 |
| 653 | 3 | 47 | 3 | 13 | 693 | 29 | 3 | 7 | 3 | 733 | — | — | 11 | 41 | 773 | 3 | 11 | 3 | 71 |
| 654 | 31 | 3 | — | 3 | 694 | 11 | 53 | — | — | 734 | 3 | 7 | 3 | — | 774 | — | 3 | 61 | 3 |
| 655 | — | — | 79 | 7 | 695 | 3 | 17 | 3 | — | 735 | — | 3 | 7 | 3 | 775 | 23 | — | — | — |
| 656 | 3 | — | 3 | — | 696 | — | 3 | — | 3 | 736 | 17 | 37 | 53 | — | 776 | 3 | 7 | 3 | 17 |
| 657 | — | 3 | — | 3 | 697 | — | 19 | — | 7 | 737 | 3 | 73 | 3 | 47 | 777 | 19 | 3 | 7 | 3 |
| 658 | — | 29 | 7 | 11 | 698 | 3 | — | 3 | 29 | 738 | 11 | 3 | 83 | 3 | 778 | 31 | 43 | 13 | — |
| 659 | 3 | 19 | 3 | — | 699 | — | 3 | — | 3 | 739 | 19 | — | 13 | 7 | 779 | 3 | — | 3 | 11 |
| 660 | 7 | 3 | — | 3 | 700 | — | 47 | 7 | 43 | 740 | 3 | 11 | 3 | 31 | 780 | 29 | 3 | 37 | 3 |
| 661 | 11 | 17 | 13 | — | 701 | 3 | — | 3 | — | 741 | — | 3 | — | 3 | 781 | 73 | 13 | — | 7 |
| 662 | 3 | 37 | 3 | 7 | 702 | 7 | 3 | — | 3 | 742 | 41 | 13 | 7 | 17 | 782 | 3 | — | 3 | — |
| 663 | 19 | 3 | — | 3 | 703 | 79 | 13 | 31 | — | 743 | 3 | — | 3 | 43 | 783 | 41 | 3 | 17 | 3 |
| 664 | 29 | 7 | 17 | 61 | 704 | 3 | — | 3 | 7 | 744 | 7 | 3 | 11 | 3 | 784 | — | 11 | 7 | 47 |
| 665 | 3 | — | 3 | — | 705 | 11 | 3 | — | 3 | 745 | — | 29 | — | — | 785 | 3 | — | 3 | 29 |
| 666 | — | 3 | 59 | 3 | 706 | 23 | 7 | 37 | — | 746 | 3 | 17 | 3 | 7 | 786 | 7 | 3 | — | 3 |
| 667 | 7 | — | 11 | — | 707 | 3 | 11 | 3 | — | 747 | 31 | 3 | — | 3 | 787 | 17 | — | — | — |
| 668 | 3 | 41 | 3 | — | 708 | 73 | 3 | 19 | 3 | 748 | — | 7 | — | — | 788 | 3 | — | 3 | 7 |
| 669 | — | 3 | 37 | 3 | 709 | 7 | 41 | 47 | 31 | 749 | 3 | 59 | 3 | — | 789 | 13 | 3 | 53 | 3 |
| 670 | — | — | 19 | — | 710 | 3 | — | 3 | — | 750 | 13 | 3 | — | 3 | 790 | — | 7 | — | 11 |
| 671 | 3 | 7 | 3 | — | 711 | 13 | 3 | 11 | 3 | 751 | 7 | 11 | — | 73 | 791 | 3 | 41 | 3 | — |
| 672 | 11 | 3 | 7 | 3 | 712 | — | 17 | — | — | 752 | 3 | — | 3 | — | 792 | 89 | 3 | — | 3 |
| 673 | 53 | — | — | 23 | 713 | 3 | 7 | 3 | 11 | 753 | 17 | 3 | — | 3 | 793 | 7 | — | — | 17 |
| 674 | 3 | 11 | 3 | 17 | 714 | 37 | 3 | 7 | 3 | 754 | — | 19 | — | — | 794 | 3 | 13 | 3 | — |
| 675 | 43 | 3 | 29 | 3 | 715 | — | 23 | 17 | — | 755 | 3 | 7 | 3 | — | 795 | — | 3 | 73 | 3 |
| 676 | — | — | 67 | 7 | 716 | 3 | 13 | 3 | 67 | 756 | — | 3 | 7 | 3 | 796 | 19 | — | 31 | 13 |
| 677 | 3 | 13 | 3 | — | 717 | 71 | 3 | — | 3 | 757 | 67 | — | — | 11 | 797 | 3 | 7 | 3 | 79 |
| 678 | — | 3 | 11 | 3 | 718 | 43 | 11 | — | 7 | 758 | 3 | — | 3 | — | 798 | 23 | 3 | 7 | 3 |
| 679 | — | — | 7 | 13 | 719 | 3 | — | 3 | 23 | 759 | — | 3 | 71 | 3 | 799 | 61 | — | 11 | 19 |

Table 1. (Continued).

| | 1 | 3 | 7 | 9 | | 1 | 3 | 7 | 9 | | 1 | 3 | 7 | 9 | | 1 | 3 | 7 | 9 |
|---|---|---|---|---|---|---|---|---|---|---|---|---|---|---|---|---|---|---|---|
| 800 | 3 | 53 | 3 | — | 840 | 31 | 3 | 7 | 3 | 880 | 13 | — | — | 23 | 920 | 3 | — | 3 | — |
| 801 | — | 3 | — | 3 | 841 | 13 | 47 | 19 | — | 881 | 3 | 7 | 3 | — | 921 | 61 | 3 | 13 | 3 |
| 802 | 13 | 71 | 23 | 7 | 842 | 3 | — | 3 | — | 882 | — | 3 | 7 | 3 | 922 | — | 23 | — | 11 |
| 803 | 3 | 29 | 3 | — | 843 | — | 3 | 11 | 3 | 883 | — | 11 | — | — | 923 | 3 | 7 | 3 | — |
| 804 | 11 | 3 | 13 | 3 | 844 | 23 | — | — | 7 | 884 | 3 | 37 | 3 | — | 924 | — | 3 | 7 | 3 |
| 805 | 83 | — | 7 | — | 845 | 3 | 79 | 3 | 11 | 885 | 53 | 3 | 17 | 3 | 925 | 11 | 19 | — | 47 |
| 806 | 3 | 11 | 3 | — | 846 | — | 3 | — | 3 | 886 | — | — | — | 7 | 926 | 3 | 59 | 3 | 13 |
| 807 | 7 | 3 | 41 | 3 | 847 | 43 | 37 | 7 | 61 | 887 | 3 | 19 | 3 | 13 | 927 | 73 | 3 | — | 3 |
| 808 | — | 59 | — | — | 848 | 3 | 17 | 3 | 13 | 888 | 83 | 3 | — | 3 | 928 | — | — | 37 | 7 |
| 809 | 3 | — | 3 | 7 | 849 | 7 | 3 | 29 | 3 | 889 | 17 | — | 7 | 11 | 929 | 3 | — | 3 | 17 |
| 810 | — | 3 | 11 | 3 | 850 | — | 11 | 47 | 67 | 890 | 3 | 29 | 3 | 59 | 930 | 71 | 3 | 41 | 3 |
| 811 | — | 7 | — | 23 | 851 | 3 | — | 3 | 7 | 891 | 7 | 3 | 37 | 3 | 931 | — | 67 | 7 | — |
| 812 | 3 | — | 3 | 11 | 852 | — | 3 | — | 3 | 892 | 11 | — | 79 | — | 932 | 3 | — | 3 | 19 |
| 813 | 47 | 3 | 79 | 3 | 853 | 19 | 7 | — | — | 893 | 3 | — | 3 | 7 | 933 | 7 | 3 | — | 3 |
| 814 | 7 | 17 | — | 29 | 854 | 3 | — | 3 | 83 | 894 | — | 3 | 23 | 3 | 934 | — | — | 13 | — |
| 815 | 3 | 31 | 3 | 41 | 855 | 17 | 3 | 43 | 3 | 895 | — | 7 | 13 | 17 | 935 | 3 | 47 | 3 | 7 |
| 816 | — | 3 | — | 3 | 856 | 7 | — | 13 | 11 | 896 | 3 | — | 3 | — | 936 | 11 | 3 | 17 | 3 |
| 817 | — | 11 | 13 | — | 857 | 3 | — | 3 | 23 | 897 | — | 3 | 47 | 3 | 937 | — | 7 | — | 83 |
| 818 | 3 | 7 | 3 | 19 | 858 | — | 3 | 31 | 3 | 898 | 7 | 13 | 11 | 89 | 938 | 3 | 11 | 3 | 41 |
| 819 | — | 3 | 7 | 3 | 859 | 11 | 13 | — | — | 899 | 3 | 17 | 3 | — | 939 | — | 3 | — | 3 |
| 820 | 59 | 13 | 29 | — | 860 | 3 | 7 | 3 | — | 900 | — | 3 | — | 3 | 940 | 7 | — | 23 | 97 |
| 821 | 3 | 43 | 3 | — | 861 | 79 | 3 | 7 | 3 | 901 | — | — | 71 | 29 | 941 | 3 | — | 3 | — |
| 822 | — | 3 | 19 | 3 | 862 | 37 | — | — | — | 902 | 3 | 7 | 3 | — | 942 | — | 3 | 11 | 3 |
| 823 | — | — | — | 7 | 863 | 3 | 89 | 3 | 53 | 903 | 11 | 3 | 7 | 3 | 943 | — | — | — | — |
| 824 | 3 | — | 3 | 73 | 864 | — | 3 | — | 3 | 904 | — | — | 83 | — | 944 | 3 | 7 | 3 | 11 |
| 825 | 37 | 3 | 23 | 3 | 865 | 41 | 17 | 11 | 7 | 905 | 3 | 11 | 3 | — | 945 | 13 | 3 | 7 | 3 |
| 826 | 11 | — | 7 | — | 866 | 3 | — | 3 | — | 906 | 13 | 3 | — | 3 | 946 | — | — | — | 17 |
| 827 | 3 | — | 3 | 17 | 867 | 13 | 3 | — | 3 | 907 | 47 | 43 | 29 | 7 | 947 | 3 | — | 3 | — |
| 828 | 7 | 3 | — | 3 | 868 | — | 19 | 7 | — | 908 | 3 | 31 | 3 | 61 | 948 | 19 | 3 | 53 | 3 |
| 829 | — | — | — | 43 | 869 | 3 | — | 3 | — | 909 | — | 3 | 11 | 3 | 949 | — | 11 | — | 7 |
| 830 | 3 | 19 | 3 | 7 | 870 | 7 | 3 | — | 3 | 910 | 19 | — | 7 | — | 950 | 3 | 13 | 3 | 37 |
| 831 | — | 3 | — | 3 | 871 | 31 | — | 23 | — | 911 | 3 | 31 | 3 | 11 | 951 | — | 3 | 31 | 3 |
| 832 | 53 | 7 | 11 | — | 872 | 3 | 11 | 3 | 7 | 912 | 7 | 3 | — | 3 | 952 | — | 89 | 7 | 13 |
| 833 | 3 | 13 | 3 | 31 | 873 | — | 3 | — | 3 | 913 | 23 | — | — | 13 | 953 | 3 | — | 3 | — |
| 834 | 19 | 3 | 17 | 3 | 874 | — | 7 | — | 13 | 914 | 3 | 41 | 3 | 7 | 954 | 7 | 3 | — | 3 |
| 835 | 7 | — | 61 | 13 | 875 | 3 | — | 3 | 19 | 915 | — | 3 | — | 3 | 955 | — | 41 | 19 | 11 |
| 836 | 3 | — | 3 | — | 876 | — | 3 | 11 | 3 | 916 | — | 7 | 89 | 53 | 956 | 3 | 73 | 3 | 7 |
| 837 | 11 | 3 | — | 3 | 877 | 7 | 31 | 67 | — | 917 | 3 | — | 3 | 67 | 957 | 17 | 3 | 61 | 3 |
| 838 | 17 | 83 | — | — | 878 | 3 | — | 3 | 11 | 918 | — | 3 | — | 3 | 958 | 11 | 7 | — | 43 |
| 839 | 3 | 7 | 3 | 37 | 879 | 59 | 3 | 19 | 3 | 919 | 7 | 29 | 17 | — | 959 | 3 | 53 | 3 | 29 |

**Table 1**. (Continued).

| | 1 | 3 | 7 | 9 | | 1 | 3 | 7 | 9 | | 1 | 3 | 7 | 9 | | 1 | 3 | 7 | 9 |
|---|---|---|---|---|---|---|---|---|---|---|---|---|---|---|---|---|---|---|---|
| 960 | — | 3 | 13 | 3 | 970 | 89 | 31 | 18 | 7 | 980 | 3 | — | 3 | 17 | 990 | — | 3 | — | 3 |
| 961 | 7 | — | 59 | — | 971 | 3 | 11 | 3 | — | 981 | — | 3 | — | 3 | 991 | 11 | 23 | 47 | 7 |
| 962 | 3 | — | 3 | — | 972 | — | 3 | 71 | 3 | 982 | 7 | 11 | 31 | — | 992 | 3 | — | 3 | — |
| 963 | — | 3 | 23 | 3 | 973 | 37 | — | 7 | — | 983 | 3 | — | 3 | — | 993 | — | 3 | 19 | 3 |
| 964 | 31 | — | 11 | — | 974 | 3 | — | 3 | — | 984 | 13 | 3 | 43 | 3 | 994 | — | 61 | 7 | — |
| 965 | 3 | 7 | 3 | 13 | 975 | 7 | 3 | 11 | 3 | 985 | — | 59 | — | — | 995 | 3 | 37 | 3 | 23 |
| 966 | — | 3 | 7 | 3 | 976 | 43 | 13 | — | — | 986 | 3 | 7 | 3 | 71 | 996 | 7 | 3 | — | 3 |
| 967 | 19 | 17 | — | — | 977 | 3 | 29 | 3 | 7 | 987 | — | 3 | 7 | 3 | 997 | 13 | — | 11 | 17 |
| 968 | 3 | 23 | 3 | — | 978 | — | 3 | — | 3 | 988 | 41 | — | — | 11 | 998 | 3 | 67 | 3 | 7 |
| 969 | 11 | 3 | — | 3 | 979 | — | 7 | 97 | 41 | 989 | 3 | 13 | 3 | 19 | 999 | 97 | 3 | 13 | 3 |

**Table 2**. Values of Some Arithmetic Functions.

| $n$ | $\phi(n)$ | $\tau(n)$ | $\sigma(n)$ |
|---|---|---|---|
| 1 | 1 | 1 | 1 |
| 2 | 1 | 2 | 3 |
| 3 | 2 | 2 | 4 |
| 4 | 2 | 3 | 7 |
| 5 | 4 | 2 | 6 |
| 6 | 2 | 4 | 12 |
| 7 | 6 | 2 | 8 |
| 8 | 4 | 4 | 15 |
| 9 | 6 | 3 | 13 |
| 10 | 4 | 4 | 18 |
| 11 | 10 | 2 | 12 |
| 12 | 4 | 6 | 28 |
| 13 | 12 | 2 | 14 |
| 14 | 6 | 4 | 24 |
| 15 | 8 | 4 | 24 |
| 16 | 8 | 5 | 31 |
| 17 | 16 | 2 | 18 |
| 18 | 6 | 6 | 39 |
| 19 | 18 | 2 | 20 |
| 20 | 8 | 6 | 42 |
| 21 | 12 | 4 | 32 |
| 22 | 10 | 4 | 36 |
| 23 | 22 | 2 | 24 |
| 24 | 8 | 8 | 60 |
| 25 | 20 | 3 | 31 |
| 26 | 12 | 4 | 42 |
| 27 | 18 | 4 | 40 |
| 28 | 12 | 6 | 56 |
| 29 | 28 | 2 | 30 |
| 30 | 8 | 8 | 72 |
| 31 | 30 | 2 | 32 |
| 32 | 16 | 6 | 63 |
| 33 | 20 | 4 | 48 |
| 34 | 16 | 4 | 54 |
| 35 | 24 | 4 | 48 |
| 36 | 12 | 9 | 91 |
| 37 | 36 | 2 | 38 |
| 38 | 18 | 4 | 60 |
| 39 | 24 | 4 | 56 |
| 40 | 16 | 8 | 90 |
| 41 | 40 | 2 | 42 |
| 42 | 12 | 8 | 96 |
| 43 | 42 | 2 | 44 |
| 44 | 20 | 6 | 84 |
| 45 | 24 | 6 | 78 |
| 46 | 22 | 4 | 72 |
| 47 | 46 | 2 | 48 |
| 48 | 16 | 10 | 124 |
| 49 | 42 | 3 | 57 |

**Table 2**. (Continued).

| $n$ | $\phi(n)$ | $\tau(n)$ | $\sigma(n)$ |
|---|---|---|---|
| 50 | 20 | 6 | 93 |
| 51 | 32 | 4 | 72 |
| 52 | 24 | 6 | 98 |
| 53 | 52 | 2 | 54 |
| 54 | 18 | 8 | 120 |
| 55 | 40 | 4 | 72 |
| 56 | 24 | 8 | 120 |
| 57 | 36 | 4 | 80 |
| 58 | 28 | 4 | 90 |
| 59 | 58 | 2 | 60 |
| 60 | 16 | 12 | 168 |
| 61 | 60 | 2 | 62 |
| 62 | 30 | 4 | 96 |
| 63 | 36 | 6 | 104 |
| 64 | 32 | 7 | 127 |
| 65 | 48 | 4 | 84 |
| 66 | 20 | 8 | 144 |
| 67 | 66 | 2 | 68 |
| 68 | 32 | 6 | 126 |
| 69 | 44 | 4 | 96 |
| 70 | 24 | 8 | 144 |
| 71 | 70 | 2 | 72 |
| 72 | 24 | 12 | 195 |
| 73 | 72 | 2 | 74 |
| 74 | 36 | 4 | 114 |
| 75 | 40 | 6 | 124 |
| 76 | 36 | 6 | 140 |
| 77 | 60 | 4 | 96 |
| 78 | 24 | 8 | 168 |
| 79 | 78 | 2 | 80 |
| 80 | 32 | 10 | 186 |
| 81 | 54 | 5 | 121 |
| 82 | 40 | 4 | 126 |
| 83 | 82 | 2 | 84 |
| 84 | 24 | 12 | 224 |
| 85 | 64 | 4 | 108 |
| 86 | 42 | 4 | 132 |
| 87 | 56 | 4 | 120 |
| 88 | 40 | 8 | 180 |
| 89 | 88 | 2 | 90 |
| 90 | 24 | 12 | 234 |
| 91 | 72 | 4 | 112 |
| 92 | 44 | 6 | 168 |
| 93 | 60 | 4 | 128 |
| 94 | 46 | 4 | 144 |
| 95 | 72 | 4 | 120 |
| 96 | 32 | 12 | 252 |
| 97 | 96 | 2 | 98 |
| 98 | 42 | 6 | 171 |
| 99 | 60 | 6 | 156 |
| 100 | 40 | 9 | 217 |

## Table 3. Primitive Roots Modulo Primes

The least primitive root $r$ modulo $p$ for each prime $p$, $p < 1000$ is given in the table.

| $p$ | $r$ | $p$ | $r$ | $p$ | $r$ | $p$ | $r$ |
|-----|-----|-----|-----|-----|-----|-----|-----|
| 2 | 1 | 191 | 19 | 439 | 15 | 709 | 2 |
| 3 | 2 | 193 | 5 | 443 | 2 | 719 | 11 |
| 5 | 2 | 197 | 2 | 449 | 3 | 727 | 5 |
| 7 | 3 | 199 | 3 | 457 | 13 | 733 | 6 |
| 11 | 2 | 211 | 2 | 461 | 2 | 739 | 3 |
| 13 | 2 | 223 | 3 | 463 | 3 | 743 | 5 |
| 17 | 3 | 227 | 2 | 467 | 2 | 751 | 3 |
| 19 | 2 | 229 | 6 | 479 | 13 | 757 | 2 |
| 23 | 5 | 233 | 3 | 487 | 3 | 761 | 6 |
| 29 | 2 | 239 | 7 | 491 | 2 | 769 | 11 |
| 31 | 3 | 241 | 7 | 499 | 7 | 773 | 2 |
| 37 | 2 | 251 | 6 | 503 | 5 | 787 | 2 |
| 41 | 6 | 257 | 3 | 509 | 2 | 797 | 2 |
| 43 | 3 | 263 | 5 | 521 | 3 | 809 | 3 |
| 47 | 5 | 269 | 2 | 523 | 2 | 811 | 3 |
| 53 | 2 | 271 | 6 | 541 | 2 | 821 | 2 |
| 59 | 2 | 277 | 5 | 547 | 2 | 823 | 3 |
| 61 | 2 | 281 | 3 | 557 | 2 | 827 | 2 |
| 67 | 2 | 283 | 3 | 563 | 2 | 829 | 2 |
| 71 | 7 | 293 | 2 | 569 | 3 | 839 | 11 |
| 73 | 5 | 307 | 5 | 571 | 3 | 853 | 2 |
| 79 | 3 | 311 | 17 | 577 | 5 | 857 | 3 |
| 83 | 2 | 313 | 10 | 587 | 2 | 859 | 2 |
| 89 | 3 | 317 | 2 | 593 | 3 | 863 | 5 |
| 97 | 5 | 331 | 3 | 599 | 7 | 877 | 2 |
| 101 | 2 | 337 | 10 | 601 | 7 | 881 | 3 |
| 103 | 5 | 347 | 2 | 607 | 3 | 883 | 2 |
| 107 | 2 | 349 | 2 | 613 | 2 | 887 | 5 |
| 109 | 6 | 353 | 3 | 617 | 3 | 907 | 2 |
| 113 | 3 | 359 | 7 | 619 | 2 | 911 | 17 |
| 127 | 3 | 367 | 6 | 631 | 3 | 919 | 7 |
| 131 | 2 | 373 | 2 | 641 | 3 | 929 | 3 |
| 137 | 3 | 379 | 2 | 643 | 11 | 937 | 5 |
| 139 | 2 | 383 | 5 | 647 | 5 | 941 | 2 |
| 149 | 2 | 389 | 2 | 653 | 2 | 947 | 2 |
| 151 | 6 | 397 | 5 | 659 | 2 | 953 | 3 |
| 157 | 5 | 401 | 3 | 601 | 2 | 967 | 5 |
| 163 | 2 | 409 | 21 | 673 | 5 | 971 | 6 |
| 167 | 5 | 419 | 2 | 677 | 2 | 977 | 3 |
| 173 | 2 | 421 | 2 | 683 | 5 | 983 | 5 |
| 179 | 2 | 431 | 7 | 691 | 3 | 991 | 6 |
| 181 | 2 | 433 | 5 | 701 | 2 | 997 | 7 |

**Table 4**.  Indices

| p | \multicolumn{16}{c}{Numbers} |
|---|---|
| | 1 | 2 | 3 | 4 | 5 | 6 | 7 | 8 | 9 | 10 | 11 | 12 | 13 | 14 | 15 | 16 |
| 3 | 2 | 1 | | | | | | | | | | | | | | |
| 5 | 4 | 1 | 3 | 2 | | | | | | | Indices | | | | | |
| 7 | 6 | 2 | 1 | 4 | 5 | 3 | | | | | | | | | | |
| 11 | 10 | 1 | 8 | 2 | 4 | 9 | 7 | 3 | 6 | 5 | | | | | | |
| 13 | 12 | 1 | 4 | 2 | 9 | 5 | 11 | 3 | 8 | 10 | 7 | 6 | | | | |
| 17 | 16 | 14 | 1 | 12 | 5 | 15 | 11 | 10 | 2 | 3 | 7 | 13 | 4 | 9 | 6 | 8 |
| 19 | 18 | 1 | 13 | 2 | 16 | 14 | 6 | 3 | 8 | 17 | 12 | 15 | 5 | 7 | 11 | 4 |
| 23 | 22 | 2 | 16 | 4 | 1 | 18 | 19 | 6 | 10 | 3 | 9 | 20 | 14 | 21 | 17 | 8 |
| 29 | 28 | 1 | 5 | 2 | 22 | 6 | 12 | 3 | 10 | 23 | 25 | 7 | 18 | 13 | 27 | 4 |
| 31 | 30 | 24 | 1 | 18 | 20 | 25 | 28 | 12 | 2 | 14 | 23 | 19 | 11 | 22 | 21 | 0 |
| 37 | 36 | 1 | 26 | 2 | 23 | 27 | 32 | 3 | 16 | 24 | 30 | 28 | 11 | 33 | 13 | 4 |
| 41 | 40 | 26 | 15 | 12 | 22 | 1 | 39 | 38 | 30 | 8 | 3 | 27 | 31 | 25 | 37 | 24 |
| 43 | 42 | 27 | 1 | 12 | 25 | 28 | 35 | 39 | 2 | 10 | 30 | 13 | 32 | 20 | 26 | 24 |
| 47 | 46 | 18 | 20 | 36 | 1 | 38 | 32 | 8 | 40 | 19 | 7 | 10 | 11 | 4 | 21 | 26 |
| 53 | 52 | 1 | 17 | 2 | 47 | 18 | 14 | 3 | 34 | 48 | 6 | 19 | 24 | 15 | 12 | 4 |
| 59 | 58 | 1 | 50 | 2 | 6 | 51 | 18 | 3 | 42 | 7 | 25 | 52 | 45 | 19 | 56 | 4 |
| 61 | 60 | 1 | 6 | 2 | 22 | 7 | 49 | 3 | 12 | 23 | 15 | 8 | 40 | 50 | 28 | 4 |
| 67 | 66 | 1 | 39 | 2 | 15 | 40 | 23 | 3 | 12 | 16 | 59 | 41 | 19 | 24 | 54 | 4 |
| 71 | 70 | 6 | 26 | 12 | 28 | 32 | 1 | 18 | 52 | 34 | 31 | 38 | 39 | 7 | 54 | 24 |
| 73 | 72 | 8 | 6 | 16 | 1 | 14 | 33 | 24 | 12 | 9 | 55 | 22 | 59 | 41 | 7 | 32 |
| 79 | 78 | 4 | 1 | 8 | 62 | 5 | 53 | 12 | 2 | 66 | 68 | 9 | 34 | 57 | 63 | 16 |
| 83 | 82 | 1 | 72 | 2 | 27 | 73 | 8 | 3 | 62 | 28 | 24 | 74 | 77 | 9 | 17 | 4 |
| 89 | 88 | 16 | 1 | 32 | 70 | 17 | 81 | 48 | 2 | 86 | 84 | 33 | 23 | 9 | 71 | 64 |
| 97 | 96 | 34 | 70 | 68 | 1 | 8 | 31 | 6 | 44 | 35 | 86 | 42 | 25 | 65 | 71 | 40 |

| p | \multicolumn{17}{c}{Numbers} |
|---|---|
| | 17 | 18 | 19 | 20 | 21 | 22 | 23 | 24 | 25 | 26 | 27 | 28 | 29 | 30 | 31 | 32 | 33 |
| 19 | 10 | 9 | | | | | | | | | | Indices | | | | | |
| 23 | 7 | 12 | 15 | 5 | 13 | 11 | | | | | | | | | | | |
| 29 | 21 | 11 | 9 | 24 | 17 | 26 | 20 | 8 | 16 | 19 | 15 | 14 | | | | | |
| 31 | 7 | 26 | 4 | 8 | 29 | 17 | 27 | 13 | 10 | 5 | 3 | 16 | 9 | 15 | | | |
| 37 | 7 | 17 | 35 | 25 | 22 | 31 | 15 | 29 | 10 | 12 | 6 | 34 | 21 | 14 | 9 | 5 | 20 |
| 41 | 33 | 16 | 9 | 34 | 14 | 29 | 36 | 13 | 4 | 17 | 5 | 11 | 7 | 23 | 28 | 10 | 18 |
| 43 | 38 | 29 | 19 | 37 | 36 | 15 | 16 | 40 | 8 | 17 | 3 | 5 | 41 | 11 | 34 | 9 | 31 |
| 47 | 16 | 12 | 45 | 37 | 6 | 25 | 5 | 28 | 2 | 29 | 14 | 22 | 35 | 39 | 3 | 44 | 27 |
| 53 | 10 | 35 | 37 | 49 | 31 | 7 | 39 | 20 | 42 | 25 | 51 | 16 | 46 | 13 | 33 | 5 | 23 |
| 59 | 40 | 43 | 38 | 8 | 10 | 26 | 15 | 53 | 12 | 46 | 34 | 20 | 28 | 57 | 49 | 5 | 17 |
| 61 | 47 | 13 | 26 | 24 | 55 | 16 | 57 | 9 | 44 | 41 | 18 | 51 | 35 | 29 | 59 | 5 | 21 |
| 67 | 64 | 13 | 10 | 17 | 62 | 60 | 28 | 42 | 30 | 20 | 51 | 25 | 44 | 55 | 47 | 5 | 32 |
| 71 | 49 | 58 | 16 | 40 | 27 | 37 | 15 | 44 | 56 | 45 | 8 | 13 | 68 | 60 | 11 | 30 | 57 |
| 73 | 21 | 20 | 62 | 17 | 39 | 63 | 46 | 30 | 2 | 67 | 18 | 49 | 35 | 15 | 11 | 40 | 61 |
| 79 | 21 | 6 | 32 | 70 | 54 | 72 | 26 | 13 | 46 | 38 | 3 | 61 | 11 | 67 | 56 | 20 | 69 |
| 83 | 56 | 63 | 47 | 29 | 80 | 25 | 60 | 75 | 54 | 78 | 52 | 10 | 12 | 18 | 38 | 5 | 14 |
| 89 | 6 | 18 | 35 | 14 | 82 | 12 | 57 | 49 | 52 | 39 | 3 | 25 | 59 | 87 | 31 | 80 | 85 |
| 97 | 89 | 78 | 81 | 69 | 5 | 24 | 77 | 76 | 2 | 59 | 18 | 3 | 13 | 9 | 46 | 74 | 60 |

Table 4. (Continued).

| p | 34 | 35 | 36 | 37 | 38 | 39 | 40 | 41 | 42 | 43 | 44 | 45 | 46 | 47 | 48 | 49 |
|---|----|----|----|----|----|----|----|----|----|----|----|----|----|----|----|----|
| | | | | | | | | | | Numbers | | | | | | |
| 37 | 8 | 19 | 18 | | | | | | | | | | | | | |
| 41 | 19 | 21 | 2 | 32 | 35 | 6 | 20 | | | Indices | | | | | | |
| 43 | 23 | 18 | 14 | 7 | 4 | 33 | 22 | 6 | 21 | | | | | | | |
| 47 | 34 | 33 | 30 | 42 | 17 | 31 | 9 | 15 | 24 | 13 | 43 | 41 | 23 | | | |
| 53 | 11 | 9 | 36 | 30 | 38 | 41 | 50 | 45 | 32 | 22 | 8 | 29 | 40 | 44 | 21 | 23 |
| 59 | 41 | 24 | 44 | 55 | 39 | 37 | 9 | 14 | 11 | 33 | 27 | 48 | 16 | 23 | 54 | 36 |
| 61 | 48 | 11 | 14 | 39 | 27 | 46 | 25 | 54 | 56 | 43 | 17 | 34 | 58 | 20 | 10 | 38 |
| 67 | 65 | 38 | 14 | 22 | 11 | 58 | 18 | 53 | 63 | 9 | 61 | 27 | 29 | 50 | 43 | 46 |
| 71 | 55 | 29 | 64 | 20 | 22 | 65 | 46 | 25 | 33 | 48 | 43 | 10 | 21 | 9 | 50 | 2 |
| 78 | 29 | 34 | 28 | 64 | 70 | 65 | 25 | 4 | 47 | 51 | 71 | 13 | 54 | 31 | 38 | 66 |
| 79 | 25 | 37 | 10 | 19 | 36 | 35 | 74 | 75 | 58 | 49 | 76 | 64 | 30 | 59 | 17 | 28 |
| 83 | 57 | 35 | 64 | 20 | 48 | 67 | 30 | 40 | 81 | 71 | 26 | 7 | 61 | 23 | 76 | 16 |
| 89 | 22 | 63 | 34 | 11 | 51 | 24 | 30 | 21 | 10 | 29 | 28 | 72 | 73 | 54 | 65 | 74 |
| 97 | 27 | 32 | 16 | 91 | 19 | 95 | 7 | 85 | 39 | 4 | 58 | 45 | 15 | 84 | 14 | 62 |

| p | 50 | 51 | 52 | 53 | 54 | 55 | 56 | 57 | 58 | 59 | 60 | 61 | 62 | 63 | 64 | 65 |
|---|----|----|----|----|----|----|----|----|----|----|----|----|----|----|----|----|
| | | | | | | | | | | Numbers | | | | | | |
| 53 | 43 | 27 | 26 | | | | | | | | | Indices | | | | |
| 59 | 13 | 32 | 47 | 22 | 35 | 31 | 21 | 30 | 29 | | | | | | | |
| 61 | 45 | 53 | 42 | 33 | 19 | 37 | 52 | 32 | 36 | 31 | 30 | | | | | |
| 67 | 31 | 37 | 21 | 57 | 52 | 8 | 26 | 49 | 45 | 36 | 56 | 7 | 48 | 35 | 6 | 34 |
| 71 | 62 | 5 | 51 | 23 | 14 | 59 | 19 | 42 | 4 | 3 | 66 | 69 | 17 | 53 | 36 | 67 |
| 73 | 10 | 27 | 3 | 53 | 26 | 56 | 57 | 68 | 43 | 5 | 23 | 58 | 19 | 45 | 48 | 60 |
| 79 | 50 | 22 | 42 | 77 | 7 | 52 | 65 | 33 | 15 | 31 | 71 | 45 | 60 | 55 | 24 | 18 |
| 83 | 55 | 46 | 79 | 59 | 53 | 51 | 11 | 37 | 13 | 34 | 19 | 66 | 39 | 70 | 6 | 22 |
| 89 | 68 | 7 | 55 | 78 | 19 | 66 | 41 | 36 | 75 | 43 | 15 | 69 | 47 | 83 | 8 | 5 |
| 97 | 36 | 63 | 93 | 10 | 52 | 87 | 37 | 55 | 47 | 67 | 43 | 64 | 80 | 75 | 12 | 26 |

| p | 66 | 67 | 68 | 69 | 70 | 71 | 72 | 73 | 74 | 75 | 76 | 77 | 78 | 79 | 80 | 81 |
|---|----|----|----|----|----|----|----|----|----|----|----|----|----|----|----|----|
| | | | | | | | | | | Numbers | | | | | | |
| 67 | 33 | | | | | | | | | | | | | | | |
| 71 | 63 | 47 | 61 | 41 | 35 | | | | | | | | Indices | | | |
| 78 | 69 | 50 | 37 | 52 | 42 | 44 | 36 | | | | | | | | | |
| 79 | 73 | 48 | 29 | 27 | 41 | 51 | 14 | 44 | 23 | 47 | 40 | 43 | 39 | | | |
| 83 | 15 | 45 | 58 | 50 | 36 | 33 | 65 | 69 | 21 | 44 | 49 | 32 | 68 | 43 | 31 | 42 |
| 89 | 13 | 56 | 38 | 58 | 79 | 62 | 50 | 20 | 27 | 53 | 67 | 77 | 40 | 42 | 46 | 4 |
| 97 | 94 | 57 | 61 | 51 | 66 | 11 | 50 | 28 | 29 | 72 | 53 | 21 | 33 | 30 | 41 | 88 |

| p | 82 | 83 | 84 | 85 | 86 | 87 | 88 | 89 | 90 | 91 | 92 | 93 | 94 | 95 | 96 |
|---|----|----|----|----|----|----|----|----|----|----|----|----|----|----|----|
| | | | | | | | | | Numbers | | | | | | |
| 83 | 41 | | | | | | | | | | | Indices | | | |
| 89 | 37 | 61 | 26 | 76 | 45 | 60 | 44 | | | | | | | | |
| 97 | 23 | 17 | 73 | 90 | 38 | 83 | 92 | 54 | 79 | 56 | 49 | 20 | 22 | 82 | 48 |

**Table 4**. (Continued).

| p | Indices | | | | | | | | | | | | | | | |
|---|---|---|---|---|---|---|---|---|---|---|---|---|---|---|---|---|
| | 1 | 2 | 3 | 4 | 5 | 6 | 7 | 8 | 9 | 10 | 11 | 12 | 13 | 14 | 15 | 16 |
| 3 | 2 | 1 | | | | | | | | | | | | | | |
| 5 | 2 | 4 | 3 | 1 | | | | | Numbers | | | | | | | |
| 7 | 3 | 2 | 6 | 4 | 5 | 1 | | | | | | | | | | |
| 11 | 2 | 4 | 8 | 5 | 10 | 9 | 7 | 3 | 6 | 1 | | | | | | |
| 13 | 2 | 4 | 8 | 3 | 6 | 12 | 11 | 9 | 5 | 10 | 7 | 1 | | | | |
| 17 | 3 | 9 | 10 | 13 | 5 | 15 | 11 | 16 | 14 | 8 | 7 | 4 | 12 | 2 | 6 | 1 |
| 19 | 2 | 4 | 8 | 16 | 13 | 7 | 14 | 9 | 18 | 17 | 15 | 11 | 3 | 6 | 12 | 5 |
| 23 | 5 | 2 | 10 | 4 | 20 | 8 | 17 | 16 | 11 | 9 | 22 | 18 | 21 | 13 | 19 | 3 |
| 29 | 2 | 4 | 8 | 16 | 3 | 6 | 12 | 24 | 19 | 9 | 18 | 7 | 14 | 28 | 27 | 25 |
| 31 | 3 | 9 | 27 | 19 | 26 | 16 | 17 | 20 | 29 | 25 | 13 | 8 | 24 | 10 | 30 | 28 |
| 37 | 2 | 4 | 8 | 16 | 32 | 27 | 17 | 34 | 31 | 25 | 13 | 26 | 15 | 30 | 23 | 9 |
| 41 | 6 | 36 | 11 | 25 | 27 | 39 | 29 | 10 | 19 | 32 | 28 | 4 | 24 | 21 | 3 | 18 |
| 43 | 3 | 9 | 27 | 38 | 28 | 41 | 37 | 25 | 32 | 10 | 30 | 4 | 12 | 36 | 22 | 23 |
| 47 | 5 | 25 | 31 | 14 | 23 | 21 | 11 | 8 | 40 | 12 | 13 | 18 | 43 | 27 | 41 | 17 |
| 53 | 2 | 4 | 8 | 16 | 32 | 11 | 22 | 44 | 35 | 17 | 34 | 15 | 30 | 7 | 14 | 28 |
| 59 | 2 | 4 | 8 | 16 | 32 | 5 | 10 | 20 | 40 | 21 | 42 | 25 | 50 | 41 | 23 | 46 |
| 61 | 2 | 4 | 8 | 16 | 32 | 3 | 6 | 12 | 24 | 48 | 35 | 9 | 18 | 36 | 11 | 22 |
| 67 | 2 | 4 | 8 | 16 | 32 | 64 | 61 | 55 | 43 | 19 | 38 | 9 | 18 | 36 | 5 | 10 |
| 71 | 7 | 49 | 59 | 58 | 51 | 2 | 14 | 27 | 47 | 45 | 31 | 4 | 28 | 54 | 23 | 19 |
| 73 | 5 | 25 | 52 | 41 | 59 | 3 | 15 | 2 | 10 | 50 | 31 | 9 | 45 | 6 | 30 | 4 |
| 79 | 3 | 9 | 27 | 2 | 6 | 18 | 54 | 4 | 12 | 36 | 29 | 8 | 24 | 72 | 58 | 16 |
| 83 | 2 | 4 | 8 | 16 | 32 | 64 | 45 | 7 | 14 | 28 | 56 | 29 | 58 | 33 | 66 | 49 |
| 89 | 3 | 9 | 27 | 81 | 65 | 17 | 51 | 64 | 14 | 42 | 37 | 22 | 66 | 20 | 60 | 2 |
| 97 | 5 | 25 | 28 | 43 | 21 | 8 | 40 | 6 | 30 | 53 | 71 | 64 | 29 | 48 | 46 | 36 |

| p | Indices | | | | | | | | | | | | | | | |
|---|---|---|---|---|---|---|---|---|---|---|---|---|---|---|---|---|
| | 17 | 18 | 19 | 20 | 21 | 22 | 23 | 24 | 25 | 26 | 27 | 28 | 29 | 30 | 31 | 32 | 33 |
| 19 | 10 | 1 | | | | | | | Numbers | | | | | | | |
| 23 | 15 | 6 | 7 | 12 | 14 | 1 | | | | | | | | | | |
| 29 | 21 | 13 | 26 | 23 | 17 | 5 | 10 | 20 | 11 | 22 | 15 | 1 | | | | |
| 31 | 22 | 4 | 12 | 5 | 15 | 14 | 11 | 2 | 6 | 18 | 23 | 7 | 21 | 1 | | |
| 37 | 18 | 36 | 35 | 33 | 29 | 21 | 5 | 10 | 20 | 3 | 6 | 12 | 24 | 11 | 22 | 7 | 14 |
| 41 | 26 | 33 | 34 | 40 | 35 | 5 | 30 | 16 | 14 | 2 | 12 | 31 | 22 | 9 | 13 | 37 | 17 |
| 43 | 26 | 35 | 19 | 14 | 42 | 40 | 34 | 16 | 5 | 15 | 2 | 6 | 18 | 11 | 33 | 13 | 39 |
| 47 | 38 | 2 | 10 | 3 | 15 | 28 | 46 | 42 | 22 | 16 | 33 | 24 | 26 | 36 | 39 | 7 | 35 |
| 53 | 3 | 6 | 12 | 24 | 48 | 43 | 33 | 13 | 26 | 52 | 51 | 49 | 45 | 37 | 21 | 42 | 31 |
| 59 | 33 | 7 | 14 | 28 | 56 | 53 | 47 | 35 | 11 | 22 | 44 | 29 | 58 | 57 | 55 | 51 | 43 |
| 61 | 44 | 27 | 54 | 47 | 33 | 5 | 10 | 20 | 40 | 19 | 38 | 15 | 30 | 60 | 59 | 57 | 53 |
| 67 | 20 | 40 | 13 | 26 | 52 | 37 | 7 | 14 | 28 | 56 | 45 | 23 | 46 | 25 | 50 | 33 | 66 |
| 71 | 62 | 8 | 56 | 37 | 46 | 38 | 53 | 16 | 41 | 3 | 21 | 5 | 35 | 32 | 11 | 6 | 42 |
| 73 | 20 | 27 | 62 | 18 | 17 | 12 | 60 | 8 | 40 | 54 | 51 | 36 | 34 | 24 | 47 | 16 | 7 |
| 79 | 48 | 65 | 37 | 32 | 17 | 51 | 74 | 64 | 34 | 23 | 69 | 49 | 68 | 46 | 59 | 19 | 57 |
| 83 | 15 | 30 | 60 | 37 | 74 | 65 | 47 | 11 | 22 | 44 | 5 | 10 | 20 | 40 | 80 | 77 | 71 |
| 89 | 6 | 18 | 54 | 73 | 41 | 34 | 13 | 39 | 28 | 84 | 74 | 44 | 43 | 40 | 31 | 4 | 12 |
| 97 | 83 | 27 | 38 | 93 | 77 | 94 | 82 | 22 | 13 | 65 | 34 | 73 | 74 | 79 | 7 | 35 | 78 |

**Table 4.** (Continued).

| p | Indices | | | | | | | | | | | | | | | |
|---|----|----|----|----|----|----|----|----|----|----|----|----|----|----|----|----|
|   | 34 | 35 | 36 | 37 | 38 | 39 | 40 | 41 | 42 | 43 | 44 | 45 | 46 | 47 | 48 | 49 |
| 37 | 28 | 19 | 1 | | | | | | | Numbers | | | | | | |
| 41 | 20 | 38 | 23 | 15 | 8 | 7 | 1 | | | | | | | | | |
| 43 | 31 | 7 | 21 | 20 | 17 | 8 | 24 | 29 | 1 | | | | | | | |
| 47 | 34 | 29 | 4 | 20 | 6 | 30 | 9 | 45 | 37 | 44 | 32 | 19 | 1 | | | |
| 53 | 9 | 18 | 36 | 19 | 38 | 23 | 46 | 39 | 25 | 50 | 47 | 41 | 29 | 5 | 10 | 20 |
| 59 | 27 | 54 | 49 | 39 | 19 | 38 | 17 | 34 | 9 | 18 | 36 | 13 | 26 | 52 | 45 | 31 |
| 61 | 45 | 29 | 58 | 55 | 49 | 37 | 13 | 26 | 52 | 43 | 25 | 50 | 39 | 17 | 34 | 7 |
| 67 | 65 | 63 | 59 | 51 | 35 | 3 | 6 | 12 | 24 | 48 | 29 | 58 | 49 | 31 | 62 | 57 |
| 71 | 10 | 70 | 64 | 22 | 12 | 13 | 20 | 69 | 57 | 44 | 24 | 26 | 40 | 67 | 43 | 17 |
| 73 | 35 | 29 | 72 | 68 | 48 | 21 | 32 | 14 | 70 | 58 | 71 | 63 | 23 | 42 | 64 | 28 |
| 79 | 13 | 39 | 38 | 35 | 26 | 78 | 76 | 70 | 52 | 77 | 73 | 61 | 25 | 75 | 67 | 43 |
| 83 | 59 | 35 | 70 | 57 | 31 | 62 | 41 | 82 | 81 | 79 | 75 | 67 | 51 | 19 | 38 | 76 |
| 89 | 36 | 19 | 57 | 82 | 68 | 26 | 78 | 56 | 79 | 59 | 88 | 86 | 80 | 62 | 8 | 24 |
| 97 | 2 | 10 | 50 | 56 | 86 | 42 | 16 | 80 | 12 | 60 | 9 | 45 | 31 | 58 | 96 | 92 |

| p | Indices | | | | | | | | | | | | | | | |
|---|----|----|----|----|----|----|----|----|----|----|----|----|----|----|----|----|
|   | 50 | 51 | 52 | 53 | 54 | 55 | 56 | 57 | 58 | 59 | 60 | 61 | 62 | 63 | 64 | 65 |
| 53 | 40 | 27 | 1 | | | | | | | Numbers | | | | | | |
| 59 | 3 | 6 | 12 | 24 | 48 | 37 | 15 | 30 | 1 | | | | | | | |
| 61 | 14 | 28 | 56 | 51 | 41 | 21 | 42 | 23 | 46 | 31 | 1 | | | | | |
| 67 | 47 | 27 | 54 | 41 | 15 | 30 | 60 | 53 | 39 | 11 | 22 | 44 | 21 | 42 | 17 | 34 |
| 71 | 48 | 52 | 9 | 63 | 15 | 34 | 25 | 33 | 18 | 55 | 30 | 68 | 50 | 66 | 36 | 39 |
| 73 | 67 | 43 | 69 | 53 | 46 | 11 | 55 | 56 | 61 | 13 | 65 | 33 | 19 | 22 | 37 | 39 |
| 79 | 50 | 71 | 55 | 7 | 21 | 63 | 31 | 14 | 42 | 47 | 62 | 28 | 5 | 15 | 45 | 56 |
| 83 | 69 | 55 | 27 | 54 | 25 | 50 | 17 | 34 | 68 | 53 | 23 | 46 | 9 | 18 | 36 | 72 |
| 89 | 72 | 38 | 25 | 75 | 47 | 52 | 67 | 23 | 69 | 29 | 87 | 83 | 71 | 35 | 16 | 48 |
| 97 | 72 | 69 | 54 | 76 | 89 | 57 | 91 | 67 | 44 | 26 | 33 | 68 | 49 | 51 | 61 | 14 |

| p | Indices | | | | | | | | | | | | | | | |
|---|----|----|----|----|----|----|----|----|----|----|----|----|----|----|----|----|
|   | 66 | 67 | 68 | 69 | 70 | 71 | 72 | 73 | 74 | 75 | 76 | 77 | 78 | 79 | 80 | 81 |
| 67 | 1 | | | | | | | | | | | | | | | |
| 71 | 60 | 65 | 29 | 61 | 1 | | | | | | | Numbers | | | | |
| 73 | 49 | 26 | 57 | 66 | 38 | 44 | 1 | | | | | | | | | |
| 79 | 10 | 30 | 11 | 33 | 20 | 60 | 22 | 66 | 40 | 41 | 44 | 53 | 1 | | | |
| 83 | 61 | 39 | 78 | 73 | 63 | 43 | 3 | 6 | 12 | 24 | 48 | 13 | 26 | 52 | 21 | 42 |
| 89 | 55 | 76 | 50 | 61 | 5 | 15 | 45 | 46 | 49 | 58 | 85 | 77 | 53 | 70 | 32 | 7 |
| 97 | 70 | 59 | 4 | 20 | 3 | 15 | 75 | 84 | 32 | 63 | 24 | 23 | 18 | 90 | 62 | 19 |

| p | Indices | | | | | | | | | | | | | | | |
|---|----|----|----|----|----|----|----|----|----|----|----|----|----|----|----|----|
|   | 82 | 83 | 84 | 85 | 86 | 87 | 88 | 89 | 90 | 91 | 92 | 93 | 94 | 95 | 96 | |
| 83 | 1 | | | | | | | | | | | | | | | |
| 89 | 21 | 63 | 11 | 33 | 10 | 30 | 1 | | | | | | Numbers | | | |
| 97 | 95 | 87 | 47 | 41 | 11 | 55 | 81 | 17 | 85 | 37 | 88 | 52 | 66 | 39 | 1 | |

# Table 5. Simple Continued Fractions for Square Roots of Positive Integers

| $d$ | $\sqrt{d}$ | $d$ | $\sqrt{d}$ |
|---|---|---|---|
| 2 | $[1;\overline{2}]$ | 53 | $[7;\overline{3,1,1,3,14}]$ |
| 3 | $[1;\overline{1,2}]$ | 54 | $[7;\overline{2,1,6,1,2,14}]$ |
| 5 | $[2;\overline{4}]$ | 55 | $[7;\overline{2,2,2,14}]$ |
| 6 | $[2;\overline{2,4}]$ | 56 | $[7;\overline{2,14}]$ |
| 7 | $[2;\overline{1,1,1,4}]$ | 57 | $[7;\overline{1,1,4,1,1,14}]$ |
| 8 | $[2;\overline{1,4}]$ | 58 | $[7;\overline{1,1,1,1,1,1,14}]$ |
| 10 | $[3;\overline{6}]$ | 59 | $[7;\overline{1,2,7,2,1,14}]$ |
| 11 | $[3;\overline{3,6}]$ | 60 | $[7;\overline{1,2,1,14}]$ |
| 12 | $[3;\overline{2,6}]$ | 61 | $[7;\overline{1,4,3,1,2,2,1,3,4,1,14}]$ |
| 13 | $[3;\overline{1,1,1,1,6}]$ | 62 | $[7;\overline{1,6,1,14}]$ |
| 14 | $[3;\overline{1,2,1,6}]$ | 63 | $[7;\overline{1,14}]$ |
| 15 | $[3;\overline{1,6}]$ | 65 | $[8;\overline{16}]$ |
| 17 | $[4;\overline{8}]$ | 66 | $[8;\overline{8,16}]$ |
| 18 | $[4;\overline{4,8}]$ | 67 | $[8;\overline{5,2,1,1,7,1,1,2,5,16}]$ |
| 19 | $[4;\overline{2,1,3,1,2,8}]$ | 68 | $[8;\overline{4,16}]$ |
| 20 | $[4;\overline{2,8}]$ | 69 | $[8;\overline{3,3,1,4,1,3,3,16}]$ |
| 21 | $[4;\overline{1,1,2,1,1,8}]$ | 70 | $[8;\overline{2,1,2,1,2,16}]$ |
| 22 | $[4;\overline{1,2,4,2,1,8}]$ | 71 | $[8;\overline{2,2,1,7,1,2,2,16}]$ |
| 23 | $[4;\overline{1,3,1,8}]$ | 72 | $[8;\overline{2,16}]$ |
| 24 | $[4;\overline{1,8}]$ | 73 | $[8;\overline{1,1,5,5,1,1,16}]$ |
| 26 | $[5;\overline{10}]$ | 74 | $[8;\overline{1,1,1,1,16}]$ |
| 27 | $[5;\overline{5,10}]$ | 75 | $[8;\overline{1,1,1,16}]$ |
| 28 | $[5;\overline{3,2,3,10}]$ | 76 | $[8;\overline{1,2,1,1,5,4,5,1,1,2,1,16}]$ |
| 29 | $[5;\overline{2,1,1,2,10}]$ | 77 | $[8;\overline{1,3,2,3,1,16}]$ |
| 30 | $[5;\overline{2,10}]$ | 78 | $[8;\overline{1,4,1,16}]$ |
| 31 | $[5;\overline{1,1,3,5,3,1,1,10}]$ | 79 | $[8;\overline{1,7,1,16}]$ |
| 32 | $[5;\overline{1,1,1,10}]$ | 80 | $[8;\overline{1,16}]$ |
| 33 | $[5;\overline{1,2,1,10}]$ | 82 | $[9;\overline{18}]$ |
| 34 | $[5;\overline{1,4,1,10}]$ | 83 | $[9;\overline{9,18}]$ |
| 35 | $[5;\overline{1,10}]$ | 84 | $[9;\overline{6,18}]$ |
| 37 | $[6;\overline{12}]$ | 85 | $[9;\overline{4,1,1,4,18}]$ |
| 38 | $[6;\overline{6,12}]$ | 86 | $[9;\overline{3,1,1,1,8,1,1,1,3,18}]$ |
| 39 | $[6;\overline{4,12}]$ | 87 | $[9;\overline{3,18}]$ |
| 40 | $[6;\overline{3,12}]$ | 88 | $[9;\overline{2,1,1,1,2,18}]$ |
| 41 | $[6;\overline{2,2,12}]$ | 89 | $[9;\overline{2,3,3,2,18}]$ |
| 42 | $[6;\overline{2,12}]$ | 90 | $[9;\overline{2,18}]$ |
| 43 | $[6;\overline{1,1,3,1,5,1,3,1,1,12}]$ | 91 | $[9;\overline{1,1,5,1,5,1,1,18}]$ |
| 44 | $[6;\overline{1,1,1,2,1,1,1,12}]$ | 92 | $[9;\overline{1,1,2,4,2,1,1,18}]$ |
| 45 | $[6;\overline{1,2,2,2,1,12}]$ | 93 | $[9;\overline{1,1,4,6,4,1,1,18}]$ |
| 46 | $[6;\overline{1,3,1,1,2,6,2,1,1,3,1,12}]$ | 94 | $[9;\overline{1,2,3,1,1,5,1,8,1,5,1,1,3,2,1,18}]$ |
| 47 | $[6;\overline{1,5,1,12}]$ | 95 | $[9;\overline{1,2,1,18}]$ |
| 48 | $[6;\overline{1,12}]$ | 96 | $[9;\overline{1,3,1,18}]$ |
| 50 | $[7;\overline{14}]$ | 97 | $[9;\overline{1,5,1,1,1,1,1,1,5,18}]$ |
| 51 | $[7;\overline{7,14}]$ | 98 | $[9;\overline{1,8,1,18}]$ |
| 52 | $[7;\overline{4,1,2,1,4,14}]$ | 99 | $[9;\overline{1,18}]$ |

# Answers to Selected Problems

## Section 1.1

3. 0
9. a) no  b) yes  c) yes  d) no  e) yes
10. a) -1  b) 0  c) 3  d) -2  e) 3  f) -4
13. 0 when $x$ is an integer, -1 when $x$ is an integer
26. 1,2,3,4,6,8,11,13,16,18

## Section 1.2

1. a) 20  b) 55  c) 385  d) 2047
2. a) 14  b) 154  c) 10  d) 168
3. a) 243  b) 720  c) 14400  d) 32768
4. a) 503284375  b) 34560  c) 0  d) 14400
5. a) $5 \sum_{i=m}^{n} a_i$  b) $- \sum_{i=m}^{n} b_i$  c) $\sum_{i=m}^{n} a_i - \sum_{i=m}^{n} b_i$  d) $3 \sum_{i=m}^{n} a_i + 4 \sum_{i=m}^{n} b_i$
6. 1, 2, 6, 24, 120, 720, 5040, 40320, 362880, 3628800
7. $2^{100} < (50!)^2 < 100! < 100^{100}$
8. a) $k^n \prod_{i=1}^{n} a_i$  b) $n! \prod_{i=1}^{n} a_i$  c) $\left[ \prod_{i=1}^{n} a_i \right]^k$
9. a) 315  b) 540  c) 90  d) 195
10. a) 17915904  b) 36288000  c) 13824  d) 0
11. a) 2352  b) 6000  c) 1320  d) 5544
21. $x = y = 1, z = 2$

## Section 1.3

2. $n(n+1)$
4. $n/(n+1)$
5. $A^n = \begin{bmatrix} 1 & n \\ 0 & 1 \end{bmatrix}$
11. $2^{n(n+1)/2}$
15. $2 - (1/2^{100})$
28. The statement for one horse does not imply the statement for two horses.
30. $2^n$
31. 65536
37. a) 55  b) 233  c) 610  d) 1597  e) 4181  f) 46368
39. $f_{2n}$
52. b) more than 500 billion years

## Section 1.4

1. a) 1  b) 50  c) 1140  d) 462 e) 120 f) 1
2. 84, 126, 210
4. $\begin{pmatrix} 200 \\ 99 \end{pmatrix} 2^{99} \, 3^{101}$
6. both are $2^{n-1}$

## Section 1.5

1. $99 = 3 \cdot 33$, $145 = 5 \cdot 29$, $343 = 7 \cdot 49$, $0 = 888 \cdot 0$
2. (a), (c), (d), (e)
3. a) 5,15  b) 17,0  c) −3,7  d) −6,2
4. $a = \pm b$
15. b) 3
20. 200, 40, 8, 1
21. 128, 18
22. 533
24. 250
39. 39,59,89,134,67,101,152,76,38,19,29,44,22,11,17,26,13,20,10,5,8,4,2,1

## Section 1.6

1. $(5554)_7$, $(2112)_{10}$
2. $(328)_{10}$, $(11111000000)_2$
3. $(8F5)_{16}$, $(74E)_{16}$
4. $(10101011110011011110111)_2$, $(110111101111010110011101101)_2$, $(1001101000001011)_2$
7. $-39, 26$
8. $(1001)_{-2}$, $(110011)_{-2}$, $(1001101)_{-2}$
16 a) 001001 b) 000000 c) 111000 d) 101100
17 a) -6 b) 2 c) -14 d) 0
20. $14 = 2 \cdot 3! + 1 \cdot 2!$, $56 = 2 \cdot 4! + 1 \cdot 3! + 1 \cdot 2!$, $384 = 3 \cdot 5! + 1 \cdot 4!$

## Section 1.7

1. $(10010110110)_2$
2. $(111110111)_2$
3. $(10110001101)_2$
4. $(1110)_2$, $(10001)_2$
5. $(3314430)_5$
6. $(3444230)_5$
7. $(4320023)_5$
8 $(22)_5$, $(313)_5$
9. $(16665)_{16}$
10. $(33EF)_{16}$
11. $(B705736)_{16}$
12. $(11C)_{16}$, $(2B95)_{16}$

18. a) 7 gross, 7 dozen, and 8 eggs   b) 11 gross, 5 dozen, and 11 eggs
  c) 3 gross, 11 dozen, and 6 eggs

## Section 1.8

1. a) yes b) no c) yes d) yes e) yes f) yes

## Section 1.9

1. a) prime b) prime c) prime d) composite  e) prime f) composite
7. 3, 7, 31, 211, 2311, 59
10. a) 24, 25, 26, 27, 28  b) 1000001! + 2,1000001! + 3,...,1000001! + 1000001
12. a)  $50 = 3+47$  b)  $98 = 79 + 19$  c)  $102 = 5 + 97$ d)  $144 = 5 + 139$
  e)  $200 = 3 + 197$  f)  $222 = 211 + 11$
13. a)  $7 = 2 + 2 + 3$  b)  $17 = 13 + 2 + 2$  c)  $27 = 23 + 2 + 2$
  d)  $97 = 3 + 8 + 89$ e)  $101 = 97 + 2 + 2$  f)  $199 = 193 + 2 + 2$
17. 53
20. a) 1,3,7,9,13,15,21,25,31,33,37,43,49,51,63,67,69,73,75,79,87,93,99
21.  $2^3$  and  $3^2$

## Section 2.1

1. a) 5  b) 111  c) 6  d) 1  e) 11  f) 2
2.  $a$
3. 1
4. 1 if  $a$  is odd and  $b$  is even or *vice versa*, 2 otherwise
17. a) 2  b) 5  c) 99  d) 3  e) 7  f) 1001
18. 66,70,105; 66,70,165; or 42,70,165
22.  $(3k+2, 5k+3) = 1$  since  $5(3k+2)-3(5k+3) = 1$
24.  $\dfrac{0}{1}, \dfrac{1}{7}, \dfrac{1}{6}, \dfrac{1}{5}, \dfrac{1}{4}, \dfrac{2}{7}, \dfrac{1}{3}, \dfrac{2}{5}, \dfrac{3}{7}, \dfrac{1}{2}, \dfrac{4}{7}, \dfrac{3}{5}, \dfrac{2}{3}, \dfrac{5}{7}, \dfrac{3}{4}, \dfrac{4}{5}, \dfrac{5}{6}, \dfrac{6}{7}, \dfrac{1}{1}$

## Section 2.2

1. a) 15  b) 6  c) 2  d) 5
2. a)  $15 = 2 \cdot 45 + (-1)75$  b)  $6 = 6 \cdot 222 + (-13)102$
  c)  $2 = 65 \cdot 1414 + (-138)666$  d)  $5 = 800 \cdot 44350 + (-1707)20785$
3. a) 1  b) 7  c) 5
4. a)  $1 = 1 \cdot 6 + 1 \cdot 10 + (-1)15$  b)  $7 = 0 \cdot 70 + (-1)98 + 1 \cdot 105$
  c)  $5 = -5 \cdot 280 + 4 \cdot 330 + (-1)405 + 1 \cdot 490$
5. 2
10. 2

## Section 2.3

1. a)  $2^2 \cdot 3^2$  b)  $3 \cdot 13$  c)  $2^2 \cdot 5^2$  d)  $17^2$  e)  $2 \cdot 3 \cdot 37$  f)  $2^8$  g)  $5 \cdot 103$  h)  $23 \cdot 43$
  i)  $2^4 \cdot 3^2 \cdot 5 \cdot 7$  j)  $2^6 5^3$  k)  $3 \cdot 5 \cdot 7^2 \cdot 13$  l)  $3^2 \cdot 11 \cdot 101$
11.  $2^{18} \cdot 3^8 \cdot 5^4 \cdot 7^2 \cdot 11 \cdot 13 \cdot 17 \cdot 19$
12. 249, 331

13. 300, 301, 302, 303, 304
16. 5,9,13,17,21,29,33,37,41,49,53,57,61,69,73,77,89,93,97,101
18. $693 = 21 \cdot 33 = 9 \cdot 77$
20. a) 24  b) 210  c) 140  d) 11211  e) 80640  f) 342657
21. a) $2^2 \ 3^3 \ 5^3 \ 7^2$ , $2^7 \ 3^5 \ 5^5 \ 7^7$  b) 1, $2 \cdot 3 \cdot 5 \cdot 7 \cdot 11 \cdot 13 \cdot 17 \cdot 19 \cdot 23 \cdot 29$
  c) $2 \cdot 5 \cdot 11$, $2^3 \cdot 3 \cdot 5^7 \cdot 7 \cdot 11^{13} \cdot 13$  d) $101^{1000}$, $41^{11} \ 47^{11} 79^{111} \ 83^{111} \ 101^{1001}$
23. 2121
24. 18,540; 36,270; 54, 180; 90, 108
36. 308, 490
38. 30, 1001
48. a) $3^3 \cdot 7 \cdot 11 \cdot 13 \cdot 37$  b) $3^2 \cdot 11 \cdot 73 \cdot 101 \cdot 137$  c) $7 \cdot 31 \cdot 151$  d) $3^2 \cdot 5 \cdot 7 \cdot 13 \cdot 17 \cdot 241$
  e) $3^2 \cdot 7 \cdot 11 \cdot 31 \cdot 151 \cdot 331$  f) $3^3 \cdot 5 \cdot 7 \cdot 13 \cdot 19 \cdot 37 \cdot 73 \cdot 109$
49. 103
55. $m = n$, $m = 4$ and $n = 2$, or $m = 2$ and $n = 4$

### Section 2.4

1. a) $3 \cdot 5^2 \cdot 7^3 \cdot 13 \cdot 101$  b) $11^3 \cdot 13 \cdot 19 \cdot 641$  c) $13 \cdot 17 \cdot 19 \cdot 47 \cdot 71 \cdot 97$
2. a) $83 \cdot 97$  b) 73  c) $17 \cdot 641$  d) $103 \cdot 107$  e) $1601 \cdot 1999$  f) $4957 \cdot 4967$
8. $17 \cdot 347$
12. a) $13 \cdot 17$  b) $41 \cdot 61$  c) $293 \cdot 3413$
13. $5 \cdot 13 \cdot 37 \cdot 109$
18. 7
19. $2^n \log_{10} 2$

### Section 2.5

1. a) $x = 33 + 5n$, $y = -11-2n$  b) $x = -300 + 13n$, y $= 400-17n$
  c) $x = 21 + 14n$, $y = -21-21n$  d) no solution
  e) $x = 889 + 1969 \ n$, $y = -633-1402n$
2. 39 French francs, 11 Swiss francs
3. 17 apples, 23 oranges
4. 18
5. a) (14-cent stamps,21-cent stamps) $= (25,0)$, $(22,2)$, $(19,4)$, $(16,6)$, $(13,8)$,
  $(10,10)$, $(7,12)$, $(4,14)$, $(1,16)$
b) no solution
c) (14-cent stamps, 21-cent stamps) $= (54,1)$, $(51,3)$, $(48,5)$, $(45,7)$, $(42,9)$,
  $(39,11)$, $(36,13)$, $(33,15)$, $(30,17)$, $(27,19)$, $(24,21)$, $(21,23)$, $(18,25)$,
  $(15,27)$, $(12,29)$, $(9,31)$, $(6,33)$, $(3,35)$, $(0,37)$
10. a) 3  b) 29  c) 242
11. a) $x = 98-6n$, $y = 1 + 7n$, $z = 1-n$  b) no solution
  c) $x = 50-n$, $y = -100 + 3n$, $z = 150-3n$, $w = n$
12. (nickels, dimes, quarters) $= (20, 0, 4)$, $(17, 4, 3)$, $(14, 8, 2)$, $(11, 12, 1)$,
  $(8, 16, 0)$
13. 9 first-class, 19 second-class, 41 standby
14. no
20. $(x,y) = (15,210)$, $(16,112)$, $(18,63)$, $(21,42)$, $(28,28)$, $(42,21)$, $(63,18)$,

$(112,16)$, $(210,15)$, $(13,-182)$, $(12,-84)$, $(10,-35)$, $(7,-14)$, $(-14,7)$, $(-35,10)$, $(-84,12)$, or $(-182,13)$

## Section 3.1

2. a) yes  b) yes  c) no  d) no  e) yes  f) yes
3. a) 1,2,11,22  b) 1,3,9,27,37,111,333,999  c) 1,11,121,1331
6. a) 9  b) 9  c) 0  d) 12  e) 4  f) 1
7. a) 1  b) 5  c) 9  d) 13

| 12. + | 0 1 2 3 4 5 |
|---|---|
| 0 | 0 1 2 3 4 5 |
| 1 | 1 2 3 4 5 0 |
| 2 | 2 3 4 5 0 1 |
| 3 | 3 4 5 0 1 2 |
| 4 | 4 5 0 1 2 3 |
| 5 | 5 0 1 2 3 4 |

| 13. - | 0 1 2 3 4 5 |
|---|---|
| 0 | 0 5 4 3 2 1 |
| 1 | 1 0 5 4 3 2 |
| 2 | 2 1 0 5 4 3 |
| 3 | 3 2 1 0 5 4 |
| 4 | 4 3 2 1 0 5 |
| 5 | 5 4 3 2 1 0 |

| 14. × | 0 1 2 3 4 5 |
|---|---|
| 0 | 0 0 0 0 0 0 |
| 1 | 0 1 2 3 4 5 |
| 2 | 0 2 4 0 2 4 |
| 3 | 0 3 0 3 0 3 |
| 4 | 0 4 2 0 4 2 |
| 5 | 0 5 4 3 2 1 |

15. a) 4 o'clock  b) 6 o'clock  c) 4 o'clock
16. 0,1,5,6
17. $a \equiv \pm b \pmod{p}$
21. $n \equiv \pm 1 \pmod 6$
24. 1,3,5,7,9,11,13,15,17,19,21,23,25
28. a) 42  b) 2  c) 18
33. a) 1  b) 1  c) 1  d) 1  e) $a^{p-1} \equiv 1 \pmod{p}$ when $p$ is prime and $p \nmid a$
34. a) $-1$  b) $-1$  c) $-1$  d) $-1$  e) $(p-1)! \equiv -1 \pmod{p}$ when $p$ is prime
38. 15621

## Section 3.2

1. a) $x \equiv 3 \pmod 7$  b) $x \equiv 2,5,8 \pmod 9$  c) $x \equiv 7 \pmod{21}$  d) no solution
   e) $x \equiv 812 \pmod{1001}$  f) $x \equiv 1596 \pmod{1597}$
2. c) $x \equiv 5 \pmod{23}$
3. 19 hours
4. $c \equiv 0,6,12,18,24 \pmod{30}$, 6 solutions
5. a) 13  b) 7  c) 5  d) 16
8. a) $(x,y) \equiv (0,5),(1,2),(2,6),(3,3),(4,0),(5,4),(6,1) \pmod 7$
   b) $(x,y) \equiv (1,1),(1,3),(1,5),(1,7),(3,0),(3,2),(3,4),(3,6),(5,1),(5,3),(5,5),$
   $(5,7),(7,0),(7,2),(7,4),(7.6) \pmod 8$
   c) $(x,y) \equiv (0,0),(0,3),(0,6),(1,1),(1,4),(1,7),(2,2),\ (2,5),(2,8),(3,0),(3,3),(3,6),$
   $(4,1),(4,4),(4,7),(5,2),(5,5),(5,8),(6,0),(6,3),(6,6),(7,1),(7,4),(7,7),(8,2)$
   $(8,5),(8,8) \pmod 9$
   d) no solution

## Section 3.3

1. $x \equiv 1 \pmod 6$
2. 21

3. 32
4. a) $x \equiv 37$ (mod 187)　b) $x \equiv 23$ (mod 30)　c) $x \equiv 6$ (mod 210)
　 d) $x \equiv 150999$ (mod 554268)
5. 204
7. 2101
8. 119
12. a) $x \equiv 28$ (mod 30)　b) no solution
14. a) $x \equiv 23$ (mod 30)　b) $x \equiv 100$ (mod 210)　c) no solution
　 d) $x \equiv 44$ (mod 840)　e) no solution
17. 0000,0001,0625,9376
25. 26 feet 6 inches

## Section 3.4

1. a) $(x,y) \equiv (2,2)$ (mod 5)　b) no solution
　 c) $(x,y) \equiv (0,2),(1,3),(2,4),(3,0)$ or $(4,1)$ (mod 5)
2. a) $(x,y) \equiv (0,4),(1,1),(2,5),(3,2),(4,6),(5,3),(6,0)$ (mod 7)　b) no solution
3. 0, 1, $p$, or $p^2$

4. $\begin{bmatrix} 0 & 1 \\ 2 & 3 \end{bmatrix}$

8. a) $\begin{bmatrix} 0 & 1 \\ 1 & 0 \end{bmatrix}$　b) $\begin{bmatrix} 3 & 1 \\ 4 & 2 \end{bmatrix}$　c) $\begin{bmatrix} 1 & 4 \\ 2 & 1 \end{bmatrix}$

9. a) $\begin{bmatrix} 4 & 4 & 3 \\ 4 & 3 & 4 \\ 3 & 4 & 4 \end{bmatrix}$　b) $\begin{bmatrix} 2 & 0 & 6 \\ 2 & 1 & 4 \\ 3 & 4 & 0 \end{bmatrix}$　c) $\begin{bmatrix} 5 & 5 & 5 & 4 \\ 5 & 5 & 4 & 5 \\ 5 & 4 & 5 & 5 \\ 4 & 5 & 5 & 5 \end{bmatrix}$

10. a) $x \equiv 0, y \equiv 1, z \equiv 2$ (mod 7)　b) $x \equiv 1, y \equiv 0, z \equiv 0$ (mod 7)
　 c) $x \equiv 5, y \equiv 5, z \equiv 5, w \equiv 5$ (mod 7)
11. a) 0 b) 5 c) 25 d) 1

## Section 4.1

1. a) $2^8$ b) $2^4$ c) $2^{10}$ d) $2^1$
2. a) $5^3$ b) $5^4$ c) $5^1$ c) $5^9$
3. a) by 3 not by 9　b) by 3, and 9　c) by 3, and 9　d) not by 3
4. a) no　b) yes　c) no　d) no
5. those with their number of digits divisible by 3, and by 9
6. those with an even number of digits
7. those with their number of digits divisible by 6 (same for 7 and for 13)
8. 11
13. $a_{2n}a_{2n-1}...a_1a_0 \equiv a_{2n}a_{2n-1}\,a_{2n-2} + \cdots + a_5a_4a_3 + a_2a_1a_0$ (mod 37),
　 37 $\nmid$ 443692, 37 $\mid$ 11092785
15. a) no b) not by 3, by 5 c) not by 5, not by 13 d) yes
16. 73 ¢

17. ?=6
18. a) incorrect  b) incorrect  c) passes casting out nines check
19. no, for example 18(c) is incorrect, but passes the check

## Section 4.2

2.  a) Friday       b) Friday       c) Monday      d) Thursday
    e) Saturday     f) Saturday     g) Tuesday     h) Thursday
    i) Monday       j) Sunday       k) Friday      l) Wednesday
    m) Wednesday

## Section 4.3

1. a)

| Team<br>Round | 1 | 2 | 3 | 4 | 5 | 6 | 7 |
|---|---|---|---|---|---|---|---|
| 1 | 7 | 6 | 5 | bye | 3 | 2 | 1 |
| 2 | bye | 7 | 6 | 5 | 4 | 3 | 2 |
| 3 | 2 | 1 | 7 | 6 | bye | 4 | 3 |
| 4 | 3 | bye | 1 | 7 | 6 | 5 | 4 |
| 5 | 4 | 3 | 2 | 1 | 7 | bye | 5 |
| 6 | 5 | 4 | bye | 2 | 1 | 7 | 6 |
| 7 | 6 | 5 | 4 | 3 | 2 | 2 | bye |

3.  a) Home teams: Round 1: 4,5, Round 2: 2,3, Round 3: 1,5, Round 4: 3,4, Round 5: 1,2

## Section 4.4

5.  558, 1002, 2174, 4

## Section 5.1

1.  $-1$
2.  1
4.  4
5.  a) $x \equiv 9 \pmod{17}$   b) $x \equiv 17 \pmod{19}$
18. 1
25. b) 52

## Section 5.2

17. $7 \cdot 23 \cdot 67$

## Section 5.3

1. a) 1,5  b) 1,2,4,5,7,8  c) 1,3,7,9  d) 1,3,5,9,11,13  e) 1,3,5,7,9,11,13,15
   f) 1,2,3,4,5,6,7,8,9,10,11,12,13,14,15,16
2. $1,3,5,...,2^m-1$
5. 11
9. a) $x \equiv 9 \pmod{14}$  b) $x \equiv 13 \pmod{15}$  c) $x \equiv 7 \pmod{16}$
11. 1
12. 1
13. $\phi(13) = 12, \phi(14) = 6, \phi(16) = 8, \phi(17) = 16, \phi(18) = 6, \phi(19) = 18, \phi(20) = 8$

## Section 6.1

1. a) yes b) no c) no d) no e) yes f) no g) no h) no i) yes
2. a) 40 b) 128 c) 720 d) 5760
3. all three equal 2592
4. a) 1, 2  b) 3, 4, 6  c) no solution  d) 7, 9, 14, and 18  e) no solution
   f) 35, 39, 45, 52, 56, 70, 72, 78, 84, 90
5. 1, 2
6. those integers n such that $8 \mid n; 4 \mid n$, and n has at least one odd prime
   factor; n has at least two odd prime factors; or n has a prime factor $p \equiv 1 \pmod 4$
7. $2^k, k = 1,2,...$

## Section 6.2

1. a) 48  b) 399  c) 2340  d) $2^{101}-1$  e) 6912
2. a) 9  b) 6  c) 15  d) 256
3. perfect squares
4. those positive integers that have only even powers of odd primes in their prime-
   power factorization
5. a) 6,11  b) 10,17  c) 14,15,23  d) 33,35,47  e) no solution  f) 44, 65, 83
6. a) 1  b) 2  c) 4  d) 12  e) 192  f) 45360
8. primes
9. squares of primes
10. products to two distinct primes or cubes of primes
11. $n^{\tau(n)/2}$
12. 73, 252, 2044
13. $1 + p^k$
14. $(p^{k(a+1)}-1)/(p^k-1)$
16. $\prod_{j=1}^{m}(p_j^{k(a_j+1)}-1)/(p_j^k-1)$

## Section 6.3

1. 6, 28, 496, 8128, 33550336, 8589869056
3. 12, 18, 20, 24, 30, 36
4. 945
23. (a) and (c) are prime

24.  (a), (b), and (d) are prime

### Section 7.1

1.  DWWDF NDWGD ZQ
2.  I CAME I SAW I CONQUERED
3.  IEXXK FZKXC UUKZC STKJW
4.  PHONE HOME
5.  12
6.  $k = 17$,  THE KEY IS SEVENTEEN
7.  9,12
9.  $C \equiv 7P + 16 \pmod{26}$
10. $C \equiv acP + bc + d \pmod{26}$
11. VSPFXH HIPKLB KIPMIE GTG
12. EXPLOSIVES INSIDE

### Section 7.2

1.  RL OQ NZ OF XM CQ KG QI VD AZ
2.  IGNORE THIS

3.  $\begin{pmatrix} 3 & 24 \\ 24 & 25 \end{pmatrix}$

4.  a) 1   b) 13   c) 26

6.  $\begin{pmatrix} 2 & 13 & 3 \\ 1 & 23 & 10 \\ 25 & 3 & 7 \end{pmatrix}$

7.  digraphic Hill cipher with enciphering matrix $\begin{pmatrix} 11 & 6 \\ 2 & 13 \end{pmatrix}$

10. $\begin{pmatrix} 5 & 2 & 0 & 0 & 0 & 0 \\ 3 & 1 & 3 & 1 & 0 & 0 \\ 2 & 1 & 3 & 1 & 0 & 0 \\ 0 & 0 & 2 & 1 & 3 & 1 \\ 0 & 0 & 2 & 1 & 2 & 1 \\ 0 & 0 & 0 & 0 & 5 & 2 \end{pmatrix}$

### Section 7.3

1.  14 17 17 27 11 17 65 76 07 76 14
2.  DO NOT READ THIS
4.  GOOD GUESS
5.  92
6.  150

### Section 7.4

1.  97, 151
2.  1453, 3019

5. 1215 1224 1471 0023 0116
6. EAT CHOCOLATE CAKE
7. a) 0371 0354 0858 0858 0087 1359 0354 0000 0087 1543 1797 0535 b) 0019 0977
   0756 0370 0343 0647 0274 0872 0821 0073 0845 0740 0000 0008 0148 0803 0415
   0458 0274 0740
8. c) 0042 0056 0481 0481 0763 0000 0051 0000 0294 0262 0995 0495 0543 0972
   0000 0734 0152 0647 0972
9. d) 1383 1812 0352 0000 1383 0130 1080 1351 1383 1812 0130 0972 1208 0956
   0000 0972 1515 0937 1297 1208 2273 1515 0000
10. 0872 2263 1537 2392

## Section 7.5

1. a) yes  b) no  c) yes  d) no
4. $18 = 2+16 = 2+3+13 = 3+4+11 = 2+5+11 = 7+11$
5. $(17,51,85,8,16,49,64)$
6. 6242382306332274
8. $(44,37,74,72,50,24)$
10. $60 = 2\cdot3\cdot10 = 2\cdot5\cdot6 = 6\cdot10$
11. $15960 = 8\cdot21\cdot95$

## Section 7.6

1. a) 3696, 2640, 5600, 385   b) 5389
2. 829

## Section 8.1

1. a) 4  b) 4  c) 6  d) 3
2. a) 3  b) 2, 3  c) 3, 7  d) 2, 6, 7, 11  e) 3, 5  f) 5, 11
4. 4
16. b) $23\cdot89$
20. 2209

## Section 8.2

1. a) 2  b) 4  c) 8  d) 6  e) 12  f) 22
4. a) 4  b) the modulus is not prime
6. 1
11. b) 6
12. c) 22, 37, 8, 6, 8, 38, 26

## Section 8.3

1. 4, 10, 22
2. a) 2   b) 2   c) 3   d) 2
3. a) 2   b) 2   c) 2   d) 3
4. a) 5   b) 5   c) 15   d) 15
5. 7, 13, 17, 19

## Section 8.4

1. $\text{ind}_5 1 = 22$, $\text{ind}_5 2 = 2$, $\text{ind}_5 3 = 16$, $\text{ind}_5 4 = 4$, $\text{ind}_5 5 = 1$, $\text{ind}_5 6 = 18$, $\text{ind}_5 7 = 19$, $\text{ind}_5 8 = 6$, $\text{ind}_5 9 = 10$, $\text{ind}_5 10 = 3$, $\text{ind}_5 11 = 9$, $\text{ind}_5 12 = 20$, $\text{ind}_5 13 = 14$, $\text{ind}_5 14 = 21$, $\text{ind}_5 15 = 17$, $\text{ind}_5 16 = 8$, $\text{ind}_5 17 = 7$, $\text{ind}_5 18 = 12$, $\text{ind}_5 19 = 15$, $\text{ind}_5 20 = 5$, $\text{ind}_5 21 = 13$, $\text{ind}_5 22 = 11$
2. a) $x \equiv 9 \pmod{23}$   b) $x \equiv 9, 14 \pmod{23}$
3. a) $x \equiv 7, 18 \pmod{22}$   b) no solution
4. $a \equiv 2, 5$, or $6 \pmod{13}$
5. $b \equiv 8, 9, 20$, or $21 \pmod{29}$
6. $x \equiv 10, 16, 57, 59, 90, 99, 115, 134, 144, 145, 149$, or $152 \pmod{156}$
7. $x \equiv 1 \pmod{22}$, $x \equiv 0 \pmod{23}$, or $x \equiv 1, 12, 45, 47, 78, 91, 93, 100, 137, 139, 144,$ $183, 185, 188, 210, 229, 231, 232, 252, 254, 275, 277, 321, 323, 367, 369,$ $386, 413, 415, 430, 459, 461$, or $496 \pmod{506}$
11. (1,2), (0,2)
13. $x \equiv 29 \pmod{32}$, $x \equiv 42 \pmod{8}$
15. (0, 0, 1, 1), (0, 0, 1, 4)
17. $x \equiv 17 \pmod{60}$
23. b) $(49938 \cdot 99876)/(4 \cdot 49939 \cdot 99877) = .24999249\ldots$

## Section 8.6

1. a) 20   b) 12   c) 36   d) 48   e) 180   f) 388080   g) 8640   h) 125411328000
2. a) 1,2   b) 3, 4, 6, 8, 12, 24   c) no solution   d) 5, 10, 15, 16, 20, 30, 40, 48, 60, 80, 120, 240   e) no solution   f) 7, 9, 14, 18, 21, 28, 36, 42, 56, 63, 72, 84, 126, 168, 252, 504
3. 65520
4. a) 11   b) 2   c) 7   d) 11   e) 19   f) 38
14. $5 \cdot 13 \cdot 17 \cdot 29$, $5 \cdot 17 \cdot 29$, $5 \cdot 29 \cdot 73$

## Section 8.7

1. 69, 76, 77, 92, 46, 11, 12, 14, 19, 36, 29, 84, 5, 2, 0, 0,...
2. 6, 13, 10, 14, 15, 1, 7, 18, 16, 6, 13, ..., period length is 9
3. 10
7. a) 31   b) 715827882   c) 31   d) 195225786   e) 1073741823
9. 1, 24, 25, 18, 12, 30, 11, 10

## Section 8.8

1. a) 8   b) 5   c) 2   d) 6   e) 30   f) 20

2. a) 2  b) 3  c) 2  d) 2  e) 5  f) 7
3. a) use spread s = 3   b) use spread s = 21   c) use spread s = 2

### Section 9.1

1. a) 1   b) 1,4   c) 1,3,4,9,10,12   d) 1,4,5,6,7,9,11,16,17
2. 1,1,−1,1,−1,−1
11. a) $x \equiv 2,4 \pmod 7$   b) $x \equiv 1 \pmod 7$   c) no solution
17. $x \equiv 1,4,11,14 \pmod{15}$
43. c) DETOUR

### Section 9.2

1. a) − 1  b) − 1  c) − 1  d) − 1  e) 1  f) 1
4. $p \equiv \pm 1 \pmod 5$
5. $p \equiv \pm 1, \pm 3, \pm 9 \pmod{28}$

### Section 9.3

1. a) 1  b) −1  c) 1  d) 1  e) −1  f) 1
2. $n \equiv$  1, 7, 11, 17, 43, 49, 53, or 59 $\pmod{60}$
3. $n \equiv 1,7,13,17,19,2937,71,83, 91,101,103,107,109,113,$ or 119 $\pmod{120}$
9. a) −1  b) −1  c) −1

### Section 10.1

1. a) $.4$  b) $.41\overline{6}$  c) $\overline{.923076}$  d) $.5\overline{3}$  e) $.00\overline{9}$  f) $.\overline{000999}$
2. a) $(.\overline{25})_8$  b) $(.2)_8$  c) $(.\overline{1463})_8$  d) $(.12\overline{5})_8$  e) $(.0\overline{52})_8$  f) $(.0\overline{2721350564})_8$
3. a) $\dfrac{3}{25}$  b) $\dfrac{11}{90}$  c) $\dfrac{4}{33}$
4. a) $\dfrac{66}{343}$  b) $\dfrac{3}{70}$  c) $\dfrac{3}{20}$  d) $\dfrac{916}{1365}$
5. $b = 2^{s_1}3^{s_2}5^{s_3}7^{s_4}$, where $s_1,s_2,s_3,$ and $s_4$ are nonnegative integers, not all zero
6. a) 2,1  b) 1,1  c) 2,1  d) 0,22  e) 3,6  f) 0,60
7. a) 1,0  b) 2,0  c) 1,4  d) 2,1  e) 1,1  f) 2,4
9. a) 3  b) 11  c) 37  d) 101  e) 41,271  f) 7,13

### Section 10.2

1. a) 15/7  b) 10/7  c) 6/31  d) 355/113  e) 2  f) 3/2  g) 5/3  h) 8/5
2. a) [1;5]  b) [3;7]  c) [0;1,1,9]   d) [0;199,1,4]  e) [−1;1,22,3,1,1,2,2]
   f) [0;5,1,1,2,1,4,1,21]

### Section 10.3

1. a) [1;2,2,2,...]  b) [1;1,2,1,2,1,2,...]  c) [2;4,4,4,...]  d) [1;1,1,1,...]
2. a) 1,3,1,5,1 b) 6,3,1,1,7 c) 0,2,6,10,14  d) 0,1,3,5,7
3. $\dfrac{312689}{99532}$
4. a) $\dfrac{2}{1},\dfrac{3}{1},\dfrac{8}{3},\dfrac{11}{4},\dfrac{19}{7},\dfrac{87}{32},\dfrac{106}{39},\dfrac{193}{71}$  b) $\dfrac{193}{71}$

18. $\dfrac{25}{8}, \dfrac{47}{15}, \dfrac{69}{22}, \dfrac{91}{29}, \dfrac{113}{36,} \dfrac{135}{43}, \dfrac{157}{50}, \dfrac{179}{57}, \dfrac{201}{64}, \dfrac{223}{71}, \dfrac{245}{78}, \dfrac{267}{85}, \dfrac{289}{92}, \dfrac{311}{99}$

## Section 10.4

1. a) $[2;\overline{1,1,1,4}]$  b) $[3;\overline{3,6}]$  c) $[4;\overline{1,3,1,8}]$  d) $[6;\overline{1,5,1,12}]$
2. a) $[\overline{1;2}]$
3. a) $(23 + \sqrt{29})/10$  b) $(-1 + \sqrt{45})/2$  c) $(8 + \sqrt{82})/6$
4. b) $[10;\overline{20}]$, $[17;\overline{34}]$, $[47;\overline{94}]$
5. c) $[9;\overline{1,18}]$, $[10;\overline{2,20}]$, $[16;\overline{2,32}]$, $[24;\overline{2,48}]$
6. c) $[6;\overline{1,5,1,12}]$, $[7;\overline{7,14}]$, $[16;\overline{1,15,1,32}]$
11. b), c), e)

## Section 10.5

1. $7\cdot 17$
2. $29\cdot 53$
3. $3119\cdot 4261$
7. $3001\cdot 4001$
8. $199\cdot 911$

## Section 11.1

1. a) 3,4,5; 5,12,13; 15,8,17; 7,24,25; 21,20,29; 35,12,37  b) 3,4,5; 6,8,10; 5,12,13;9, 12,15; 15,8,17; 12,16,20; 7,24,25; 15,20,25; 10,24,26; 21,20,29; 18,24,30; 30,16,34; 21,28,35; 35,12,37; 15,36,39; 24,32,40
8. $x = (m^2 - 2n^2)/2$, $y = mn$, $z = (m^2 + 2n^2)/2$ where $m$ and $n$ are positive integers, $x = (2m^2 - n^2)/2, y = mn, z = (2m^2 + n^2)/2$ where $m$ and $n$ are positive integers, $m > n/\sqrt{2}$, and $n$ is even
9. $x = (m^2 - 3n^2)/2, y = mn, z = (m^2 + 3n^2)/2$ where $m$ and $n$ are positive integers, $m > \sqrt{3}\, n$, and $m \equiv n \pmod 2$

## Section 11.3

1. a) $19^2 + 4^2$  b) $23^2 + 11^2$  c) $37^2 + 9^2$  d) $137^2 + 9^2$
2. a) no  b) yes  c) yes  d) yes  e) no  f) yes  g) no  h) no  i) yes
3. a) $5^2 + 3^2$  b) $9^2 + 3^2$  c) $10^2 + 0^2$
   d) $21^2 + 7^2$  e) $147^2 + 7^2$  f) $448^2 + 352^2$
5. a) $1^2 + 1^2 + 1^2$  b) impossible  c) $3^2 + 1^2 + 1^2$
   d) $3^2 + 3^2 + 0^2$  e) impossible  f) impossible
9. a) $10^2 + 2^2 + 1^2 + 0^2$  b) $510 = 22^2 + 4^2 + 1^2 + 1^2$
   c) $238 = 14^2 + 5^2 + 4^2 + 1^2$  d) $3570 = 46^2 + 34^2 + 17^2 + 3^2$
10. a) $2^2 + 1^2 + 1^2 + 0^2$  b) $3^2 + 1^2 + 1^2 + 1^2$  c) $4^2 + 2^2 + 1^2 + 0^2$
    d) $9^2 + 2^2 + 2^2 + 0^2$  e) $9^2 + 3^2 + 3^2 + 0^2$  f) $23^2 + 5^2 + 1^2 + 0^2$

## Section 11.4

1. a) $x = \pm 2, y = 0$ ; $x = \pm 1, y = \pm 1$   b) no solution   c) $x = \pm 1, y = \pm 2$
2. a) $x = \pm 3, y = \pm 1$   b) no solution   c) $x = \pm 5, y = 0$;   $x = \pm 13, y = \pm 8$
3. a) $x = 70, y = 13$   b) $x = 9801, y = 1820$
5. $x = 1520, y = 273$ ; $x = 4620799, y = 829920$; $x = 42703566796801$,
    $y = 766987012160$
6. a), d), e), g), h) yes   b), c), f) no
7. $x = 6239765965720528801, y = 79892016576262330040$

# Bibliography

## BOOKS

### Number Theory

1. W. W. Adams and L. J. Goldstein, *Introduction to Number Theory*, Prentice-Hall, Englewood Cliffs, New Jersey, 1976.

2. G. E. Andrews, *Number Theory*, W. B. Saunders, Philadelphia, 1971.

3. T. A. Apostol, *Introduction to Analytic Number Theory*, Springer-Verlag, New York, 1976.

4. R. G. Archibald, *An Introduction to the Theory of Numbers*, Merrill, Columbus, Ohio, 1970.

5. I. A. Barnett, *Elements of Number Theory*, Prindle, Weber, and Schmidt, Boston, 1969.

6. A. H. Beiler, *Recreations in the Theory of Numbers*, 2nd ed., Dover, New York, 1966.

7. E. D. Bolker, *Elementary Number Theory*, Benjamin, New York, 1970.

8. Z. I. Borevich and I. R. Shafarevich, *Number Theory*, Academic Press, New York, 1966.

9. D. M. Burton, *Elementary Number Theory*, Allyn and Bacon, Boston, 1976.

10. R. D. Carmichael, *The Theory of Numbers and Diophantine Analysis*, Dover, New York, 1959 (reprint of the original 1914 and 1915 editions).

11. H. Davenport, *The Higher Arithmetic*, 5th ed., Cambridge University Press, Cambridge, 1982.

12. L. E. Dickson, *Introduction to the Theory of Numbers*, Dover, New York 1957 (reprint of the original 1929 edition).

13. U. Dudley, *Elementary Number Theory*, 2nd ed., Freeman, New York, 1969.

14. H. M. Edwards, *Fermat's Last Theorem,* Springer-Verlag, New York, 1977.

15. C. F. Gauss, *Disquisitiones Arithmeticae,* Yale, New Haven, Connecticut, 1966.

16. A. A. Gioia, *The Theory of Numbers*, Markham, Chicago 1970.

17. E. Grosswald, *Topics from the Theory of Numbers*, 2nd ed., Birkhäuser, Boston, 1982.

18. H. Gupta, *Selected Topics in Number Theory*, Abacus Press, Kent, England, 1980.

19. R. K. Guy, *Unsolved Problems in Number Theory*, Springer-Verlag, New York, 1981.

20. G. H. Hardy and E. M. Wright, *An Introduction to the Theory of Numbers*, 5th ed., Oxford University Press, Oxford, 1979.

21. L. Hua, *Introduction to Number Theory*, Springer-Verlag, New York 1982.

22. K. Ireland and M. I. Rosen, *A Classical Introduction to Modern Number Theory*, Springer-Verlag, New York, 1982.

23. E. Landau, *Elementary Number Theory*, Chelsea, New York, 1958.

24. W. J. LeVeque, *Fundamentals of Number Theory*, Addison-Wesley, Reading, Massachusetts, 1977.

25. W. J. LeVeque, editor, *Reviews in Number Theory* [1940-1972], and R. K. Guy, editor, *Reviews in Number Theory* [1973-1983], six volumes each, American Mathematical Society, Washington, D.C., 1974 and 1984, respectively.

26. C. T. Long, *Elementary Introduction to Number Theory*, 2nd ed., Heath, Lexington, Massachusetts, 1972.

27. G. B. Matthews, *Theory of Numbers*, Chelsea, New York (no date).

28. I. Niven and H. S. Zuckerman, *An Introduction to the Theory of Numbers*, 4th ed., Wiley, New York, 1980.

29. O. Ore, *An Invitation to Number Theory*, Random House, New York, 1967.

30. O. Ore, *Number Theory and its History*, McGraw-Hill, New York, 1948.

31. A. J. Pettofrezzo and D. R. Byrkit, *Elements of Number Theory*, Prentice-Hall, Englewood Cliffs, New Jersey, 1970.

32. H. Rademacher, *Lectures on Elementary Number Theory*, Blaisdell, New York 1964, reprint Krieger, 1977.

33. P. Ribenboim, *13 Lectures on Fermat's Last Theorem*, Springer-Verlag, New York, 1979.

34. J. Roberts, *Elementary Number Theory*, MIT Press, Cambridge, Massachusetts, 1977.

35. M. R. Schroeder, *Number Theory in Science and Communication*, 2nd ed., Springer-Verlag, Berlin, 1986.

36. D. Shanks, *Solved and Unsolved Problems in Number Theory*, 2nd ed., Chelsea, New York, 1978.

37. H. S. Shapiro, *Introduction to the Theory of Numbers*, Wiley, New York, 1983.

38. J. E. Shockley, *Introduction to Number Theory*, Holt, Rinehart, and Winston, 1967.

39. W. Sierpiński, *Elementary Theory of Numbers*, Polski Akademic Nauk, Warsaw, 1964.

40. W. Sierpiński, *A Selection of Problems in the Theory of Numbers*, Pergammon Press, New York, 1964.

41. W. Sierpiński, *250 Problems in Elementary Number Theory*, Polish Scientific Publishers, Warsaw, 1970.

42. H. M. Stark, *An Introduction to Number Theory*, Markham, Chicago, 1970; reprint MIT Press, Cambridge, Massachusetts, 1978.

43. B. M. Stewart, *The Theory of Numbers*, 2nd ed., Macmillan, New York, 1964.

44. J. V. Uspensky and M. A. Heaslet, *Elementary Number Theory*, McGraw-Hill, New York, 1939.

45. C. Vanden Eyden, *Number Theory*, International Textbook, Scranton, Pennsylvania, 1970.

46. I. M. Vinogradov, *Elements of Number Theory*, Dover, New York, 1954.

47. H. N. Wright, *First Course in Theory of Numbers*, Wiley, New York, 1939.

**Number Theory with Computer Science**

48. A. M. Kirch, *Elementary Number Theory: A Computer Approach*, Intext, New York, 1974.

49. D. G. Malm, *A Computer Laboratory Manual for Number Theory*, COMPress, Wentworth, New Hampshire, 1979.

50. D. D. Spencer, *Computers in Number Theory*, Computer Science Press, Rockville, Maryland, 1982.

**Factorization and Primality Testing**

51. J. Brillhart, D. H. Lehmer, J. L. Selfridge, B. Tuckerman, S. S. Wagstaff, Jr., *Factorizations of $b^n \pm 1$, $b = 2,3,5,6,7,10,11,12$ up to high powers*, American Mathematical Society, Providence, Rhode Island, 1983.

52. E. Kranakis, *Primality and Cryptography*, Wiley-Teubner, Stuttgart, West Germany, 1986.

53. C. Pomerance, *Lecture Notes on Primality Testing and Factoring*, Mathematical Association of America, Washington, D.C., 1984.

54. H. Riesel, *Prime Numbers and Computer Methods for Factorization*, Birkhäuser, Boston, 1985.

**History**

55. E. T. Bell, *Men of Mathematics*, Simon & Schuster, New York, 1965.

56. C. B. Boyer, *A History of Mathematics*, Wiley, New York, 1968.

57. D. M. Burton, *The History of Mathematics*, Allyn and Bacon, Boston, 1985.

58. L. E. Dickson, *History of the Theory of Numbers*, three volumes, Chelsea, New York, 1952 (reprint of the 1919 original).

59. H. Eves, *An Introduction to the History of Mathematics* , 5th edition, Saunder, Philadelphia, 1983.

60. M. Kline, *Mathematical Thought from Ancient to Modern Times*, Oxford University, New York, 1972.

61. U. Libbrecht, *Chinese Mathematics in the Thirteenth Century, The Shu-shu chiu-chang of Ch'in Chiu-shao:'* MIT, 1973.

62. W. Scharlau and H. Opolka, *From Fermat to Minkowski, Lectures on the Theory of Numbers and its Historical Development*, Springer-Verlag, New York, 1985.

63. A. Weil, *Number Theory: An approach through history from Hummurapi to Legendre*, Birkhäuser, Boston, 1984.

**Cryptography**

64. H. Beker and F. Piper, *Cipher Systems*, Wiley, New York, 1982.

65. B. Bosworth, *Codes, Ciphers, and Computers*, Hayden, Rochelle Park, New Jersey, 1982.

66. D. Chaum, R. L. Rivest, A. T. Sherman, eds., *Advances in Cryptology - Proceedings of Crypto 82*, Plenum, New York, 1983.

67. D. Chaum, ed., *Advances in Cryptology - Proceedings of Crypto 83*, Plenum, New York, 1984.

68. D. E. R. Denning, *Cryptography and Data Security*, Addison-Wesley, Reading, Massachusetts, 1982.

69. W. F. Friedman, *Elements of Cryptanalysis*, Aegean Park Press, Laguna Hills, California, 1978.

70. A. Gersho, ed., *Advances in Cryptography*, Dept. of Electrical and Computer Engineering, Univ. Calif. Santa Barbara, 1982.

71. D. Kahn, *The Codebreakers, the Story of Secret Writing*, Macmillan, New York, 1967.

72. A. G. Konheim, *Cryptography: A Primer*, Wiley, New York, 1981.

73. S. Kullback, *Statistical Methods in Cryptanalysis*, Aegean Park Press, Laguna Hills, California, 1976.

74. R. C. Merkle, *Secrecy, Authentication, and Public Key Systems*, UMI Research Press, Ann Arbor, Michigan, 1982.

75. C. H. Meyer and S. M. Matyas, *Cryptography: A New Dimension in Computer Data Security*, Wiley, New York, 1982.

76. G. J. Simmons, ed., *Secure Communications and Asymmetric Cryptosystems*, AAAS Selected Symposium Series 69, Westview Press, Boulder, Colorado, 1982.

77. A. Sinkov, *Elementary Cryptanalysis*, Mathematical Association of America, Washington, D.C., 1966.

78. H. C. Williams, ed., *Advances in Cryptology - CRYPTO '85*, Springer-Verlag, Berlin, 1986.

**Computer Science**

79. K. Hwang, *Computer Arithmetic: Principles, Architecture and Design*, Wiley, New York, 1979.

80. D. E. Knuth, *Art of Computer Programming: Semi-Numerical Algorithms* Volume 2, 2nd ed., Addison Wesley, Reading Massachusetts, 1981.

81. D. E. Knuth, *Art of Computer Programming: Sorting and Searching*, Volume 3, Addison-Wesley, Reading, Massachusetts, 1973.

82. L. Kronsjö, *Algorithms: Their Complexity and Efficiency*, Wiley, New York, 1979.

83. J. H. McClellan and C. M. Rader, *Number Theory in Digital Signal Processing*, Prentice-Hall, Englewood Cliffs, New Jersey, 1979.

84. M. A. Soderstrand, et. al., editors, *Residue Number System Arithmetic: Modern Applications in Digital Signal Processing*, IEEE Press, New York, 1986.

85. N. S. Szabó and R. J. Tanaka, *Residue Arithmetic and its Applications to Computer Technology*, McGraw-Hill, 1967.

**General**

86. H. Anton, *Elementary Linear Algebra*, 3rd ed., Wiley, New York, 1981.

87. E. Landau, *Foundations of Analysis*, 2nd ed., Chelsea, New York, 1960.

88. K. H. Rosen, *Discrete Mathematics and its Applications*, Random House, New York, 1988.

89. W. Rudin, *Principles of Mathematical Analysis*, 2nd ed., McGraw-Hill, New York 1964.

## ARTICLES

**Number Theory**

90. L. M. Adleman, C. Pomerance, and R. S. Rumely, "On distinguishing prime numbers from composite numbers," *Annals of Mathematics*, Volume 117 (1983), 173-206.

91. J. Brillhart, "Fermat's factoring method and its variants," *Congressus Numerantium*, Volume 32 (1981), 29-48.

92. J. D. Dixon, "Factorization and primality tests," *American Mathematical Monthly*, Volume 91 (1984), 333-353.

93. J. Ewing, " $2^{86243} - 1$ is prime," *The Mathematical Intelligencer*, Volume 5 (1983), 60.

94. J. E. Freund, "Round robin mathematics," *American Mathematical Monthly*, Volume 63 (1956), 112-114.

95. R. K. Guy, "How to factor a number" *Proceedings of the Fifth Manitoba Conference on Numerical Mathematics*, Utilitas, Winnepeg, Manitoba, 1975, 49-89.

96. A. K. Head, "Multiplication modulo n," *BIT*, Volume 20 (1980), 115-116.

97. P. Hagis, Jr., "Sketch of a proof that an odd perfect number relatively prime to 3 has at least eleven prime factors," *Mathematics of Computations*, Volume 46 (1983), 399-404.

98. J. C. Lagarias and A. M. Odlyzko, "New algorithms for computing $\pi(x)$," Bell Laboratories Technical Memorandum TM-82-11218-57.

99. H. P. Lawther, Jr., "An application of number theory to the splicing of telephone cables," *American Mathematical Monthly*, Volume 42 (1935), 81-91.

100. D. H. Lehmer and R. E. Powers, "On factoring large numbers," *Bulletin of the American Mathematical Society*, Volume 37 (1931), 770-776.

101. H. W. Lenstra, Jr., "Primality testing," *Studieweek Getaltheorie en Computers*, 1-5 September 1980, Stichting Mathematisch Centrum, Amsterdam, Holland.

102. G. L. Miller, "Riemann's hypothesis and tests for primality," *Proceedings of the Seventh Annual ACM Symposium on the Theory of Computing*, 234-239.

103. L. Monier, "Evaluation and comparison of two efficient probablistic primality testing algorithms, *Theoretical Computer Science*, Volume 11 (1980), 97-108.

104. C. Pomerance, "Recent developments in primality testing," *The Mathematical Intelligencer*, Volume 3 (1981), 97-105.

105. C. Pomerance, "The search for prime numbers," *Scientific American*, Volume 247 (1982), 136-147.

106. M. O. Rabin, "Probabilistic algorithms for testing primality," *Journal of Number Theory*, Volume 12 (1980), 128-138.

107. H. Riesel, "Modern factorization methods," *BIT* (1985), 205-222.

108. R. Rumely, "Recent advances in primality testing," *Notices of the American Mathematical Society*, Volume 30 (1983), 475-477.

109. D. Slowinski, "Searching for the 27th Mersenne prime," *Journal of Recreational Mathematics*, Volume 11 (1978/9), 258-261.

110. R. Solovay and V. Strassen, "A fast Monte Carlo test for primality," *SIAM Journal for Computing*, Volume 6 (1977), 84-85 and erratum, Volume 7 (1978), 118.

111. S. S. Wagstaff and J. W. Smith, "Methods of factoring large integers," in *Number Theory, New York, 1984-1985*, Lecture Notes in Mathematics, Volume 1240, Springer-Verlag, Berlin, 1987, 281-303.

112. H. C. Williams, "An overview of factoring," in *Advances in Cryptology, Proceedings of Crypto 83*, Plenum, New York, 1984, 87-102.

113. H. C. Williams, "The influence of computers in the development of number theory," *Computers and Mathematics with Applications*, Volume 8 (1982), 75-93.

114. H. C. Williams, "Primality testing on a computer", *Ars Combinatorica*, Volume 5 (1978), 127-185.

115. M. C. Wunderlich, "Implementing the continued fraction algorithm on parallel machines," *Mathematics of Computation*, Volume 44 (1985), 251-260.

## Cryptography

116. L. M. Adleman, "A subexponential algorithm for the discrete logarithm problem with applications to cryptography," *Proceedings of the 20th Annual Symposium on the Foundations of Computer Science*, 1979, 55-60.

117. M. Blum, "Coin-flipping by telephone - a protocol for solving impossible problems," *IEEE Proceedings, Spring Compcon.*, 133-137.

118. W. Diffie and M. Hellman, "New directions in cryptography," *IEEE Transactions on Information Theory*, Volume 22 (1976), 644-655.

119. D. R. Floyd, "Annotated bibliographical in conventional and public key cryptography," *Cryptologia*, Volume 7 (1983), 12-24.

120. J. Gordon, "Use of intractable problems in cryptography," *Information Privacy*, Volume 2 (1980), 178-184.

121. M. E. Hellman, "The mathematics of public-key cryptography," *Scientific American*, Volume 241 (1979) 146-157.

122. L. S. Hill, "Concerning certain linear transformation apparatus of cryptography," *American Mathematical Monthly*, Volume 38 (1931), 135-154.

123. A. Lempel, "Cryptology in transition," *Computing Surveys*, Volume 11 (1979), 285-303.

124. R. J. Lipton, "How to cheat at mental poker," and "An improved power encryption method," unpublished reports, Department of Computer Science, University of California, Berkeley, 1979.

125. R. C. Merkle and M. E. Hellman, "Hiding information and signatures in trapdoor knapsacks," *IEEE Transactions in Information Theory*, Volume 24 (1978), 525-530.

126. S. Pohlig and M. Hellman, "An improved algorithm for computing logarithms over $GF(p)$ and its cryptographic significance," *IEEE Transactions on Information Theory*, Volume 24 (1978), 106-110.

127. M. O. Rabin, "Digitalized signatures and public-key functions as intractable as factorization," MIT Laboratory for Computer Science Technical Report LCS/TR-212, Cambridge, Massachusetts, 1979.

128. R. L. Rivest, A. Shamir, and L. M. Adleman, "A method for obtaining digital signatures and public-key cryptosystems," *Communications of the ACM*, Volume 21 (1978), 120-126.

129. A. Shamir, "A polynomial time algorithm for breaking the basic Merkle-Hellman cryptosystem," *Proceedings of the 23rd Annual Symposium of the Foundations of Computer Science*, 145-152.

130. A. Shamir, "How to share a secret," *Communications of the ACM*, Volume 22 (1979), 612-613.

131. A. Shamir, R. L. Rivest, and L. M. Adleman, "Mental Poker," *The Mathematical Gardner*, ed. D. A. Klarner, Wadsworth International, Belmont, California, 1981, 37-43.

# Index

Absolute least residues, 112
Abundant integer, 205
Additive function, 193
Additive inverse, 4
Affine transformation, 211
Algorithm, 49
   continued fraction factoring, 390
   division, 35-37
   Euclidean, 76-85
   for addition, 49-50
   for base $b$ expansion, 41-42
   for division, 58
   for matrix multiplication, 60
   for modular exponentiation, 116
   for modular multiplication, 120
   for multiplication, 57-58
   for subtraction, 50-51
   least-remainder, 84
Amicable pair, 205
Approximation,
   best rational, 372
   by rationals, 370
Arithmetic function, 185
Arithmetic, fundamental
   theorem of, 85-87
Arithmetic mean, 26
Arithmetic progression,
   primes in, 90
Associative law, 4
Automorph, 134

Babylonians, 1
Balanced ternary expansion, 45
Base, 43
Base $b$ expansion,
   of an integer, 43
   of a real number, 343-345

Basis step, 15
Best rational approximation, 372
Big-$O$ notation, 55-58
Binary coded decimal notation, 46
Binary notation, 43
Binomial coefficient, 27
Binomial expression, 27
Binomial theorem, 30-32
Biorhythms, 134
Bit operation, 55
Bit, 43
Block cipher, 218
Borrow, 50

Caesar, Julius, 153,209
Caesar cipher, 209-210
Calendar, 152-157
   Gregorian, 153
   International Fixed, 157
   Julian, 153
   perpetual, 152-157
Cancellation law, 4
Cantor expansion, 46
Card shuffling, 171
Carmichael number, 174,284-285
Carry, 49
Casting out nines, 151-152
Chang Ch'iu-chien, 109
Character cipher, 209
Chessboard, 27
Chinese, ancient, 127,172
Chinese remainder theorem, 126-133
Cicada, periodic, 94
Cipher,
   block, 218
   Caesar, 209-210
   character, 209

digraphic, 218
exponentiation, 224-225
Hill, 218
iterated knapsack, 242
knapsack, 240
monographic, 209
polygraphic, 218
product, 216
public-key, 230
Rabin, 234
RSA, 231
substitution, 209
transposition, 224
Vigenère, 217
Ciphertext, 208
Class, congruence, 112
Closure property, 4,5
Clustering, 161
Coconut problem, 121
Coefficient, binomial, 27
Coin flipping, 308
Collatz conjecture, 40
Collision, 161
Common key, 227
Common ratio, 16
Commutative law, 4
Complete system of residues, 112
Completely additive function, 193
Completely multiplicative
    function, 185
Complexity, computational, 55
Composite, 61
Computational complexity, 55
  of addition, 57
  of division, 58
  of Euclidean algorithm, 79
  of matrix multiplication, 60
  of modular exponentiation, 117
  of multiplication, 57-58
  of subtraction, 57
Computer arithmetic, 129-130
Congruence, 110-146
    linear, 122
    of matrices, 138
Congruence class, 112
Congruent to, 111
Conjecture,
    Collatz, 40
    Goldbach, 65

Conjugate, 377
Constant, Kaprekar, 47
Continued fraction, 353-392
    factoring with, 389-392
    finite, 353-362
    infinite, 363-375
    periodic, 375-389
    purely periodic, 383
    simple, 354
Convergent, 357
Covering set of congruences, 134
Cryptanalysis, 208
    of character ciphers, 213
    of block ciphers, 221
Cryptography, 208
Cryptology, 208-249
    definition of, 208
Cubic residue, 276

Database, 244
Decimal fraction, 343-353
Deciphering, 208
Deciphering key,
    for RSA cipher, 231
Decryption, 208
Deficient integer, 205
Descent, proof by, 399
Diabolic square, 146
Digraphic cipher, 218
Diophantine equation,
    linear, 104-109
    nonlinear, 393-419
Diophantus, 105
Dirichlet, G. Lejeune, 90
Dirichlet product, 192
Dirichlet's theorem on primes in
    arithmetic progression, 90
Distributive law, 4
Divide, 35
    exactly, 93
Divisibility, 35-40
Divisibility tests, 147-152
Division, 58
Division algorithm, 35-37
Divisor, 35
    greatest common, 70
Double hashing, 161
Dozen, 54
Draim factorization, 102-103

Duodecimal notation, 54

Eggs, 132
Electronic poker, 229
Enciphering, 208
Enciphering key,
  for RSA cipher, 231
Encryption, 208
Equation,
  diophantine, 104
  Pell's, 413-419
Eratosthenes, 52
Eratosthenes, sieve of, 52-53,65
Euclid, 61,76
  game of, 84
Euclidean algorithm, 76-85
Euler, Leonhard, 180,406
Euler phi-function, 180-183,186-190
Euler pseudoprime, 334-342
Euler's criterion, 301
Euler's factorization, 103-104
Euler's theorem, 182-183
Even number, 37
Exactly divide, 93
Expansion,
  balanced ternary, 45
  base b, 43
  Cantor, 46
  continued fraction, 353-392
  periodic base b, 347
  periodic continued function, 375-389
  terminating decimal, 346
±1-exponent, 292
Exponentiation cipher, 224-225

Factor, 35
Factor base, 392
Factor table, 422-428
Factorial function, 11,19,20
Factorization, 97-104
  Draim, 102-103
  Euler, 103-104
  Fermat, 99-100
  prime-power, 86
  speed of, 98
  using continued fractions, 389-392
Faltings, G., 402
Farey series, 75

Fermat, Pierre de, 99,399,406
Fermat factorization, 99-100
Fermat number, 100-102,312,321-322
Fermat prime, 100-101
Fermat quotient, 171
Fermat's last theorem, 399
Fermat's little theorem, 168-169
Fibonacci, 20
Fibonacci numbers, 20-22,78-80
  generalized, 26
Fibonacci pseudo-random number
    generator, 292
Four squares, sums of, 406-410
Fractional part, 343
Frequencies,
  of digraphs, 221
  of letters, 213
  of polygraphs, 221
Function,
  additive, 193
  arithmetic, 185
  completely additive, 193
  completely multiplicative, 185
  Euler phi, 180-183
  factorial, 11,19-20
  greatest integer, 6
  hashing, 160
  Liouville's, 193
  Möbius, 192
  multiplicative, 185
  number of divisors, 194
  strongly multiplicative, 191
  sum of divisors, 194
Fundamental theorem of
    arithmetic, 85-87

Game of Euclid, 84
Gauss, Karl Friedrich, 63,110,317
Gauss' generalization of
    Wilson's theorem, 171
Gauss' lemma, 304
Generalized Fibonacci number, 26
Generalized Riemann hypothesis, 177
Geometric mean, 26
Geometric series,
  sum of infinite, 344
Geometric progression, 16
Goldbach, Christian, 65-66

Goldbach's conjecture, 65-66
Greatest common divisor, 70-85,87-88
   of two integers, 70
   of more than two integers, 72
Greatest integer function, 5
Greeks, ancient, 199

Hadamard, J., 63
Hanoi, tower of, 26
Harmonic series, 23
Hashing, 164
   double, 161
   quadratic, 314
Hashing function, 160
Hexadecimal notation, 43,44
Hilbert, David, 410
Hilbert prime, 93
Hill cipher, 218

Identity elements, 4
Identity matrix modulo $m$,
Inclusion-exclusion, principle of, 34
Incongruent, 111
Index of an integer, 267
Index of summation, 9
Index system, 276
Induction, mathematical, 15-27
Inductive step, 15
Infinite descent, 399
Infinite simple continued fraction, 363-375
Infinitude of primes, 61-62
Integer,
   abdundant, 205
   deficient, 205
   palindromic, 151
   powerful, 93
   square-free, 93
International fixed calendar, 157
Inverse of an arithmetic function, 192
Inverse modulo $m$, 123
Inverse of a matrix modulo $m$, 140
Inversion, Möbius, 192
Involutory matrix, 144
Irrational number, 6,91-92,350-351
   quadratic, 376

Jacobi, Carl G.J., 324
Jacobi symbol, 324
Jigsaw puzzle, 24

$k$-abundant number, 205
$k$-perfect number, 205
Kaprekar constant, 47
Key,
   common, 227
   deciphering, 231
   enciphering, 231
   for hashing, 160
   master, 246
   public, 230
Knapsack cipher, 239
Knapsack problem, 237
Kronecker symbol, 332
$k$th power residue, 271

Lagrange, Joseph, 167,406
Lagrange interpolation, 259
Lagrange's theorem
   (on continued functions), 378
Lagrange's theorem
   (on polynomial congruences), 255-256
Lamé, Gabriel, 79
Lamé's theorem, 79
Law of quadratic reciprocity, 315-324
Leap year, 153-154
Least common multiple, 88
   of two integers, 88
   of more than two integers, 95
Least nonnegative residue, 112
Least-remainder algorithm, 84
Legendre, Adrien-Marie, 301
Legendre symbol, 300
Lemma, Gauss', 304
Linear combination, 71
   greatest common divisor as a, 71,81-83
Linear congruence, 122-126
Linear congruential method, 288
Liouville's function, 193
Logarithms modulo $p$, 226
Lucas-Lehmer test, 202
Lucky number, 68

Magic square, 146
Master key, 246
Mathematical induction, 15-27
   second principle, 18
Matrix, involutory, 144
Matrix multiplication, 60
Maximal $\pm 1$-exponent, 293

Mayans, 1
Mean,
    arithmetic, 26
    geometric, 26
Mersenne, Marin, 201
Mersenne number, 201,313
Mersenne prime, 201-204
Method of infinite descent, 399
Middle-square method, 287-288
Miller's test, 175
Minimal universal exponent, 282
Möbius function, 192
Möbius inversion, 192
Modular exponentiation, 116
    algorithm for, 116
Monographic cipher, 209
Monkeys, 121,132
Multiple precision, 49
Multiplication, 57-58
    matrix, 60
Multiplicative function, 185-186
Multiplicative knapsack problem, 243-244
Mutually relatively prime, 73

Nim, 47
Nonresidue, quadratic, 299
Notation,
    big-$O$, 55
    binary, 43
    binary coded decimal, 46
    decimal, 43
    duodecimal, 54
    hexadecimal, 43
    octal, 43
    one's complement, 46
    product, 11
    summation, 9
Number,
    abundant, 205
    Carmichael, 174
    deficient, 205
    even, 37
    Fermat, 100-102
    Fibonacci, 20-22
    generalized Fibonacci, 26
    irrational, 6,91-92
    $k$-abundant, 205
    $k$-perfect, 205
    lucky, 68

Mersenne, 201
    odd, 37
    perfect, 199
    pseudo-random, 287-292
    rational, 6
    superperfect, 206
    triangular, 24
    Ulam, 9
Number of divisors function, 194

Octal notation, 43
Odd number, 37
One's complement representation, 46
Operation, bit, 55
Order of an integer, 249

Pairwise relatively prime, 73
Palindromic integer, 151
Partial quotient, 354
Partial remainder, 52
Pascal, Blaise, 29
Pascal's triangle, 29-30
Pell's equation, 413-419
Pepin's test, 321
Perfect number, 199
Period,
    of a base $b$ expansion, 347
    of a continued fraction, 375
Periodic base $b$ expansion, 347
Periodic cicada, 94
Periodic continued fraction, 375-389
Perpetual calendar, 152-157
Plaintext, 208
Poker, electronic, 227-229
Polygon, regular, 102
Polygraphic cipher, 218
Powerful integer, 93
Pre-period, 348
Primality test,
    probabilistic, 177
    using primitive roots, 277-280
Prime, 61-69
    definition of, 61
    Fermat, 99-100
    Hilbert, 93
    in arithmetic progressions, 90
    infinitude of, 61-62
    Mersenne, 201
    relatively, 70

Wilson, 171
Prime number theorem, 63,97
Prime-power factorization, 86
Primitive Pythagorean triple, 394
Primitive root, 251
Principle of inclusion-exclusion, 34
Principle of mathematical induction, 15-27
 second, 18
Probabilistic primality test, 177
Probing sequence, 162
Problem,
 coconut, 121
 hundred fowls, 109
 knapsack, 237
 multiplicative knapsack, 243
 Waring's, 410
Product, Dirichlet, 192
Product cipher, 216
Progression,
 geometric, 16
Property,
 reflexive, 111
 symmetric, 111
 transitive, 111
 well-ordering, 6
Pseudo-random number generator,
 Fibonacci, 292
 linear congruential, 288
 middle-square, 287-288
 pure multiplicative, 289
Pseudo-random numbers, 287-292
Pseudoconvergent, 375
Pseudoprime, 172-180
 Euler, 334-342
 strong, 176-178
Public-key cipher, 230
Pure multiplicative congruential
 method, 289
Purely periodic continued fraction, 383
Pythagoras, 393
Pythagorean theorem, 393
Pythagorean triple, 393

Quadratic hashing, 314
Quadratic irrational, 376
Quadratic nonresidue, 299
Quadratic reciprocity law, 315-324
Quadratic residue, 299

Quotient,
 Fermat, 171
 partial, 354

Rabbits, 26
Rabin's cipher system, 234,314
Rabin's probabilistic primality
 test, 177-178
Radix, 43
Ratio, common, 16
Rational number, 6,91
Read subkey, 245
Reciprocity law,
 for Jacobi symbols, 327-329
 quadratic, 315-324
Recursive definition, 19
Reduced residue system, 181
Reduced quadratic irrational, 383
Reflexive property, 111
Regular polygon,
 constructability, 102
Relatively prime, 70
 mutually, 73
 pairwise, 73
Remainder, 36
Repunit, 151
Residue,
 cubic, 276
 $k$th power, 271
 least nonnegative, 112
 quadratic, 299
Residues,
 absolute least, 112
 complete system of, 112
 reduced, 181
Root of a polynomial modulo $m$, 255
Round-robin tournament, 158-160
RSA cipher system, 231,254

Second principle of
 mathematical induction, 18
Seed, 288
Shadows, 246
Shift transformation, 211
Shifting, 51
Sieve of Eratosthenes, 52-53
Signature, 234
Signed message, 234
Simple continued fraction, 354

Solovay-Strassen probabilistic
    primality test, 341-342
Splicing of telephone cables, 292-298
Spread of a splicing scheme, 296
Square,
    diabolic, 146
    magic, 146
Square-free integer, 93
Strong pseudoprime, 176-178
Strongly multiplicative
    function, 191-192
Subkey,
    read, 245
    write, 245
Substitution cipher, 209
Succinct certificate of primality, 280
Sum, telescoping, 13
Sum of divisors function, 194
Sums of squares, 402-410
Summation notation, 9
Sun-Tsu, 127
Super-increasing sequence, 239
Superperfect number, 206
Symbol,
    Jacobi, 324
    Kronecker, 332
    Legendre, 300
Symmetric property, 111
System of congruences, 126-134,135-146
System of residues,
    complete, 112
    reduced, 181

Team,
    Away, 159
    Home, 159
Telephone cables, 292-298
Telescoping sum, 13
Terminating base $b$ expansion, 346
Test,
    divisibility, 147-152
    Lucas-Lehmer, 202
    Miller's, 175
    Pepin's, 321
    primality, 172,277-280
    probalistic primality, 177
Theorem,
    binomial, 30-32
    Chinese remainder, 126-133

Dirichlet's, 90
Euler's, 182-183
Fermat's last, 399
Fermat's little, 168-169
fundamental, of arithmetic, 85
Lagrange's (on continued
    fractions), 378
Lagrange's (on polynomial
    congruences), 255-256
Lamé's, 79
prime number, 63
Wilson's, 166-167
Threshold scheme, 246
Tournament, round-robin, 158-160
Tower of Hanoi, 26
Transitive property, 111
Transposition cipher, 224
Triangular number, 24
Triangle,
    Pascal's, 29-30
    Pythogrean, 393
Trichotomy law, 5
Twin primes, 65
    application to hashing, 162
Two squares, sums of, 402-406

Ulam number, 9
Universal exponent, 282

Vallée-Poussin, C. de la, 63
Vigenère cipher, 217
Von Neumann, John, 287

Waring's problem, 410
Well-ordering property, 6
Wilson, J., 166
Wilson prime, 171
Wilson's theorem, 166-167
    Gauss' generalization of, 171
Winning position in nim, 47
Word size, 48
Write subkey, 245

Year end day, 157
Year, leap, 153-154

Zeller, Rev., 156

# List of Symbols

| | |
|---|---|
| $[x]$ | Greatest integer, 6 |
| $\Sigma$ | Summation, 9 |
| $\Pi$ | Product, 11 |
| $n!$ | Factorial, 11 |
| $f_n$ | Fibonacci number, 20 |
| $\binom{m}{k}$ | Binomial coeficient, 27 |
| $a \mid b$ | Divides, 35 |
| $a \nmid b$ | Does not divide, 35 |
| $(a_k a_{k-1}...a_1 a_0)_b$ | Base $b$ expansion, 43 |
| $w$ | Computer word size, 48 |
| $O(f)$ | Big-$O$ notation, 55 |
| $\pi(x)$ | Number of primes, 63 |
| $(a,b)$ | Greatest common divisor, 70 |
| $(a_1,a_2,...,a_n)$ | Greatest common divisor (of $n$ integers), 72 |
| $\mathscr{F}_n$ | Farey series of order $n$, 75 |
| $[a,b]$ | Least common multiple, 88 |
| $\min(x,y)$ | Minimum, 88 |
| $\max(x,y)$ | Maximum, 88 |
| $p^a \mid\mid n$ | Exactly divide, 93 |
| $[a_1,a_2,...,a_n]$ | Least common multiple (of $n$ integers), 95 |
| $F_n$ | Fermat number, 100 |
| $a \equiv b \pmod{m}$ | Congruent, 111 |
| $a \not\equiv b \pmod{m}$ | Not congruent, 111 |
| $\bar{a}$ | Inverse, 123 |
| $\mathbf{A} \equiv \mathbf{B} \pmod{m}$ | Congruent (matrices), 138 |
| $\bar{\mathbf{A}}$ | Inverse (of matrix), 140 |